Brain and intelligence in vertebrates

Brain and intelligence in vertebrates

E. M. MACPHAIL
Senior Lecturer in Psychology,
University of York

CLARENDON PRESS · OXFORD
1982

Oxford University Press, Walton Street, Oxford OX2 6DP

London Glasgow New York Toronto
Delhi Bombay Calcutta Madras Karachi
Kuala Lumpur Singapore Hong Kong Tokyo
Nairobi Dar es Salaam Cape Town
Melbourne Auckland
and associates in
Beirut Berlin Ibadan Mexico City Nicosia

Published in the United States
by Oxford University Press, New York

British Library Cataloguing in Publication Data
Macphail, E. M.
Brain & intelligence in vertebrates.
1. Psychobiology 2. Brain
3. Nervous system–Vertebrates
I. Title
596'.0188 QP360

ISBN 0-19-854550-9
ISBN 0-19-854551-7 Pbk

Set in Great Britain by Oxford Print Associates
Printed in Hong Kong

Preface

The original impetus for writing this book lay in my early efforts to place my research work in a wider context. That work concerned the role of avian forebrain structures in learning, and led naturally to an interest in non-mammalian brains, and to following through the relevance of studies using non-mammals to the intelligence of mammals including, of course, man. I found that ideas concerning the functional organization of non-mammalian forebrains were changing rapidly, but that the relevant research was published in a wide range of journals, many of which were not widely read by psychologists. I also found that, although comparative studies of learning did appear in the standard psychological journals, there was no detailed recent survey of them available, and no attempt to relate them to neurological information. In this book I have tried to bring together neurological and behavioural findings relevant to intelligence in the various classes of vertebrates, in the hope of making information derived from diverse disciplines more readily accessible to all those with an interest in intelligence.

Although this book is intended as a contribution to comparative psychology, it should be noted that it does not deal with invertebrates nor (at least directly) with the evolution of either the brain or intelligence. My major concern has been to organize and describe the findings obtained from each group in as clear a way as I could, and to avoid speculations (such as those involved in surmising the course of evolution) for which little hard evidence was available. I have not, however, avoided speculation altogether; indeed, one of the unexpected outcomes of the hours spent in libraries involved in preparing my material has been that I have come to a very different view of the nature of intelligence in vertebrates from the one I held some years ago. My current view is, of course, represented throughout the book, but I hope that, although it may have coloured the presentation of some areas of research, it will not interfere with the utility of the survey to readers who wish to be given a sufficiently full account of the experimental results available to enable them to provide their own interpretations. I shall not be too downcast if my views are not adopted by all (or even many) who read this book,but I shall be disappointed if those who do not share those views find the book on that account less useful.

Two apologies are no doubt in order. First, to those who find some of the behavioural sections too complex. I can only say that theories of animal learning have never been easy to follow, and that, in order to grasp the essence of current controversies, we must go rather deep into theory. The second apology concerns errors which I have assuredly made, but have not yet detected, in presenting results from fields in which I am far from expert: here, I must hope that the virtue of having a single relatively consistent view outweighs the evident disadvantage of lack of direct research experience in many of the areas surveyed.

Many friends and colleagues have read drafts of the book and have given me comments, and, equally important, encouragement. Drafts of individual chapters were read by Peter Bailey, Bob Boakes, Anthony Dickinson, Charles Hulme, Bundy Mackintosh, and Neil Thomson; drafts of the entire manuscript were read by Jane Mitchell and Steve Reilly, my research associates at York, and by Richard Morris and Mick Rugg, and I am very grateful to all of them for their time and advice. Finally, Geoff Hall and Nick Mackintosh were both obliged to spend many hours discussing drafts with me, and made substantial contributions to the overall organization of the book: I am very glad to acknowledge my indebtedness to them both.

I should also like to express my gratitude to those friends who have borne the burden of typing for me: to Jenny Bailey, who typed almost all the final draft, and to June Edwards, Sue Medd, Jenny Debenham, and Jane Moor, who produced the references and indexes. My wife Kate typed all the early drafts, skilfully deflected children's demands on my time, and gave me constant support; my debt to her cannot be put into words.

York,
August 1981

E. M. M.

Contents

1. The comparative approach to intelligence 1

Introduction. Assessment of intelligence in animals. Classification of learning tasks. Goals of comparative studies. The classification of animals: differing degrees of relatedness. Vertebrates: their place in evolution. Organization of the book.

2. The physiological analysis of intelligence 25

Brain and intelligence. The vertebrate brain. Concluding comments.

3. Fish 44

Classification of fish: relationship to land vertebrates. Forebrain organization in fish: deviant nature of teleost forebrain. Sensory and motor organization in fish forebrain. Habituation in fish. Classical conditioning in fish. Instrumental conditioning in fish. Complex learning in fish. Interspecies differences in learning in fish. Forebrain lesions and learning in fish. Summary and conclusions.

4. Amphibians 114

Classification of amphibians. Structure of amphibian forebrain. Sensory and motor organization in amphibian forebrain. Habituation in amphibians. Classical conditioning in amphibians. Instrumental conditioning in amphibians. Forebrain lesions and learning in amphibians. Summary and conclusions.

5. Reptiles 136

Classification of reptiles. Structure of reptilian forebrain. Sensory and motor organization in reptilian forebrain. Habituation in reptiles. Classical conditioning in reptiles. Instrumental conditioning in reptiles. Complex learning in reptiles. Forebrain lesions and learning in reptiles. Summary and conclusions.

6. Birds 168

Classification of birds. Structure of avian forebrain. Sensory and motor organization in avian forebrain. Habituation in birds. Classical conditioning in birds. Instrumental conditioning in birds. Complex learning in birds. Interspecies comparisons in birds. Brain mechanisms and learning in birds. Summary and conclusions.

7. Mammals 237

Classification of mammals. Brain size in mammals. Forebrain organization in mammals. Comparative studies of learning in mammals. Summary and conclusions.

8. Language and intelligence 290

Introduction. Linguistic capacities of non-human mammals. Cerebral lateralization. The hippocampal formation and memory. Memory in non-human mammals. Summary and conclusions.

9. Intelligence in vertebrates: two hypotheses 330

Introduction. Phylogenetic considerations. General adaptiveness. Differences in brain size. Vertebrate brain organization. General intelligence. Implications.

References 344

Author index 387

Subject index 398

Acknowledgements

I should like to thank the following scientists for permission to use material from their publications: M. E. Bitterman, R. L. Boord, E. C. Crosby, N. Geschwind, R. C. Gonzalez, W. Hodos, H. J. Jerison, P. J. Livesey, N. J. Mackintosh, P. D. MacLean, H. J. Markowitsch, R. C. Miles, R. Nieuwenhuys, R. G. Northcutt, R. E. Passingham, H. Stephan, J. M. Warren, W. I. Welker.

The following organizations have kindly allowed me to reproduce material: Academic Press; the American Association for the Advancement of Science; the American Museum of Natural History; the American Psychological Association; the American Society of Zoologists; Elsevier/North Holland Biomedical Press; Holt, Rinehart and Winston; S. Karger Verlag; Lea and Febiger; the New York Academy of Sciences; Pergamon Press; Plenum Publishing Corporation; University of Chicago Press; University of Illinois Press.

1. The comparative approach to intelligence

Introduction

Throughout the history of Western thought there have been two contrasting interpretations of the behaviour of animals: some, like Descartes, have assumed that animals are qualitatively distinct from humans, that they are mere machines, devoid of reason or consciousness, having no mental life; others, Hume for example, have assumed a continuity between the mental processes of animals and man, arguing that the differences seen are essentially quantitative in nature. It is difficult to see what experimental observations might be relevant to the question of whether an animal does or does not enjoy consciousness, and the dispute between these two points of view was conducted with a minimum of reference to the facts of animal behaviour; this remained true until the late nineteenth century when, following the publication (in 1871) of Darwin's *The Descent of Man,* the impact upon the argument of the theory of evolution became clear. Just as the physical form of the body evolved, so must those capacities responsible for human mental life have evolved: the notion that there could be a sudden discontinuity between man and beast became much more difficult to sustain, and the search for evidence of continuity in mental life throughout evolution began. The fruits of that search constitute the field of enquiry now known as comparative psychology.

The early comparative psychologists saw the conscious mind as the proper object of psychological enquiry: direct access was possible to only one mind (one's own), so that the properties of other minds (human or infrahuman) must be inferred. What rules, then, were to be used to infer from behavioural observations to mental events? An early proposal (Romanes 1882) was that one criterion for the possession of mind is the ability to learn; in Romanes' view, the evolution of mind was reflected in the evolution of learning capacity – 'The lower down we go in the animal kingdom, the less capacity do we find for changing adjustive movements in correspondence with changed conditions' (Romanes 1882, p. 8). Romanes' work, however, met criticism on two grounds. First, from Lloyd Morgan who, while sharing Romanes' general views on the primarily introspective nature of psychology, saw the dangers inherent in excessive anthropomorphism. Romanes, for example, had written: 'if we observe an ant or a bee apparently exhibiting sympathy or rage, we must either conclude that some psychological state resembling that of sympathy or rage is present, or else refuse to think about the subject at all' (Romanes 1882, p. 9). Not unnaturally the proposal that ants and bees might feel sympathy made many comparative psychologists uneasy, and this disquiet was crystallized in Lloyd Morgan's celebrated canon: 'In no case may we interpret an action as the outcome of the exercise of a higher psychical faculty, if it can be interpreted as the outcome of the exercise of one which stands lower in the psychological scale' (Morgan 1894, p. 53). Although this principle still retains its importance today, and is particularly relevant to the analysis of problem-solving in animals, the issue to which it was originally directed, the inference of types of conscious processes in animals, is no longer of central importance to psychologists; the question, for example, of whether an animal (or even another human being) feels, say, pleasure or pain, has reverted once more to the philosophers.

Psychologists abandoned their efforts to provide an analysis of states of consciousness largely as a result of the attacks on introspectionism in the early years of this century, by biologists (Loeb, for example) and psychologists—Watson in particular. The essence of Watson's position is that there is no dichotomy between behaviour and consciousness, that thoughts and images, for example, are segments of behaviour on a par with limb movements, and to be studied in the same objective way. Now since that time, few comparative psychologists have concerned themselves with attempts to infer the mental life of animals; whether this is because most psychologists accept Watson's position, and see it as a resolution of the 'mind-body' problem, or because, although rejecting his psychological stance, they nevertheless agree that observations are not relevant to questions of consciousness, need not concern us. What is pertinent here is that since the early days of this century, comparative psychology has been concerned with the behavioural capacities of animals—with what they can and cannot *do*. Theoretical inferences have been made, not to the conscious experience of the animals, but to the types of internal processing required to produce the observed behavioural output.

A second criticism directed at Romanes was that his behavioural evidence was largely anecdotal in character, and so generally unreliable. The case against the uncritical use of anecdotal evidence was made out by Thorndike, in a style that is hardly likely to be bettered: 'In the first place, most of the books [on comparative psychology] do not give us a psychology, but rather an *eulogy*, of animals. They have all been about animal *intelligence*, never about animal *stupidity*. . . . Human folk are as a matter of fact eager to find intelligence in animals. Dogs get lost hundreds of times and no one ever notices it or sends an account of it to a scientific magazine. But let one find his way from Brooklyn to Yonkers and the fact immediately becomes a circulating anecdote' (Thorndike 1898, pp. 3-4).

As a result of such arguments, there has since been fairly general agreement that the proper subject matter of comparative psychology is overt behaviour, and that the proper study of behaviour requires the use of objective methods of observation and experimentation. Agreement on these basic matters has not, however, resolved the question whether there is continuity in the processes controlling behaviour in animals and man. Thorndike's experiments led him to believe that the essence of all intelligent behaviour lay in the formation of associations, and that the differences in intellectual capacity between species reflected quantitative differences in the number, delicacy, complexity, permanence, and speed of formation of associations. 'Amongst the minds of animals that of man leads, not as a demigod from another planet, but as a king from the same race' (Thorndike 1911, p. 294). Thorndike failed to obtain any evidence of the capacity to reason in monkeys, but argued that 'rational connections are, in their basic causation, like any others, the difference being in what is connected', so that 'the denial of reasoning need not mean, does not to my mind, any denial of continuity between animal and human mentality'. Maier and Schneirla, on the other hand, concluded from their analysis of experimental data that the ability to form associations and the ability to reason were distinct, and that the ability to reason required a well-developed cortex (found only in mammals).

One of our primary aims in assessing the evidence relevant to the intellectual capacities of various vertebrates will be to determine to what extent that evidence bears on the quantitative/qualitative or continuity/discontinuity controversy. The ultimate goal, however, is not simply to resolve that controversy, but rather to provide an explanatory account of intelligent behaviour in man and in animals; the goal, then, is to elucidate the nature of intelligence. The stance adopted by a particular psychologist on the continuity question generally reflects his view of the nature of intelligence. Most of those who believe that there are no major qualitative differences between vertebrate species argue that all intellectual acitvity proceeds according to the laws of associative learning and have, in this sense, a relatively unstructured view of intelligence: the only information-processing mechanisms involved in logical (as opposed to sensory, motor, or storage) processes are those required for the formation (and, perhaps, the dissolution) of associations. The opponents of continuity naturally tend to be somewhat more diverse in their views, but have in common the belief that intelligence is structured, that intelligent behaviour is the product of the interaction of a number of devices, each having distinct properties. As we shall see, these contrasting views of the nature of intelligence imply important differences over which types of experimental approach are likely to prove most fruitful.

Assessment of intelligence in animals

It is clearly important that, before attempting to compare the intellectual capacities of various species, we should have in mind a definition of intelligence. We shall not, however, attempt a formal definition here, and in case this should smack of evasion, some justification may be provided by citing the precedent set by Darwin, who wrote, in the chapter on Instinct in *On the Origin of Species:* 'I will not attempt any definition of instinct. It would be easy to show that several distinct mental actions are commonly embraced by this term; but every one understands what is meant, when it is said that instinct impels the cuckoo to migrate and to lay her eggs in other birds' nests' (Darwin 1866, p. 148). A similar view is taken here: we need not provide the definition for a word whose meaning is already well known. Moreover, to adopt any strict definition would run the risk of imposing some theoretical bias at the outset, in particular as regards the question whether intelligence is some unitary capacity, or better seen as a complex of capacities, each of which might be quite independent of the others. It may be tempting to argue that intelligent behaviour must involve some complex behaviour, such as reasoning, and that, therefore, consideration of simpler types of behaviour is not relevant to the analysis of intelligence. The conclusion, however, supposes that reasoning involves mechanisms other than those involved in 'simple' learning, and this is an empirical matter, not one to be answered by appealing to the ways in which the word 'intelligence' is generally used. It may well be that 'complex' behaviour, such as reasoning, does involve mechanisms not engaged in 'simpler' behaviour—but this is an issue which can only be settled by evidence; to avoid ruling out such evidence, our notion of intelligence must at this stage be left open and flexible. In place, then, of a formal definition, it should be sufficient to indicate the sorts of tasks in which intelligence may play an important role: these include the entire range of learning tasks used by experimental psychologists, from habituation and elementary association formation to complex problem-solving, language analogues, and so on. Intelligence, in other words, is held to manifest itself in all those situations in which subjects are required to adapt to novel circumstances; a common feature of many such situations is that successful adaptation involves the detection of and appropriate response to regularities in the environment. This is the case as much with 'simple' association formation as with 'complex' concept formation: whether successful performance in all these situations involves the same mechanisms will remain, at this stage, an open question.

Our use of the term intelligence is, then, intended to correspond reasonably well to its common, non-specialist, use: it should be noted, however, that it stands contrasted with another use current in the branch of experimental psychology known as artificial intelligence. The goal of workers in this field is to produce a detailed account of the information-processing that underlies behaviour, however simple that behaviour might seem on the surface. One area of particular concern is that of visual perception, and the problems that arise in attempting to devise a machine that can see emphasize the extreme complexity of perception. However, we shall not apply the word 'intelligent' to processes solely on the grounds that they involve complex information-processing: the meaning used here implies a generality and flexibility of application that may well not be true of at least some of the complex devices within the central nervous system. The processes of perception appear to involve something akin to problem-solving, but whether such processes enjoy any use outside perception is unclear, and until such use is shown, perceptual capacities will be assumed to be independent of intelligence. A further example may serve to emphasize the notion of general application implied by the present definition of intelligence. There is currently no evidence from laboratory experiments that amphibians can master complex learning tasks—but frogs do make use of sun-compass orientation: that is, they orient towards 'home' by observing the position of the sun, by calculating, from the use of an internal clock, some measure of the difference between that position and the position occupied by the sun at home at that time of day, and finally, by calculating, from that measure, the direction in which home lies. Is this evidence of otherwise undetected intellectual capacity in frogs? The answer, of course, lies in the matter of definition: if by intelligence we had meant simply the capacity to carry out complex information-processing, then, of course, sun-compass orientation would have implied intelligence. On the other hand, as we wish to restrict our use of intelligence to more general capacities, sun-compass orientation does not of itself demonstrate intelligence, although it may stimulate efforts to discover other situations in which the relevant processing mechanisms may be brought into play.

As the preceding paragraph indicates, the decision to concentrate on mechanisms of general intelligence will lead to the exclusion from consideration of a number of capacities, such as navigation, imprinting, and poison-aversion learning, which are demonstrated in restricted contexts. These are capacities which have been the subjects of intensive investigation by ethologists and psychologists, and it may be worth considering whether there are grounds for supposing that such specific capacities might not be relevant to performance in the more general contexts in which we have declared an interest.

One possibility, which has been raised for at least some of these specific capacities (see, for example, Hoffman and Ratner's 1973 account of imprinting), is that they are not in fact as specific as they might seem but instances of the operation of general learning mechanisms responsible for virtually all behavioural adaptations. Now if this account is valid for any of the 'specific' capacities then we shall, of course, be considering the mechanisms responsible for the behaviour concerned when we discuss performance in standard learning tasks; no exclusion will in fact occur, and there would seem little value in pursuing the properties of those mechanisms in what might be misleadingly narrow contexts.

A second, very different, possibility is that there are specific capacities which, although having evolved to deal with restricted contexts, are available for at least some more general applications; in this case, interpretation of performance in conventional tasks should take account of the contribution to that performance that may have been made by (relatively) context-specific mechanisms. In its most exaggerated form, this view claims that 'general intelligence' is a chimera; that what we see in all examples of adaptive behaviour is the operation of mechanisms which evolved to cope with relatively specific environmental demands, but which have become to some degree emancipated so that they can be brought to bear on a wide range of problems (see, for example, Rozin 1976). According to this notion, it could be that by not examining 'specialized' forms of learning, we exclude from consideration those tasks which would exhibit the optimal operation of the capacities whose performance we are in fact exploring in our more general tasks.

The principal difficulty facing evaluation of the notion that general intellectual capacity is a composite of contributions from diverse specialized capacities is that it is vague; the proposal does not show how a given specific capacity might contribute to performance in some more general context, nor how to assess the existence or otherwise of any such contribution. Taken at face value, the suggestion would appear to be that any comparative study of intelligence should consider all the behavioural capacities of the species considered—patterns of social behaviour, of rearing the young, of feeding, aggression, territoriality, and so on—but this without any indication of how such discussions would be relevant to the central issue. Such a procedure is clearly unsatisfactory, and in any case impractical. In the absence of specific proposals concerning possible contributions from specialized capacities, we shall be unable to consider such capacities further. This is not, of course, to dismiss the theoretical possibility of such contributions, and there is one general implication of the hypothesis that should be borne in mind. If we assume, reasonably enough, that not all species possess the same complement of specialized mechanisms, and that such mechanisms do contribute to a wide range of tasks, then we might expect to find qualitative differences between species in performance of standard tasks. Any discovery of qualitative differences should, then, cause us to look at differences between the specialized capacities of the species concerned, in case they might account for the qualitative contrast observed. Should that procedure also be successful, we should have both some evidence that the overall hypothesis might be valid, and, perhaps, indications of how to devise more general techniques for evaluating the contribution of specialized mechanisms to performance in standardized contexts.

The burden of the above argument is, then, not that 'general intelligence' should be assumed to be in some sense unitary, rather than a collection of specific capacities, but rather that, since we have no means of inferring from (apparently) context-specific capacities to corresponding general capacity, we should examine (apparently) general capacities initially, to see whether that analysis suggests factors which could be derived from context-specific capacities.

It is appropriate to emphasize at this point that by intelligence is meant, not the intelligent behaviour of an organism, but its capacity or potential for such behaviour; we cannot therefore conclude from the fact that a particular species is rarely if ever seen to perform intelligently in its natural life that it is of 'low' intelligence—its natural habitat may simply not provide an appropriate situation for the demonstration of its intelligence. Although it may seem to us that animals that are well adapted to a relatively monotonous environment, and that survive essentially by use of species-specific responses to restricted classes of stimuli, have no need of intelligence, and so would not have evolved such a superfluous capacity, the conclusion would be entirely speculative: efforts to determine the limitations of the animal's capacities would have to be made in 'unnatural' situations, to see how far novel demands might be met.

Our interest, then, is to be in the behaviour of animals in situations which make demands upon the subject that allow it to demonstrate the behavioural flexibility that is implied by intelligence. How are such situations to be designed? There are, unfortunately, no rules that enable us to say *a priori* what tasks are required: their design depends basically on the ingenuity of the experimenter, on his intuitions concerning the organization of intelligent behaviour, and on the compatibility between the apparatus and his subjects. The experimenter's intuitions are of importance in the following way. If the task he designs makes very few demands on intelligence, and is, therefore, 'easy', then animals of a wide range of intellectual capacity might perform at a similar level in it—there would be, that is, a 'ceiling effect'. If, on the other hand, the task is too 'difficult', then only a few exceptionally intelligent creatures might succeed in it, the remainder appearing equal owing to a floor effect. Ideally, a number of tasks of varying levels of difficulty seem to be required: the problem is, how can the difficulty of a task be assessed in advance—what is it that makes one task more or less difficult than another? Task difficulty is determined by the organization of the intelligence of the species in question: it depends, that is, entirely on how the relevant information is stored and processed. Given a valid account of such information-processing we could of course design tasks that would make varying demands upon intelligence—but then at least one of the objects of the exercise is to obtain such an account, and, at present, *a priori* assessment of task difficulty depends on hypotheses that are far from commanding universal assent. The design of the tasks, then, will be guided by an experimenter's 'hunches' concerning intelligence; their success as assessors of intelligence is, on the other hand, an entirely empirical matter, although it is far from easy to set out criteria for success. Many problems in the interpretation of intelligence tests for animals are best considered individually, in the context of a particular test, but there are some general difficulties which may appropriately be considered here.

Suppose that a task has been devised which does obtain different rates of learning for various species of animals—may we assume that the species are now ranked in intelligence? Clearly, we cannot. Obvious alternative explanations are that the task merely distinguishes between their sensory capacities (as, for example, in the acquisition of a visual pattern discrimination), or between their motor skills (as, for example, in learning to fit one object into another), or reflects differences in motivation or incentive—some species may find the reward offered more attractive than others. It seems obvious that such 'contextual variables' (Bitterman 1965*a*) would indeed generate species differences in many learning situations—how could we rule out the possibility that one of them was responsible for a given difference in any situation? One proposal for overcoming this problem is known as the method of 'systematic variation', and was advocated by Bitterman (1965*a*): this technique involves the use of varying levels of a contextual variable in a particular task. Suppose, for example, that pigeons have acquired a particular discrimination more rapidly than goldfish, and we are concerned to know whether this is in fact because the pigeons are more highly motivated. We should then run groups of pigeons and goldfish at different levels of motivation (by, for example, varying the duration of food deprivation) on the same discrimination: if neither group shows much variation in performance with variations in motivation, and if there is no sign of an overlap between the groups at any point, then it is not unreasonable to conclude that the observed difference is not brought about by differences in motivation. At this point, of course, the method would have to be applied to other confounded contextual variables, which should in turn be varied. Moreover, variations in the level of one variable should be tested at various levels of other variables, in case there might be interactions between the effects of such variables; for example, variations in size of reward might have little effect at one level of deprivation, but a marked effect at another level. Even then, a conclusion as to absence of effects of contextual variables would be unsafe, as there is no finite catalogue of such variables. The method of systematic variation, taken at its face value, clearly makes heavy demands on both time and subjects, and cannot guarantee, however extended the series of variations, that there is not still some relevant but untested variable present in a situation.

This conclusion is not, however, as damning as might at first sight appear; the conclusion is, after all, only that, by using this technique, it is not possible to exclude all logically possible artefacts. On the other hand, the technique can be used to rule out any apparently plausible source of artefact—there is, indeed, no alternative, as one cannot 'equate' the sensations of members of different species. Once again, the implication is that we must wait upon results, and then consider what interpretations are plausible rather than all those that are logically possible.

As the method of systematic variation is so time-consuming, experimenters have attempted to devise tasks in which contextual variables may seem unlikely to play an important role in generating species differences. One popular technique has been to use situations in which the measure is not the absolute number of trials or errors in the acquisition of a task, but some relative measure that depends on a baseline score made by the species. For example, an animal might be taught to choose stimulus A rather than stimulus B in a discrimination task, the number of errors being recorded; the reward value of the two stimuli is now reversed, and the number of errors made in the course of learning to choose B rather than A is recorded, and expressed as a percentage of the errors made in acquisition. It might be supposed that such a relative score (the Reversal Index of Rajalakshmi and Jeeves 1965) would be less liable to contamination by contextual variables than would raw acquisition data, the grounds for this supposition being, of course, that the contextual variables, present presumably in both acquisition and reversal, should cancel themselves out. However, this is evidently not a watertight conclusion: for example, reversal is a different task from acquisition—it may, therefore, employ a mechanism that is not employed in acquisition, and that mechanism might be sensitive to contextual variables that are of little significance in acquisition; alternatively, it might be simply that reversal is more difficult than acquisition (or vice versa), and that the effects of some contextual variable vary with task difficulty. This technique, then, can give no guarantee of success in providing an uncontaminated measure of intelligence, but this is not to deny its possible utility: if a test using a 'relative' score provides results that appear to form a meaningful pattern, lending itself to plausible theoretical analysis, then it will have been successful at least to the extent that further empirical analysis, employing, no doubt, the method of systematic variation, will be justified.

There is one additional important problem that requires comment. The situations used by experimental psychologists expose animals to environments very different from those that would be encountered in the wild; they require unnatural responses (key-pecking, lever-pressing) to unnatural stimuli (monochromatic light, electric shocks). Poor performance by an animal might reflect, not a low level of intelligence, but rather that the situation is not one that allows expression of the adaptive capacities available in more natural environments. Experimental psychologists have become increasingly aware of this problem, and are very conscious of the need for compatibility between the behavioural demands of their situations and the species-specific behaviour patterns of their subjects. For example, it is extremely difficult to train pigeons to peck a lit key to escape or avoid trains of electric shocks (Hineline and Rachlin 1969); but this probably reflects the fact that key-pecking is, in pigeons, a most unlikely response to either fear or pain. It is not, in fact, difficult to train pigeons to avoid shocks by moving from one compartment to another in a shuttle-box (Macphail 1968) or by pressing a treadle with their feet (Foree and LoLordo 1970).

This problem may be exaggerated: the situations preferred by psychologists have become popular precisely because the species with which they are used perform efficiently in them, and do not require extensive (and tedious) training. Provided that animals are not grossly maladapted to the test situation, it does not seem that the objection has much force, except perhaps as a caveat to be borne in mind where subjects fail to perform up to expectation. Moreover, there is no alternative, unless the comparative analysis of intelligence is to be abandoned: it has already been argued that passive observations of animals in the wild do not necessarily reveal the full extent of their capacities—unnatural demands, therefore, must be made. Provided that the situations used are carefully designed, and that a variety of situations are used, there is no reason why invaluable information should not be gained in the laboratory.

Classification of learning tasks

In a number of the chapters that follow, experimental results are considered under one of four headings—habituation, classical conditioning, instrumental learning, and complex learning. It should be emphasized here that these labels refer to the nature of the procedures used by the experimenter; the allocation of results to separate headings is not intended to imply that the different procedures tap distinct learning mechanisms—that, in other words, they are different 'types of learning'. They may, of course, be different types of learning, and indeed one of our major interests will be to see whether some species succeed in giving evidence of learning in one procedure, but not in another—a finding which would give support to the view that different mechanisms are involved in the various procedures.

Habituation

The term habituation refers to the process whereby the magnitude of an unconditioned (unlearned) response (e.g. a change in heart-rate) to an originally novel stimulus declines with repeated presentations of the stimulus. One way in which habituation may differ critically from (and be 'simpler' than) the other procedures is in requiring no (or at least no obvious) association-formation: the subject has, as it were, only to register that a particular stimulus has occurred previously in order to perform adequately. It makes sense, then, to begin the sections on learning by considering habituation first, as it has some claim to be a 'simple' form of learning. Our interest will be not only in whether habituation occurs in given sets of species, but also in whether there is evidence for comparability of mechanisms underlying the phenomenon in the various species in which it is observed.

Classical and instrumental conditioning

Our major concern in the sections on these two types of conditioning will be to determine whether each type is obtained in the class of animals under review. In recent years, the distinction between classical and instrumental learning has attracted much theoretical interest, and we shall consider at this point some of the problems which apply generally to the distinction, independent of its investigation in any particular species.

In classical, or Pavlovian conditioning, the occurrence of a reinforcer (an unconditioned stimulus, or UCS, normally of biological significance, that reliably elicits a particular response, the unconditioned response or UCR) is contingent upon a prior occurrence of some arbitrary stimulus (the conditioned stimulus or CS); the CS, then, acts as a signal that the UCS is imminent, and nothing that the animal does can affect the occurrence of the UCS. That the animal has detected the contingency between CS and UCS is shown by the occurrence of some conditioned response (CR) to the presentation of the CS alone; the CR generally, but not invariably, closely resembles the UCR, and appropriate controls are, of course, required to show that the CR is not obtained in the absence of the CS-UCS contingency.

In instrumental conditioning, the occurrence of reinforcers is contingent upon the production of some specified response by the animal; the detection of that contingency is demonstrated by a change in the spontaneous rate of occurrence of that response (by an increase, where the reinforcer is a reward, or by a decrease, where punishment is used). As with classical conditioning, controls are necessary to show that the rate changes obtained are indeed due to the contingency imposed. Successful acquisition of instrumental conditioning is prima facie evidence of the capacity to form associations involving a response as one of the terms in the association.

The major complexity that arises from these two superficially simple procedures is the possibility that learning in a classical procedure might be mediated by a covert instrumental contingency, and vice versa for an apparently instrumental procedure. Consider a typical classical procedure using an appetitive reinforcer, say pairings of a bell with food; suppose that the subject, after some trials, salivates reliably to presentations of the bell alone. At first sight, this would seem to indicate the detection of the bell–food contingency (and so be an instance of classical conditioning); however, it could be the case either that salivation improves the quality of the food reward, or that the subject has formed the 'superstition' that salivation produces the food. In either case, the operative contingency would be that between the response (salivation) and the reinforcer—that is, an instance of instrumental conditioning. Similarly, an animal that freezes in response to a bell preceding an inescapable shock might be detecting, not the explicit classical contingency, but a covert instrumental contingency—freezing at shock onset might, for example, reduce the aversiveness of the shock. Similar problems arise in instrumental procedures. Suppose a hungry rat is rewarded with a food pellet each time it presses a bar; exposure to the stimuli near the bar will now regularly precede the reward. If we suppose (plausibly enough) that rewards elicit approach as a UCR, it may be that stimuli near the bar will elicit approach (and incidentally standing on the bar) as a classically conditioned response; similarly, an animal may desist from a response that is punished, not because it detects the contingency between the response and the punisher but because it withdraws from the stimuli (associated with the response) that are now regularly paired with the aversive reinforcer.

There is one control procedure that can be used to help distinguish between classical and instrumental interpretations of a given instance of conditioning, this procedure being known as omission training. Quite simply, in this control procedure, if the target response occurs on the presentation of the relevant stimulus, the reinforcer is omitted on that trial. Now if the conditioned response is classical in origin, this procedure should not have a dramatic effect—some decline in response probability should occur as a consequence of the occurrence of trials on which the CS is not followed by the UCS, but as soon as response probability declines, of course, the CS–UCS pairings are re-established and so obtain the CR once more. If on the other hand the response has been instrumentally conditioned then, if the UCS is a reward, the response should rapidly and permanently disappear, and if the UCS is a punisher, then the response should be more strongly established (as a total omission of a punisher is a more powerful reinforcer than its reduction).

It should be pointed out at this early stage that the results of experiments using omission training will not always have an unequivocal interpretation and that, in particular, it will be considerably more difficult to establish instrumental as opposed to classical conditioning (see Dickinson and Mackintosh 1978 and Mackintosh and Dickinson 1979, for discussions of this issue). It may, for example, be agreed that where a response is established or maintained in an omission procedure with an appetitive UCS, then that response is of classical origin; since the response is never followed by the UCS, it can hardly be supposed that the animal forms the superstition that responding obtains the UCS, or that responding is established because it improves the UCS.

It is not, however, clear what outcome of an omission training experiment could unequivocally demonstrate instrumental conditioning, and this is due to the fact that the imposition of response–reinforcer contingencies almost inevitably alters stimulus–reinforcer contingencies. Suppose, for example, that responding for an appetitive UCS is abolished by an omission procedure. While this result could be taken to indicate the detection by the animal of a response–reinforcer contingency (that responding leads to omission of the reward), it is also amenable to a 'classical' interpretation; this is because, as Mackintosh and Dickinson (1979) argue, the relationship between CSs and UCSs is altered by an omission schedule so that, for example, a stimulus complex containing a view of the CS but perceived at a distance from the response site (on trials, that is, on which no CR is to be emitted) is now perfectly correlated with UCS delivery, whereas the complex perceived near the response site (immediately preceding a response) is no longer correlated with the UCS. The 'distant' but not the 'near' stimulus complex might now elicit CRs of some kind but these will not be scored as such since they will not occur at the specified target site.

Similar considerations apply to omission procedures using aversive stimuli (avoidance tasks). An animal might learn to, say, run from one end of an alley to the other more rapidly if that response avoided shock than if the same number of shocks were delivered, irrespective of what response occurred; but now, whereas for animals in the avoidance condition the stimuli at the start but not the end of the runway are correlated with shock, there is no such differential correlation for animals in the classical condition, who are shocked irrespective of their responses, and therefore of their position, in the alley. It is clear that a classical account of the difference in performance between the two conditions is entirely plausible.

It appears that, to obtain convincing demonstrations of instrumental conditioning, we need to find responses which are not only influenced appropriately by omission training procedures but which also do not significantly alter the stimulus complex. We shall, in the relevant sections of Chapters 3 to 7, introduce and discuss such examples, where they are available, and this introduction to the issue may serve as early warning that the discussions may become somewhat involved. We shall in those discussions accept evidence from studies using omission procedures of an effect of the response–reinforcer contingency as *prima facie* evidence for instrumental learning, but will not be able to regard such evidence as necessarily conclusive.

Given that it will be a complicated task to distinguish between classical and instrumental conditioning, it is appropriate to emphasize here that it is an issue which is of considerable relevance to comparative psychology. It was argued above that, in instrumental conditioning, an association was formed in which a response was one of the terms. Now one school of associationist psychologists (Thorndike and Hull, for example) have taken the view that the relevant association is between the CS and the response (an S–R bond), this bond being strengthened by the ensuing UCS. This account, then, supposes that the arrival of the CS elicits the CR, there being no reason to suppose that the animal 'expects' any UCS consequent upon the CR. There are a number of reasons for rejecting this account, at least when applied to mammals, one reason being that downward shifts in the value of the UCS appear to affect behaviour much more rapidly than would be expected if the mechanism for such effects lay solely in the (gradual) change in strength of competing S–R bonds. As we shall see in Chapter 3, the effects of reward shifts in fish appear to be much less dramatic and this contrast between fish and mammals has led Bitterman (e.g. 1975) to propose that whereas fish are 'pure' S–R animals, mammals have evolved other mechanisms which allow the formation of expectancies, whose non-fulfilment leads to rapid behavioural adjustment. In Bitterman's view, then, instrumental learning, conceived of as S–R learning, is a basic or early form of learning, to which other forms of learning have been added as intelligence evolved.

An alternative view of instrumental learning (e.g. Mackintosh 1974) is that it involves an association between a response and a reinforcer, an association for which the CS acts as a discriminative stimulus; in other words, the animal learns that in a given context a particular response obtains a particular reinforcer. A difficulty with this view is that it does not provide a mechanism for performance of the response; the fact that an animal associates a given response with food does not in itself explain why it performs that response (and, in particular, why it might perform it when hungry but not when satiated). It seems that such an account must formally acknowledge not only the expectation of food given the CR, but also the desire to obtain the food. Mackintosh and Dickinson (1979) have put forward the view that response–reinforcer associations should be regarded as propositional in form, and that they lead to responses when co-ordinated with what they describe as 'imperative premises', the co-ordination proceeding by the use of rules of 'imperative inference'. In their words: 'If exposure to a contingency between wheel running and food establishes the propositional association "The response of wheel running produces food", and if a state of food deprivation engages an imperative premise of the form "Perform any response that will produce food", the rules of imperative inference will permit the derivation of the instruction "Perform the response of wheel running"' (Mackintosh and Dickinson 1979, p. 166).

To view instrumental responding as an instance of imperative inference suggests that it is a complex process, and Mackintosh and Dickinson in fact go on to contrast it with what they regard as the much simpler relationship between the nature of the association and the production of response in classical conditioning. According to Mackintosh and Dickinson, the association formed in classical conditioning takes the form of an excitatory link between the representations of the CS and UCS: 'When two events are associated, presentation of one activates a representation of the other' (Mackintosh and Dickinson 1979, p. 167). This account implies, then, that classical conditioning is a simpler process than instrumental conditioning so that it is reasonable to expect that classical conditioning may have evolved before instrumental conditioning.

These two accounts of instrumental learning, as involving either S–R or R–S associations, generate opposing views concerning the relative 'simplicity' of classical and instrumental conditioning. Any evidence for the prior evolution of one or other of these two types of learning would, then, throw light on not only the evolution of intelligence but also the true nature of instrumental conditioning.

Complex learning

We shall discuss under this heading all those subjects which do not fall easily into the preceding categories (habituation, classical conditioning, and instrumental conditioning), and for which a reasonable body of comparative data is available; whether the forms of learning concerned are more complex in the sense of requiring mechanisms beyond those required for the former categories is, of course, an open question. Since there is no wish to exclude data which, while they might reveal contrasts between groups, do not favour any currently popular hypothesis, the topics discussed will form a heterogeneous collection. Some sections will discuss particular types of experimental procedure such as serial reversal learning, probability learning, double alternation, learning-set formation, which may have diverse theoretical interpretations; other sections focus on a theoretical issue, considering a body of experiments, using diverse procedures specifically aimed at that issue (e.g. the sections on the mechanisms of attention, and on memory). There will, of course, be some 'cross-talk' between the various sections—experiments on serial reversal learning are relevant, for example, to discussions of attention—but the organization to be used has been adopted in the hope that a reasonable degree of unity within a given section can be achieved, although the organizing principle—procedure or issue—has been varied to attain that end.

It is, of course, to be expected that we shall encounter in surveys of complex learning marked differences in the level of performance of various groups, and it may be helpful here to outline the rationale for the way in which these differences will be viewed.

Where a particular phenomenon can be demonstrated in one group, but not in another, this will be taken as *prima facie* evidence for a qualitative difference between those groups; it will not, of course, be taken as good evidence for such a difference unless accounts in terms of plausibly relevant contextual variables are ruled out through systematic variation.

Where a phenomenon is demonstrated in each of two groups, then it will be assumed that there is no qualitative difference between the groups on the task concerned; this is not, of course, a necessary conclusion—two groups of animals might solve the same problem and yet be using quite different mechanisms for its solution—but it is the most parsimonious account of such an outcome. Where the levels of performance of two groups differ, an effort must be made to decide whether that difference reflects some quantitative difference in intellectual capacity, or the differential effects of contextual variables. It will be apparent that this is bound to be an extremely difficult question: it is hardly likely that the level of performance of any group will not be affected by contextual variables, so that the gap between the performances of two groups will almost inevitably be capable of reduction by manipulating some such variable. Suppose, for example, that one species achieves an accuracy of, say, 90 per cent correct on a given type of problem across a wide range of values of contextual variables, whereas another species achieves comparable accuracy only at optimal levels of, say, deprivation, reward size, discriminability of the stimuli, and so on—does the fact that the gap can be closed in that one case indicate that its appearance in all other cases is due, not to differences in intelligence, but to differences in processing of motivational and sensory information? There is no simple answer to this question, and each case must be examined individually to establish the most plausible and parsimonious interpretation. Where there is further external evidence to indicate major differences, in, for example, the sensory processing concerned, this may provide support for a 'contextual' interpretation; where there is further evidence for a comparable quantitative difference between the same species in some other task posing quite different contextual demands, then support for an 'intellectual' interpretation will be strengthened.

Where evidence for differences in intellectual capacity is obtained we shall, of course, attempt to interpret them, to specify the nature of the mechanism responsible for the differences. There have been surprisingly few suggestions concerning mechanisms which might be responsible for species differences in intelligent performance. Examples of hypotheses which will receive detailed attention in the chapters that follow are, for qualitative differences: that some species can, and others cannot, form expectancies; that humans alone can acquire language; and, for quantitative differences: that species vary in the stability of their attention to the stimulus dimensions relevant to problem solution; that species vary in the strength of memory traces set up by certain environmental events. It should be noted at the outset that serious consideration will also be given to the possibility that there are, excluding humans, neither qualitative nor quantitative differences in intellectual capacity amongst the various groups of vertebrates.

Goals of comparative studies

Formidable difficulties arise in both the design and interpretation of comparative studies of intelligence, so raising the question whether the information to be gathered will justify the effort expended. What would be gained, for example, by establishing a rank-order for intelligence of various species of animals?

One role of such a ranking would be in providing empirical evidence relevant to certain theories of progressive brain evolution. Jerison, for example, who defines 'biological intelligence' as 'the capacity to construct perceptual worlds in which sensory information from various modalities is integrated as information about objects in space and time' (Jerison 1976, p. 101), argues that the mean level of intelligence of reptiles and fish is considerably lower than that in birds and mammals, and that, amongst mammals, intelligence is most highly developed in the primates. Jerison also claims that intelligence is correlated with relative brain size (a concept that will be discussed further in Chapters 2 and 9) and provides (Jerison 1973) extensive data on relative brain sizes of both living species and fossils. Now this latter claim provides a ranking of animals that should agree with a ranking based on intelligence, and although it is not clear exactly how we should attempt to measure the perceptual worlds of animals, Jerison appears to accept that the data provided by experimental psychologists are appropriate as estimates of biological intelligence. He writes: 'In birds the fixed-action pattern is the typical behavioral mode, and biological intelligence may be a little-used capacity. Yet experimental procedures showing that birds are well within the mammalian range of competence in performing standardized—albeit "unnatural"—learning tasks seem to affirm the judgement, based on relative brain size, that birds and mammals are at comparable grades of biological intelligence' (Jerison 1976, p. 99).

There are, as we shall see, a number of rival theories concerning which brain measure is most appropriate for the estimation of intelligence, and it can be seen that one way of deciding between such theories would be to see which is best able to predict the actual ranking of species in intelligence.

The idea of producing rankings of intelligence arises naturally from those relatively unstructured views of the nature of intelligence which suppose that species differences are essentially quantitative in character. However, a more structured view allows further roles for comparative studies. If intelligent behaviour is the outcome of the interaction of a considerable number of information-processing mechanisms, then the task of the psychologist is to elucidate the properties of the various mechanisms. Suppose that in fact different species do not all possess the same complement of mechanisms, as, for example, would be the case if the evolution of intelligence involved the evolution of new devices, then, by contrasting the performance of various species, it might be possible to infer some at least of the properties of a mechanism that was possessed by one species, and not another. That is, qualitative differences between the intelligent behaviour of various species could serve to expose the logical structure of that behaviour.

Whether species differences will in fact point to the properties of the mechanisms used depends, of course, on the differences that actually exist between the species tested; there is no need to prejudge whether there are or are not any such qualitative distinctions. Moreover, it seems premature to argue that qualitative differences invalidate attempts to compare different species: Lockard (1971), for example, queries whether comparative studies are of any value, since the learning capacities of different species have evolved to be ideally adapted to the various ecological niches of the species, and, given that all species occupy different niches, no species have similar learning capacities. Of course, if different species had in fact no mechanisms in common, there would in practice be little hope of inferring the properties of any of the mechanisms through contrasts in behaviour: but if the species to be compared had some, but not all, mechanisms in common, a real possibility of such inferences would exist. The species to be considered in this book are vertebrates, and many of them in fact perform in apparently similar ways over a relatively wide range of behavioural tasks, so that there are at least grounds for supposing some common behavioural capacities.

In case it may seem over-optimistic to expect that we should be able to deduce the properties of internal logical devices simply, as it were, by 'subtracting' the capacities of one species from those of another, it should be pointed out that comparative studies may serve another, related, but somewhat more restricted, purpose. Suppose that a behavioural theory explains two or more apparently disparate behavioural phenomena as being due in each case to the operation of the same set of mechanisms—then, other things being equal, different species ought, according to the theory, to show either all or none of these phenomena. This argument is clearly made out by Sutherland and Mackintosh (1971), who also provide concrete instances of its application: the force of the argument depends, of course, upon how much is concealed in the phrase 'other things being equal'—it is possible, for example, that a superficially similar behavioural phenomenon has in fact a different causation in different species, so that the apparent conclusion need not hold good. What is clear is that comparative research of this kind, while unlikely to provide conclusive evidence for a behavioural theory, may well nevertheless generate strong additional support for a theory already possessing a sound empirical basis.

The classification of animals: differing degrees of relatedness

Classification of raw materials is a necessary first step in any scientific enterprise. This is because the basic scientific method is that of induction. Where observations indicate that a given rule dependably describes the behaviour of a particular set of individuals, then that rule is generalized to predict the behaviour of other, similar, individuals, as yet unobserved. To establish whether or not two individuals are similar requires some principle of classification: there are, however, an infinite number of ways of classifying individuals—in the case of animals, we could classify according to, for example, size, colour, lifespan, number of legs, ability to fly, and so on. None of these principles of classification, clearly, is correct or incorrect in any absolute sense, and each might indeed be the most appropriate principle for some purposes. But for scientific purposes, a classificatory principle is required that promises the most general application, a principle that gives a rational order to the huge range of types of individuals known.

One such principle that obtained wide acceptance before the nineteenth century was the so-called 'Scala naturae' or 'Great chain of being'. This principle supposed that each individual object in the universe could be assigned to a particular level in a series that began with the simplest and most inert objects, rose through the plants to simple animals, on to the highest mortal form, man, and culminated in heavenly beings such as angels. There were no gaps in this chain, so that no types had ever become extinct, and no new types were created;

apparent gaps were due to 'inessential' properties, and debate naturally ensued over which were in fact the essential properties of individuals. The problems apparent in this system need not concern us here: what is important to note is that it provides a justification for the notion that some types are 'higher' than others, and this is a notion which persists despite the general rejection of the principle.

It was, of course, the theory of evolution by natural selection that provided the principle required to supplant the notion of the scala naturae: the various living groups of animals (and plants) have all evolved from earlier forms, and so may be classified in a way that reflects their ancestry, and, in that sense, their degree of relatedness. There are a number of obvious but important observations that should be made concerning the evolutionary principle. First, no living group of animals has evolved from any other living group—to say, for example, that birds evolved from reptiles does not at all imply that any species of bird evolved from any living reptilian species, and it should be emphasized that reptiles have had just as much time in which to evolve since birds first appeared as have the birds themselves. Second, evolution implies change, but it does not necessarily imply 'progress'; although in general evolution has proceeded in the direction of increasing complexity, evolutionary change in any particular organ may involve degeneration just as it may involve increased complexity and sophistication. Third, the theory of evolution provides no rationale for regarding any one species as 'higher' than any other; it may or may not be proper to conclude that, in some context, amoebae are lower than, say, rats: but the fact that rats evolved from single-celled organisms can provide no part of any justification for such a conclusion. Evolution provides us with a rationale for the classification of animals, and an explanation of the similarities and differences between the groups that we observe, but it does not imply a linear path from simple to complex, or from lower to higher, and so provides no parallel to the scala naturae.

Although evolutionary theory provides the essential rationale for a scientific classification of animals, there remains, of course, the problem that there generally exists no direct evidence for the relationships between groups of animals; the fossil record contains many large gaps, and very few new species have arisen in the historical era. How then may animals be arranged in groups to reflect their affinities? Two techniques of classification may be contrasted here: on the one hand, classification of individuals may proceed according to whether or not they possess a particular character; on the other hand, individuals may be classified together according to how many of the total range of characters they possess in common—according, that is, to their 'general resemblance'. This latter method of classification, the 'natural' as opposed to 'artificial' methods, gains theoretical support from the knowledge that evolution progresses through the action of natural selection on the variations endemic in a population: environmental pressures will select for more extreme modification in some characters rather than in others, but which characters will be so modified cannot be predicted *a priori*—therefore, to detect relationships, all characters should be taken into account. Furthermore, where two types of animal occupy very similar ecological niches, then characters specifically related to that niche are unlikely to be reliable for diagnostic purposes, as they may have evolved independently, from unrelated ancestral forms, in response to similar pressures (the concept of convergence).

The natural method of classification distributes animals into groups of varying size, depending upon the amount of homogeneity there is within a group: cats, for example, resemble tigers more than they do bears, but cats and bears resemble each other more closely than do either resemble hippopotami. The various degrees of similarity, which are generally taken to reflect differing degrees of phylogenetic affinity, have resulted in the use by taxonomists of different ranks of classification. There are six obligate ranks in the animal kingdom, and these are, in order of increasing within-rank heterogeneity, the species, genus, family, order, class, and phylum; many other ranks are used in addition to these, but a taxonomist is not necessarily required to assign a given specimen to any but the obligate ranks. The classification is hierarchical so that, for example, a particular class of animals belongs to one and only one phylum, and may consist of one or more mutually exclusive orders, which in turn consist of one or more mutually exclusive families, and so on. In naming animals, use is made of the binomial system of nomenclature, which was first employed consistently by Linnaeus in the mid-eighteenth century. Each scientific name consists first of the name of the genus to which the species belongs, and, second, of the name of the particular species; where there is no danger of ambiguity, the first (generic) name may be abbreviated to its first one or two letters.

The lowest of the obligate ranks, the species, is the only rank for which anything like a formal definition may be given: Mayr, for example, defines species as 'groups of actually (or potentially) interbreeding natural populations which are reproductively isolated from other such groups' (Mayr 1969, p. 412). All living men belong to the same species (*Homo sapiens*), and this species, along with a number of extinct species, form the genus *Homo*. The group of species that comprise the genus *Homo* resemble each other more than any of them resembles any other species outside the group, and this is the basis for the establishment of *Homo* as a genus. This genus in turn is part of the family Hominidae, which belongs to the order Primates. This order contains all types of apes and monkeys, as well as such animals as lemurs and lorises, and so is obviously much more heterogeneous than the genus *Homo*. The primates in turn belong to the class Mammalia, whose members include the duck-billed platypus, the opossum, whales, elephants, rhinoceroses, and so on. The mammals are but one of the seven classes that go to form the subphylum Vertebrata, a division of the phylum Chordata.

It can be seen that, whereas the separation of one species from another is not arbitrary (although there are, certainly, difficulties in many cases), the separation of the higher ranks one from another depends upon what appear to be undefined notions of discontinuities in a range of the members of a lower rank. To what extent, it may be asked, are the higher ranks 'real', and what relation do they bear to phylogenetic affinities? This is a distinctly contentious issue, a detailed examination of which is, fortunately, not required here, as the number of vertebrate species on which behavioural data have been collected is so small that we need consider only coarse distinctions. For our purposes, it will be adequate to assume that, the lower the rank of the lowest group to which two species belong, the more recently their lines of descent have diverged.

This survey of the hierarchical classification of the animal kingdom may serve to emphasize the dramatic increase in heterogeneity within groups as we ascend the taxonomic ranks. On the assumption that such heterogeneity reflects differing phylogenetic histories, it becomes clear that the relationship between species that belong to different classes (using class in the technical sense), albeit within the same phylum, is very distant. In the chapters that follow we shall frequently examine differences between species that belong to different classes, and, although the term 'species-differences' will be used, as is the usual custom, the true nature of the contrast that is being drawn should always be borne in mind. Although all animals to be discussed are vertebrates, they are not all closely related.

Vertebrates: their place in evolution

The vertebrates are a subphylum of the phylum Chordata, which in turn contains but a small proportion (about 3 per cent) of the known species in the animal kingdom; the other 97 per cent belong to one or other of the 20 or so invertebrate (or, more properly, non-chordate) phyla. In terms of numbers of both individuals and species, many invertebrate groups have been vastly more successful than the vertebrates—there are, for example, more than 850 000 known species of insects, which form just one class within the phylum Arthropoda. The vertebrate skeletal structure, however, allows the greatest individual size and weight, and has led to the emergence of man. The fact that man is a vertebrate, and so more closely related to other vertebrates than to any other group of animals, is, of course, a central reason for the restriction of this book's concern to vertebrates: there are, however, other reasons. First, intelligent behaviour, our primary interest, is, at least to the naive observer, more evident in vertebrates in general than in invertebrates; second, and perhaps this is related to the first, there are far more extensive data available in the psychological literature on vertebrates than on invertebrates; third, although, as will be seen, there is considerable variation within vertebrates in organization of the central nervous system, the basic structure of the nervous system, a hollow dorsal tube, is common to all vertebrates, and, indeed, to all chordates. The basic plans of the nervous systems of all living non-chordates are strikingly different from that of chordates, so that no very meaningful comparisons may be drawn between their brain structures. No attempt will be made here to compare vertebrate with invertebrate as regards brain or intelligence, but it should be emphasized that the omission of invertebrates is not intended to imply that either the brain or the intelligence of any invertebrate is either lower or simpler than that found in any vertebrate.

Although the way in which chordates evolved from their invertebrate ancestors is far from certain, the most widely accepted view is that they evolved from the stock that gave rise to the modern echinoderms (star-fishes, sea-urchins, and sea-lilies, for example); the evidence that leads to this somewhat surprising conclusion relates primarily to similarities between the larval stages of echinoderms and certain chordates.

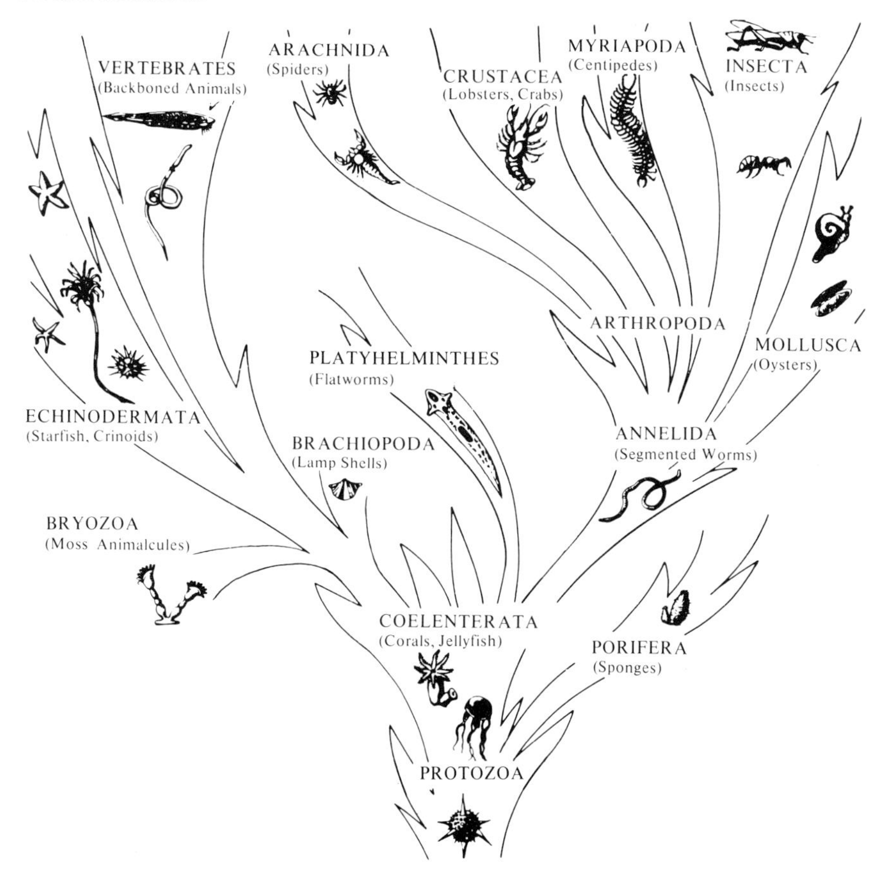

Fig. 1.1. A pictorial family tree of the animal kingdom. (From Romer 1966).

Vertebrates did not evolve, then, from the stock that gave rise to either of the two classes of invertebrates that have frequently been regarded as 'higher' invertebrates – the insects (highly regarded for their sophisticated social organization) or the cephalopods (whose ability to learn fairly complex discriminations is well documented). Figure 1.1 presents a simplified tree to show the probable relationships between the major animal phyla; this figure should serve to emphasize two cardinal features of the present discussion – first, the great 'distance' between vertebrates and most invertebrates, and, second, the fact that no modern major group of animals is the ancestor of any other modern group.

Table 1.1. *Subdivision of the three most recent geologic eras, showing the approximate date of the beginning of each period or epoch.* (After Romer and Parsons 1977.)

Era	Period	Epoch	M a ago	Events in biological evolution
CAINOZOIC (Age of mammals)	Quaternary	Pleistocene	1.8	Modern species of mammals and their forerunners; decimation of large mammals; widespread glaciation
	Tertiary	Pliocene	5.5	Appearance of many modern genera of mammals
		Miocene	22.5	Rise of modern subfamilies of mammals; spread of grassy plains; evolution of grazing mammals
		Oligocene	36	Rise of modern families of mammals
		Eocene	53.5	Rise of modern orders and suborders of mammals
		Paleocene	65	Dominance of archaic mammals
MESOZOIC (Age of reptiles)	Cretaceous		135	Dominance of angiosperm plants; extinction of large reptiles and ammonites by end of period
	Jurassic		195	Reptiles dominant on land, sea, and in air; first birds; archaic mammals
	Triassic		225	First dinosaurs, turtles, ichthyosaurs, plesiosaurs, mammals; cycads and conifers dominant

Table 1.1 (*cont.*)

Era	Period	Epoch	M a ago	Events in biological evolution
PALAEOZOIC	Permian		280	Radiation of reptiles, which displace amphibians as dominant group; widespread glaciation in southern hemisphere
	Carboniferous		345	Fern and seed fern coal forests; sharks and crinoids abundant – radiation of amphibians; first reptiles
	Devonian		394	Age of fishes; first trees, forests, and amphibians
	Silurian		440	Invasion of the land by plants and arthropods; archaic fishes
	Ordovician		500	Appearance of vertebrates (ostracoderms); brachiopods and cephalopods dominant
	Cambrian		570	Appearance of all major invertebrate phyla and many classes; dominance of trilobites and brachiopods; diversified algae

In any discussion concerned with evolution, time is an essential ingredient. The dates at which various groups of animals first appeared are basically established by determining the ages of the rocks in which their fossils are first detected, so that the geologic time scale is employed. Geologists generally divide the time since the formation of the earth (some 4600 million years ago) into five eras, of which only the three most recent provide evidence for the existence of vertebrates (although most invertebrate phyla probably evolved during the fourth most recent, Proterozoic, era, which began 1200 million years ago). The three recent eras are divided into periods, and the periods of the most recent era are further subdivided into epochs – Table 1.1 shows the duration of each of these stages, and summarizes the major events in evolution that occurred during them. More detailed accounts of the times at which the various groups of vertebrates appeared will be given, with reference to geologic time, as each group is introduced in subsequent chapters.

Organization of the book

The discussion thus far has been designed to provide an introduction to one of our major areas of concern—the comparative approach to the analysis of intelligence; Chapter 2 will serve to introduce a second major area—the physiological approach to intelligence.

Chapters 3 to 6 survey comparative and physiological data relevant to the intelligence of (in this order) fish, amphibians, reptiles, and birds. From what has already been said, it will be apparent that this ordering is in no sense intended to parallel increasing levels of complexity—the ordering may be seen as essentially traditional rather than logical. A further traditional feature of the chapters, which is inevitable given the nature of the great majority of studies reported to date, is that comparisons are drawn primarily between some species of the class being discussed and some mammalian species (usually the rat). The rationale given in this chapter for comparative studies clearly does not imply that such contrasts are likely to be of more interest than contrasts that do not involve mammalian species. However, the emphasis is, as has been said, imposed by the data available, and reflects no doubt partly the fact that not all comparative psychologists are motivated by the objectives laid out here, and partly a certain natural anthropocentric orientation. Intelligence, rightly or wrongly, has for centuries been regarded as either peculiar to or best developed in mankind, and it is perhaps for this reason that mammalian species—more closely related to man than species of any other vertebrate class—should function as reference points in so many comparative studies.

Chapter 7 considers contrasts between mammalian species (excluding humans) in brain organization and in intellectual performance. Contrasts between the intelligence of non-human mammals and humans will be considered in Chapter 8, in which particular attention will be paid to the acquisition of language; Chapter 8 will also discuss human brain organization and its relevance to language and memory. The final chapter sets out two hypotheses concerning the nature of vertebrate intelligence which have emerged from the surveys of Chapters 3 to 8; a number of potential objections to these hypotheses are considered and their wider implications are briefly discussed.

2. The physiological analysis of intelligence

Brain and intelligence

Behaviour is a consequence of brain activity: it therefore makes good sense to look to differences between brains for an explanation of differences in behaviour. There are two striking types of variation in brains of vertebrates, the first, in overall size of brain, and the second, in the relative size and differentiation of the various brain regions. In this section we shall consider these two types of variation, taking first, variations in brain size.

Brain size and intelligence

The assumption that the major differences between the intellectual capacities of species are quantitative rather than qualitative implies an interest in quantitative differences between brains as opposed to qualitative differences in brain organization. So, for example, Jerison writes: 'a notion such as total information-processing capacity is, in my judgment, more likely to be productive of an understanding of a dimension of evolution and differentiation', and 'reorganisations of the brain would be no more than the neural equivalents of species-specific behaviour patterns' (Jerison 1973, p. 81). If the organization of the brain is irrelevant, then total brain size seems an obvious candidate for use as an estimate of total information-processing capacity. The possibility that the anatomical differentiation of the brain into discrete regions might not be functionally significant at least as regards general intelligence gains some support from Lashley's concepts of cerebral mass action and equipotentiality (Lashley 1929). Lashley concluded from a series of lesion studies using rats that for complex tasks (such as maze learning) the entire rat neocortex was equipotential, and that the various parts of the neocortex somehow facilitated each other's action so that the more cortex was available the more efficient was performance. Now while these notions do provide some backing for the use of brain size as a statistic relevant to intelligence, it should be noted that Lashley's proposals referred not to the entire brain, but to the neocortex specifically: the extent to which Lashley's ideas can be maintained with regard to the neocortex will be discussed in Chapter 7.

One aspect of organization that is relevant to the quantitative approach is the relation of neuron density to brain size. In mammalian cortex, at least, the number of neurons per unit volume (that is, neuron density) declines with increase in brain size (e.g. Bok 1959). So that, on the assumption that the neuron is the functional unit relevant to information-processing, it would seem that a correction should be applied to brain size data to scale down large brains. However, Jerison points out that just as neuron density declines with increasing brain volume, so neural connectivity increases—neural connectivity being estimated from the length of dendritic trees of cortical neurons (Bok 1959): Fig. 2.1 shows this reciprocal relationship clearly. Jerison now proposes that the activity of a neuron is proportional to the length of its dendritic tree, and that, therefore, activity per unit volume of brain is independent of brain size.

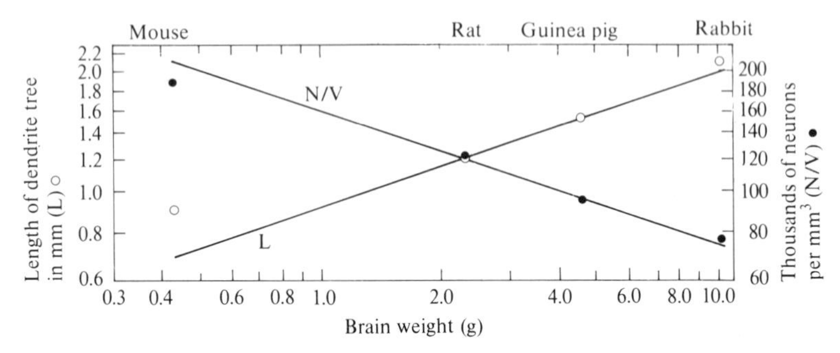

Fig. 2.1. Neuron density and dendritic tree length, both as a function of brain weight. Lines fitted by eye to have slopes of $-\frac{1}{3}$ and $\frac{1}{3}$ respectively. (From Jerison 1973.)

Two comments are worth making here. First, independence of activity and brain size is deduced from the assumption of a linear relationship between dendritic tree length and activity; there is simply no evidence to support (or refute) such an assumption. Second, the data for both density and connectivity were obtained from mammalian cortex; whether similar generalizations hold for other classes of vertebrates, and for non-cortical mammalian brain, is also unknown.

These reservations concerning the use of brain size as a measure are to some extent counterbalanced by one of its advantages, namely, that brain size can be estimated for extinct species from examination of endocranial casts (endocasts), which consist of the mineralized deposits that replaced the soft brain tissues within the skull. Endocasts occasionally reveal some surface detail, but nothing, of course, of the internal brain organization of fossil species. Analysis of endocast data is most relevant to those concerned with the evolution of brain size, but will not be of central interest here, as we can say very little of the behaviour or intelligence of extinct species.

Figure 2.2 provides a good introduction to the arguments advanced by Jerison, who has made out the most detailed case for the use of brain size as a significant statistic. The figure shows brain and body weights for 198 vertebrate species, plotted on log-log co-ordinates. The data were obtained from Crile and Quiring (1940), and are for the largest specimen of each species, where more than one specimen of a given species was available.

Three salient facts emerge from Fig. 2.2. First, man's brain is not the heaviest mammalian brain; second, there does appear to be an orderly relationship between the logarithms of brain size and body weight, implying that the two are related by a power function; third, the data points for reptiles and fish appear to be lower than those for birds and mammals. These conclusions are emphasized in

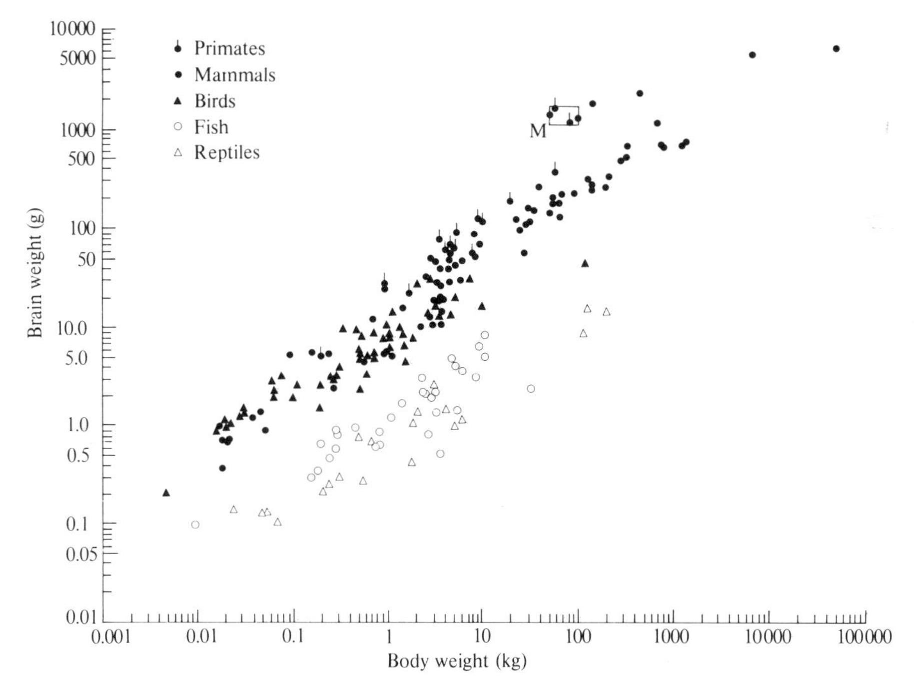

Fig. 2.2. Brain and body weight of 198 vertebrate species. Rectangle M contains the full range of data on living men from the Crile and Quiring (1940) sample; four points on the borders of the rectangle are the heaviest and lightest individuals. (From Jerison 1969.)

Fig. 2.3 (also taken from Jerison), which shows the data from birds and mammals contained within one polygon, and those from fish and reptiles within another, non-overlapping, polygon (the polygons shown here were drawn to contain all the points shown in Fig. 2.2, although only selected points are shown in Fig. 2.3).

Now it is, of course, not possible to decide how brain size might be related to intellectual capacity without making some assumptions about how vertebrates are ranked in intelligence, and that is a question which we have yet to examine; it will not be premature, however, to introduce one notion commonly adopted by workers interested in the problem, which is that man is the most intelligent vertebrate. If this proposal—or even the weaker proposal that man is at least as intelligent as any other vertebrate—is accepted, then it is clear, from the fact that man's brain is not the largest vertebrate brain, that absolute brain size does not determine intellectual capacity. An alternative suggestion might be that intelligence is determined by relative brain size, in the sense of the simple ratio of brain to body weight; but, as will become clear in what follows, brain weight in vertebrates does not in general appear to increase linearly with body weight, so that heavy vertebrates have proportionally smaller brains than light vertebrates, and many small mammals have, in terms of simple ratios, relatively larger brains than man. If we are to maintain both that brain size is critically relevant to intelligence and that man is the most intelligent vertebrate, we must, then, take body size into account and yet this cannot be done by adopting the simple ratio measure.

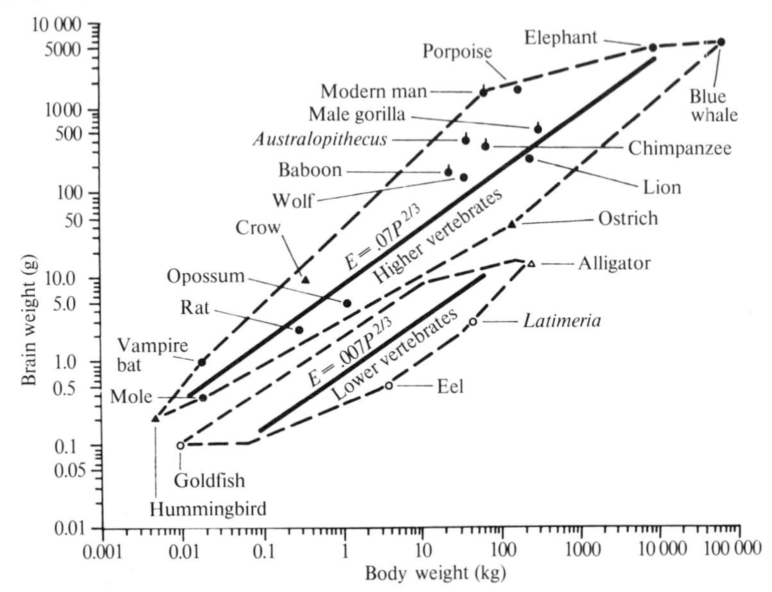

Fig. 2.3. Selected data from Fig. 2.2 to illustrate brain: body relations in familiar species. Data on the fossil hominid, *Australopithecus*, and on the living coelacanth, *Latimeria*, are added. The lines enclosed within the polygons are visually fitted. (From Jerison 1969).

One obvious solution to the problem is to evade it by comparing animals of the same body weight but different brain size, as in this case no matter what correction is appropriate to adjust brain size for body weight, the prediction clearly must be that the species having the larger brain will be the more intelligent. This proposal allows some fairly gross predictions to be made from the data shown on Fig. 2.3. That figure shows that none of the reptile or fish specimens had brains as large as those of any of the birds or mammals that were of a comparable size; bird and mammal brains are roughly ten times larger than those of fish or reptiles in species of comparable body weights. It therefore seems reasonable to assume that, if gross brain size is in any way correlated with intelligence, birds and mammals should be in general more intelligent than reptiles or fish. Further, if we assume that intelligence does not vary with body size in either group, then we may not unreasonably compare, say, a fish and a mammal of different body weights.

An alternative solution is to attempt to derive an appropriate correction for body size variations, and this brings up what is the critical question—if the increase in brain weight that accompanies increases in body weight does not necessarily increase intelligence, then what is the function of the 'extra' brain matter? The most plausible explanation is that the increased brain size of larger animals is required to cope with the increasing somatic or vegetative demands that are inevitable consequences of increases in body size; this proposal, however, does not tell us anything of the detailed quantitative correction that should be made to allow for this 'somatic' factor.

Some estimate of the quantitative effect of the somatic factor may be obtained by assuming that intelligence does not vary systematically with body weight; such an assumption seems plausible (although some writers, notably Rensch (1956) have argued that in fact intelligence in mammals does increase regularly with body weight). This notion would allow us to say that the regular increase in brain weight with increase in body weight, seen in Figs. 2.2 and 2.3, must represent the effect of the somatic factor; the cluster of brain size points at any one body weight represent, on this view, deviations of individual species in intellectual capacity from the average capacity of, say, birds or mammals, the same average level being maintained across the whole range of body sizes. If we could quantify the component due to the somatic factor, then it would be possible to quantify the extent to which the brain of some species deviates above or below the average, and so, to rank species of different body weights in terms of a measure of relative brain size.

The science of relating quantitatively the size of one part of the body to another is known as allometry, and it is this technique that we shall now examine. Figure 2.3 shows lines fitted to the data points within each polygon; the equations for the lines have the general form $E = kP^a$, where E is brain weight, P is body weight, and k and a are constants; this equation, proposed by Snell in the late nineteenth century, is known as the equation of 'simple allometry' (Gould 1966). The constant a determines the slope of the line, and is of central importance as it quantifies the supposedly somatic component in the growth of the brain. Fairly small variations in the value of a result in considerable variations in the ranking of relative brain size, so that its precise value is of some significance; how, then, is the exact value of a to be established? One method would be to take the best-fitting line for all the data points available. This method does not seem satisfactory, since a would have to be revalued with the publication of each new brain-body weight data point, and the selection of data available at any one time would necessarily appear somewhat arbitrary. More satisfactory is to adopt a value having some independent rationale, and this is the solution chosen by Jerison (1973). The lines shown on Fig. 2.3 are not best-fitting lines (the slopes for which in fact differ somewhat for fish, reptiles, birds and mammals), but take $a = \frac{2}{3}$. Jerison's selection of $a = \frac{2}{3}$ follows from the hypothesis, designed to account for the regular trends shown in Figs. 2.2 and 2.3, that it is the body surface that is the major determinant of the amount of brain required for somatic functions and that, as the body surface increases with the square of the length of an animal, whereas the volume increases with the cube of the length, then brain volume should indeed increase with body size, with an exponent of $\frac{2}{3}$.

Once an acceptable value of a has been agreed, it can be seen that brain weight is determined by two factors—P (body weight) and k. As k is the sole determinant of brain volume, once body size has been allowed for in this way, it was named the 'index of cephalization' by Dubois (1897). The allometric equation, then, gives us one way of establishing the relative capacity of brains of different species of differing body weights: by entering in the equation the absolute values for the weights of the brains and bodies of two species, and using an acceptable value of a, the value of k can be deduced for each species. If k, the index of cephalization, is larger in one species than in another, it may be concluded that the former species is, in relative terms, better endowed with brain matter than the latter. For the data shown in Fig. 2.3, the higher the value of k, the greater is the vertical displacement of a data point above the fitted lines. Figure 2.3 shows that, assuming $a = \frac{2}{3}$, modern man has the highest index of all the species plotted, that the porpoise is not far behind, and that both have considerably higher indices than those of either the elephant or the blue whale, each of which possesses, in absolute terms, a very much larger brain. The fact that man enjoys the highest cephalization index gives support to the idea that relative brain size measured in this way may be a critical determinant of intelligence.

The problem with this conclusion is, of course, that it rests heavily on assumptions that require empirical verification: it assumes that any two species having different brain sizes but the same body weight invariably have different intellectual capacities, and that the purely somatic factor is exactly quantified by the proposal that the constant a in the allometric equation has a value of $\frac{2}{3}$—that any individual deviations from the points obtained by using such a value inevitably represent not, for example, an increased or decreased capacity for somatic information processing, but changes in general intelligence. These matters are all in fact controversial. Dubois, for example, arrived at a value posed, the same intelligence, and fitting lines to their brain-body weight data points. Other authors have assumed still other values for a, and the problem of deciding on an appropriate value of a becomes very real where comparisons between closely related species are to be drawn, as, in general, brain volumes appear to increase much less than would be expected from $a = \frac{2}{3}$ as body sizes of such species increase (Pilbeam and Gould 1974).

Alternative solutions have been proposed for the problem of estimating the proportion of brain that may be assumed to be devoted to purely bodily function; if we assume, for example, that spinal cord functions are purely bodily functions, we could perhaps estimate relative intelligence by comparing ratios of brain to spinal cord (as suggested by Krompecher and Lipak 1966). Most of these alternative solutions have been applied to one class of animals only, and do not have the generality of Jerison's account; the important point to note at this introductory stage is simply that all of these theories rest on assumptions concerning the appropriate allowance for differences in body size, and that these assumptions, while generally enjoying some measure of intuitive plausibility, are far from being proven facts. Evidence concerning species differences in intelligence is clearly critical for selecting between these theories; Jerison's case, like all versions of the simple allometric equation, must suppose a gross difference between mammals and birds on the one hand, and reptiles and fish on the other. Other workers have produced encephalization rankings for species of birds and species of mammals, and these rankings should, if they have any behavioural validity, correlate with rankings obtained in objective tests of intelligence.

Although the emphasis here has been on the necessity for behavioural investigations to provide an empirical basis for hypotheses concerning factors involved in gross brain size, it should also be pointed out that, should there emerge a method of ranking in brain size that does successfully predict intellectual capacity, there would be major implications, in turn, for theories of intelligence. For such an outcome would support the notion that structural reorganization of the brain is not relevant to increasing intelligence, which would in turn imply that the major variations between the intellectual capacities of vertebrates are quantitative and not qualitative in nature.

Brain organization and intelligence

Differences in organization of brains are of primary interest to those who consider that there are qualitative behavioural differences between species. Bitterman, for example, has argued that the capacities of fish and reptiles on the one hand, and of mammals on the other, are qualitatively different, and that this difference reflects the absence of neocortex in fish and reptiles. However, it has to be noted that gross differences in anatomical organization do not by any means imply corresponding differences in function: whereas monkeys, for example, have a highly developed visual neocortex, pigeons have none, but that of itself provides no guarantee that there are any important differences between the species in visual perception. Moreover, marked differences in behavioural capacity do not imply that correspondingly striking anatomical differences will be found: one cannot tell from the study of human brains anything of the differences amongst their former possessors in personality, intellect, skills, memories, and so on.

If, on the other hand, it could be shown that a particular brain structure in one species does play an important role in a certain class of behavioural tasks, and that another species, not possessing that structure, performs poorly on those tasks, it may very reasonably be argued that the poor performance of the latter species does indeed reflect the absence of the structure in question. It might seem that such a conclusion would be of minor interest to psychologists, who would be satisfied simply by the information that the two species differed in their behavioural capacities. But, as has been suggested, it is not at all easy to show that differences in behavioural performance do reflect differences in information-processing rather, than, say, in motivation. A case made out on purely behavioural grounds for a significant species difference may receive good support from anatomical and physiological data that provide plausible explanations of the difference. The task of establishing the role of various brain structures in complex behaviour has long been a central preoccupation of physiological psychologists, and it is appropriate to introduce the major techniques and a discussion of their rationale at this point.

The lesion technique has been the method most favoured by physiological psychologists in their attempts to localize brain mechanisms involved in intelligent behaviour. If we wish to test the hypothesis that a given area is necessary for a particular behavioural function, then we may remove that area, and test to see whether that function is still available to the animal. Alternatively, if we simply wish to know what role a given brain area might play, that area may be destroyed, and a series of tasks administered to assess what capacities are now impaired.

There have been a number of objections raised concerning the utility of this approach. An obvious question is, of what interest is it to psychologists that a given function is localized in, say, the occipital lobe and not in the frontal lobe? One rejoinder to this has already been advanced, namely, that if one species possesses a brain structure not possessed by another species, then evidence that the structure plays a given role in behaviour may support the contention that there is a significant difference in the organization of behaviour between the two species. There are, of course, further advantages of localization per se: first, analyses of the nature of the inputs and outputs of the region may help generate hypotheses about the way in which the mechanism is engaged in behaviour; second, more detailed exploration of the area, using a variety of techniques, may provide a finer-grain analysis of its behavioural function.

A second objection to the lesion technique is nicely stated by Gregory (1961), in a consideration of the brain as 'an engineering problem'. Gregory compares ablation of a brain area to removal of a component from a complex machine and argues that: 'If a component is removed almost anything may happen: a radio set may emit piercing whistles or deep growls, a television set may produce curious patterns, a car engine may back-fire or blow up or simply stop' (Gregory 1961, p. 320). Gregory claims that such dysfunctions are likely to be meaningless unless we first of all understand the principles governing the behaviour of the machine—and such an understanding is precisely what is being sought by physiological psychologists. Although persuasively stated, there is a quite unjustified assumption concealed in Gregory's argument which is that the dysfunctions produced by brain lesions will in fact be analogous to growls and whistles. This is surely very much an empirical question; until the experiments have been carried out we simply do not know whether the behavioural changes induced by brain lesions will be meaningful or not.

What the effects of lesions are must depend on the relationship between anatomy and function: it is not impossible that at least some discrete brain areas perform discrete functions, so that removal of one region might selectively remove one behavioural capacity. To hope for such an outcome may, however, be regarded as a particularly optimistic view; an alternative view is simply that the dysfunctions obtained may 'reveal the natural fracture lines of behaviour' (Thomas, Hostetter, and Barker 1968, p. 234). The dysfunctions may, for example, show that certain classes of behaviour seem to belong together in that lesions produce 'syndromes' of behavioural deficits, no one of which is seen in isolation. It should be borne in mind that the behaviour seen in lesioned animals is a consequence of activity in mechanisms possessed by normal animals; unless we suppose that, following a lesion, the remaining intact brain regions are performing abnormally, we may regard the differences in behaviour of normal as compared with that of lesioned animals as due to the superimposition of some capacity in normals upon the capacities of the lesioned subjects. Analyses of the differences in behaviour may, then, enable us to define the properties of the mechanism or mechanisms that are available in the normal but not the lesioned animals, and this in turn may be an important contribution to behavioural theory. For the emphasis now is not on localization of function so much as on description of function. We are using brain lesions not to analyse brain function as such, but rather to elucidate behavioural organization. If we wish to obtain a catalogue of discrete behavioural functions involved in intelligence, then, depending upon how function and anatomy are related in the brain, the lesion technique may provide some entries for that catalogue.

The primary concern of this section has been to argue that only empirical data can decide whether brain damage will or will not provide useful information, and it is appropriate, therefore, to describe some of the evidence that gives encouragement to physiological psychologists.

Perhaps the most striking example of a relatively discrete behavioural deficit brought about by localized brain damage is that seen following hippocampal damage in man, first described for Scoville and Milner's patient H.M., and since explored in a number of other patients (see Chapter 8 for a more detailed discussion). In brief, H.M. appears unaffected in general intelligence but suffers from a severe anterograde amnesia; events that occurred prior to his operation are generally well recalled, whereas events occurring subsequently are retained for a few seconds only, while overt rehearsal is possible. Very little information is stored for any length of time following hippocampal damage, and rather sophisticated testing techniques are required for detection of that storage. Although there is, as we shall see in Chapter 8, currently no universally accepted interpretation of this phenomenon, one common account is that the capacity for transferring information from a short-term to a long-term memory store is lost: this interpretation implies that brain damage has in this instance produced a discrete effect in a relatively complex process, and, what is of particular significance here, H.M.'s dysfunction is used by psychologists as evidence in support of the controversial claim that there are in fact discrete short-term and long-term memory stores (e.g. Baddeley 1972). The case of H.M. is brought forward here to illustrate two points: first, lesion-induced dysfunctions need not be so bizarre as to be meaningless, and second, psychologists do in fact use such dysfunctions as evidence that is clearly relevant to behavioural theory.

It should be pointed out that the effects of brain lesions need not be as discrete as those seen in H.M. to be theoretically useful. Lesions of a number of structures in the mammalian 'limbic system' tend to disrupt behaviour in situations where either shocks are given as punishments or anticipated rewards are omitted, while having little or no effect on the learning of new tasks for, say, food reward. An interpretation of such findings is that response to punishment and to non-reward is mediated by a common system; this notion has been used by some psychologists to support the hypothesis that fear and frustration are similar, closely related emotions (e.g. Gray 1971).

It is, then, possible to obtain, by the lesion method, disruptions of behaviour that seem amenable to meaningful interpretation, so that progress may indeed be made in attempting to describe functions of various brain regions. Such progress will be valuable both from a comparative viewpoint, and in the wider field of behavioural theory.

This discussion has centred on the use of the lesion technique; similar arguments are, however, relevant to the two other major techniques of physiological psychology, stimulation and recording. Brain stimulation (in unanaesthetized, free-moving animals, via permanently implanted electrodes) presumably brings about the simultaneous discharge of many thousands of nerve cells, a distinctly unphysiological event; but there is reason to suppose that in some areas the behaviour obtained by such stimulation in fact mimics that obtained by naturally occurring activity in the area—in, for example, the lateral hypothalamus, where lesions and stimulation have basically opposite effects on appetitive behaviour. In other areas, e.g. the frontal cortex, brain stimulation appears to produce effects similar to those found following lesions of the area—the stimulation, that is, appears to 'jam' the activity of that region; this latter effect is to be expected in areas whose activity may be critically 'patterned', as might reasonably be supposed to be the case for most regions involved in the control of intelligent behaviour. To anticipate the evidence, our interest in this technique will be limited and largely restricted to its use in analysing the mechanisms of reinforcement, an interest which derives from the fact that stimulation of some areas appears to be rewarding, and of others, punishing.

Variations in type of electrode allow recordings to be taken from a single cell, from a small group of cells, or from many thousands of cells simultaneously. Recording from one or a few nerve cells has been of central importance in the analysis of peripheral sensory and motor systems, and in neuroanatomical studies, where functional connections may be established by showing which cells show short-latency responses to localized stimulation of distant brain regions; however, little information has yet been obtained with this technique that will be of interest here. This may be because, as has been suggested, there is complex patterning of activity in regions concerned with intelligent behaviour, so that no one cell can provide a meaningful picture of the integrated behaviour of a mechanism. Recording from large populations of cells (the electroencephalographic, or EEG, technique) has, on the other hand, provided some interesting data, and this too is not unexpected. Suppose that, according to some hypothesis, a given brain region is involved in a particular behavioural capacity, and that this capacity is, according to behavioural theory, supposed to be used at one stage only in performance of a task. If it could be shown that the gross activity of the area changed significantly from some baseline level during performance of the relevant stage, support would be obtained for both the physiological hypothesis and the behavioural theory.

In recent years, the above techniques of physiological psychology have been supplemented by pharmacological techniques. Drugs may be used to disrupt (temporarily or permanently) specific systems of neurons having, say, the same neurotransmitter in common; conversely, drugs may be used to stimulate similarly restricted neuronal systems. The advantage of these pharmacological procedures, as compared to the corresponding traditional techniques for lesioning and stimulation, lies, of course, in their selectivity: their use allows one to disrupt (or stimulate) some but not all of the neurons in a particular area, thus allowing the possibility of, for example, the fractionation of a syndrome produced by a gross lesion at a given site. Their disadvantage is that it is more difficult to restrict their site of action so as, for example, to avoid side-effects consequent upon the activation of peripheral neuronal systems of no direct interest. As pharmacological techniques have in fact been little used in studies relevant to the comparative psychology of intelligence, we shall have little to add, apart from observing that the rationale for their use is the same as that for the use of the more traditional lesion and stimulation techniques.

The vertebrate brain

Introduction to the vertebrate central nervous system

The preceding section, although providing a rationale for the use of the lesion (and other) techniques, provides no guide as to which parts of the brain might most profitably be explored and that is the issue to which attention will now be directed. As a preliminary, however, some introduction to the basic anatomy of the vertebrate nervous system is required.

It is widely agreed that the most fruitful way to view the basic plan of the vertebrate brain is to consider its embryonic development, the course of which is similar in all vertebrates. The central nervous system is derived from an infolding of dorsally placed ectodermal tissue which gradually develops from a groove, through the dorsal attachment of the walls of the groove, into a hollow tube. The anterior end of this tube expands more rapidly than the caudal end and, with its expansion, the three major anatomical subdivisions of the anterior part become evident: the most anterior section, the forebrain (or prosencephalon) is a swelling anterior to the ventral cerebral (or mesencephalic) flexure, the neighbouring region, the midbrain (or mesencephalon) being in turn separated from the final brain region, the hindbrain (or rhombencephalon) by a constriction of the tube at the isthmus; the remainder of the neural tube develops into the spinal cord. Further expansion of the forebrain results in the appearance of two subdivisions of that structure, the endbrain (telencephalon) and the between-brain (diencephalon); a constriction of the hindbrain divides that region into metencephalon and myelencephalon. Figure 2.4 shows stages in the formation of the brain in the human embryo.

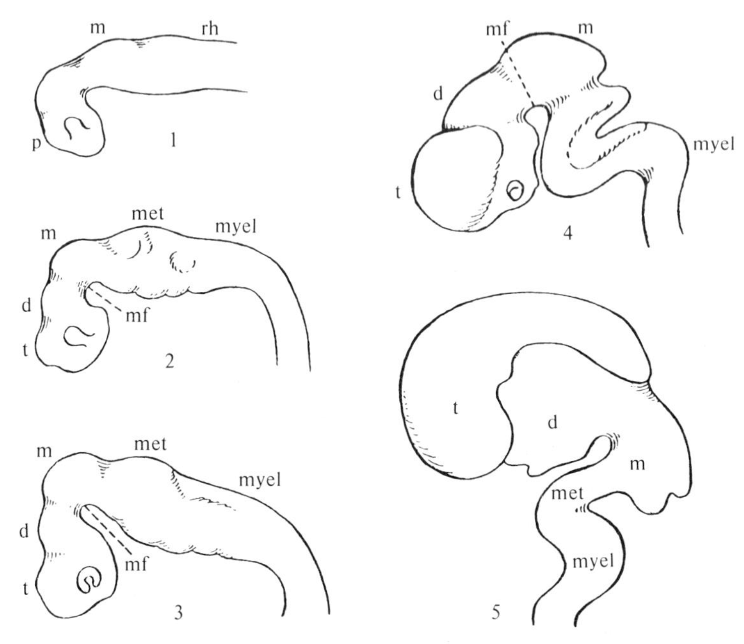

Fig. 2.4. Five stages in the development of the human brain. (From Everett 1965.)

Key: d – diencephalon; m – mesencephalon; mf – mesencephalic flexure; met – metencephalon; myel – myelencephalon; rh – rhombencephalon; t – telencephalon.

The diencephalon, or between brain, is divided into three parts, the epithalamus, the thalamus, and the hypothalamus. The mesencephalon consists of two regions, a roof (tectum) and floor (tegmentum), the former region being primarily sensory, and the latter, motor in function. The medulla is a structure that extends through the whole length of the rhombencephalon, being again essentially motor in the function of its ventral region, and sensory dorsally and laterally. The metencephalon, or more anterior region of the rhombencephalon, consists of, in addition to part of the medulla, the dorsally placed cerebellum and, in mammals, of the pons, a structure intimately related to the cerebellum.

The three major brain divisions (forebrain, midbrain, and hindbrain) are each related to different components of the cranial nerves, the sensory and motor nerves that supply primarily the head region. The forebrain is related to the olfactory nerve, as fibres from the olfactory bulb terminate in the telencephalon. The midbrain is related to the optic nerve, as fibres from the retina terminate in the tectum (optic nerve fibres also terminate in the thalamus of the forebrain, and the relative distribution of fibres to the midbrain and forebrain differs markedly amongst the vertebrate classes). There are also, in the floor of the midbrain, the nuclei of two of the nerves that control eye movements, the oculomotor and trochlear nerves. Finally, the medulla contains the nuclei of the nerves that control the great majority of muscular movements in the head region and receives the sensory roots of nerves relaying information from sense modalities other than vision and smell; such systems include those in the ear, concerned with balance and hearing; taste, receptors for which may be widely distributed over the body surface; the lateral-line system, present in fishes and larval amphibians, which is sensitive to water displacement and which, in some species, contains electroreceptive elements; and the common skin senses, such as touch, pressure, heat, and cold.

The forebrain as the site of major interest

We shall concentrate, in the chapters that follow, on the differences between forebrain organization that are found between vertebrate species, and there are three reasons for this emphasis on the forebrain. First, there is a traditional view of the evolution of the central nervous system that supposes that the forebrain evolved along with increasing intellectual capacity; second, most relevant investigations by physiological psychologists have explored forebrain rather than other brain regions; and third, there is indeed good evidence that, in mammals at least, forebrain damage does affect performance in a variety of learning tasks.

The traditional view of brain evolution in vertebrates may be briefly summarized as follows: in primitive vertebrates, behaviour was essentially reflexive in character, and reflexes were in general segmentally arranged, so that a stimulus to a particular region of the body surface tended to elicit muscular responses in that region. This primitive organization is reflected in the spinal cord of extant vertebrates. The anteriorly directed locomotion of primitive vertebrates favoured a concentration of sensory receptors at the anterior end of the body, this being especially true of those receptors concerned with stimuli (chemical, visual, or mechanical) arising from distant sources. These aggregates of sensory cells brought about the development of parts of the anterior nervous system devoted almost exclusively to the analysis of information within one sensory modality – the forebrain for olfaction, the midbrain for vision. Information from these regions flowed down to motor centres in the hindbrain. According to the traditional view, a condition similar to this may be found in fishes.

There are clear advantages to be gained from integrating information obtained from the different senses, and these advantages were increasingly exploited as vertebrates evolved; the telencephalon became the primary area of integration, and information from non-olfactory senses was now conveyed from the lower centres of the mid- and hindbrain via relay stations in the dorsal thalamus which accordingly is a minor structure in fishes, but a major brain area in mammals. The transfer of information from 'lower' to 'higher' centres resulted in the gradual take-over, by the higher telencephalic centres, of processes originally carried out by the lower centres, a process known as 'encephalization', and this was one factor leading to a progressively higher ratio of forebrain to mid- and hindbrain size. A second factor yielding telencephalic enlargement was a by-product of the capacity developed to integrate information from different senses—a shift from the dependence of behaviour on simple local reflexes to dependence upon learning, and so to the development of intelligence.

The traditional view, then, incorporates two major ideas: first, that brain and intelligence in vertebrates developed along the lines described, and second, that steps in that evolutionary process can be seen in living vertebrates, fish being in this sense lower than mammals (amphibians, reptiles, and birds occupying progressively higher intermediate steps). These are independent suppositions, so that one may very well accept the account of brain evolution without supposing that proof of that process is to be found in living species, of whatever class. The difficulty with rejecting the latter source of evidence is, of course, that no other primary evidence is available: fossilized animals provide only very gross clues to the organization of brain structures, and effectively no evidence relevant to the intelligence of the species. We shall, in subsequent chapters, encounter many findings that bear on ideas derived from the 'traditional view', in particular, on the notion that the telencephalon of 'lower' vertebrates is exclusively concerned with olfaction, on the notion that a progressive encephalization of sensory function (particularly of vision) occurs, and finally, of course, on the notion that as we go from fish to amphibians to reptiles, birds and mammals, we ascend a scale of intelligence. To anticipate the evidence, we shall find difficulties with many of these ideas: what should be borne in mind, however, is that difficulties found in the analysis of living species do not necessarily disprove the traditional account of brain evolution, many aspects of which retain a plausibility that is relatively independent of the conditions obtaining in living species.

Basic anatomy of the vertebrate forebrain

Although the fundamental anatomical and functional organization of the central nervous system is similar in vertebrate embryos of all classes, and although marked similarities in the adult organization of mid- and hindbrain structures can be discerned throughout vertebrates, there are striking differences in adult forebrain structures which, as we go from fishes to mammals, grow increasingly large relative to mid- and hindbrain structures. The increase in relative size of forebrain is accompanied by an increased differentiation of structures within the forebrain, leading to difficulties in deciding correspondences between forebrain structures in the various vertebrate classes. Before looking at the particular problems associated with the forebrain, it is important to discuss the senses in which two structures in different species may be said to correspond. Biologists traditionally distinguish between two major types of correspondence, namely, homology and analogy. Structures are said to be homologous if they are derived, through modification brought about by evolutionary processes, from the same structure in some common ancestor; thus the wing of a bat and the hand of a man are homologous. Structures are analogous if they are not homologous but do serve similar functions: the retina of the octopus is analogous to that of the goldfish (while the retina of the goldfish is homologous with that of man). It is important to note that two structures may be homologous while having very different functions: to take a well-known example, the mammalian middle-ear bones evolved from the jaw bones of some ancestral reptile, and are homologous, therefore, with the jaw bones of living reptiles.

Establishing homologies for soft tissues which do not form fossils poses obvious problems: the nature of those tissues in the supposed common ancestor cannot be established, and no intermediate steps in the series of modifications leading to structures in living species are demonstrable. Comparative neuroanatomists have relied heavily on two methods of inferring homologies. First, they have studied the embryological differentiation of the nervous system, and supposed that structures deriving from corresponding parts of the neural tube are homologous; second, they have compared the relative positions and fibre connections of nuclear masses in the adult brains of different species. Some anatomists have suggested that the functions of the structures might be taken into account: but we have already seen that there is no necessary relation between homology and function, so that correspondence in function, while it may corroborate conclusions reached from evidence of another type, is not convincing in itself in the establishment of homologies.

Embryological data, while being the most convincing evidence available, are of limited use particularly in establishing homologies between relatively small differentiated areas: embryological studies convince us that the forebrains of all vertebrates are indeed homologous, but provide rather less clear-cut implications for homologies to be drawn involving structures within, say, the dorsal telencephalon.

The basic evidence that has been used to establish brain homologies is the relative position and connections of nuclear masses. Relative position is, of course, a good general guide, but, again, imprecise particularly as regards problems of boundaries and subdivisions within brain areas. Reliance on fibre connections as a source of evidence suffers from two problems: first, the fibre connections that have been supposed by the early anatomists to exist have not infrequently been shown to be inaccurate by more recent work, using new experimental neuroanatomical techniques; second, there is no reason why two areas that are indeed homologous should not have very different fibre connections—if their functions had diverged over the course of their independent evolution, then their connections would almost inevitably have changed.

There are clearly profound difficulties in establishing homologies for brain regions amongst the various classes of vertebrates; for our purposes, this is not a serious problem—our main concern is to use physiological evidence in support of a psychological enquiry into the organization of intelligence, and that enquiry has no direct interest in the establishment of homologies of any kind. What the difficulties do mean is that we must exercise caution in accepting a suggestion that a brain area in one species 'corresponds to' a brain area in another species. The temptation to do so is particularly strong when the two areas have been given the same name: one should bear in mind that although giving two structures the same name implies homology, homology is far from guaranteed, and that, even if homology does exist, the areas may functionally be very different, and our prime interest is in the function, not the derivation, of structures in living species.

Given these caveats, we may now introduce the main regions of the vertebrate telencephalon, and their nomenclature. The major distinction is between those structures that are supposed to have derived from dorsal portions of the enlarged anterior neural tube, named pallial, and those ventrally derived, the ventromedial septal region, the ventral and lateral striatal regions and rostroventral structures, such as the olfactory tubercle and the anterior olfactory nucleus, associated with the entry of the olfactory tract into the telencephalon. The term pallium (mantle), which is interchangeable with cortex (bark), was introduced because in mammals pallial structures have enlarged laterally, finally curving ventrally so as to envelop the entire forebrain, thus forming its outer covering; the term striatum was introduced because the ventral telencephalic structures of mammals are striated by fibres of passage on their path from the thalamus to the cortex (no such striations are seen in the striatal regions of non-mammalian vertebrates).

In some vertebrate classes, both pallial and striatal regions may be subdivided into three major subdivisions, prefixed by paleo-, archi-, or neo-, these prefixes indicating the presumed order of evolutionary origin of the subdivisions (paleo- being the oldest). It should be noted here that these prefixes were introduced early in this century at a time when, as we shall see in subsequent chapters, prevailing ideas about the evolution of forebrain regions were very different from those current today; our use of terms including those prefixes should not, then, be taken to indicate any support for their evolutionary implications. In mammals, the paleocortex is represented by the piriform cortex, derived from lateral pallium, and the archicortex, by the medially derived hippocampus; the neocortex derives from pallium lying between the paleo- and archipallium. In mammals, paleostriatum is represented by the globus pallidus, archistriatum by the amygdala, and neostriatum by the caudate nucleus-putamen complex. So, where a structure in a non-mammal is named, say, archistriatum, the implication is that it is homologous to the mammalian amygdala; while it is not our concern to scrutinize closely such proposed homologies, they do help to give some meaning to the anatomical terms, and provide a framework within which to consider ways in which the forebrain may have evolved. They also, of course, provide guides as to where to look for structures that might play corresponding roles in intelligent behaviour in different species.

Sensorimotor organization of forebrain: rationale and methods

Although the central concern here is with the role of brain structures in intelligence, it is of interest to know, not only the anatomical appearance and description of a brain (with its attendant implied homologies) but also, something of the involvement of the various regions in sensory and motor processes. This interest arises from a further traditional (and plausible) view, namely, that brain areas not involved in sensory or motor processing are prime candidates for involvement in integrative processing (thus the 'association areas' of mammalian neocortex were so called not because of any positive evidence that they were concerned in association formation, but because there was evidence that they were not involved in sensory or motor processes).

Evidence for sensorimotor involvement may be obtained by either physiological or anatomical techniques. Of these, the physiological approach provides the most direct proof of involvement: stimulation or destruction of a motor area may obtain or disrupt movement; similarly, recording from a sensory area can be used to demonstrate that stimulation in a given modality obtains activity in that area, and lesions of a sensory area may show impaired sensory discrimination. However, a large body of evidence concerning sensory and motor organization in non-mammals is anatomical, and awaits confirmation by physiological techniques. Much of this evidence, moreover, was obtained in the first four decades of this century, using 'normal' as opposed to 'experimental' material. In this context, normal refers to tissue obtained from an animal that has been subjected to no special procedures while living; in normal material, fibres are traced from sensory receptors to their destinations by patient scrutiny of one stained section after another. This method is liable to error: it is extremely difficult to distinguish between the various fibre components of a given bundle of fibres, so as to be able to say, of a given group of fibres seen to terminate in one region, what the origin of that group is. It is even more difficult to say whether all the fibres from one source terminate in a given region, or whether some (or all) are in fact fibres of passage through that region. The last three decades have seen important developments in histological techniques which now allow confident determination of the origin, course, and termination of fibre bundles. These techniques involve subjecting animals to experimental procedures prior to histological examination. For example, Nauta and Gygax (1954) developed a technique for selectively staining degenerating axons; thus a brain area may be lesioned and, after sufficient survival time (usually a few days—it depends, among other things, on the length of the axons concerned) has been allowed to enable degeneration to proceed, the animal's brain may be sectioned and stained to show the course of axons originating in the lesion site. Refinements of this technique enable clear identification of the final terminations of axons, in addition to the course followed by the axons. Other experimental techniques enable the tracing back of pathways from their terminations to their cells of origin; for example, if horseradish peroxidase is injected into a region it is taken up by axon terminals in that region and transported back into the cell bodies from which the axons originate; selective staining of the tissue a few days after injection enables these cell bodies to be visualized. These anatomical techniques, using 'experimental' material, have been responsible for considerable advances in our detailed knowledge of neuronal pathways in the brain; in particular, they have clearly shown that many earlier conclusions, drawn using normal material, were inaccurate. It is therefore advisable to regard accounts of forebrain organization based solely on analyses of normal material with some caution, and it will in fact emerge that such a cautious approach to traditional views of forebrain organization in non-mammals is well justified.

Concluding comments

The rationale advanced for physiological psychology has marked similarities to that proposed for comparative psychology: it has been argued that both disciplines may serve to expose the logical structure of intelligent behaviour, in one case, by comparing individuals of one species differing in having either intact or damaged brains, and in the other, by comparing individuals of different species. Each approach relies on contrasts, the one making use of experimentally induced differences, the other using experiments made by nature (Boycott and Young 1950). It can be seen that other disciplines too may provide comparably useful contrasts: developmental psychology, for example, allows contrasts between behavioural capacities available at different stages of life; behavioural genetics generates contrasts between strains selectively bred for differential performance in restricted behavioural situations.

It has also been argued that whether either of these approaches will be fruitful is currently an open question, depending, for example, on what qualitative differences (if any) do exist amongst the intellectual capacities of various species, and on the relationship between anatomy and function in the central nervous system. Optimism concerning the utility of either approach may be related to a theorist's position on the continuum outlined in the first section of Chapter 1: if there are many diverse mechanisms involved in intelligence, then it is reasonable to suppose that not all the mechanisms are common to all species, and, similarly, that at least some of these mechamisms may be related in an orderly way to certain brain regions. So, for example, physiological psychologists like Lashley (1929) and comparative psychologists like Maier and Schneirla (1935) have argued that the principles of associative learning fail to provide adequate explanations of all forms of intelligent behaviour; conversely, learning theorists in the Thorndikian tradition have assumed that there are no important species differences, and view physiological experimentation with marked reservation. On species differences, Hull writes: 'the natural-science theory of behavior being developed by the present author and his associates assumes that all behavior of the individuals of a given species, and that of all species of mammals, including man, occurs according to the same set of primary laws, (Hull 1945, p. 56). Similarly, Skinner writes: '. . . I may say that the only differences I expect to see revealed between the behavior of rat and man (aside from enormous differences of complexity) lie in the field of verbal behavior' (Skinner 1938, p. 442). Hull, on the subject of neurophysiology, believed that 'Unfortunately, the knowledge of neurophysiology has not yet advanced to a point where it is of much assistance in telling us how the nervous system operates in the determination of important forms of behavior' (Hull 1951, p. 5). Once more, Skinner's view coincides with Hull's: 'I venture to assert that no fact of the nervous system has as yet told anyone anything new about behavior' (Skinner 1938, p. 425). Neither Hull nor Skinner positively rules out qualitative species differences or the possibility of physiological methods providing useful data: each makes it clear that, whatever his assumptions may be, these are questions that may be decided only by experimental evidence, and this is something on which we may surely all agree.

One of the ultimate goals of experimental psychology is the explanation of the human capacity to reason; that capacity evolved, and reflects the evolution of our brains. In what follows, we shall examine the evidence relevant to the proposal that study of brain and intellegence in non-human vertebrates may eventually throw light on our own thought processes.

3. Fish

Classification of fish: relationship to land vertebrates

Our interest in this and the corresponding sections of the next four chapters will be in the relationships between living groups of vertebrates, and in the time at which these groups ceased to share a common ancestry. We shall rely mainly on the account given by Romer (1966; Romer and Parsons 1977) of vertebrate relationships, and it should be borne in mind that classification of animals and speculation on their phylogeny are controversial matters, and that there is as yet no general agreement on the details of vertebrate evolution (see, for example, Stahl's 1974 review). However, differences between rival accounts are not so drastic as to affect our somewhat gross conclusions: although Nelson (1976), for example, offers a rather different classification of fish, his proposals do not alter what will be the central conclusion of the discussion that follows, namely that teleosts—the most widely investigated subjects of psychological experiments using fish—are far removed from the types of fish that gave rise to terrestrial vertebrates.

The fossil record suggests that fish, the first true vertebrates, appeared some 450-500 million years ago, during the Ordovician period. The bewildering variety of species that has evolved since that time may be divided into three separate classes, the Agnatha, the Elasmobranchiomorphi, and the Osteichthyes: the relationships between those classes and between fish and the other vertebrate classes are summarized in Fig. 3.1.

The earliest fossils are of ostracoderms, which belong to the class Agnatha (jawless vertebrates); most ostracoderms were small animals, possessing a striking external bony skeleton, but all are now extinct. The only living agnathous species are cyclostomes (lampreys and hagfishes); the lamprey's (jawless) mouth acts as a sucker by which it attaches itself to living fish, whose flesh it devours by using a specialized rasped tongue. The hagfish preys on dead or dying marine animals, usually invertebrates, using a similar rasped tongue. Cyclostomes do not possess a bony skeleton and if, as seems likely, they evolved from ostracoderms, they are, in that respect, degenerate; on the other hand, ostracoderms probably did not possess the rasped tongue, so that in that respect modern cyclostomes may be seen as advanced when compared to the most primitive vertebrates. Although cyclostomes are living representatives of the most primitive class of vertebrates, it was not from fishes like modern cyclostomes that more advanced fishes evolved, but rather, like modern cyclostomes themselves, from some primitive forms that remain unknown.

Representatives of the second class of fish, the Elasmobranchiomorphi, first appeared early in the Devonian period; these early fossils were of placoderms, a subclass of elasmobranchiomorphs. Placoderms had a bony outer armour, which gave them a rather forbidding appearance; unlike ostracoderms, they had jaws and paired fins, albeit of a primitive type. Although none of these creatures resembled any modern fish, it was probably from some (unknown) species of

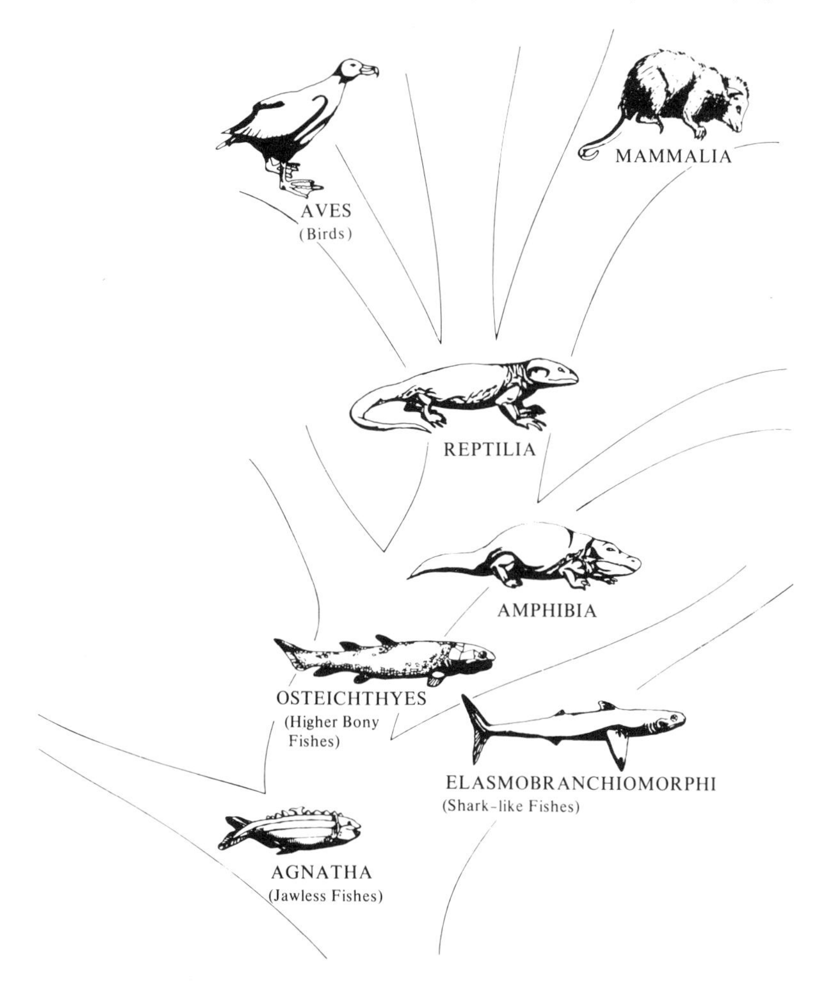

Fig. 3.1. A simplified family tree of the classes of vertebrates.
(From Romer and Parsons 1977.)

placoderm that the two living groups of elasmobranchiomorphs, the elasmobranchs (the sharks, skates, and rays) and the holocephalians (the chimaeras, or rat fishes) evolved. Elasmobranchs and holocephalians are cartilaginous fish, and it was originally thought that boneless fish were ancestral to all vertebrates. However, it is now known that the earliest jawless fish were bony (as were placoderms) and the absence of bone is seen as a degenerate feature. Sharks will be of most interest here, as there are some data available on their learning capacity, and as they do seem to be genuinely primitive vertebrates.

The earliest elasmobranch fossils, dating from the latter part of the Devonian, were of a shark-like creature belonging to the Order Cladoselachii. Although fossils of modern sharks do not appear until the Jurassic (195 million years ago), these fossils are remarkably similar to cladoselachian fossils, and, according to Young (1962), there are no clear signs of modern sharks being in any way of a 'higher' type than their Devonian ancestors. Modern sharks, then, appear to be no different from sharks that existed 195 million years ago, and, indeed, little different from their ancestors of 350 million years ago.

Although sharks are representatives of very early vertebrates, they in no way represent a form from which land vertebrates evolved. Different types of ostracoderms gave rise, on the one hand, to the elasmobranchiomorphs and, on the other, to the third class of fish, the Osteichthyes, or bony fishes. This class, in terms of numbers, is by far the most successful of the classes of fish: whereas there are some 600 species of living cartilaginous fishes, there are more than 20 000 species of living bony fish. The Class Osteichthyes is further divided into two subclasses, the Actinopterygii, or ray-finned fishes, and the Sarcopterygii, or fleshy-finned fishes; the earliest bony fish fossils date from the early Devonian (long before the earliest elasmobranch fossils) and both subclasses were already in existence at that stage.

Evolution within the actinopterygians is a complicated matter; the earliest actinopterygians belonged to the superorder Chondrostei, living examples of which are *Polypterus* (the African bichir), the sturgeons, and the paddlefish. From some long extinct forms of the Chondrostei evolved the Superorder Holostei, living representatives of which are the gar-pike and the bow fin. Finally, the Superorder Teleostei in turn evolved from extinct holostean forms, first appearing in the late Triassic (about 200 million years ago). The vast majority of living fish are teleosts, and it is to teleosts that comparative psychologists have devoted almost their entire research effort on fish. It is important to emphasize here that teleosts are representatives of the latest stage in the long history of evolution of actinopterygians, a history which has proceeded quite independently of that in sarcopterygians, for some 400 million years; it is particularly important because it is from sarcopterygians that terrestrial vertebrates evolved. Figure 3.2 summarizes the relationships between the various groups of bony fishes.

Sarcopterygians are divided into two orders, the Crossopterygii, of which the only living representative is the coelacanth (*Latimeria*), and the Dipnoi (lung fishes). Living forms of both of these orders show marked deviations from early sarcopterygian fossils; however, it was from crossopterygian stock that the first amphibians evolved, so that the coelacanth and lung fishes are the closest that we can come to assessing the nature of that ancestral stock. Although the brain anatomy of both coelacanths and lung fishes has been investigated, virtually nothing is known of the capacity of either group for intelligent behaviour.

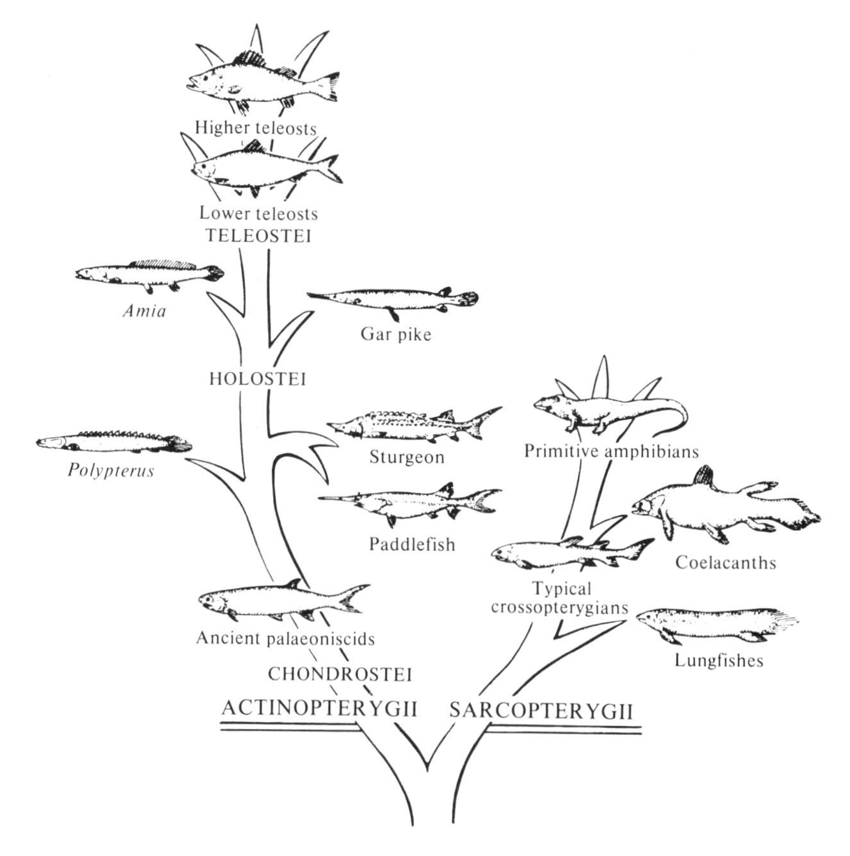

Fig. 3.2. A simplified family tree of the bony fishes, showing their relations to one another and to the amphibians. (From Romer and Parsons 1977.)

Forebrain organization in fish: deviant nature of teleost forebrain

In this section we shall consider the gross anatomical appearance of the forebrain of various groups of fish, noting the divisions within the telencephalon that can be seen in histological material. The functional organization of the brain will be considered in the succeeding section, and data on differences between the groups in brain size will be given in the section on interspecies differences (p. 101). It will be recalled from Chapter 2 (p. 28) that in general the fish brain is some ten times smaller than that of birds and mammals of comparable body-weight.

Tc the vast variety of types of fish there corresponds a similar diversity in forebrain organization. The gross anatomical appearance of the forebrains of many species is reasonably well known, although, as we shall see, relatively little is known of the connections of the fibre systems of those brains, and even less about the role of any of the parts of the brains in behaviour. Three types of brain will be considered here: a cyclostome brain, as an example of the most primitive vertebrate forebrain known, an elasmobranch brain, and a teleost brain, as the behavioural studies to be reviewed use almost exclusively species of these latter two groups.

It will be recalled that the nervous system of all vertebrate embryos consists of side walls of nervous tissue connected dorsally and ventrally by membranous tissue. The ways in which the rostral portions of this tube grow and alter their shape differ in important ways between the major groups of fishes, and this results in strikingly different types of gross forebrain organization.

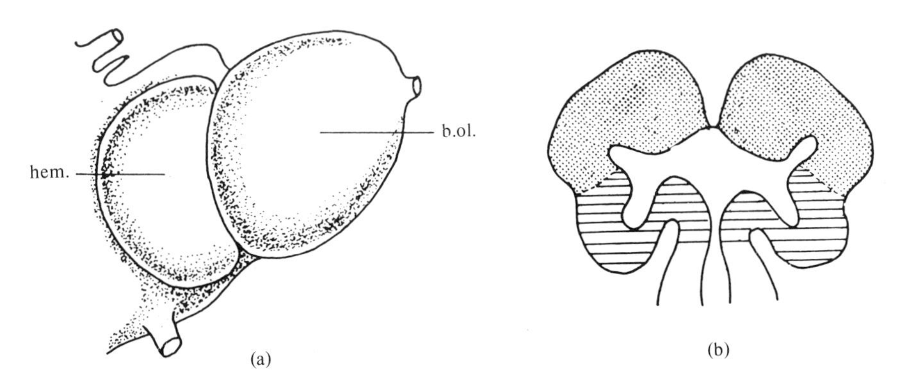

Fig. 3.3. (a) Lateral view of the lamprey forebrain. Hem. – cerebral hemisphere; b. ol. – olfactory bulb. (b) Diagrammatic horizontal section through the lamprey forebrain. Stippled parts, olfactory bulbs; striped parts, cerebral hemispheres. (From Nieuwenhuys 1967.)

Figure 3.3 illustrates the external anatomy of the lamprey forebrain. It can be seen that the olfactory bulbs are larger than the cerebral hemispheres, and that the lateral ventricle on each side runs from the hemisphere to the bulb. The shape of this brain has come about through a bulging, laterally, dorsally, and ventrally, of the walls of the neural tube, to form separate paired structures, each having an internal ventricle. This process, involving the thickening and hollowing out of the walls of the neural tube, is known to neuroanatomists as evagination, and is the basic mechanism by which the hemispheres are formed in all vertebrates barring actinopterygians. It is evagination in the rostral direction that brings about paired separate structures: Fig. 3.3 shows that virtually all the lamprey paired structures consist of the olfactory bulbs, the cerebral hemispheres being almost entirely unpaired; the unpaired portion of the telencephalon is known as the telencephalon medium.

Although Heier (1948) distinguished 12 different areas within the lamprey hemispheres, Nieuwenhuys (1967) could locate few clear-cut cytoarchitectonic boundaries, and found only four distinct zones, which he labelled the primordium hippocampi (the pallial zone of the telencephalon medium), the evaginated part of the pallium (Heier's primordium pallii dorsalis and primordium piriforme), the corpus striatum, and the preoptic nucleus. Other regions identified by Heier include both a medial and a lateral septal nucleus, and an anterior olfactory nucleus lying at the boundary of the olfactory bulbs and the telencephalon. Although the pallium possesses two layers of cells, neurons are in general scattered diffusely throughout both bulbs and hemispheres, showing little localized differentiation.

The picture which emerges, then, of the lamprey forebrain is of a relatively undifferentiated structure; on the other hand, if we accept Heier's description, then the major telencephalic divisions are already evident in these vertebrates and, in particular, there is already to be found a region—the primordium pallii dorsalis—which may have some claim to be homologous with mammalian neocortex.

There is, as one would expect of a group of more than 600 species, considerable variation in the forebrains of cartilaginous fish; Fig. 3.4 shows two very different shark brains. The basic principles of forebrain organization appear to be similar throughout the class, and Fig. 3.5 shows a section from a shark's brain. In the elasmobranch brain, the olfactory bulbs are clearly separable from the hemispheres, to which they are linked by hollow olfactory nerves, through which the lateral ventricles communicate with the olfactory ventricles. A substantial amount of the forebrain is paired, representing, compared with cyclostomes, a more marked rostral hemispheric evagination. The forebrain is distinctly divided, both medially and laterally, by a cell-free zona limitans; structures dorsal to this zone are termed pallial, and those ventrally placed, subpallial.

According to Northcutt (1977) the pallial region of the elasmobranchs shows three main longitudinal divisions into a lateral, a dorsal, and a medial pallium; the dorsal pallium possesses two laminae, the inner lamina developing caudally into a structure known as the central nucleus. Northcutt argues that these three pallial divisions may be homologues of the major regions of the mammalian cortex, the lateral region corresponding to the piriform cortex, the dorsal to the neocortex, and the medial, to the hippocampal cortex; although the last homologue is supported by topographical considerations alone, there is support also from the connections of the regions for the proposed homologues of the lateral and dorsal pallii, and those connections will be discussed in the following section.

The major divisions of the subpallial region of the elasmobranch forebrain are a lateral striatal primordium, a medial septal area (containing, according to Northcutt, both a lateral and a medial septal nucleus), and a ventral region, named by Northcutt the area superficialis basalis, and by Nieuwenhuys (1967) the olfactory tubercle. Northcutt also describes a preoptic area in the telencephalon medium, and suggests that regions that may be homologues of the pallial and subpallial divisions of the mammalian amygdala can be found, respectively, in the caudal pallium and the area superficialis basalis.

In comparison with the lamprey, then, the shark shows two 'advanced' characteristics: first, a greater proportion of the forebrain consists of paired hemispheres, and second, there is more differentiation of regions within the forebrain. Moreover, the elasmobranch forebrain shows (rather less controversially than lampreys) six major divisions, which seem generally comparable to those

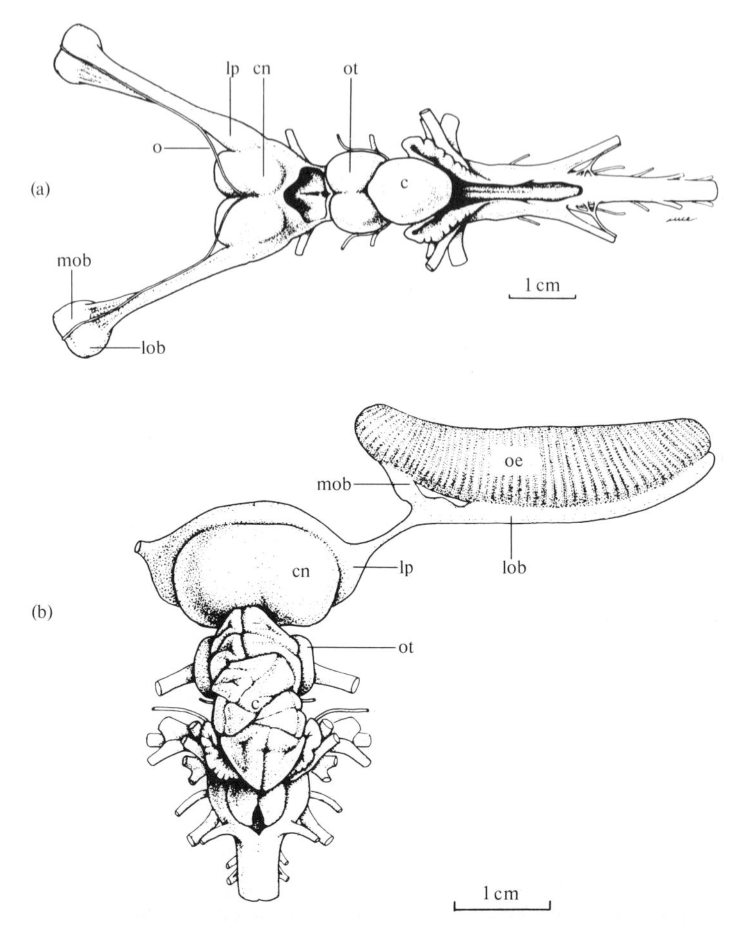

Fig. 3.4. Dorsal views of the brains of two sharks, *Notorynchus maculatus* (a) and *Sphyrna tiburo* (b), illustrating the range of shark brain variation. (From Northcutt 1977.)

Key: c – cerebellum; cn – central nucleus of telencephalon; lob – lateral division of the olfactory bulb; lp – lateral pallium; mob – medial division of the olfactory bulb; o – terminal nerve; oe – olfactory epithelium (sac); ot – optic tectum.

seen in the amphibian forebrain, although it will not be possible to postulate homologies between those divisions and structures in other vertebrates until many more experimental anatomical and physiological data are available. If in fact the homologies proposed currently are valid, the implication, given that cartilaginous fishes are not ancestral to amphibians, is either that these divisions did exist in some common ancestor, and so are very ancient, or that they have evolved in parallel, suggesting that the sixfold division is called for by similar environmental demands made upon the forebrain.

The teleost forebrain stands in striking contrast to the brains of all other non-actinopterygian vertebrates. Although there is considerable variation in the gross external morphology of teleost brains, all teleost forebrains show an unusual feature in their internal organization, which is that the hemispheres possess no lateral ventricles; according to Nieuwenhuys (1967) the teleostean hemispheres arise, not from a process of evagination, but from a process known as eversion. During the embryonic development of the teleostean forebrain, the dorsal parts

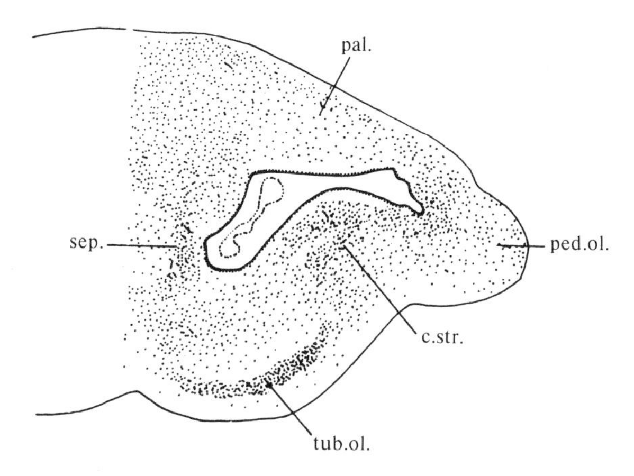

Fig. 3.5. Transverse hemisection through the telencephalon of the shark *Scyliorhinus caniculus.* (From Nieuwenhuys 1967.)
Key: c. str. – corpus striatum; pal. – pallium; ped. ol. – olfactory peduncle; sep. – septum; tub. ol. – olfactory tubercle.

of the walls of the neural tube begin to thicken and move apart, so stretching the roof membrane connecting them, while the ventral parts of the walls remain essentially static. Eventually, the dorsal parts, greatly thickened, recurve to envelop the ventral areas, which show much less enlargement; the forebrain, therefore, shows a wide, thin, membranous roof and no internal ventricles; Fig. 3.6 contrasts the processes of eversion and evagination.

The forebrain may be divided into pallial and subpallial regions, separated by the zona limitans, and Nieuwenhuys (1967) distinguishes three subpallial regions, a ventral nucleus (possibly the homologue of the lateral septal nucleus), a dorsal nucleus (homologue of part of the olfactory tubercle and of the striatal region), and a lateral nucleus (homologue of part of the olfactory tubercle and of the medial septal nucleus); Fig. 3.7 is a cross-section of the brain of a stickleback, showing the main pallial and subpallial regions.

Nieuwenhuys divides the pallial region into a central area, consisting of scattered large neurons, and a peripheral area, having three fields, which he terms the medial, dorsal, and lateral pallial nuclei. There is considerable controversy

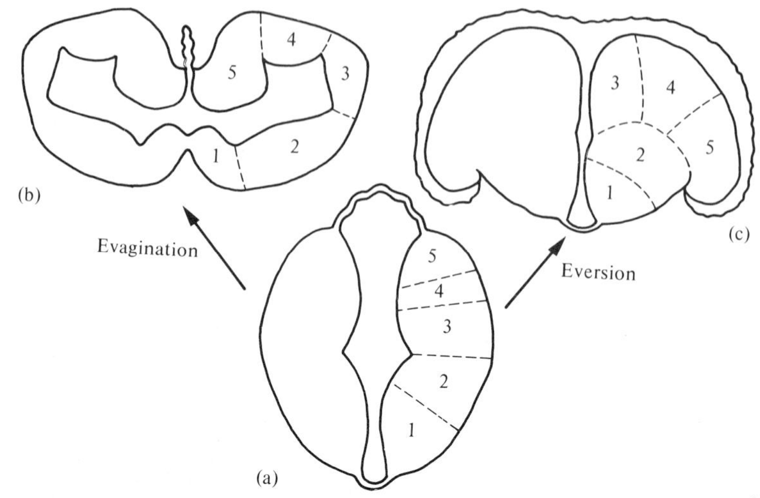

Fig. 3.6. Diagrammatic representation of the contrasting processes of evagination and eversion. (From Scalia and Ebbesson 1971.) This figure shows a hypothetical way in which the topology of the fundamental divisions of the embryonic telencephalon (a) might be preserved after evagination (b) or eversion (c).

Key: 1 – septal region; 2 – striatum; 3 – lateral pallium; 4 – dorsal; 5 – medial pallium.

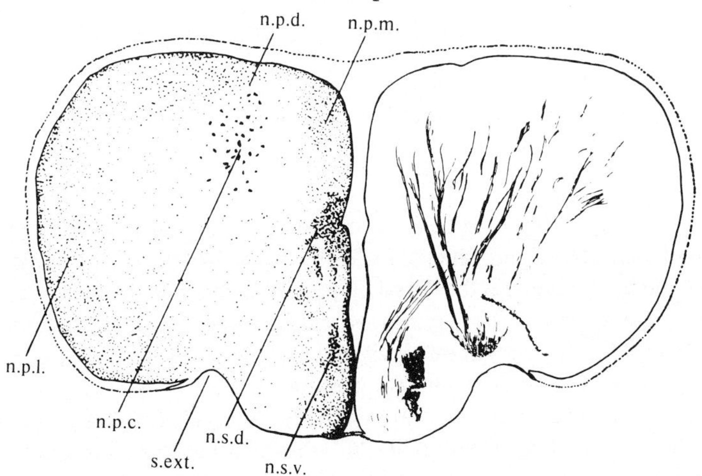

Fig. 3.7. Transverse section through the telencephalon of the stickleback, *Gasterosteus aculeatus.* Cell bodies are shown on the left, myelinated fibres on the right. (From Nieuwenhuys 1967.) The lateral subpallial nucleus is found, dorsal to the external sulcus, in regions posterior to that of this section.

Key: n.p.c. – central pallial nucleus; n.p.d. – dorsal pallial nucleus; n.p.l. – lateral pallial nucleus; n.p.m. – medial pallial nucleus; n.s.d. – dorsal subpallial nucleus; n.s.v. – ventral subpallial nucleus; s. ext. – external sulcus.

over possible homologues in other vertebrate lines for these regions: Nieuwenhuys argues that there is no rational way of establishing such homologies, and bases this argument on two grounds. First, the relative position of parts of the pallial division is of little significance, for purposes of establishing homologies, owing to the idiosyncratic mode of development in teleosts. Second, no pallial divisions are found in the chondrostean *Polypterus,* a living representative of the ancient palaeoniscoids from which the first holosteans, and finally teleosts, evolved; the division into three fields, that is, occurred entirely within the actinopterygian line, wholly independent of any divisions occurring in the sarcopterygian line that eventually gave rise to terrestrial vertebrates.

For the vast majority of investigations of behavioural capacities of fish, psychologists have chosen teleosts. What is fairly clear from both this and the preceding section is that, if such investigations were intended to throw light on the evolutionary development of intelligence, this was the least appropriate choice; on the other hand, of course, if there are any qualitative contrasts between groups of animals, it would seem most likely that they should emerge in comparisons between teleosts and terrestrial vertebrates.

Sensory and motor organization in fish forebrain

Early anatomists believed (and reflected this belief in their naming of brain regions) that olfactory processing was the primary role of fish forebrains; Herrick, for example, writes: 'In all fishes the greater part of the forebrain is dominated physiologically by the olfactory system, this entire region being designated the nosebrain. In the dogfish the nosebrain comprises about one-third of the entire brain' (Herrick 1924, p. 222). This view was based on interpretation of anatomical data, and that interpretation, it now appears, was heavily influenced by presuppositions concerning what the role of the fish forebrain should be. The evidence for this conclusion comes mainly from work on elasmobranchs, but we may first briefly consider the possibility of non-olfactory sensory information reaching the lamprey telencephalon.

Heier (1948) described a projection from the retina direct to the dorsal thalamus, and a projection from the dorsal thalamus to pallial regions of the telencephalon. Although this was based on inspection of normal material and it is, as stated in Chapter 2, extremely difficult to trace the paths of nerve fibres in such material, the projection from the retina to the thalamus has been confirmed, using modern experimental anatomical techniques (e.g. Northcutt and Przybylski 1973; Vesselkin, Ermakova, Repérant, Kosareva, and Kenigfest 1980). A further difficulty facing the proposal that the cyclostome forebrain is entirely olfactory is raised by Aronson (1970), who has pointed out that although ammocaetes (the larval form of lampreys) have poorly developed olfactory organs, their telencephalon shows no material difference from that of adult lampreys.

In recent years Ebbesson and his colleagues, using experimental neuroanatomical techniques, have provided valuable new data on sensory afferents to both teleost and elasmobranch telencephalon, the elasmobranch studies being more detailed at this stage. Using various species of sharks, it has been shown, first, that secondary olfactory fibres (whose cell bodies are in the olfactory bulb) project to a laterally situated area, near the attachment of the olfactory peduncle to the hemisphere (Ebbesson and Heimer 1970); this projection, which is entirely ipsilateral, indicates a more restricted distribution of bulb fibres than that suggested by earlier accounts, based on normal material. Ebbesson (1972) has since shown that the lateral olfactory telencephalic area itself projects to only restricted parts of the telencephalon (mainly ipsilaterally, to a ventrolateral telencephalic region); that is, the tertiary olfactory connections also do not involve large portions of the telencephalon. This account of restricted olfactory function in the shark telencephalon is supported by electrophysiological work carried out by Bruckmoser and Dieringer (1973), who showed that forebrain-evoked potentials, in response to electrical stimulation of the olfactory mucosa or bulb, were restricted to a small lateral area of ipsilateral forebrain.

Further work has shown that there are direct afferents from the thalamus to the shark telencephalon (Schroeder and Ebbesson 1974), the largest bundle of which terminates in the contralateral central telencephalic nucleus. The thalamus, in turn, receives afferents from both the (contralateral) retina (Graeber and Ebbesson 1972; Smeets 1981) and the spinal cord (Ebbesson 1972), and so could act as a sensory 'relay station', comparable to mammalian thalamus (except that, in the shark, the major thalamotelencephalic pathway is crossed—goes, that is, to the contralateral hemisphere). The possibility that the thalamus does indeed act as a visual relay station gains support from three further observations: first, there are short-latency evoked potentials to electrical stimulation of the optic tract in the ipsilateral posterior central telencephalic nucleus (Cohen, Duff, and Ebbesson 1973); second, sharks entirely deprived of their optic tectum are capable of performing visual discriminations requiring both brightness and pattern vision, although their performance in these situations may be somewhat poorer than that of normals (Graeber, Ebbesson, and Jane 1973); third, there is evidence, (Graeber 1978) that bilateral lesions of the posterior telencephalon in sharks produce visual dysfunction.

The studies reviewed above, when contrasted with Herrick's account, justify Ebbesson's statement that 'it is evident that the shark brain must be considered as poorly understood up to the advent of the silver impregnation techniques for tracing degenerating pathways' (Ebbesson 1972, p. 121). It is, of course, equally true that the shark telencephalon is still poorly understood; we have, for example, no information on the possible involvement of the forebrain in the control of movement and know little of the telencephalic representation of other senses (although there are electrophysiological data to show that both electrosensory and acoustic information reaches the telencephalon—Platt, Bullock, Czéh, Kovecević, Konjević, and Gojković 1974; Bullock and Corwin 1979).

Unfortunately, our knowledge of teleost brain organization is even more fragmentary, if we regard, as we surely must, the tracings of pathways by the earlier anatomists with suspicion. Scalia and Ebbesson (1971) have followed the destinations of secondary olfactory fibres in the moray eel (*Gymnothorax funebris*) from the bulb to three areas in the hemispheres, one in the lateral wall of the hemispheres, the second and third in a posterior and a ventromedial area, respectively. They point out that in mammals also, secondary olfactory fibres project primarily to three sites—the prepiriform cortex, the amygdaloid complex, and the olfactory tubercle; Scalia and Ebbesson add that to establish the comparability of these areas in mammals and teleosts, further study, in particular of the projections of the areas in teleosts, would be required. We can, then, say that in teleosts, as in elasmobranchs, the secondary olfactory fibres project to limited telencephalic regions; in the case of teleosts, however, we have no modern accounts of the projections of tertiary olfactory fibres.

Relatively little work has been carried out on the central pathways of vision in teleosts. Most reports (e.g. Ebbesson 1968; Campbell and Ebbesson 1969; Sharma 1972; Repérant and Lemire 1976) agree that the retinal projection is entirely crossed (goes, that is, entirely to contralateral destinations), although Voneida and Sligar (1976) have reported some ipsilateral fibres. The majority of the retinofugal fibres run to the superficial layers of the tectum but there are also, as in elasmobranchs (Smeets 1981) and, as we shall see, in land vertebrates, direct retinal projections to the hypothalamus, the pretectum, and a variety of sites in the thalamus, as well as to the accessory optic nuclei, the nuclei of the basal optic root (Finger and Karten 1978). There is currently no evidence that any of the thalamic nuclei to which the retina projects in turn project to the telencephalon; however, there is a projection from the optic tectum to a nucleus in the thalamus (variously termed the nucleus rotundus or nucleus prethalamicus) and there is some evidence that there may be a projection from this nucleus to the telencephalon (Ebbesson and Vanegas 1976; Vanegas and Ebbesson 1976; Ito, Morita, Sakamoto, and Ueda 1980). This latter pathway may parallel that in mammals from the superior colliculus via the pulvinar or lateral posterior nucleus of the thalamus to extrastriate visual cortex, which is one of the two major routes to the telencephalon for visual information (to which much attention will be paid in later chapters).

A recent report (Finger 1980) which used both anterograde and retrograde tracing techniques provides anatomical evidence for a projection to the telencephalon of a component of the lateral line system in a teleost, the bullhead catfish (*Ictalurus nebulosus*). The lateral line system, found only in fish and larval aquatic amphibians, contains both mechanoreceptive organs (of two types, each sensitive to displacement of water in their vicinity), and (in some species) electroreceptive organs (again, of two types, each sensitive to specific changes in local electric fields). The catfish system contains both a mechanoreceptive and an electroreceptive component, and these obtain independent representation throughout the central nervous system (Knudsen 1977; Finger 1980).

Finger's report concerns the mechanoreceptive component of the system, cells of which project from a primary sensory area in the rostral medulla to an area in the (contralateral) mesencephalon known as the torus semicircularis, and from this region to a complex of nuclei in the (ipsilateral) ventral thalamus, which Finger calls the mechanoreceptive thalamic zone. This zone in turn projects to a restricted part of the telencephalon which appears to correspond to Nieuwenhuys' central pallial nucleus.

Finger points out the overall similarity between the organization of the system he describes in the catfish and that of the auditory system in terrestrial vertebrates (which will be discussed in subsequent chapters); points of similarity are the general location and number of synapses in the system, as well as the point at which decussation (crossing of the midline) occurs. These parallels are of particular interest in the light of the fact that the sensory receptors of the mechanoreceptive component are hair cells, just as is the case in vertebrate auditory systems; however, auditory sensitivity in catfish is represented independently of both components of the lateral line system (Knudsen 1977), and it is not yet known whether auditory information reaches the telencephalon in these fishes. Finger's report, although it awaits physiological confirmation, does, then, strongly suggest that relatively specific sensory information from a nonolfactory modality projects to a restricted telencephalic site, and that the overall organization of that projection resembles that of one of the systems found in terrestrial vertebrates.

Virtually nothing further is known of the representation of other senses in the teleost telencephalon. Although teleosts without forebrains are, as would be expected, incapable of olfactory discriminations (see Segaar's 1965 review), there do not appear to be any other major sensory deficits in such animals. They may therefore differ from elasmobranchs in the role played by the telencephalon in vision; it will be recalled that Graeber (1978) described visual deficits after posterior telencephalic damage in sharks, and this finding is backed up by the fact that, whereas sharks are capable of visual discriminations after loss of the optic tectum, tectal-ablated teleosts appear to be completely blind (Dijkgraaf 1949; Iwai, Saito, and Tsukahara 1970).

Although stimulation of the teleost forebrain elicits, from some sites, behavioural patterns, and although damage to the forebrain may alter the temporal organization of certain activities (see Segaar 1965), there is no evidence that the telencephalon is involved in the organization of movement, as opposed to the possibility that it is involved in its initiation. Evidence for a general lack of major motor or sensory deficits following hemispheric damage in teleosts will be found as the experiments whose primary concern is with the effects of lesions on learning are reviewed.

We may summarize this section by saying that as yet little is known of sensorimotor involvement in the fish forebrain; the few reliable data currently available indicate that there is no reason to suppose that the fish telencephalon is wholly taken up with sensory and motor analysis, that, in other words, there may well be ample 'space' for higher-level processes.

Habituation in fish

A review by Peeke and Peeke (1973) shows that habituation is obtained in a variety of response systems in fish. The responses studied may be classified according to the type of stimulus that initially elicits them, and Peeke and Peeke quote evidence of habituation in responses to five types of stimulus: (i) the general stimuli of a novel environment, where the initial response of arrested movement (sinking to the bottom and remaining still) wanes (Welker and Welker 1958, using the marine species *Eucinostomus gula*); (ii) specific stimuli that presumably indicate the presence of a non-aquatic predator, e.g. light taps on the side of a tank, which initially elicit a 'tail-flip' response in goldfish, *Carassius auratus* (Rodgers, Melzack, and Segal 1963); (iii) food or prey stimui where either the 'food' is worthless (e.g. a cotton ball, Herrick 1924) or the prey is unobtainable, e.g. brine shrimp protected by a transparent tube from goldfish or paradise fish (*Macropodus opercularis*) where the initial response is biting (Peeke and Peeke 1972); (iv) an unresponsive female, where the initially elicited courtship behaviour of the male wanes (Baerends 1957, using the male guppy, *Lebistes reticulatus*); (v) the sight of a conspecific, where the initial response consists of aggressive behaviour (Baenninger 1966, using male Siamese fighting fish, *Betta splendens*).

The reports summarized above describe habituation in a number of different species, but all of these are teleosts, and there do not appear to be any definitive reports of habituation in either agnathans or elasmobranchs; there are reports (Birnberger and Rovainen 1971; Wickelgren 1977) of phenomena that resemble habituation in lampreys, but the data are insufficient to rule out the possibility that the waning of response observed is due to fatigue (perhaps of interneurons), rather than to a learning process. There are, on the other hand, no reports of failures to obtain habituation in either of those classes, and so no reason to suppose that there is (or is not) any difference between the various living classes of fish as regards the occurrence of habituation.

Analytical studies have shown that fish exhibit many of the phenomena associated with habituation that have been found in other vertebrate (and invertebrate) species. The rate at which habituation proceeds, for example, varies with the response system measured: Rodgers *et al.* (1963) found that, whereas the goldfish tail-flip response to taps declined rapidly, the orienting response also elicited by the taps was markedly more persistent. Responses show spontaneous recovery following habituation, gradually returning to approximately the levels seen initially: Clayton and Hinde (1968) examined spontaneous recovery of the gill-cover extension response (a component of the aggressive display) in Siamese fighting fish that had been exposed to mirrors (which elicit the display) continuously for ten days. The mirror was removed for periods of from 15 minutes to four days, and the fish showed, on its replacement, recovery of the gill cover extension response to 25 per cent of its original initial level after a 15 minute recovery period, rising to 67 per cent of the original level after a four day recovery period. Presentation of a novel stimulus (which may be a change in some parameter of the training stimulus) can restore the response abruptly to its initial level (the phenomenon of dishabituation): so, for example, Rodgers *et al.* showed that the tail-flip response, habituated to weak taps, could be dishabituated by presenting either stronger taps or a visual stimulus. As a final example, repeated or continuous presentations of a stimulus may obtain not only decremental effects on responding (habituation) but also incremental effects (sensitization): in the Clayton and Hinde (1968) study, the rate of gill-cover extension responses to the mirror increased over the first few minutes following its original exposure, declining thereafter to a somewhat lower level, and eventually falling over the course of a few days to about 10 per cent of the original level.

Habituation is, then, widespread amongst teleosts, and associated phenomena, such as spontaneous recovery, dishabituation, and sensitization are also observed. Parametric comparisons between fish and other groups of animals are complicated first by the fact that the absolute rate of habituation varies, in fish as in other animals, with the response system concerned and second, by the absence of agreement on the details of parameters applicable to any one species. Thompson and Spencer (1966) did draw up a list of nine parametric characteristics of habituation, and argued that these characteristics reflected the interaction of two processes, one decremental, the other, incremental; Hinde (1970), on the other hand, has challenged many of the Thompson and Spencer generalizations, concluding that multiple short-term and long-term incremental and decremental processes are involved. The proponents of each view take their examples more or less indiscriminately from a wide variety of animals, including invertebrates and spinal mammals (mammals whose spinal cord has been severed from the brain). Thompson and Spencer believe that their two basic processes, habituation and sensitization, are widespread throughout the animal kingdom; Hinde holds that such a conclusion is premature until detailed comparisons of the parametric effects of wide ranges of dependent variables have been carried out. Both may agree, however, that there is as yet no convincing evidence of differences in response to repeated stimulation that can be ascribed to a difference of species rather than of, say, response system, and this general conclusion may be applied to fish as compared to other vertebrates. No suggestion that habituation in fish might differ from habituation in any other vertebrate has been made and, accordingly, no detailed comparisons of fish with other species have been made experimentally. There are clear overall similarities between fish and other groups, and so no reason currently to suppose that there are any important differences between the processes of habituation in fish and other vertebrate species.

Classical conditioning in fish

Our goals in this section will be to establish first, whether fish can form associations between stimuli and reinforcers, and second, whether there appear to be any distinctions to be drawn between the way such associations are formed in fish and the methods of association formation in other groups of vertebrates.

The first part of this section will survey the results obtained in fish using standard classical conditioning procedures, and will compare those results with the performance of mammals in comparable situations. The latter part of the section will turn to the substantive question whether the successful conditioning obtained using these procedures necessarily implies that fish can detect contingencies between stimuli and reinforcers, and so, are capable of 'genuine' classical conditioning.

Although the great majority of experiments have used teleosts, there are some data on representatives of the other two classes of fish, and these will be described first.

Razran (1971) summarizes a number of experiments by Russian workers which point to the conclusion that lampreys may not be capable of classical conditioning. Although pairings of CSs and UCSs may produce CRs in lampreys, it appears that these are a consequence of sensitization rather than association-formation; for example, no CRs are found using CSs which do not themselves at some higher intensity elicit a UCR similar to the target CR. A similar conclusion is reached by Wickelgren (cited by Rovainen 1979), who found as many CRs in a group of lampreys given independent random presentations of light flashes and electric shocks as in a group given pairings of light and shock.

There is one well-controlled study (Gruber and Schneiderman 1975) of classical conditioning in sharks. The animals used were lemon sharks (*Negaprion brevirostris*), and the response measured was movement of the nictitating membrane, a third eyelid that is present in sharks, in most amphibians, in reptiles, and birds, and in some mammals. The CS was light onset, and the UCS, electric shock. The sharks showed rapid acquisition of the CR, averaging more than 95 per cent CRs after two days' training (100 pairings each day); the rate of learning is similar to that seen in conditioning the rabbit's nictitating membrane, using a tonal CS, and a puff of compressed nitrogen to the eyeball as UCS (Gormezano, Schneiderman, Deaux, and Fuentes 1962). The sharks' performance resembled that of the rabbits in three other ways: first, during acquisition, modal CR latency decreased gradually; second, in extinction (when CSs were no longer followed by UCSs) the decline in percentage CRs was gradual; and third, spontaneous recovery was observed—that is, over the series of extinction sessions, performance at the start of each day was somewhat higher than that seen at the end of the preceding day.

One group in the Gruber and Schneiderman study was subjected to the so-called 'random control procedure': these subjects received the same number of CSs and UCSs as the experimental subjects, but there was in this case no contingency between their occurrence. This control is now regarded as the most appropriate way to show that CRs occur as a result of the contingency between CS and UCS and not, for example, as a consequence of simply presenting the CS against a background of UCS occurrences (Rescorla 1967). There were virtually no CRs in this control group, so that we may conclude that the CRs seen in the experimental group were indeed a consequence of the CS–UCS contingency.

Although some early studies (e.g. Harlow 1939) suggested that CRs in teleosts might be simply artefacts, not due to the contingency between CS and UCS, there is by now good evidence that genuine CRs may be established using both aversive and appetitive USCs. Controlled studies using shock UCS and general body movement as CR have been carried out, for example, by Bitterman (1964*a*) and Overmier and Curnow (1969), both using goldfish; control groups in both of these studies received, in place of paired presentations of CS and UCS, a series of CS alone and UCS alone presentations, in random order. Further evidence against the possibility of artefact comes from studies that show that the CR is elicited by a stimulus paired with the UCS and not by a stimulus not so paired. A study by Thompson and Sturm (1965) used as an appetitive UCS exposure of a mirror to Siamese fighting fish; these fish consistently react with an aggressive display to their own images, and will in fact learn to work to obtain mirror exposure (Thompson 1963), so that the mirror does function as a reward. Elements of the aggressive display were successfully conditioned to a red light that preceded its exposure, and a second experiment showed that where a green light was paired with a mirror exposure, and a red light presented alone, CRs developed to the green light, but not to the red light. Similarly, successful discrimination between coloured CSs has been established, using shock UCS in goldfish, by Overmier and Starkman (1974).

Most studies of classical conditioning in teleosts have used visual CSs, but a variety of alternative CRs have been scored. Otis, Cerf, and Thomas (1957), for example, measured changes in breathing and heart-rate in goldfish; Scarborough and Addison (1962) measured 'forward darting' in golden shiner minnows (*Notemigonus crysolecus*); Geller (1963) showed that goldfish, trained to produce a steady rate of response (lever presses) for food rewards, suppressed that response during a 3-min presentation of a flashing light that preceded, on 50 per cent of its occurrences, an unavoidable shock (a phenomenon known as 'conditioned suppression').

Both Woodard and Bitterman (1974) and Bottjer, Scobie, and Wallace (1977) have shown that where the transillumination of a response key with coloured light (for 20 seconds) regularly preceded the delivery of food from a nearby feeder, goldfish developed the CR of approaching and 'nosing' the lit response key; this procedure parallels a similar procedure developed for pigeons (with pecking as the CR) by Brown and Jenkins (1968), a technique known as 'autoshaping'–Squier (1969) also reports successful autoshaping in three other teleost species (the Oscar, *Astronotus ocellatus*; mouthbreeders, *Tilapia mossambica*; and mullet, *Mugil cephalus*). The phenomenon of autoshaping has been of considerable theoretical interest, as it appears to demonstrate conditioning of skeletal ('voluntary') responses in a classical procedure.

In general, the rates of acquisition of CRs in fish appear comparable to those obtained in mammals. Any suggestion that there might be a significant difference in acquisition performance would require variations in all those factors known to affect acquisition in mammals. Such factors include (and this is not an exhaustive list) CS intensity, UCS intensity, deprivation level, and type of CR monitored.

As acquisition rates do fall within the general range of those found in mammals, and as those rates are influenced by so many different variables, it is hardly surprising that no effort has been made to see whether, in any particular situation, a species difference of any important kind exists. It is not, of course, possible to conclude that there is no difference between fish and mammals in rate of acquisition, but it does seem sufficiently unlikely that the research effort required for the necessary systematic variation has not so far seemed justified.

Comparability of acquisition rates does not, of course, necessarily imply identity of the mechanisms employed; a more analytical approach might be to see whether the factors that are critical to one group of animals have a similar relevance in other groups. One such variable, known to be of importance in mammals, is inter-stimulus interval – the time elapsing, that is, between CS onset and UCS onset. Noble, Gruender, and Meyer (1959) measured CRs to a shock UCS in green sailfin mollies (*Mollienesia latipinna*), and found an 'optimal' ISI of 2 s, where ISIs tested ranged from 0.5 to 4 s; this suggested to them that fish might differ in some significant way from mammals, as the optimum ISI for mammals was believed to be approximately 0.5 s. Subsequent work has, however, cast doubt on this suggestion: first, Klinman and Bitterman (1963) failed to obtain the Noble *et al.* result in either mollies or goldfish, finding instead a tendency for a 0.5 s ISI to result in more efficient conditioning than a 2 s ISI; a subsequent study by Bitterman (1964*a*), using goldfish, found that CRs increased steadily as ISI was increased from 0 to 1 s, and remained at about the same level up to an ISI of 27 s. Second, work with mammals has shown that there is no unique ISI for optimizing classical conditioning in that class: optimal ISIs depend, for example, on the response involved. Whereas the optimal ISI for conditioning rabbit nictitating membrane is 0.2 s, variations in ISI from 0.2 to 4 s have very little effect on conditioning of rabbit jaw movements, where UCS is an intraoral squirt of cold water (Gormezano 1972).

A second variable whose effect on conditioning (and extinction) has been explored in fish is the probability of occurrence of UCS following CS occurrence – the effects, that is, of partial as opposed to consistent reinforcement (e.g. Gonzalez, Longo, and Bitterman 1961); the utility of these studies for comparative purposes is severely limited by the fact that the effects of partial reinforcement on classical conditioning in other groups of animals are inconsistent and poorly understood (Mackintosh 1974). Partial reinforcement of instrumental responding does have, in mammals, more reliable effects, particularly in extinction, so that contrasts between fish and mammals in response to partial reinforcement will be analysed in a later section, where instrumental responding will be under consideration.

Although there appears to have been only one study using teleosts (Wolach, Breuning, Roccaforte, and Solhkhan 1977) which has used the ideal baseline control, the random control procedure, the results of that study, along with the controls that have been used in a number of other studies have provided generally convincing evidence that teleosts do acquire conditioned responses that depend upon the contingency between the CS and UCS. This of itself, however, does not necessarily indicate that teleosts form associations between CSs and UCSs: as argued in Chapter 1, alternative possibilities are that the CRs obtained occur either because they in some way modify the UCS to the benefit of the fish, or because the fish adopt the 'superstition' that their CRs in some way are responsible for the arrival of the UCSs. If these alternatives cannot be ruled out, then it remains possible that fish may not be capable of true classical conditioning, and may form only associations involving responses—may, that is, show only instrumental conditioning. This issue assumes particular theoretical relevance in the light of Bitterman's proposal (e.g. Brandon and Bitterman 1979) that fish are 'pure S–R reinforcement animals' (p. 61), capable of forming bonds between stimuli and responses, but not between stimuli and other stimuli (such as reinforcers); that proposal will be explored more fully in context of the experiments that originally led to its formulation (see p. 86), and we shall restrict our discussion here to relevant evidence obtained using standard classical procedures.

That responding by fish in classical training situations might modify the UCS is by no means implausible, at least in the case of aversive UCSs: Woodard and Bitterman (1971*a*) have shown that the amount of current actually transmitted through a fish's body depends critically on the orientation of the fish to the electrodes in the tank; therefore, the subjective strength of the shock could be reduced by adopting some optimal orientation prior to shock onset. This is a possibility that must be noted, but which cannot be further assessed on the evidence available, as no studies of the actual orientation of fish at UCS onset have been reported. It was suggested in Chapter 1 that, to resolve difficulties of this kind, omission training procedures should be used. Where the UCS is aversive, omission training is, of course, an avoidance procedure: when the animal responds to the CS the UCS is avoided. Evidence will be presented in a later section to show that in fish avoidance training does produce more responses than does classical training—suggesting an instrumental component in responding. That demonstration does not allow us to conclude whether the responding (at a lower rate) seen under the classical procedure is truly classical or instrumental (the consequence, possibly, of some response-induced reduction in shock). So, where avoidance training indicates the existence of instrumental responding, the outcome of omission training using appetitive UCSs become theoretically important; for in this procedure, responding to the CS prevents a reward (rather than avoiding a punishment), and if sustained responding under omission training occurs, the implication is that the response is under control of the classical contingency.

A recent study (Brandon and Bitterman 1979) compared the performance of 'master' goldfish subjected to an omission procedure in a food-reinforced autoshaping situation with that of yoked controls which received the same number of CS alone and CS–UCS trials, but whose responses had no effect on UCS delivery. The master fish responded at a lower rate than the yoked subjects (suggesting that responding was indeed modified by its consequences—that an instrumental contingency was effective) but, what is more relevant here, the master (omission-trained) fish maintained a good level of response thoughout. This latter finding gives strong support to the notion that the classical contingency was also effective; since no response is ever followed by food presentation, the fish could not have formed the 'superstition' that responding produced food. Brandon and Bitterman, however, believing on other grounds (which we shall discuss in later sections) that goldfish do not form classical associations, explored an alternative account of the maintained responding, namely, that it might be an instance of sensitization, that the background of UCS deliveries might elicit responding in a non-associative manner. They examined this possibility by having goldfish perform a discrimination between two CSs, successively presented in random order, one CS being associated with omission training, the other, with simple extinction: if occasional UCS deliveries elicit non-associative responding, the rate of response to the two stimuli should be the same, whereas if classical associations have been formed, the omission stimulus should obtain more responding than the extinction stimulus. The results showed that where the inter-stimulus interval was relatively short and the CS duration relatively long, there was actually less responding to the omission than to the extinction stimulus, whereas when the inter-stimulus interval was long, and CS duration short, the omission stimulus showed more trials on which one or more responses occurred than did the extinction stimulus. The former finding suggests an instrumental component in performance: in omission training, responses are 'punished' by non-delivery of food, and in this condition, the punishment is more effective in eliminating responding than is the simple absence of food. The latter finding, of course, implicates a classical contingency, for the reasons advanced above. Why the difference between the conditions? The most likely answer seems to be that the optimal conditions for classical association-formation in an autoshaping procedure require long inter-stimulus intervals and short CSs—since, in these conditions, the CS is a more effective signal for the imminent arrival of food. It may be noted that a similar pattern of contrasting effectiveness of omission training and extinction in reducing responding, varying with inter-stimulus interval and CS duration, appears to be found in pigeons (cf. Weiler 1971; Woodard, Ballinger, and Bitterman 1974); the evidence for a classical component in autoshaping is probably better in pigeons than in any other group of non-human animals, so that the overall pattern of Brandon and Bitterman's (1979) findings gives good support to the view that goldfish do form classical associations.

One difficulty facing the assumption of classical association formation in fish is that there is one classical training procedure which is effective in mammals, but has so far been unsuccessful with fish, namely sensory preconditioning (SPC). In this procedure two neutral stimuli, CS_1 and CS_2, are paired (CS_1 preceding CS_2), in the first stage of training; the second stage of training pairs CS_2 with a UCS, so that CS_2 elicits a CR; the third stage tests whether an association was formed between CS_1 and CS_2 in the first stage, by now presenting CS_1. Successful SPC is demonstrated where the originally ineffective CS_1 now produces a CR. Since neither CS originally produces a salient UCR, and since neither is of motivational significance, it seems unlikely that the association between them is mediated by associations involving responses, or that any such associations could reflect the consequences of responses. In other words, animals capable of only instrumental conditioning would not be expected to show successful SPC.

A recent report by Amiro and Bitterman (1980) failed to detect SPC using a procedure very similar to one in which second-order conditioning was successfully demonstrated. Second-order conditioning is procedurally rather similar to SPC, except that the first two stages of training are reversed—the neutral CS is paired from the outset with a CS that has already been trained and elicits a CR. Since the tasks make comparable sensory, motor and motivational demands, it seems reasonable to conclude that the failure to obtain SPC is due to the nature of the association to be formed (it may be added that Sergeyev, cited by Razran 1971, p. 198, also reports failure to obtain SPC in fish). Amiro and Bitterman (1980), following an analysis by Rescorla (1977), propose that second-order conditioning is mediated by stimulus-response (instrumental) associations, whereas SPC involves, as most workers would agree, stimulus-stimulus (classical) associations. The grounds for regarding second-order conditioning as instrumental are too complex to be discussed here, but it is clear that the analysis does agree well with the proposal that fish are 'pure S–R' animals, incapable of classical association-formation. Although the failure to obtain SPC does accord with the view that fish cannot form classical associations, it is the only piece of evidence that we have seen so far which does support that view, and it would be premature at this stage to conclude that there are no conditions under which fish would show SPC.

The overall conclusion of this survey is, then, that fish, or at least elasmobranchs and teleosts, are capable of detecting contingencies between stimuli and reinforcers and that in general their performance appears no less efficient than in other vertebrates. This conclusion should of course be qualified by mention of the failure to date to obtain classical conditioning in lampreys, or to obtain SPC in teleosts.

Instrumental conditioning in fish

Our concern in this section will be to establish whether fish are capable of forming associations involving their own responses, and whether such associations are formed with either more or less facility than in other groups of vertebrates.

Procedures using positive reinforcers

Almost all studies of instrumental conditioning in fish have employed teleosts; Clark (1959) did report an experiment in which two lemon sharks and three nurse sharks (*Ginglymostoma cirratum*) were trained to push a target with their snouts to obtain food rewards, but the procedure was poorly controlled, and it is very possible that his effects were a result of classical rather than instrumental conditioning. A variety of reinforcers have proved successful with teleosts: the most common reinforcer has been food reward, and other reinforcers include visual exposure to a conspecific (Bols 1977), access to a model of a symbiotic cleaner fish (Losey and Margules 1974), opportunity to regulate water temperature (Rozin and Mayer 1961), and increases in oxygenation of water (Van Sommers 1962). Not all of these reports present detailed acquisition data, but in general it appears that acquisition is rapid, and no suggestion that there might be any significant difference between fish and mammals in simple instrumental tasks has been made. Any such suggestion would, of course, have to be backed up by data which would inevitably require reliance upon the method of systematic variation.

There remains, however, the question whether successful performance in fact demonstrates that true instrumental conditioning, involving learning about responses, has occurred, or whether, as has been suggested previously (in Chapter 1), such conditioning might be classical in origin—a variety of autoshaping, where appetitive UCSs are concerned. We saw in the preceding section that there is evidence, from the Brandon and Bitterman (1979) report, for an instrumental component in performance of goldfish using a food-reinforced autoshaping technique: first, subjects yoked to master subjects on an omission schedule showed higher rates of response than the master fish; second, under one condition, omission training was more effective than extinction in reducing response rate. Each of these findings suggests that the fish detected the consequence (food omission) of a response and so were sensitive to response-reinforcer contingencies. Unfortunately, as we saw in Chapter 1, such results do not conclusively demonstrate instrumental conditioning, since the performance of the CR concerned systematically altered the stimulus complex so that, in an omission procedure, the stimuli near the manipulandum would have become differentially correlated with non-reward and might therefore have been avoided through detection of that (classical) contingency.

There is one report using an appetitive UCS which provides support of a different kind for the existence of instrumental learning in fish, and which may not be open to the criticisms outlined above. Van Sommers (1962) placed single goldfish in a tube which had across its front end a photobeam, interruptions of which constituted responses. Water flowed through the tube from the front, and the level of oxygen in the water could be varied, so that increased oxygenation might act as a reward. The fish were run in a discrimination such that, when a red light was on, each response obtained oxygenated water for 15 s, after which deoxygenated water was presented, and when a green light was on, water was deoxygenated until 20 s had passed *without* a response, at which point, oxygenated water was presented for 15 s. The three subjects showed good discrimination, responding at a much higher rate in red than in green components, and so obtained substantial numbers of rewards in both components. Now this situation is difficult to analyse in terms of purely classical conditioning: the fish in the tube will invariably face forwards, so that the stimuli at the front end should in each type of component regularly precede occurrence of reward—nevertheless these stimuli are approached in one component, and not in the other.

Teleost fish, then, will perform efficiently for a variety of positive reinforcers, and it seems probable that they do indeed learn to form associations involving responses.

Procedures using aversive stimuli

All the experiments to be described in this section used teleost subjects, with electric shocks as the punishing stimulus. The procedures may be divided into two broad categories: first, those in which the fish are required to learn to perform some response to avoid shocks, and, second, those in which the requirement is to inhibit some response to avoid shock. The former category is known as active avoidance, the latter, as passive avoidance, or punishment.

The great majority of experiments on active avoidance in fish have used shuttle-boxes; these are tanks divided into two compartments by a barrier designed so that, although the fish may shuttle freely between the compartments, they are unlikely to remain midway across the barrier for any length of time (this is achieved by allowing a narrow gap either above or below the barrier, or by making a small hole in the barrier). A standard design is to present a CS (usually light onset) for, say, 10 s, after which, unless the fish has responded by moving from one compartment into the other, shock is delivered until a response does occur. A response occurring within 10 s of CS onset both terminates the CS and avoids the shock, and is scored as an avoidance response; responses occurring after shock onset terminate both CS and shock, and are termed escape responses. Using designs of this kind, reliable avoidance responding has been established in a variety of teleosts, including goldfish, Siamese fighting fish, guppies, and Beau gregory (*Pomacentrus leucostictus*) (e.g. Otis and Cerf 1963; Werboff and Lloyd 1963; Wodinsky, Behrend, and Bitterman 1962). Differences in rates of acquisition have been found between some of these species; Wodinsky *et al.*, for example, found that goldfish showed superior acquisition to Beau gregory where the CS-UCS interval was long (more than 5 s) but were inferior when the interval was short (2.5 s); Otis and Cerf (1963) found slower acquisition in Siamese fighting fish than in goldfish. The authors of these reports do not, however, suppose that any difference in intelligence is indicated: Otis and Cerf suggest that the differences may find their explanation in the contrasting ways the two species react to sudden intense stimuli—that such 'species-specific defence responses' are of considerable importance in acquisition of avoidance learning is well established in mammals (Bolles 1970). Apart from the Wodinsky *et al.* report on the effects of varying CS-UCS interval, no effort has been made to vary other factors that might be relevant.

A number of factors have been shown to influence active avoidance performance in fish: these include CS-UCS interval (e.g. Behrend and Bitterman 1962), UCS intensity (Bintz 1971), CS modality (Jacobs and Popper 1968), CS location (Gallon 1972), individual versus group training (Warren, Bryant, Petty, and Byrne 1975), and the nature of the target response (shuttling versus lever-bumping, Rakover 1979). The results of these studies show a wide range of rates of acquisition, overlapping extensively with those reported for various species of mammals, so that there is currently no suggestion of any significant species differences either within fishes or between fish and any other group of vertebrates.

An important question concerning the nature of this acquisition was raised by Woodard and Bitterman (1973*a*), who had noted in an earlier study of classical conditioning in a shuttle-box (Woodard and Bitterman 1971*b*) that simple pairing of a CS and UCS obtained frequent shuttling, although that response was in fact without any effect on UCS delivery; in other words, as we have seen, goldfish respond to fear-inducing stimuli with an increase in movement—could this apparently classical conditioning effect account for the shuttling seen in instrumental avoidance situations? Their 1973 report was concerned with this possibility. Four groups of goldfish were run under the following conditions: for one group, responses had no effect; for a second group, responses both terminated the CS and avoided the UCS; for a third group, responses avoided the UCS but did not terminate the CS; for the fourth group, responses terminated the CS but did not avoid the UCS. No significant differences were seen amongst the four groups in terms of the number of avoidance responses that occurred (and all of the groups averaged more than 70 per cent responses by the end of training). Such a finding carries the theoretical implication that although fish can form associations between stimuli and aversive reinforcers, they cannot form associations involving their own responses where those responses avoid the aversive stimuli; it may, of course, be remarked that this notion is quite the reverse of that currently held by Bitterman—that fish are pure S–R animals, capable of only instrumental learning. Scobie and Fallon (1974), however, repeated the experiment, with minor procedural changes, and found statistically reliable effects of both avoiding the UCS and terminating the CS, the best performance being obtained in the group whose responses both terminated CS and avoided the UCS, and the worst performance, in the group whose responses were without any effect. They point out that, although Woodard and Bitterman did not find reliable inter-group differences, the ordering of the group means in that experiment was identical to that in the replication, and conclude that real group differences in the Woodard and Bitterman experiment were obscured by uncontrolled sources of variance. The Scobie and Fallon study, then, provides evidence that goldfish can indeed detect the consequences of their responses, and so are capable of instrumental conditioning.

Further evidence for an instrumental component in avoidance by fish is provided by Mandriota, Thompson, and Bennett (1968), who used as a target CR an elevation in rate of electric organ discharge by a mormyrid fish, *Gnathonemus*. Their study compared the performance of master fish, whose CRs avoided the UCS (electric shocks) and terminated the CS (a light) with that of yoked controls, given the same number of CS and UCS deliveries, whose responses were without effect. The master fish showed reliably more CRs than the yoked control fish and, since the same classical (CS–UCS) contingency was operative for both, the implication is that the instrumental contingency (CRs avoid the UCS) was indeed effective in the master fish.

It should be noted that neither of the experiments described above appears amenable to a classical interpretation since the CRs did not in either case systematically affect the stimulus complex: in the Scobie and Fallon (1974) design, all parts of the apparatus would have been equally associated with shock, so that CRs could not move the fish away from a region differentially associated with shock, and the changes in the organ discharge rate in the Mandriota *et al.* (1968) study could have had no effect on the stimulus complex. They do, therefore, provide some support for the notion that instrumental learning is obtained in teleosts.

There is one active avoidance procedure that deviates in an important way from those that we have been considering, and that is free operant or Sidman avoidance. In this task, no CS is presented by the experimenter, and a brief aversive UCS occurs at a fixed rate (determined by the shock-shock interval) unless a response occurs, in which case the next train of shocks is postponed for a fixed time (the response-shock interval). Two reports using this technique in shuttle-boxes have shown that free operant avoidance responding can be established in goldfish, that the rates of response exceed those of yoked controls that receive the same temporal distribution of shocks but whose responses have no effect, and that responding varies, as should be expected, with the response-shock interval (Behrend and Bitterman 1963; Pinckney 1968). No detailed analyses of the temporal distributions of responses are currently available, but there is no evidence to date which might suggest that there is any difference between the performance of fish and any other group of vertebrates on this task.

Passive avoidance or punishment procedures depend on the existence of some response which, in the absence of punishment, would reliably occur; such responses may be either 'spontaneous' or the result of explicit training. One interesting observation made in studies using this technique is that goldfish are capable of 'one-trial' learning. Riege and Cherkin (1971) describe a procedure in which goldfish swam spontaneously against a current to a calm-water well; following one experience of being shocked on entering the well, the fish showed greatly increased latencies on a succeeding trial. A control group, given shocks in a different apparatus, showed that this effect was not due to any debilitating effect of the shocks. It is not clear from this report precisely what the fish had learned—the shock might have been associated with the response (which was, accordingly, inhibited) or with the well (which was accordingly, avoided) or simply with the apparatus (resulting, perhaps, in general suppression or 'freezing'); nevertheless, it is a demonstration that fish can form at least one type of association in one trial, and so reinforces the general impression that simple association formation is efficient in fish.

There have been a number of reports on the effect of punishing aggressive display in the Siamese fighting fish. Adler and Hogan (1963) presented a mirror for a maximum of 2 min, with 10 min inter-presentation intervals, and recorded occurrences of gill membrane extension. For four experimental fish, shocks were delivered, and the mirror removed, as soon as a response occurred; for the four controls, stimulus duration and shocks were yoked to those of control fish, there now being no contingency between their responses and shocks. In this experiment, both experimental and control subjects showed marked suppression of responding, although the controls showed more rapid recovery when shocks were no longer delivered. It is, of course, possible that the controls were, at least on some occasions, shocked while performing the response, and this might account for the minimal difference seen between the groups. Fantino, Weigele, and Lancy (1972) carried out a similar experiment, shocking experimental subjects for each gill membrane response, and yoked controls, on the same number of occasions, but in a quasi-random sequence. Fantino *et al.* did not interrupt mirror presentation when responses occurred, and, perhaps as a consequence of this procedural change, found in their experiment that whereas all three experimental subjects abandoned membrane extension, the three control fish showed no decline. This result provides *prima facie* evidence that fish can detect a contingency between responses and shocks: an alternative interpretation, however, is that the mirror, which would inevitably be viewed by the experimental fish immediately preceding each delivery of shock, came to act as a CS for shock, and that the resulting CR—movement of some kind—interfered with the membrane extension response—that, in other words, classical rather than instrumental conditioning was obtained.

Conclusions

The conclusions to be drawn from this section are, first, that fish (or, at least, teleosts) are capable of genuine instrumental conditioning using both appetitive and aversive UCSs, and, second, that their performance in simple instrumental tasks differs in no significant respect from that of mammals.

Complex learning in fish

Any modern survey of comparative studies of learning in vertebrates must inevitably take as its starting point the work carried out over the last two decades or so by Bitterman and his colleagues. Their experiments and hypotheses, frequently highly controversial, have been responsible, directly or indirectly, for the great majority of comparative studies using non-mammalian vertebrates.

Bitterman's early experiments suggested that two experimental procedures (serial reversal and probability learning) revealed qualitative differences in intellectual functioning amongst vertebrate classes. Table 3.1 summarizes the conclusions to which those experiments pointed; the table classifies the performance of various vertebrates as either 'rat-like' or 'fish-like' in either spatial or visual versions of serial reversal and probability learning (to anticipate details that follow in subsequent sections, 'rat-like' in the reversal task means that subjects show progressive improvement across a series of reversals, whereas 'fish-like' means no improvement; rat-like in the probability task means maximizing (or non-random matching), and fish-like means random matching; in the visual versions, the correct stimulus is to be identified by visual characteristics, such as colour, alone, whereas in spatial versions, the correct stimulus is identified by its position alone).

Bitterman's own account of the evidence is revealing: 'As we go up the evolutionary scale new modes of adjustment appear earlier in spatial than in visual settings' (Bitterman 1965*a*, p. 97). That is, the performance of rats is not simply different from that of fish—it is an advance reflecting the superior standing of mammals in terms of the evolutionary development of intelligence. The following sections will be devoted primarily to analysing evidence from these and other tasks, evidence which, while it has forced Bitterman to abandon the early classificatory scheme, has not altered his view that the intellect of fish is qualitatively different from, and inferior to, that of mammals.

Table 3.1. *Bitterman's classification of animals from different vertebrate classes according to whether their performance resembles that of the rat or the fish.*

	Spatial problems		Visual problems	
Animal	Reversal	Probability	Reversal	Probability
Rat	Rat	Rat	Rat	Rat
Pigeon	Rat	Rat	Rat	Fish
Turtle	Rat	Rat	Fish	Fish
Fish	Fish	Fish	Fish	Fish

Serial reversal learning

In the standard serial reversal paradigm, an animal is trained, first, to choose one of two simultaneously presented stimuli (S+) in preference to the other (S–); then, after either achieving a certain criterion or having experienced a predetermined number of trials, the reinforcement value of the two stimuli is reversed, and the opposite discrimination is now formed. An indefinite number of such reversals may be carried out, so that the asymptotic performance level of an animal may be determined.

This situation was introduced into comparative psychology by Bitterman and his colleagues who 'decided to study reversal-learning on the hunch that something like flexibility would prove to be an important functional correlate of phylogenetic level' (Bitterman, Wodinsky, and Candland 1958, p. 95). This notion, allied to Bitterman's often-stated view that fish are 'lower' and 'simpler' than mammals, gained support from a number of early studies, which showed that, whereas rats improved over a series of reversals, error scores on later reversals being lower than those on early reversals, fish did not; failures to improve were found in different species of fish (e.g. in African mouthbreeders, *Tilapia macrocephala,* in Paradise fish, and in goldfish—Behrend, Domesick, and Bitterman 1965; Warren 1960*a*) in a series of reports from different laboratories.

Further experiments, however, have shown that fish can, after all, show improvement in serial reversals. Mackintosh and Cauty (1971) tested rats, pigeons, and goldfish in a series of spatial reversals (for all subjects, both side keys were white, and the left key was correct in original acquisition, the right key correct on the first reversal, and so on). Now although the fish showed significantly less improvement over the series than the pigeons or the rats, they did, nevertheless, show significant improvement. Mackintosh and Cauty point out that a procedural difference between the conventional rat and pigeon testing situation and that normally used (up to that time) for fish was that, whereas 'rats and pigeons usually receive reinforcement from a magazine located close to the response keys', fish are usually 'reinforced by food dropped onto the surface of the water—often near the back of their tank' (Mackintosh and Cauty 1971, p. 281). In their study, Mackintosh and Cauty used a device which delivered paste food to a magazine located midway between the two response keys and, as has been said, obtained serial reversal improvement.

That training conditions are indeed critical determinants of serial reversal performance in fish was further demonstrated in a series of experiments reported by Engelhardt, Woodard, and Bitterman (1973). In one experiment, three response keys were available in the tank: the third of these was at the rear of the tank, and was lit with white light to indicate the start of a trial. When the subjects (goldfish) struck the rear key, it was turned off, and the two keys at the fromt of the tank were illuminated, one with S+, the other, with S− (the stimuli were red versus green light). Choice of S− extinguished both keys, and initiated the inter-trial interval; choice of S+ obtained 6 s access to liquid food, pumped through a nipple in the centre of the S+ key. The experiment compared one condition, in which the correct key remained illuminated with the S+ colour throughout the reinforcement cycle, with another, in which the S+ colour changed to white light at the start of the cycle: fish showed substantial improvement across the reversal series in both conditions, but performance was better when the key remained illuminated with the S+ colour. A further experiment compared performance in those two conditions with performance where the rear key was used, not to initiate trials, but as a feeder (closely approximating the conditions of some of the earlier experiments); when the rear key was illuminated with white light during the reinforcement cycle very little serial reversal improvement was obtained—and this is what would be expected from previous work. However, when the key target was illuminated with the S+ colour throughout the reinforcement cycle, performance (of three subjects) was 'quite as good' as that of the animals given food at the correct front key, even when that key remained coloured through reinforcement. In other words, the separation of response and reward *does* have an adverse effect, but this effect can apparently be overcome by presenting the S+ colour at the reward site.

Although it has now been shown that fish are capable of progressive improvement in both spatial (Mackintosh and Cauty 1971) and visual (Engelhardt *et al.* 1973) serial reversals, it may nevertheless, of course, be the case that their performance does differ in some significant way from that of rats (the overall amount of improvement shown by fish in the Mackintosh and Cauty study was much less, for example, than that shown by the rats), and it is to specific theoretical proposals concerning the causes of possible species differences in performance that we must now turn.

Bitterman initially assumed that improvements involved some (unspecified) 'higher order process' that was absent in fish; in 1967, however, he concluded that the process responsible for improvement was 'after all, a simple one—that improvement in reversal results from decrements in retention which are produced by proactive interference' (Gonzalez, Behrend, and Bitterman 1967, p. 519). In detail, his argument was that, as a series of reversals progressed, rats tended to forget overnight which stimulus was S+, and which S−, this forgetting in turn being due to the interference of the memories of older reversals with that for the most recently completed reversal (proactive interference). This forgetting would naturally improve performance as animals would not begin later reversals, as they had the early ones, with a series of errors to the most recent S+; goldfish, on the other hand, did not show overnight forgetting, and so showed no improvement. It should be pointed out that the suggestion is not necessarily that fish have 'better' memories than rats—it could be that fish have severely restricted memory stores, so that there is no 'space' both for memories of older reversals and for that of the current discrimination. In other words, memory for the current reversal might, in an animal with a restricted store, entirely replace the memory of any previous discrimination, and so avoid the possibility of proactive interference.

An alternative account has been provided by Mackintosh (1969*a*), whose argument is that fish have less stable attention than rats. This hypothesis is based on a theory of discrimination learning (Sutherland and Mackintosh 1971) which proposes that, in solving a discrimination, an animal has to learn two things: first, what to attend to, or which dimension (e.g. colour, position, brightness) is relevant; and second, which points along that dimension (e.g. blue versus yellow) are reinforced, and which not. In the terminology of Sutherland and Mackintosh, animals have to learn, first, to switch in the relevant analyser, and second, to build up response strengths at appropriate points along the dimension selected by that analyser. Mackintosh claims that a major source of improvement in rat serial reversal improvement is due to the gradual strengthening across the series of the relevant analyser; rats performing, for example, a brightness discrimination, become gradually less likely to lapse into position habits as a consequence of errors once a number of reversals have been experienced—they learn to persist in attending to the relevant dimension, despite non-reward. The analysers of fish, on the other hand, are according to this view more labile, and tend to be switched out in response to errors despite extensive training in tasks with a constant relevant dimension.

These two proposals, which are not mutually exclusive, will recur as themes in explanations of tasks to be described in later sections; at this point, relevant evidence drawn from reversal studies will be discussed. Experiments by Gonzalez, Behrend, and Bitterman (1967), and by Mackintosh and Cauty (1971) compared overnight forgetting during serial reversal performance by goldfish and pigeons (which, like rats, also show improvement over a reversal series). Both experiments agreed in showing that pigeons came to perform roughly at chance on the first trial of each day, and so showed marked overnight 'forgetting' of S+. Gonzalez *et al.* also showed that, if pigeons were given a choice between the stimuli 20 min after the end of a daily session, 72 per cent of choices were S+; choices made, on the other hand, after 24 and 48 h were 48 and 51 per cent S+, respectively. Mackintosh and Cauty found that, at asymptote, their pigeons were performing at 80 per cent correct over the last ten trials of each daily reversal, but showed only 58 per cent choices of the former S+ on the first trial of each day; their experiment also included a group of rats, who showed an overnight drop in S+ choice from 93 to 55 per cent. The experiments also agree in showing relatively little overnight forgetting in goldfish; Mackintosh and Cauty, for example, found that their goldfish declined overnight from 73 per cent S+ choices over the last ten trials of each day to 65 per cent former S+ choices on the first trial of the following day. The experiments, however, do not necessarily indicate that goldfish forget less than pigeons or rats. The problem stems from the fact that, in both studies, reversals were run, not to criterion, but for a fixed number of trials; the level of accuracy achieved at the end of each reversal was markedly lower in the fish than in the pigeons and rats, so that the 'amount of room for forgetting' (Mackintosh and Cauty 1971, p. 282) was greater in the rats and pigeons.

There were no significant differences between the percentages of Trial 1 former S+ choices amongst the three species used by Mackintosh and Cauty, and so it is quite possible that, in each case, performance declined overnight to a 'floor' level, slightly above chance, so that any animal that achieved accurate performance by the end of a reversal would appear to show a larger drop in retention overnight. Gonzalez *et al.* did not present Trial 1 data for goldfish, but their relative retention scores are entirely consistent with such an interpretation.

Analysis of an experiment from the Engelhardt *et al.* (1973) report throws further light on the sources of serial reversal improvement in fish. In this experiment, which dispensed with the rear target, the fish were required to choose between two coloured keys, which also served as feeders following correct choices. The animals were given 40 trials a day for five days on each reversal, and Fig. 3.8 compares performance in ten-trial blocks for each of the first two days of early and late reversals (for a group for which the S+ colour remained on the chosen key during the reinforcement cycle). This figure illustrates two important features of the subjects' performance: first, the goldfish did show overnight forgetting—performance, in late reversals, over the first ten trials of the second day of a reversal was markedly inferior to that over the last ten trials of the preceding day; second, the major difference between early and late reversals lay, not in the error scores over the first ten trials of Day 1 of a reversal, but in the *rate* of within-reversal improvement—especially on Day 1. This latter observation has important implications: it suggests that, after a number of reversals, goldfish are more efficient at modifying their behaviour as a result of the outcome of the early choices of each day, and this in turn suggests that the only safe measure of overnight forgetting preceding the start of a reversal is performance on Trial 1 of each day—performance on subsequent trials will presumably be modified as a consequence of the outcome of that trial; indeed, the Woodard,

Schoel, and Bitterman (1971) study of reversal learning in pigeons and goldfish in a situation in which stimuli were presented singly rather than simultaneously, shows that, at asymptote, very rapid improvements in discrimination are seen over the first five trials of each reversal (see Fig. 2 of their report). It will be recalled that the Mackintosh and Cauty (1971) report found no significant differences between fish, pigeons, and rats on Trial 1 performance after a series of reversals. We may, then, conclude that goldfish do show overnight forgetting, that there is no evidence to suggest that they forget either more or less than rats,

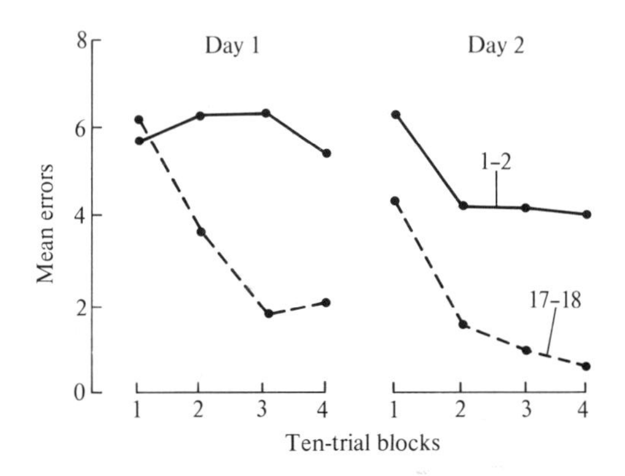

Fig. 3.8. Within-day performance of goldfish. (From Engelhardt *et al.* 1973.) For explanation see text.
Key: 1-2 – performance averaged over the first two reversals of the experiment; 17-18 – performance averaged over the 17th and 18th reversals.

and that this forgetting does not appear to be the major source of improvement in their serial reversal performance.

If overnight forgetting is not the prime source of serial reversal improvement, what factor can explain the beneficial transfer from early to late problems? One proposal, to which reference has already been made, is that of Mackintosh (e.g. 1969*a*), namely, that in early problems subjects learn which dimension is relevant, and attend to that dimension preferentially in later problems. Mackintosh's proposal forms part of a more general thesis, the essence of which is that rats, pigeons, and fish show quantitative differences in stability of attention (rats having the most, and fish the least, stable attention), and that these differences are reflected in quantitative differences in performance in a variety of situations, including serial reversal and probability learning tasks. In support of this claim, Mackintosh (1969*a*) advances evidence to show that rats and birds show improvements in serial reversal performance that cannot be attributed to increased forgetting, and, in the case of rats, positive transfer from a reversal task to a probability learning task having the same relevant dimension (which supports the view that the two tasks do indeed share some common mechanism).

The evidence that relates directly to stability of attention in fish is not drawn from studies of serial reversal performance (and will be considered in a later section), but two relevant observations may be made here. Each concerns the Mackintosh and Cauty (1971) report of performance on serial reversals in which no salient irrelevant stimuli were present (both response keys were white) and in which, therefore, selective attention should have played a minor role. First, the goldfish (like the rats and the pigeons) did show a significant improvement across the series, and this is an improvement which cannot be attributed wholly to an increase in forgetting. Although we are not given Trial 1 data for the early reversals in the series, it is clear from the data that are presented (on the accuracy of early as opposed to late reversals) that the likelihood of the fish choosing the former S+ on Trial 1 in early reversals cannot have exceeded by much (if at all) that on the final ten reversals (when the Trial 1 probability of a former S+ choice was 65 per cent). This suggests, therefore, a source of serial reversal improvement in fish other than increased forgetting or decreases in attention to salient but irrelevant stimuli. The second observation is that the fish (and the pigeons) both performed at a poorer level than did the rats, even though there were no significant differences between them in Trial 1 choices over the final ten reversals; in other words, in a procedure in which selective attention was of minor importance, and in which differential forgetting was not observed, rats were still more efficient than fish. Before, therefore, considering Mackintosh's claim that there is a quantitative difference, not attributable to contextual variables alone, between fish and rats in serial reversal performance, we should note that differences in selective attention will not in any case explain all of the differences between rats and fish in serial reversal performance. Some other factor or factors must be involved, both in the improvement seen in fish, and in the difference between fish and rats; this is an issue to which we shall return having considered the general question of quantitative performance differences between fish and rats.

There is no doubt that the majority of reports of serial reversal in rats show asymptotic error scores considerably lower than those seen in fish; many studies (e.g. Pubols 1957; North 1959; Theios 1965) find substantial numbers of reversals containing only one error – a level of accuracy superior to anything so far reported in fish. Moreover, the early series of failures to obtain any improvement across reversals in fish suggests that, at the least, such improvement is restricted to a narrower range of conditions than is the case for rats; this impression is strengthened by Mackintosh and Cauty's (1971) report that when rats were trained with a magazine in the back wall of the training chamber (similar to the situation commonly used in Bitterman's early studies), the rats, although showing more errors (nearly twice as many) than those of rats trained in a conventional situation, nevertheless did show a 'highly significant' improvement (Mackintosh and Cauty 1971, p. 281). Mackintosh has also cited data to show that conditions which obtain similar original acquisition scores in rats and fish still show greater serial reversal improvement in rats than in fish (Fig. 3.9). Is it reasonable, then, to conclude that the similarity of acquisition error scores indicates that the relevant contextual variables have been, in some sense, equated, and that the residual quantitative differences do, therefore, indicate quantitative differences in the efficiency of some information-processing mechanism involved in reversal improvement? The answer must be, it is not, for two reasons: first, in principle—reversal performance may involve some mechanism not involved in acquisition, and this mechanism might be sensitive to contextual variables not relevant to acquisition—therefore even if it was accepted that contextual variables had been

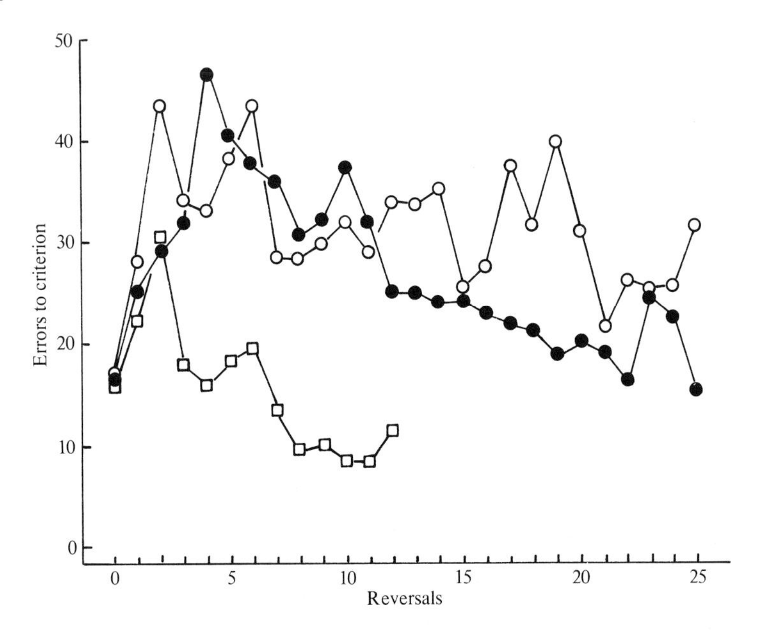

Fig. 3.9. Performance of rats (open squares), pigeons (filled circles), and fish (open circles) trained on a series of visual reversals. (From Mackintosh 1969*a*.) Rat data from Gonzalez, Roberts, and Bitterman (1964); pigeon data from Stearns and Bitterman (1965); fish data from Behrend *et al.* (1965).

equated in acquisition, there need be no implication that they had been equated for reversal; second, on empirical grounds—Gonzalez, Berger, and Bitterman (1966*a*) have shown that different training procedures may obtain, in the same species, similar original acquisition error scores, but very different serial reversal performance (Fig. 3.10).

Given that it is not possible to equate contextual variables, and that fish serial reversal performance does vary (in a currently unpredictable manner) with contextual variables, is it, then, reasonable to assume that the failure to obtain efficient reversal performance and, in particular, one-trial reversals in fish is due to contextual variables? We may approach this question by returning to a consideration of what other factor (besides forgetting and selective attention) might be concerned in serial reversal improvement.

One potential source of improvement in serial reversals is a strategy such as win-stay, lose-shift (persist in responding to a stimulus that has just been rewarded, avoid a stimulus that has just been non-rewarded). The advantage of such a strategy is that it emancipates the animal from control by the previous reinforcement-history of the stimulus concerned; the problem need no longer be solved gradually through increases and decreases in associative strength of the stimuli concerned from whatever levels they possessed before the shift in reinforcement-value. Could part of the difference between rats and fish lie in the ability of rats, but not fish, to adopt strategies?

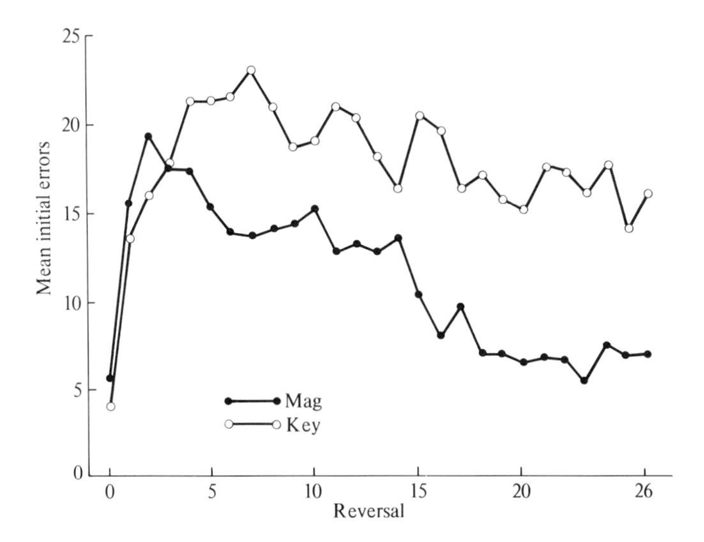

Fig. 3.10. Habit reversal in the pigeon as a function of training procedure. The Key group chose between keys of different colours, the Mag group chose between food magazines of different colours. (From Gonzalez *et al.* 1966*a*.)

Probably the best evidence for the use of strategies is the demonstration of positive transfer from one set of discriminations for which a strategy would be relevant to another set, using different discriminanda, to which an identical strategy would also be relevant. Evidence of this kind is in fact available for both birds (Kamil, Jones, Pietrewicz, and Mauldin 1977; see also p. 210) and primates (Schusterman 1962; see also p. 272), but is not available for either rats or fish. The existence, however, of one-trial reversals in rats is good evidence in itself for the formation of such strategies, since it does not seem plausible that such a level of efficiency could be obtained through gradual changes in stimulus-specific associative strengths. Is the absence of one-trial reversals in fish then sufficient to indicate an inability to use strategies?

The fact that fish have not shown one-trial reversals does not in itself show that strategies were not in use: the Kamil *et al.* (1977) report on win-stay, lose-shift strategies in blue jays (*Cyanocitta cristata*) showed that such strategies had been adopted during serial reversal training, despite the fact that, at the end of that training, Trial 2 performance on each reversal was barely above the 50 per cent correct level. It is, therefore, possible that fish do use such strategies on serial reversals, albeit inefficiently.

If fish do not adopt strategies, how is the improvement seen in the Mackintosh and Cauty (1971) study to be explained? Three relevant proposals have been made. Bitterman (1968) has suggested that one source of improvement in serial reversal performance might be due to 'the general effects of practice (known since the turn of the century)' (p. 100). Mackintosh and Little (1969*a*) make two suggestions, the first of which, similar to Bitterman's, is that one source might be 'habituation to the experimental conditions and apparatus, or perhaps adaptation to the disrupting (because frustrating) consequences of non-reinforcement' (p. 344). Mackintosh and Little's second suggestion rests on the assertion that all testing situations contain irrelevant stimuli, whether salient or not, and that even the improvement seen in spatial reversals in an unchanging stimulus complex could be due in part to a reduction in attention to features of the situation which are not relevant to correct choices.

Unfortunately, the three proposals described above lead to no clear predictions (except, perhaps, that serial reversal improvement should always be obtained—which it is not), and we shall find it difficult to decide whether we should or should not assume the use of strategies by fish. On the one hand, there is evidence, from improvements not attributable to forgetting or (in any obvious way) to selective attention, that they do use strategies; on the other hand, there are alternative explanations of such improvements, explanations which seem capable neither of proof nor disproof, and fish do not show one-trial reversals. We might take the view that, since the existence of strategies in fish is not proved, it is most parsimonious to assume that they do not exist; but, equally, we might conclude that since their existence in fish is not disproved, it is most parsimonious to assume that fish employ the same set of mechanisms as do rats, and that no qualitative differences are demonstrated by serial reversal performance. One factor that may tilt the balance in favour of the latter view is that there is evidence, as we shall see in the section on reward shifts, that goldfish appear to be less sensitive than rats to the after-effects of reward and non-reward, and to make less use of such after-effects as stimuli. This in turn raises the possibility that a contributing factor to the inefficiency of goldfish in serial reversal may be the relative lack of salience for fish of the stimuli generated by reward and non-reward per se. An animal that is less sensitive to the distinction between the after-effects of reward and those of non-reward is clearly likely to be less efficient in its use of win-stay lose-shift strategies. Therefore, one difficulty facing the notion that fish use hypotheses, the absence of one-trial reversals, may be accounted for by assuming quantitative differences in the use of after-effects, differences for which we shall see good independent support.

Although, therefore, no confident decision may be reached over the capacity of fish to use strategies, the view tentatively taken here will be that to assume such a capacity provides a good account of the data and is to be preferred to evidently untestable alternative accounts.

It may be useful to summarize here the evidence obtained from the use of the serial reversal paradigm with teleosts. First, contrary to some early reports, fish can show serial reversal improvement. Second, fish do show overnight forgetting in this situation, although it is not clear (since relevant Trial 1 data have not been reported) whether the amount of this forgetting increases as the number of reversals experienced increases, and so not clear whether forgetting accounts for some of the improvement, or whether proactive interference occurs in the task. Third, fish show progressive improvement, as do rats, in the rate of within-reversal learning: part of this improvement might be due to the fish learning to attend to the relevant dimension, although no evidence directly pertinent to this issue is currently available; more generally, this improvement provides some support for the view that fish may adopt strategies. Fourth, although the amount of improvement seen in fish is generally less than that seen in rats, there is good evidence that contextual variables associated with the apparatus have marked effects on fish performance, and, as we shall see, good reason to suppose that other variables (the after-effects of reward and non-reward) might affect the efficiency of strategies. Finally, there is as yet no evidence that there are factors which have qualitatively different effects on fish and rats. More generally, the conclusions that may be drawn seem to be that there is no convincing evidence now of any qualitative species difference in performance in this situation, at least where fish are concerned (although the possibility that there may be quantitative differences remains open), and that the problem of identifying and allowing for contextual variables is, perhaps, even more daunting than had been foreseen earlier (in Chapter 1).

Probability learning

The second procedure which, according to Bitterman (1965*a*), may reveal qualitative differences in learning capacity is probability learning. This is a discrimination in which reinforcement is available for response to one of two stimuli on every trial, but where the probability of reinforcement is higher for response to one of the two stimuli (S+) than it is to the other (S−). Typically, S+ is reinforced on 70 per cent of the trials, S−, on the remaining 30 per cent. Early studies by Bitterman and his colleagues (Bitterman, Wodinsky, and Candland 1958) suggested that whereas rats in this task tended to 'maximize', i.e. to select S+ on 100 per cent of trials, fish (African mouthbreeders) tended to 'match', i.e. to select S+ and S− with probabilities corresponding to their respective probabilities of reinforcement. Adoption of the maximizing strategy over, say, 100 trials of a 70:30 problem would yield immediate reinforcement on 70 trials, while the matching strategy would succeed on only 58 (70×0.7 plus 30×0.3), so that the fish performance does seem in that sense inferior to that of the rats.

Before discussing further results obtained from this task, one important procedural issue must be clarified: matching is obtained in fish, or indeed in any group of vertebrates so far tested, only where every trial ends in reward—where, that is, the initial choice of the non-rewarded stimulus on a given trial, whether it be S+ or S–, is followed by either correction trials (where both stimuli are re-presented with the same reinforcement condition) or guidance trials (where only the stimulus to be rewarded is re-presented); non-correction procedures obtain maximizing in all species tested to date. Matching and maximizing refer to initial choice performance only, and ignore performance on correction trials. This qualification is, on reflection, hardly surprising: if an animal matched on a non-correction procedure, then it would obtain, over 100 trials of a 70:30 problem, 49 rewards for responses to S+, as opposed to nine for responses to S–. In the correction or guidance procedures, of course, reinforcement is obtained in such a series on 70 occasions for responses to S+, and on 30 occasions, for responses to S–, whatever strategy (maximizing, matching, or some other strategy) is employed. Where matching occurs, then, it reflects the ratio of reinforcements actually obtained for responses to the stimuli; *in parenthesi,* it may be added that this matching of choice probabilities to ratios of reinforcement occurs only in a context in which *choices* have been inconsistently rewarded—the mere presentation of one rewarded stimulus on twice as many occasions as that of another rewarded stimulus does not lead in mouthbreeders to a preference for that stimulus (Behrend and Bitterman 1961).

Mackintosh (1969*a*) challenged Bitterman's view that fish were qualitatively inferior to rats in probability matching tasks on a number of grounds: first, rats do not in most studies select S+ on *every* trial (although they do come much nearer to maximizing than matching); second, fish sometimes exceed matching performance, and third, birds appear to perform at a level somewhere between that of fish and rats, and so could not reasonably be described as either 'fish-like' or 'rat-like'. This final proposition will be discussed in a later chapter on birds. Mackintosh's first two points have by now received strong experimental support: it is clear that under certain conditions, rats may actually match rather than maximize (e.g. Weitzman 1967; Bitterman 1971), and also that fish may maximize rather than match (e.g. Woodard and Bitterman 1973*b*). Matching may, however, be obtained from a group of animals for a variety of reasons: for example, if, on a 70:30 problem, four of a group of ten animals maximized, and six showed no preference between the stimuli, then the mean group performance would show 70 per cent selection of S+; moreover, matching in individuals might occur as a result of some systematic strategy, such as selecting the stimulus most recently reinforced (reward following). Random probability matching—matching, that is, in individuals that does not appear to be a consequence of systematic strategies, and where choice on a trial does not depend on any property of the immediately preceding trial—has at present been shown only in fish (e.g. Behrend and Bitterman 1966; Woodard and Bitterman 1973*b*), and never in mammals (e.g. Mackintosh 1970; Weitzman 1967; Wilson, Oscar, and Bitterman 1964*a, b*;

Bitterman 1971). The significance of this observation, however, is obscured by the fact that, as in mammals, non-random matching does sometimes occur in fish (Mackintosh, Lord, and Little 1971), and there are at present no suggestions concerning what might be the critical factors responsible for the occurrence of random matching; it remains possible, therefore, that the manipulation of some variable may obtain random matching in mammals—and as no attempt has been made to use the method of systematic variation in probability learning with mammals, this is a possibility that cannot lightly be dismissed.

There is then very little reason to suppose that there are qualitative differences between fish and other vertebrates in probability learning tasks. Mackintosh (1969*a*) takes the view that there are, nevertheless, quantitative differences in performance, and his claim should now be discussed. According to Mackintosh, quantitative differences in probability learning reflect quantitative differences in stability of attention: in order to maximize, subjects must consistently attend to the relevant dimension, despite the fact that, even when attention is consistently maintained, reward will be inconsistent. So, animals (such as fish) whose stability of attention is supposed to be weak in the face of inconsistent reinforcement should tend to perform below the optimal (maximizing) level. It is true that the studies reported to date show that maximizing is much more common in rats than it is in goldfish; however, there is one report (Woodard and Bitterman 1973*b*) of a tendency to maximize in goldfish, and it is clear that contextual variables are, as in serial reversal learning, critical. Weitzman (1967), for example, ran rats in a probability learning task using either a conventional rat jumping stand apparatus or an apparatus built to be closely similar (in relative size, type of stimuli, and so on) to the apparatus usually used with goldfish; in the conventional apparatus, the rats maximized, but in the fish-type apparatus, the rats matched. We have, then, a familiar problem: it seems that fish maximize in a considerably more restricted range of situations than do rats, but it remains possible that some (perhaps fairly trivial) procedural innovation might considerably extend the range for fish, and so eliminate the apparent distinction.

Even if it were to be accepted that fish are less likely to maximize than rats, there remains the question whether the quantitative difference is in fact due to attentional differences; direct evidence on this question will be discussed in a later section, but there are two relevant reports in the probability literature. First, Woodard and Bitterman (1973*b*) found matching in goldfish in a visual (red-green) probability learning task, and point out that the apparatus was the same as that in which Engelhardt *et al.* (1973) obtained progressive improvement in serial reversal of a red-green discrimination by goldfish. Woodard and Bitterman argue that if the phenomena of maximizing and serial reversal improvement are both due to the operation of the same mechanism, then, if contextual variables are held constant, one ought to obtain either both or neither but not, as is in fact the case, one (serial reversal improvement) without the other (maximizing). Second, a study of Mackintosh *et al.* (1971) compared performance of pigeons and goldfish on a visual (red-green) probability learning task. Pigeons tended to maximize on the problem, showing 93 per cent choices of the majority (70 per cent) stimulus; the goldfish matched, showing almost exactly 70 per cent choice of the majority stimulus. Analysis of errors showed, for both groups, strong position preferences: 79 per cent of pigeon errors were for the minority stimulus in a preferred position, as were 68 per cent of the goldfish errors. These percentages indicate, presumably, that position (the salient irrelevant dimension) was controlling behaviour on error trials to a lesser extent in goldfish than in pigeons, which, then, raises the question, why are goldfish so much worse than pigeons? Neither of these objections to Mackintosh's view is particularly strong—it is not difficult to imagine various ways in which the findings could be accommodated within his account—but they do indicate that Mackintosh too can provide only an incomplete theoretical account of the results obtained using the probability learning paradigm.

The results reviewed in this and the preceding section suggest that neither of the experimental procedures concerned is in fact adequate to distinguish categorically between the intellectual capacities of groups of vertebrates. Bitterman's more recent experimental work has encouraged him to develop a new account of a qualitative intellectual difference between fish and rats, and it is to studies relevant to that view that the following section is directed.

Effects of reward shifts

The feature common to the experiments to be described in this section is that they examine the response of fish to shifts in the conditions of reward. The basic example of such a task is simple extinction—the shift from the consistent delivery of UCSs or rewards to their total absence, a principal effect of which is the reduction or abolition of the relevant conditioned response. There are, of course, a wide variety of factors that affect rates of extinction in the rat, the animal for which most data are available; like rats, fish do extinguish responding following removal of reinforcers, and no attempt has been made to date to show that there are any gross species differences in rates of simple extinction. There have, however, been suggestions that some variables might have contrasting effects on extinction in rats and fish, and that these contrasts reflect important differences in the ways in which rewards influence behaviour.

One such variable is amount of acquisition training: Gonzalez and Bitterman (1967) have claimed that 'resistance to extinction in the fish is more sensitive than in the rat to sheer frequency of reinforcement' (p. 163). The evidence for this proposal is, however, not strong. Gonzalez, Eskin, and Bitterman (1961) exposed mouthbreeders to a series of acquisition and extinction sessions, finding that, like rats, mouthbreeders showed a progressive decline in extinction scores, but that, unlike rats, they were influenced by the number of acquisition sessions interpolated between succeeding extinction sessions, so that extinction was slowest, the more interpolated sessions had been administered. A subsequent study (Gonzalez, Holmes, and Bitterman 1967*a*) directly compared goldfish and rats, giving subjects of both species daily sessions that were alternately acquisition

(30 reinforcers, one for each response) and extinction (15 min duration). Both groups showed declines in the overall level of extinction responding, but the rats showed a much lower level, the difference between the groups being most marked in the first few minutes of each extinction session (Fig. 3.11). Subsequent tests showed that, whereas rats' extinction scores were identical whether an acquisition session was interpolated or not, goldfish extinction scores were very much lower in the absence of interpolated training: there was in this experiment, however, no effect of the number (one, three, or six) of interpolated sessions (as there

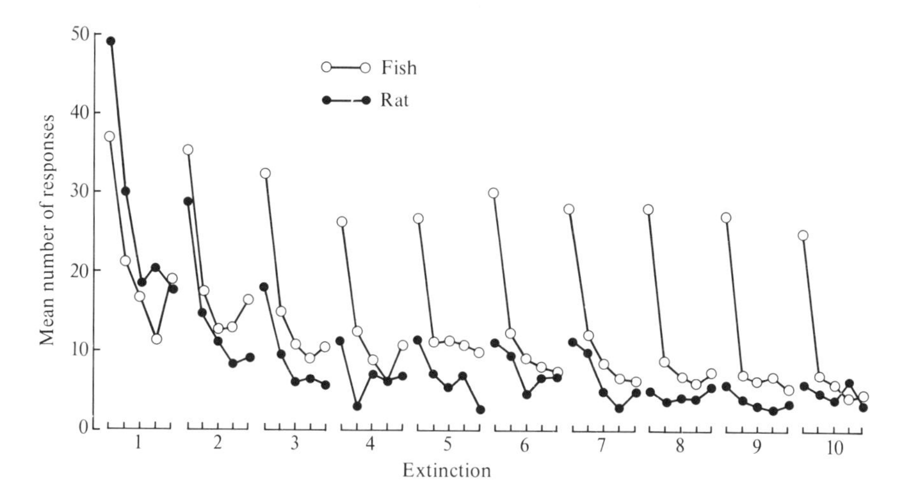

Fig. 3.11. Course of extinction in fish and rat plotted in terms of mean number of responses for successive 3-min periods in each of ten 15-min extinction sessions separated by one of more retraining sessions. (From Gonzalez *et al.* 1967*a*.)

had been in the mouthbreeder study). These studies may best be interpreted as showing that whereas rats do learn to use the delivery or non-delivery of rewards for the first few responses of a session as a cue indicating whether acquisition or extinction is in force, goldfish do not, and so simply build up habit strength (in acquisition sessions) or reduce it (in extinction sessions). This finding is reminiscent of the suggestion made earlier (p. 78) to account for the relatively poor performance of goldfish on serial reversal tasks, namely, that they are less able than rats to use the outcome of a trial (reward or non-reward) as a cue associated with a change in response: this is a notion which will be further developed in later parts of this section.

There does not appear to be any significant contrast between rat and fish sensitivity to frequency of reward in simple 'one off' extinction. Gonzalez, Holmes, and Bitterman (1976*b*) showed that in a procedure in which fish paced their own striking of a target, and were consistently reinforced, resistance to extinction increased with the number of rewards, and increased resistance to extinction with increased training is also generally found in rats in similar free-operant situations (for a review, see Mackintosh 1974, p. 423). Zych and Wolach (1973), on the other hand, using a water alley with goldfish, found more resistance to extinction after few (56) as opposed to many (196) consistently reinforced acquisition trials; in rats also, an inverse relationship between amount of training and resistance to extinction is generally found after consistent reinforcement in alleys (Mackintosh 1974, p. 424).

There is rather more convincing evidence that magnitude of reinforcement may differentially affect extinction in rats and fish. A large number of studies have shown that, following consistent reinforcement in alleys, rats extinguish more rapidly, the larger the reward (Mackintosh 1974, p. 427). In lever-pressing tasks, the effect of reward magnitude is less clear: Marx (1967), for example, found a similar inverse relationship between amount of reward and resistance to extinction in rats trained under high drive, but also found that, when trained under low drive, rats showed a positive relationship between reward size and resistance to extinction. Experiments with goldfish have, in contrast, invariably found slower extinction after large rewards in both alleys (Mackintosh 1971; Gonzalez, Potts, Pitcoff, and Bitterman 1972) and free-operant tasks (Gonzalez and Bitterman 1967; Gonzalez, Holmes, and Bitterman 1967*b*). The theoretical implications of these findings will be considered after relevant data from another procedure, partial reinforcement, have been presented.

In rats, the rate of extinction in instrumental tasks is heavily influenced by the schedule of reinforcement during acquisition. A large range of experiments have shown that irregular (partial) reinforcement, usually on 50 per cent of trials, leads to slower extinction than is seen after consistent reinforcement, and this phenomenon is known as the partial reinforcement extinction effect (PREE). Some early experiments by Bitterman and his colleagues (Gonzalez, Milstein, and Bitterman 1962; Wodinsky and Bitterman 1960) failed to obtain a PREE in experiments which used goldfish and mouthbreeders in both classical and instrumental procedures. In all of these early studies, the consistently reinforced control group received the same number of trials as the partially reinforced group, which in consequence obtained half as many reinforcers as the controls (the trials-equated procedure); that this is critical was shown by a number of subsequent studies which equated the number of reinforcers obtained by the groups and found a PREE in both classical and instrumental tasks (Gonzalez, Eskin, and Bitterman 1962, 1963). Since then, Gonzalez and Bitterman (1967) have shown that, using a high (seven tubifex worms) but not a low (one worm) reward, a PREE can be obtained in goldfish in a 'trials-equated' procedure; Zych and Wolach (1973), using a 'relatively small reward' (three brine shrimps) demonstrated a PREE in an equated trials procedure following extensive (196) but not brief (56 trials) training. These latter two results suggest that there may, after all, be little substantive difference between rats and fish in this procedure: although a PREE is found in rats involving small rewards and few trials, the magnitude of the PREE may be increased by increases in both reward magnitude (Wagner 1961) and amount of training (Wilson 1964). There is, however, one other pertinent variable—inter-trial interval (ITI); in rats, a PREE can be obtained using a 24 h ITI (Wagner 1961) but no such effect has been obtained in fish, even in those studies (Gonzalez, Behrend, and Bitterman 1965; Schutz and Bitterman 1969; Boitano and Foskett 1968) which have used the reinforcer-equated procedure. Bitterman (1975) has laid considerable emphasis on the failure to obtain a PREE in fish with spaced (i.e. 24 h ITI) training; however, the absence of the PREE in those studies may not have been due to the long ITI. In no case were control groups run at a shorter ITI to show that, with the same reward, the same number of training trials, the same species, and the same apparatus, a PREE would have appeared. Similar reservations apply to the failure of Longo and Bitterman (1960) to obtain the PREE in mouthbreeders, where the reward was five food pellets, and the ITI, 24 h. Gonzalez, Eskin, and Bitterman (1962) did obtain a PREE using the same species, in the same apparatus, with a smaller food reward (one pellet), but a brief (a few seconds) ITI—however, the partially reinforced fish in the Longo and Bitterman study received in acquisition 30 rewards in all, as opposed to the 60 received by control subjects, and in the Gonzalez *et al.* study, both groups received 100 reinforcers each. Until a closely controlled study of the effects of ITI on the PREE in fish is carried out, caution must be exercised, particularly as there have been failures to obtain a spaced-trial PREE in rats: Gonzalez and Bitterman (1969), for example, found no PREE in rats trained with a small reward with trials spaced 1 h apart.

Fish, then, appear to show a PREE in a narrower range of situations than do rats: they are sensitive to magnitude of reward, to amount of training, and, perhaps, to ITI. Do these results encourage theoretical speculation? The answer to such a question depends upon the theoretical analysis provided for the phenomena in rats (theoretical analyses being almost invariably directed at results obtained in that species alone). Bitterman (1975) argues that the massed trial PREE and the spaced trial PREE are two phenomena having distinct causes. The massed trial effect is, he believes (following Sheffield 1950), due to a sensory carry-over of reinforcement in continuously reinforced animals, which forms an important part of the stimulus complex controlling responding, so that its absence (in extinction) drastically alters that stimulus complex, and obtains weak responding. The partially reinforced subjects, in contrast, are accustomed to obtaining rewards in the absence of reinforcement-produced sensory carry-over, and so find extinction less different from acquisition than do continuously reinforced subjects, and extinguish more slowly. It should be emphasized that an important part of this account is the assumption that the sensory carry-over has a rapid decay, and so does not play a role in the control of responding with spaced trials.

The spaced-trial PREE is, according to Bitterman (e.g. Gonzalez and Bitterman 1969), caused by the operation of a 'contrast mechanism', which is also responsible for both the inverse relation between reward size and resistance to extinction, and the phenomenon of successive negative contrast (which will be discussed in a later part of this section). The onset of extinction is, through the action of the contrast mechanism, a particularly disruptive event to an animal that confidently expects a large reward; partial reinforcement prevents the formation of confident expectations, and so, with large rewards, partially reinforced animals extinguish more slowly than consistently reinforced animals. The formation of expectations is independent of ITI, so that a spaced-trial PREE is obtained; the contrast mechanism takes effect only where large rewards are involved, so that the spaced trial PREE is dependent on large reward. The fact that the massed-trial PREE can be obtained with small reward indicates that some mechanism other than that of contrast is responsible, although Gonzalez and Bitterman (1969) agree that '. . . contrast may contribute to the PREE in massed as well as in spaced trials' (p. 100).

To account for the differences between rats and fish in response to reward shifts, Bitterman has proposed that the contrast mechanism does not operate in the fish, which 'gives no indication of learning about reward, which is prerequisite, it seems, to contrast' (Gonzalez and Bitterman 1969, p. 102). The essence of Bitterman's proposal is that fish, unlike rats, do not come to 'anticipate' reinforcers, or to form expectancies of any kind; fish do not, then, experience thwarted expectancies, the disruptive consequences of which can be seen in rats. Bitterman does not speculate further on the nature of the contrast mechanism, although he appears to rule out (Gonzalez and Bitterman 1969) one widely accepted mechanism, the emotional response of frustration to the non-arrival of anticipated rewards (e.g. Amsel 1962).

In Bitterman's view, then, the fish is a somewhat simpler animal than the rat, and one which behaves according to the stimulus-response reinforcement principle, first formulated by Thorndike. According to this principle '. . . the role of reward is simply to connect responses to stimuli; large rewards produce stronger connections than small rewards, and strong connections produce better performance than weak connections' (Bitterman 1975, p. 700). If fish possess a mechanism of this kind, and no other mechanism, then they can form only associations involving responses, and cannot form associations in which both terms are stimuli, cannot, that is, form classical associations. Rats, on the other hand, are capable of forming associations between stimuli and reinforcers, and this capacity enables them to anticipate rewards, so that in them we find important consequences of non-arrival of anticipated rewards.

Shifts in reward in massed trial designs should produce comparable effects in rats and fish since, in massed trials, after-effects play an important role in both species; shifts in reward in spaced trial designs, on the other hand, should produce differential performance, as after-effects play no role, and only rats experience disappointed expectations.

It remains to be shown how Bitterman's account applies to the data reviewed above. First, fish in alleys extinguish more slowly after large than after small rewards; rats show the reverse effect. In Bitterman's view, the inverse relation between reward and resistance to extinction arises in rats because the bigger reward elicits a larger response to non-reward, which severely disrupts performance, and hastens extinction of responding; in fish, larger rewards merely serve to forge stronger connections, which in turn take longer to counteract by non-reward (no mechanism for the reduction in response strength being specified).

The massed-trial PREE is explicable in terms of after-effect theory, and is found in both rats and fish; the spaced-trial PREE, a consequence of the contrast mechanism, has been found only in rats (although it will be recalled that there are as yet no very good grounds for supposing that it cannot be obtained in fish).

The proposal that fish do not learn to anticipate rewards does, then, seem to fit the data in a general way, but there remain problems of detail: for example, although a massed-trial PREE does occur in fish, the conditions for its appearance appear rather stringent. Gonzalez and Bitterman (1967) failed to obtain a massed-trial PREE in goldfish using a small reward, although the number of reinforcements was equated across groups. Such observations have led Mackintosh (1971) to propose that there are quantitative, but not qualitative, differences between rats and fish in response to reward: specifically, Mackintosh proposes that, in contrast to rats, '. . . in fish, after-effects of reward and non-reward form a relatively small part of the stimulus complex controlling instrumental performance' (Mackintosh 1971, p. 226). As a reasonably direct test of this notion, Mackintosh explored goldfish performance in a single-alternation experiment. Single-alternation is a type of partially reinforced schedule in which rewarded and non-rewarded trials occur in strict alternation; rats in such procedures typically come to respond rapidly on trials on which reinforcement is available, and to withhold responses in the intervening non-reinforced trials, and this performance is presumed to depend upon the use of contrasting after-effects of rewarded and non-rewarded trials. Mackintosh failed to obtain any sign of patterning (responding more rapidly on rewarded than on non-rewarded trials) in goldfish when given 41 trials a day, with a 20 s ITI. Mackintosh goes on to describe a further experiment in which fish were given only two trials a day: each pair of trials was on alternate days NR (a non-rewarded trial followed by a rewarded trial) and RN, so that the outcome of the first trial predicted the type of the second trial. Patterning was demonstrated for the second trial of each day in this design, and Mackintosh concludes therefore that the goldfish are capable of conditioning to the after-effects of reward and non-reward, but that they 'discriminate poorly between more and less recent after-effects' (Mackintosh 1971, p. 231). An alternative interpretation grants that the failure to obtain patterning in the multi-trial study is due to interference between trials, but supposes that such interference is the result of the after-effects being relatively weak, and able to support conditioning only in optimal circumstances. Gonzalez (1972) provided evidence for this view by demonstrating that, in a multi-trial situation, goldfish did show patterning, but only where large (ten worms) and not small (one worm) rewards were used (and even then, good patterning emerged only after some 900 trials). These data suggest, then, that it may be simplest to assume that the after-effects of rewards of the size conventionally used in goldfish are weaker than those of rats given rewards within the conventional range. Before leaving the single alternation paradigm, it should be pointed out that Mackintosh's 1971 studies provide very good evidence that proactive interference does occur in fish: it is difficult to conceive any other explanation for the failure of patterning in his multi-trial experiment than that some stored after-effect of the outcomes of early trials interfered with the after-effects of the later trials.

Mackintosh can, then, account for the relative difficulty of obtaining even a massed-trial PREE in fish by agreeing with Bitterman that the phenomenon is due largely to the role played by after-effects in continuously reinforced animals, and further assuming that, in fish, such after-effects play a small role in the control of behaviour. Can Mackintosh, however, use this notion to explain the two phenomena most amenable to Bitterman's account, namely, the differential effects of reward size on extinction in rats and fish, and the absence of the spaced-trial PREE in fish? Once again, the answer to this question depends critically upon the explanation adopted for those effects in rats. Mackintosh in each case takes a view that is quite different from that of Bitterman, believing that after-effect theory provides an adequate account of both phenomena. Mackintosh does not, however, restrict the term 'after-effect' to sensory consequences of rewards (the usage of Sheffield and Bitterman), but instead broadens its reference to that employed by Capaldi (e.g. Capaldi 1967). According to Capaldi, both reward and non-reward produce discriminable after-effects, which may be regarded rather as memories for reward or non-reward, and which persist, as do other memories, for a considerable time—certainly for more than 24 h. This version of after-effect theory sees the inverse relation between reward size and resistance to extinction as being due, not to an increased disruptive consequence of non-arrival of a large anticipated reward, but to the fact that the absence of large rewards causes larger changes in the stimulus complex than the absence of small rewards, and obtains therefore less responding; similarly, the spaced-trial PREE may be accounted for, like the massed-trial PREE, by supposing that partially reinforced subjects have, unlike continuously reinforced subjects, been rewarded in the presence of the after-effects of non-reward and so have acquired habit strength to stimulus complexes similar to those encountered in extinction. The assumptions that smaller rewards set up weaker after-effects, and that there is some decay over time of these effects, enable after-effect theory to accommodate the fact that the spaced trial PREE (which is generally smaller than the massed-trial effect) is dependent upon the use of a large reward.

The arguments relevant to Mackintosh's view are complex, and a detailed presentation of his case can be found in Mackintosh (1974). This is not the place to attempt a resolution of conflicting theoretical accounts of behaviour in rats; our strategy must be to note that there are opposing views on the nature of, for example, the spaced-trial PREE, and to see to what extent any proposed theoretical interpretation applies consistently across the range of phenomena found in fish. The evidence to be discussed in the remainder of this section, which concerns contrast effects, has an important bearing on the question of the relative merits of the approaches of Bitterman and Mackintosh to the interpretation of the whole range of reward shift effects in fish.

The procedures considered up to this point have concerned extinction – the total removal of reward. Our interest will now focus on shifts between one reward magnitude and another, non-zero, reward magnitude. It is a well-established phenomenon in rats that when animals accustomed to performing for a high reward are shifted to a lower reward, their performance (e.g. speed of running in an alley) falls sharply, to a level below that of rats accustomed to working for that low level throughout training. This phenomenon, associated with the name of Crespi (1942) is variously known as the 'depression effect' or the successive negative contrast effect (successive NCE). The NCE is powerful evidence against the operation of a 'strict reinforcement principle' in rats: low reward should not reduce at all the strength of a connection previously established by a high reward, much less reduce it below the strength of a connection established by low reward. Many theorists have taken the NCE as evidence, then, that reward serves in rats, not merely to strengthen connections between stimuli and responses, but to establish expectancies of rewards of a particular magnitude; according to this view, the depression effect must in some way reflect the disappointment of established expectancies, there being no other mechanism that could reduce response strengths below that established by consistent low reward.

A number of experimenters, using a variety of types of apparatus, have failed to obtain a successive NCE in goldfish (Lowes and Bitterman 1967; Mackintosh 1971; Gonzalez *et al.* 1972; Wolach, Raymond, and Hurst 1973; Gonzalez, Ferry, and Powers 1974). Indeed, using spaced (24 h ITI) trials, both Lowes and Bitterman (1967) and Gonzalez *et al.* (1972) found that a downshift in reward produced no decrement at all in the performance of their fish; fish rewarded with the lower amount, without any experience of the higher amount, performed at a significantly lower level, in both experiments, than fish working for the same amount but having previously experienced the higher amount. These are striking results – particularly as, in the Gonzalez *et al.* 1972 study, the post-shift phase lasted 36 days without any detectable decline after a change from a 40 worm reward to a one worm reward. The data do, of course, provide good support for the notion that a strict reinforcement principle does operate in fish; however, there are two contrary reports, by Raymond, Aderman, and Wolach (1972) and Wolach *et al.* (1973), each of which used spaced trials and found a fairly rapid adjustment to a tenfold reduction in reward magnitude, performance dropping, in each report, to that maintained by a group trained throughout for the lower reward. The Raymond *et al.* (1972) report suffered from the possibility that the fish, swimming an alley, might have seen the food reward from the alley, as the goal box was baited before the start of each trial. However, the Wolach *et al.* (1973) experiment excluded this possibility, and it is not currently possible to provide any convincing resolution of the conflict between this report, and those of Lowes and Bitterman (1967) and Gonzalez *et al.* (1972).

Where highly massed trials (ITI 20 s or less) have been used, there is general agreement that a downshift in performance does occur (Gonzalez *et al.* 1974; Gonzalez and Powers 1973; Cochrane, Scobie, and Fallon 1973); Mackintosh

(1971) used an ITI of 3 to 5 min, and found some evidence of a gradual decline in alley speeds following a fivefold reduction in reward magnitude, although the decline did not reach statistical significance. It appears, then, that at very short ITIs, significant reductions in performance may be obtained reliably, but that the effect is very much less consistent at longer ITIs. It is also generally accepted that a major contribution to this decline is the change in after-effects occasioned by the shift in reward; Mackintosh uses the fact that he failed to obtain a significant downshift using a change in reward that is more than sufficient to obtain marked downshifts in rats in situations with comparable ITIs (Di Lollo and Beez 1966) as support for his view that after-effects play a less prominent role in the control of behaviour in fish than is the case with rats.

Although, as we have seen, a number of reports have found downshifts in performance as a consequence of reductions in reward magnitude, the only investigators to have found a significant successive NCE in fish are Breuning and Wolach (1977, 1979); these workers used a classical training procedure with a short (average 55 s) ITI, in which the target CR was general activity, with light as CS and food delivery as UCS. In the other reports, the performance of the fish accustomed to large reward declined, on the introduction of small reward, to the level of performance of fish trained throughout for small reward, but not below that level (the defining feature of the NCE). The significance of the elusiveness of the successive NCE may be considered along with evidence from another procedure, which examines a simultaneous NCE.

When rats are trained to respond to two discriminable stimuli, one of which (S+) signals large reward, and the other (S−), small reward, performance on S− trials is consistently found to be below that of control animals for whom all trials use only the small reward. As S+ and S− trials are intermixed in a given session, and both large and small rewards are experienced on each day of training, this phenomenon is known as the simultaneous NCE. Experiments conducted with goldfish have clearly demonstrated a simultaneous NCE (Cochrane *et al.* 1973; Gonzalez and Powers 1973; Burns, Woodard, Henderson, and Bitterman 1974).

The fact that, in goldfish, a simultaneous NCE is readily observed, but a successive NCE can, apparently, rarely be obtained, has been taken by many authors (Bitterman 1975; Gonzalez and Powers 1973; Mackintosh 1974) as evidence to support the view that simultaneous and successive NCEs have different causes; there is, however, no general agreement on the nature of the causes. Bitterman (1975) holds that the successive NCE (at least with spaced trials) depends upon the operation of the contrast mechanism, and so depends upon the ability, absent in fish, to 'learn about rewards'; Gonzalez and Powers (1973) agree upon the role played by expectations, as they 'concur with the most commonly held view that it [the successive NCE] is the product of an emotional reaction to the discrepancy between anticipated and encountered amounts of reward' (p. 97). Mackintosh, on the other hand, believes that the successive NCE can be explained in terms of the change in after-effects occasioned by the shift in reward, a change which alters the stimulus complex to one which commands weak habit-strength, and so obtains responding at a very low level. Once again, the relevant arguments are involved, and depend almost entirely upon data from rat studies; they cannot, therefore, be presented in detail here, although we shall consider some of the relevant rat data when making an overall assessment of the rival accounts of reward shift effects in fish.

The causes of simultaneous NCE are equally contentious. Bitterman holds the distinctly idiosyncratic opinion that 'if the reward value of a small amount of food is less immediately after experience with a large amount than after experience with a small amount, weaker connections should be established in the experimental animals than in the controls' (Bitterman 1975, p. 709); this he characterizes as an 'interpretation in terms of sensory contrast' (p. 706). Regrettably, we are given no guide, apart, perhaps, from the use of the word 'sensory' as to how this mechanism might operate; there seems, moreover, no reason why the *sensory* response to a given magnitude of food should be less following ingestion of a larger amount of food some 10 to 20 s previously (10 to 20 s being the range of ITIs used in the three successful demonstrations of the simultaneous NCE in fish). A higher level perceptual contrast would be most plausible if we supposed that the animals had learned to anticipate rewards of a given magnitude—but this is precisely the assumption that Bitterman does not make. Moreover, there is good evidence that such a sensory contrast account does not apply to rats: Ludvigson and Gay (1967), for example, have shown that a simultaneous NCE obtained when using a single start-box and two discriminable alleys as S+ and S− is virtually abolished when two clearly discriminable start-boxes are used. The implication is, then, that it is not enough for the small rewards to occur in a context of large rewards—they must occur in association with background cues that are common to both S+ and S− trials. It therefore seems that (in rats, at least) the simultaneous NCE is due, not to a perceptual reduction in size of the small reward, but to the arrival of a small reward in a context that generates expectations of large rewards.

Gonzalez and Powers (1973), who agree with Bitterman that fish do not form expectancies, believe that the differential reinforcement of the two stimuli results in the development of 'inhibitory properties' by S−, these being 'basic products of discrimination'. The inhibitory properties subtract from the excitation that accrues to S− in the normal way, on the receipt of each reward. Mackintosh also argues that the simultaneous NCE is due to inhibitory processes involving S−, but holds that 'the effectiveness of inhibitory conditioning is a function of the prevailing expectation of reinforcement' (Mackintosh 1974, p. 34). Mackintosh, in other words, holds that the development of inhibition to S− is a consequence of the expectation of a large reward, this in turn the result of associations between contextual cues in the situation, and the larger reward. The difficulty in assessing Gonzalez and Powers' position is that they do not provide a detailed account of the processes that they assume to be involved in discrimination learning. It is fair to say that most theorists currently concerned with conditioned inhibition agree with one central tenet of the influential Wagner–Rescorla model of classical conditioning (e.g. Rescorla and Wagner 1972), namely, that inhibition is generated only in situations in which the stimulus complex predicts the arrival of a reinforcer larger in magnitude than that which arrives.

The conflict between the Mackintosh and Bitterman accounts has now taken a somewhat ironic turn. In discussing the effects of reward size on extinction, the spaced-trial PREE, and the successive NCE, Mackintosh has opposed Bitterman's interpretation, not by objecting to Bitterman's most important claim, that fish do not form expectancies, but by arguing that expectancies are not critical factors in those phenomena in rats. However, the simultaneous NCE, the one effect in which Bitterman does not claim that expectancies are involved, is the sole case in which Mackintosh supposes that they are, and so, becomes a critical phenomenon. If Mackintosh's account of the first three phenomena is accepted, it may nevertheless be the case that fish do not (and rats do) form expectancies, as Bitterman claims: however, if Mackintosh's account of the simultaneous NCE is accepted, then fish do form expectancies, and Bitterman's thesis must be rejected.

This review of reward shift phenomena in fish has shown two major advantages for Mackintosh's position, one empirical, the other, theoretical. Whether or not Mackintosh's general view on the role of after-effects is accepted, there are good grounds for the proposal that after-effects play a minor role in fish behaviour: first, it will be recalled that, both in serial reversal learning and when given a series of acquisition and extinction sessions, fish appear to be less efficient than rats in the use of reward or non-reward as a cue; second, the massed-trial PREE is found, in fish, only in a limited range of circumstances, particularly where small rewards are used (since small rewards do not activate Bitterman's contrast mechanism, small reward PREEs should, if after-effects played equal roles, be generally equivalent in the two species); third, although single alternation patterning can be established in fish, this requires extended training and large rewards, the after-effects of smaller rewards being, presumably, too weak. Mackintosh's theoretical advantage lies in the fact that he has provided (Mackintosh 1974) explicit accounts of his theoretical analyses of the phenomena, which make his position very clear. This was of particular importance in the discussion of simultaneous contrast—perhaps the most critical phenomenon reviewed in this section, as regards the presence or absence of expectations in fish. As both Mackintosh and Bitterman agree that the formation of expectancies involves the detection of stimulus-reinforcer contingencies, the general question whether fish are capable of classical conditioning is of central relevance; it will be recalled now that the balance of evidence strongly suggested that fish do show classical conditioning (although the failure to establish SPC in fish did cast some doubt on that conclusion).

It is obvious that the theoretical interpretations of the relevant phenomena in rats play a crucial role in determining the relative plausibility of the rival accounts applied to the fish data. Throughout this section, it has been argued that detailed consideration of the arguments for and against current interpretations—based, as they are, largely on rat data—would be out of place. However, some relevant data will be presented here, as they have 'comparative' implications, which derive from a prediction of Mackintosh's account, namely, that where rats in reward shift procedures experience weak after-effects of reward and non-reward, they should behave like fish. Two types of reward that, according to Mackintosh, produce poorly discriminable after-effects are sucrose solutions, and delayed rewards. As direct evidence for his proposal, Mackintosh quotes studies which show that single alternation in rats does not produce patterning when either different concentrations of sucrose are alternated (Likely, cited by Mackintosh and Lord 1973), or immediate and delayed reward are alternated (Cogan and Capaldi 1961; Burt and Wike 1963). Simultaneous NCEs are found (in rats) where either different sucrose solutions (Flaherty, Riley, and Spear 1973) or immediate versus delayed reward (Mackintosh and Lord 1973) are used; the same experiments, using the same reward parameters, found no successive NCEs with shifts in either sucrose solutions or from immediate to delayed reward. The contrast phenomena, then, parallel those found in goldfish using reward versus non-reward. Finally, Likely, Little, and Mackintosh (1971) showed a positive relationship between sucrose concentration and resistance to extinction in an experiment in which other groups of rats, exposed to differing amounts of Noyes food pellets, showed the customary inverse relationship. There is, it seems clear, a good overall agreement between the pattern of results established in goldfish, and that reported in rats working for either various sucrose concentrations or immediate and delayed rewards, which in turn reinforces Mackintosh's overall analysis.

The studies on rats using sucrose rewards are relevant to another issue, namely, whether the differential reliance on after-effects should be regarded as a difference in intellect, the weaker after-effects found in fish reflecting, perhaps, a reduced memory capacity, or whether the difference is due simply to contextual variables. Since the behaviour of rats can be made to approximate to that of fish simply by changing the nature of the reinforcer concerned, and since there is no way of deciding which type of rat food reinforcer is more properly comparable to the rewards given to goldfish, it appears to be most parsimonious to conclude that the weaker after-effects observed in fish are associated with the nature of the reinforcer used, a 'contextual variable', rather than with any basic difference in memory mechanisms.

This section has generally shown Mackintosh's position in a somewhat more favourable light than that of Bitterman, this to some extent being a consequence of the detailed elaboration of theory and data that Mackintosh has provided. It may restore the balance somewhat to bring out a weakness in Mackintosh's position—a weakness common to most proposals of quantitative as opposed to qualitative differences, which is, that the proposal can accommodate a wide variety of potential experimental outcomes, and tends to set up rather imprecise predictions. For example, Mackintosh's account does not *predict* that there will be, in goldfish, no spaced-trials PREE, no spaced-trials successive NCE, and a positive relationship between reward size and resistance to extinction: what it predicts is that it will be relatively (compared to rats) less likely that such phenomena will be obtained in fish, without specifying precisely under what conditions they will occur. The theory, that is, is in most instances applied *ad hoc*, and so accommodates rather than predicts the more striking experimental outcomes.

There is also an imprecision as to the nature of the differences between the after-effects of immediate dry food reward in the rat and those of sucrose solutions and delayed rewards (in rats) or of food reward in fish. This is most clearly shown in the case of delayed reward. We have seen that rats find difficulty in adjusting to single alternation between immediate and delayed rewards, and this was taken to show that the after-effects of immediate and delayed reward were similar. This same interpretation was used to account for the absence of a successive NCE following a shift from immediate to delayed reward. However, resistance to extinction is often found to be greater following delayed as opposed to immediate reward, and to account for this result, Mackintosh (1974, p. 432) advances the hypothesis that 'a delayed reward . . . sets up traces or after-effects relatively similar to those encountered in extinction'. It is, of course, not at all implausible that immediate and delayed reward set up similar after-effects, or that delayed reward and extinction have elements in common; however, it can be seen that, since delayed reward can be regarded as similar to both reward and non-reward, an arbitrary choice can be used to accommodate rather too many possible outcomes.

The arguments above are directed, not so much against the validity of Mackintosh's account as against the tendency to overvalue the consistency of that account applied to a wide variety of findings. It remains the case that Mackintosh's approach allows a consistent and plausible account of all the phenomena described. Moreover, his analysis of the simultaneous NCE, which does not appeal to after-effect theory, is particularly convincing; it is an analysis which contradicts the central tenet of Bitterman's account, and which, until successfully countered, sways the argument over reward shift effects in fish strongly in Mackintosh's favour.

Mechanisms of attention in fish

Our concern in this section will be to determine whether there are any significant differences between the behaviour of fish and that of other vertebrates (principally rats) in situations which, according to some theorists, involve the use of selective attention in at least some species. Selective attention will be used here in the sense advanced by Sutherland and Mackintosh (1971), and refers to the notion that animals may attend to, and store information about, stimuli defined by their position on some one analysing dimension at the expense of attending to those on other dimensions.

Reference has already been made (in the sections on serial reversal and probability learning) to Mackintosh's proposal that selective attention (the 'switching in of analysers') is less stable in fish than in rats, and this proposal may now be examined in more detail.

The first step in Mackintosh's account consists in the observation that fish show generally less improvement in serial reversal than birds, which, in turn, show less improvement than rats; similarly, fish are less likely to maximize than rats. Mackintosh's second step is to argue that learning to attend consistently to the relevant analyser despite intermittent non-reinforcement is critical to achieving both good serial reversal improvement and maximizing. Evidence for this claim is provided by showing that, in the case of rats and birds, following serial reversal or probability learning, subjects have indeed learned to attend to the relevant dimension; the evidence takes the form of showing better transfer, following either type of training, to other discriminations for which the same dimension is relevant rather than irrelevant. Mackintosh's conclusion is that the ordering of fish, birds, and rats in ability in the serial reversal and probability situations reflects an ordering in the element common to those situations, namely, in the stability of attention.

As the interest at this point centres on fish, we shall not examine the data on birds here; the argument that follows is in any case directed at the lack of certainty concerning the (larger) contrast between fish and rats, so that the bird data would not be seen as particularly helpful. We have already seen that fish have in general performed more poorly than rats in both the serial reversal and probability learning tasks, but our conclusion was that there could not be, in either case, much confidence that the difference reflected anything more than a failure to balance contextual variables across the species. In particular, it was argued in the relevant section that it appeared that poor stability of attention could not account completely for the relatively low performance of fish, so that the overall rankings of fish, pigeons, and rats, for example, need not indicate differences in attention, rather than differences in other factors contributing to their performance.

Direct evidence that either serial reversal improvement or maximizing is due, at least in part, to attentional processes is derived entirely from bird and mammal experiments: no corresponding transfer experiments have yet been carried out in fish. It will be recalled that fish do show, in the course of serial reversal training, an increase in rate of within-problem learning (Engelhardt *et al.* 1973), and it was suggested that this might be due to attentional processes; it should be emphasized here that no direct test of this suggestion has been made. The scant evidence drawn from fish that is relevant to Mackintosh's proposal provides at best very weak support, and will be described here very briefly. Two reports (Warren 1960*a*—using paradise fish; and Mackintosh, Mackintosh, Safriel-Jorne, and Sutherland 1966—using goldfish) compared performance in a reversal of a simultaneous discrimination following, in one (non-overtrained) group, acquisition training to some criterion, with that in an overtrained group, which were given extra training trials on the same original discrimination after satisfying the acquisition criterion. Overtrained rats in comparable procedures show more rapid reversal than non-overtrained animals, and this somewhat counter-intuitive phenomenon, the overtraining reversal effect, is amenable to analysis in attentional terms on the assumption that overtraining further strengthens the relevant analyser, so that it is not switched out by the non-rewards that occur at the start of reversal training (Sutherland and Mackintosh 1971). Both Warren (1960*a*) and Mackintosh *et al.* (1966) agree that the overtraining reversal effect is not obtained in fish; in other words, according to attentional theory, analysers in fish remain unstable despite previous overtraining. A second experiment of Mackintosh *et al.* (1966) showed that the so-called 'dip' effect (D'Amato and Jagoda 1960), a phenomenon obtained in non-overtrained but not in overtrained rats, an effect that is, again, amenable to a (highly involved) attentional analysis (Sutherland and Mackintosh 1971) is obtained in both non-overtrained and overtrained fish—a result that confirms that overtraining does not selectively strengthen analysers in fish. Now while these experiments suggest that, if a mechanism of attention does exist in fish, that mechanism is unstable compared to the rat mechanism, they can also be interpreted as evidence that fish simply do not have a mechanism of selective attention: the fish results, that is, can be analysed in terms of a simple 'reinforcement history' account, without appeal to differential strengthening of 'analysers' and 'responses'.

In a series of papers, Bitterman and his colleagues have argued that, in fact, fish do not possess any mechanism of attention, that they do not learn to attend to one dimension at the expense of others (Tennant and Bitterman 1973, 1975; Couvillon, Tennant, and Bitterman 1976). Much of their evidence bears on the question whether discrimination training to stimuli differing along one dimension shows better transfer to new discriminations involving stimuli drawn from the same dimension (intradimensional shifts—IDS) than to new discriminations involving stimuli from other dimensions (extradimensional shifts—EDS). Attention theory expects superior IDS over EDS performance since, at the outset of an IDS, the appropriate analyser is already firmly switched in. Superior IDS over EDS performance has in fact been shown in human children and adults (see Wolff 1967), in rats (Shepp and Eimas 1964), and in monkeys (Shepp and Schrier 1969). The IDS versus EDS design provides, according to Mackintosh, 'perhaps the best evidence that transfer between discrimination problems may be based partly on increases in attention to relevant dimensions and decreases in attention to irrelevant dimensions' (Mackintosh 1974, p. 597). Bitterman reports, however, that he has failed to find, in goldfish or carp, any advantage of IDS over EDS learning—which leads him to reject the notion that there are any mechanisms of attention in fish.

The designs of the two Bitterman IDS versus EDS experiments (Tennant and Bitterman 1973, 1975) are fairly complex, and each report suffers from the difficulty that its design differs in respects that may be important from any of the designs used to demonstrate superiority of IDS over EDS in mammals. Both studies, for example, use successive presentation of the various stimuli on a single key, discrimination being established by contrasting rates of response to S+ with those to S−, whereas all the mammalian studies to date have used simultaneous presentation of the two stimuli. Although it is true that attentional theory would expect superior IDS performance with successive as with simultaneous presentation, it is not easy to conclude that a real species difference exists on the basis of a situation that has not been reasonably closely replicated in the two species concerned. A further difficulty is that, in each report, the experimental subjects were not experimentally naive, as has generally been the case in the corresponding mammalian studies; again, although it is true that failure to find superior IDS learning in experienced mammals would pose problems for attentional theory, it would clearly be more convincing to use naive fish, if a fish/mammal contrast is to be established. The IDS versus EDS paradigm suffers, moreover, another problem, when applied to fish, and that is, that attentional theory, as modified by Mackintosh to accommodate his own fish data, does not necessarily predict superiority of IDS training. This is because, as the animal is shifted from one discrimination to the next, errors are inevitable; we have already seen that, according to Mackintosh, analysers in fish are unstable in the presence of errors. It is therefore to be expected that the relevant analyser may be switched out at an early stage of even IDS shifts, and so abolish the advantage of ID over ED training. Although this analysis is consistent with attentional theory it carries, of course, the disadvantage that analyser theory can now 'explain' any result except a superiority of EDS over IDS, and so, at least in this experimental design, lacks any predictive power.

A third experiment (Couvillon *et al.* 1976) also compared ID versus ED training, but did not use a shift design, and did use naive subjects. The design of this study, which examines the concurrent acquisition of two discriminations, is complex, and, to give the flavour of the type of design that Bitterman has used, will be detailed below; before doing so, it may be added that the general features of the design—the use, for example, of a single key, and, in particular, the use of colour and line angle as the two dimensions of the stimulus display—were common to the two IDS versus EDS experiments discussed previously.

In the Couvillon *et al.* (1976) study, a single target key could be illuminated to show any one of eight stimuli, each of which consisted of a white line on a coloured background. The stimuli were arranged in two sets, so that red and yellow were combined with either horizontal or vertical lines (giving four stimuli), and green and blue with either 45° or 135° diagonals (giving the other four stimuli). For each of four groups of goldfish, three per group, four of these stimuli were positive, and four negative. Each of 24 daily sessions contained two 3-min exposures of each of the eight stimuli, in balanced order; responses to S+ were reinforced irregularly, with on average one reward a minute, while responses to S− obtained no reinforcers. For two groups of animals (the ID groups), the stimuli that were S+ and S− from the red and yellow set (S+ was, in fact, for one group, red, and for the other, the vertical line) differed along the same dimension as the S+ and S− for the blue and green set (S+ being green for the former, and one of the diagonals for the latter group). For the two ED groups, the S+ and S− stimuli from the red and yellow set differed along a different continuum from the S+ and S− of the blue and green set. The major outcome was that there was no difference in rates of acquisition of the discriminations between the ID and ED groups; Bitterman holds that the need (in the ID groups) to use only one analyser throughout should confer an advantage (according to attentional theory) over the ED group, and that the result, therefore, indicates that attentional theory does not apply to fish. Several comments on the design do, however, seem warranted. First, a familiar objection, no design remotely similar to this has been used in any other species—therefore, there is no guarantee that the fish behaved differently from any other species in this study; second, there were only three subjects per group, and the positive and negative stimuli were not counterbalanced—a negative result, therefore, is not wholly convincing; third, the very complexity of the design may have encouraged the fish to treat each stimulus as unique (there being no indication to the subject of the experimenters' logical division into two sets of stimuli); fourth, perhaps related to the third, performance overall was poor—after the 24 sessions, discrimination ratios (ratio of S+ responses to total responses) hovered around the 0.8 level—further training might yet have revealed some inter-group difference.

The studies reviewed to this point have shown that two of the phenomena that have provided strong evidence of the operation of attentional processes in other animals (the ORE and the ID over ED advantage) have not been found in fish. There are, however, two further effects, also widely believed to be due to attentional mechanisms, that have been demonstrated in fish, and these are, overshadowing and blocking (Tennant and Bitterman 1975; Wolach *et al.* 1977). If two novel stimuli, A and B, are presented simultaneously, and paired with a UCS until the AB compound obtains a CR, subsequent tests of A and B alone may be conducted, to see to what extent conditioning has occurred to the two elements of the compound. Tests of this kind in mammals have shown that if stimulus A is more salient than stimulus B, then A will obtain a larger CR than B, when each is tested in isolation; moreover, and this is the phenomenon of overshadowing, B will obtain a smaller CR than that seen in a control group for whom B alone was paired with the UCS. According to attention theory, attention is captured by the more salient stimulus A, so that little is learned about stimulus B. Blocking is a similar phenomenon: if stimulus A is paired alone with the UCS for a number of trials, and is then presented in compound with an equally or more salient stimulus B, the compound being paired with the same UCS, subsequent tests of the stimuli separately show that less is learned about B than in a control group trained from the beginning with the compound. Attentional theory supposes that the initial pairing of A with the UCS results in the subject attending to A at the expense of B in the compound training phase, so reducing the amount learned about B.

In an experiment using carp (*Cyprinus carpio*) Tennant and Bitterman (1975) showed that successive discrimination training with two relevant cues (colour and line orientation) resulted in less learning about the line orientation stimuli than did similar training with only line orientation relevant; colour cues, that is, overshadowed line orientation cues. In another experiment of a similar design Tennant and Bitterman (1975) also showed that pretraining with either line angle or colour cues blocked subsequent learning about cues from the other dimension in goldfish. Similarly, Wolach *et al.* (1977) have found both overshadowing and blocking using visual and auditory stimuli in a classical conditioning task. Although the designs used vary slightly from those conventionally used with mammals, the parallel is clear enough for the conclusion that blocking and overshadowing, as established in rats, do occur in fish. Tennant and Bitterman, having found overshadowing in carp, subsequently failed to find any superiority of IDS over EDS in those same animals, and concluded that the proper explanation of overshadowing and blocking cannot, therefore, be one of selective attention to dimensions. It may be noted, however, that these procedures (unlike IDS) are ones in which attending to the relevant dimension does not have to be maintained through blocks of errorful performance: in blocking, for example, the relevant analyser stays switched in when the new cues are introduced precisely because it is successfully enabling solution of the discrimination. Attention theory, therefore, may be slightly discomfited by the absence of IDS superiority in fish, but can accommodate that fact alongside successful demonstrations of both overshadowing and blocking without undue strain.

As is, however, almost inevitable, the demonstration of overshadowing and blocking in fish does not prove the applicability of attentional theory to these animals. There is currently lively debate in the literature concerning the proper explanation of the phenomena. A principal rival account, for example, supposes that the stimuli do not compete for attention so much as for a limited amount of associative strength that a given UCS can support. According to this theory (Rescorla and Wagner 1972), UCSs do not increase associative strength unless they are surprising—unless, that is, they are not already predicted by the stimuli present. So in blocking, the pretraining phase establishes stimulus A as a good predictor of the UCS; therefore, when stimulus B is introduced, the AB compound already predicts the UCS, and so, no change in the habit strength of B (or A) is seen as a result of the pairing of AB and the UCS. This controversy is very much alive and all that is appropriate here is to note that there are two rival theoretical accounts of the phenomena: if we accept the Mackintosh interpretation of overshadowing and blocking, then it is reasonable to conclude that selective attention is seen in fish, but that analysers are unstable, so that no ORE is obtained, and intra-dimensional training procedures confer no advantage; if we accept the Rescorla and Wagner interpretation (an interpretation which does not predict IDS over EDS superiority), then there is currently no experimental finding that requires the postulation of selective attention in fish.

Concluding comments

Our survey of analytical studies of complex learning in fish (or, to be more exact, in teleosts) is now complete although, of course, not exhaustive and it may be worthwhile to emphasize that a large number of less analytical studies have been excluded: studies that, for example, report a single failure to obtain a phenomenon in fish, where no effort has been made to vary even one potentially relevant contextual variable to make it more plausible that a general inability has been uncovered. It has become clear in the survey that this is a genuinely important consideration in comparative work, and a broad conclusion that emerges is that the significance of such negative outcomes in comparative psychology depends on first, the theoretical account given to positive outcomes in other species, and second, the thoroughness with which the search for the phenomenon has been pursued.

It has clearly emerged that the attempt to decide the critical question—are there qualitative differences between the intellectual capacities of fish and other vertebrates?—in fact hinges on the theoretical interpretation of phenomena, such as negative contrast effects, that have been most extensively explored in rats. If we were certain about the causes of negative contrast in rats then we could make a positive decision between rival interpretations of differences in performance between fish and rats. It might seem, therefore, that the most rational procedure is to work exclusively with rats until a complete account is available, and then to test other species having now an opportunity of interpreting differences with confidence. This solution (which appears indeed to have been favoured by animal learning theorists of recent decades) may be rejected on two grounds. First there is, of course, the bizarre possibility that in fact the intelligence of the rat differs from that of all other vertebrates: that is, unless we do some comparative psychology, it will remain possible that we are establishing the capacities of one creature, the laboratory rat, that may have little or nothing in common with other creatures. The laboratory rat is an animal of restricted intrinsic interest, and it is important to be certain that at least some of the principles that emerge from its study do generalize to other animals (and, in fact, the survey of fish performance clearly supports the view that rats and fish do have much in common, which, if not surprising, is a valuable and reassuring conclusion). The second ground, already discussed in Chapter 1, is that one purpose of comparative psychology is to contribute, by work on one species, to the theoretical analysis of phenomena seen in another species. Work with fish has provided specific examples of how such contributions might be made: for example, any theory of negative contrast that can accommodate the fact that simultaneous but not successive NCEs are routinely obtained in fish is to be preferred to a theory that supposes that both phenomena are dependent on identical mechanisms; similarly, if future research, using techniques that reasonably approximate to those conventionally used with rats consistently concludes that intra-dimensional shift superiority does not occur in fish, then accounts of overshadowing and blocking (which do occur in fish) that suppose that these phenomena are independent of intra-dimensional shift superiority will gain good support.

Inter-species differences in learning in fish

Contrasts between the behavioural capacities of various species of fish would be of interest for a number of reasons: first, they might yield insights into the organization of the behaviour in question; second, they might yield a 'ranking' of fish in intelligence, which could be tested against some other ranking in, say, relative brain size; third, qualitative differences might be related to the presence or absence of a particular brain structure or type of forebrain organization. From the preceding review it will, however, have become clear that there have been virtually no attempts to contrast learning capacities between different groups of fish, and so, no data to suggest differential capacities are available.

Given the amount of work required to produce even plausible (as opposed to conclusive) evidence for species-differences in behavioural capacity, the lack of

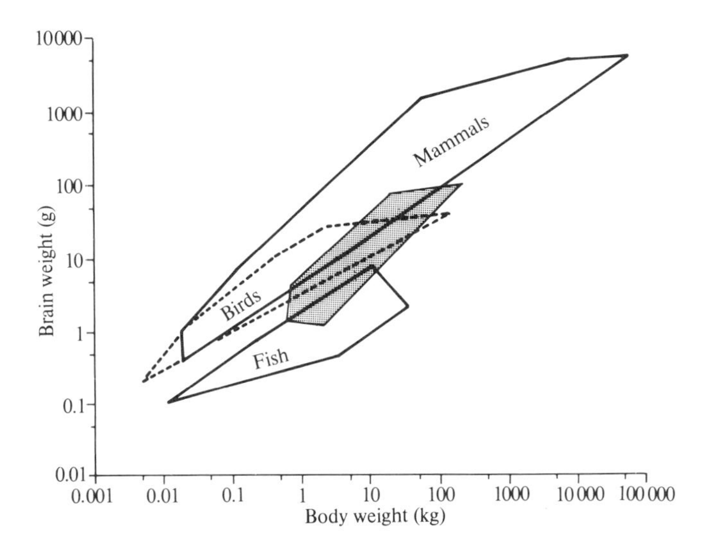

Fig. 3.12. Brain- and body-weight data for elasmobranchs (stippled polygon) superimposed on Jerison's (1969) data for fish, birds, and mammals. (From Northcutt 1977.)

such data is not surprising, although it is regrettable, particularly as no between-class comparisons have yet been made. Not only are there, as we have seen, striking differences in forebrain organization amongst representatives of the three classes of fish, but there appear also to be marked differences in relative brain size between these classes. Ebbesson and Northcutt have produced data to show 'that the lamprey, *Petromyzon*, possesses a brain-body ratio that is lower than that of any other living vertebrate' (Ebbesson and Northcutt 1976, p. 120); on the data currently available, the lamprey brain is five times smaller than that of any other fish of comparable weight. Similarly, Northcutt (1977) has produced further data, summarized in Fig. 3.12, on elasmobranchs (both sharks and rays) to show that their brain-body weight ratios overlap extensively with those recorded for birds and mammals, and are generally considerably higher than those recorded for bony fish. Jerison (1973), commenting on the large size of elasmobranch brains, argues that the elasmobranchs invaded a new ecological niche early in their evolutionary development; it is, in his view, the invasion of new niches that creates the environmental demands that result in selection for improved biological intelligence. Jerison would, then, expect sharks to be more intelligent than teleosts; whether they are is an interesting question to which, regrettably, no answer is available at present. Similarly, although, as we have seen, classical conditioning has not been found in lampreys (and reports of habituation in lampreys are far from conclusive), it is still premature to conclude with any confidence that cyclostome intelligence is of a lower order than that of teleosts.

Forebrain lesions and learning in fish

The experiments that are relevant to this section have two features in common: first, they all involve teleosts, and, second, they all examine the effects of total forebrain ablation; although a number of experimenters report that some forebrain tissue remains, no systematic effort has been made to see whether the deficits seen are due to invasion of any critical area.

The reports agree that classical conditioning is unaffected by total telencephalic ablation. Overmier and Curnow (1969), for example, showed that forebrainless goldfish acquired a classically conditioned CR (general movement) to a red light CS as rapidly as sham-operated controls (animals given identical operative procedures except that no brain tissue was removed). Similarly, Bernstein (1961, 1962) found that forebrainless goldfish acquired a discrimination between stimuli of differing brightness or of different colour as efficiently as normals; in this case, the CR was cardiac deceleration. Overmier and Savage (1974) showed that forebrainless goldfish were also unimpaired in a 'trace' classical conditioning situation; in trace procedures, the CS terminates some time before the UCS onset, so that conditioning is presumed to depend upon some 'trace' set up by the CS. In the Overmier and Savage study, the CS (a red light) came on for 5 s, and was followed by a 5 s interval before UCS onset. The UCS in all of these studies was electric shock and there have been to date no reports of the effects of forebrain lesions in classical conditioning procedures using any other reinforcer. One conclusion that is strengthened by these data is that telencephalic ablation does not cause major visual loss in teleosts.

The data available on simple appetitively reinforced instrumental conditioning are somewhat less clear. Savage has argued (Savage 1969*a*; Savage and Swingland 1969) that where the site of response and the site of reward are spatially and temporally contiguous, forebrainless fish learn to approach stimuli associated with food at the same rate as normals: as evidence for this proposition, Savage (1969*a*) has shown that forebrainless goldfish learned as rapidly as normals to swim through a small hole from one compartment to another to gain access to food cups that actually contained food, but were much slower than normals to acquire a simple response (interruption of a photobeam) when the food reward was 6 inches (15 cm) away from the photocell used to detect responses. In this latter task, forebrainless fish did eventually achieve the same asymptotic speed of response as normals. Hale (1956), however, found that forebrainless green sunfish (*Lepomis cyanellus*) lost a preoperatively acquired habit of swimming from one compartment to another for food reward, and were very slow to re-acquire the habit; this may, of course, reflect a species difference, or perhaps some unintentional delay in Hale's procedure, in which food was manually dropped once the response had occurred. Flood and Overmier (1971) tested alley acquisition in goldfish given either ten or thirty trials a session; forebrainless fish swam slower than controls (which were either unoperated, or had had the olfactory tract sectioned) at an early stage of acquisition (over trials 31 to 40), but reached criterion in the same number of trials, and were indistinguishable at asymptote. These data do suggest, then, that, at least in goldfish, asymptotic performance of simple instrumental responses is not affected by forebrain removal, and this in turn suggests that the lesions do not cause gross motor or motivational deficits.

There have been several reports of simultaneous discrimination performance by forebrainless fish in positively reinforced instrumental behaviour. Two of these reports (Savage 1969*a*; Flood 1975) found no deficit in lesioned subjects in the acquisition of shape discriminations, and in each of these studies, the food reward was contained in cups located immediately below the shapes that were to be discriminated. A third report (Warren 1961) found impaired acquisition of a T-maze, in which both position and brightness were relevant, in forebrainless paradise fish; in this experiment, correct choices were reinforced by brine shrimps dropped into the compartment. The differences between the outcomes of the experiments may have been due to species differences, or to some delay occasioned by the technique of dropping food used by Warren. It appears, then, that given optimal training conditions, forebrainless goldfish are capable of the efficient acquisition of both simple instrumental responses and responses involving visual discriminations. The data are clearly insufficient for generalization beyond goldfish at this stage.

A few experiments have studied more complex tasks. Both Warren (1961), using paradise fish, and Frank, Flood, and Overmier (1972), who used goldfish, found that forebrainless fish were less efficient than normals at reversals of a simultaneous position discrimination in, respectively, a T-maze and a Y-maze. Warren also investigated (using the same fish that later acquired and reversed the T-maze problem) performance in a relatively complex closed-field maze, originally designed to measure intelligence in rats (Rabinovitch and Rosvold 1951). This maze uses a set of barriers of various lengths that can be inserted and removed by the experimenter, so giving a range of problems in one box-like piece of apparatus. Warren tested his fish on 12 such 'umweg' problems, finding that the forebrainless subjects were poorer than their unoperated controls on six of the problems, were indistinguishable from controls on five, and superior to controls on one. The occurrence or otherwise of a deficit was not related to the number of errors made by the controls, and so did not seem to depend on task difficulty. There is, then, some fragmentary evidence that forebrainless teleosts tend to be impaired in performance of food-reinforced complex tasks.

There are a considerable number of papers dealing with avoidance behaviour following forebrain removal, and the great majority agree that forebrainless fish show gross impairments of avoidance responding. Most studies have used the conventional fish shuttle-box situation, and found severe impairments across a variety of conditions: deficits have been seen, for example, using as CSs tones, buzzers, and lights (Overmier and Starkman 1974; Savage 1969*b*), with a wide range of ISIs (from 2.5. to 20 s, e.g. Kaplan and Aronson 1967; Overmier and Starkman 1974), and these deficits are seen in both acquisition and retention of preoperatively acquired responses (e.g. Savage 1969*b*; Kaplan and Aronson 1967). Hainsworth, Overmier, and Snowdon (1967) have also shown a disruption in forebrainless fish of both acquisition and retention of a one-way avoidance response, in which fish, instead of shuttling from one compartment to another to avoid shock, had to learn to swim invariably from one (start) compartment to the other (safe) compartment. Although most studies used goldfish, the Kaplan and Aronson (1967) experiment used African mouthbreeders. Finally, Gordon (1979) has reported that forebrainless goldfish fail to acquire an aversion to the taste of food paired with a poison (lithium chloride); unfortunately, Gordon does not report control data to show that the aversion found in intact fish was specific to the paired taste rather than a general reduction in food intake and does not in any case show that the lesioned fish suffered no loss in the sensory capacity for taste discrimination.

The disruption of avoidance behaviour, although severe, is not total, and a number of authors have shown that forebrainless fish do eventually learn to avoid (e.g. Overmier and Starkman 1974; Hainsworth *et al.* 1967); however, even where avoidance is acquired, performance does not seem to be stable: Kaplan and Aronson (1967) reported that their forebrainless mouthbreeders showed much more variability than controls once they had learned to avoid, and were inefficient in that they did not, as controls did, learn to wait during the ITI beside the hole through which they swam to avoid shocks; Hainsworth *et al.* (1967) found that forebrainless goldfish were both very slow to reach criterion of avoidance in their one-way procedure, and very fast to extinguish once shocks were discontinued.

Two studies report avoidance tasks in which forebrainless goldfish did not show a deficit: one of these (Savage 1968*a*) found poor shuttle avoidance acquisition where 10 or 20 trials a day were given, but no disruption with 30 trials a day. The significance of Savage's result will be discussed at a later stage in this section. The second study (Dewsbury and Bernstein 1969) looked at two variables, ISI and the presence of some kind of barrier between the two compartments. Dewsbury and Bernstein argued that some experimenters had used very short ISIs (Kaplan and Aronson, for example, used a 2.5 s ISI), which gave the fish very little time in which to perform the response; they further argued that there might be a sensorimotor impairment of some kind that interfered with the efficient swimming of forebrainless fish from one compartment to another. The shuttleboxes conventionally used do normally have some obstacle—either a barrier with a hole in it, or a barrier allowing a small gap either above or below it—between the two compartments; the combination of a short ISI and a difficult response might, then, produce an apparent deficit in learning. Dewsbury and Bernstein therefore ran an experiment in which there was an adequate (10 s) ISI, and in which the two compartments, one black, one white, had no intervening obstacle. They found no impairment in the acquisition of avoidance responding by forebrainless fish, although there was a large (but not total) retention loss in a group given preoperative training. As there are now a number of studies showing severe deficits in forebrainless fish using ISIs of 10 s and more (e.g. Savage 1968*a*, 1969*b*; Overmier and Starkman 1974; Hainsworth *et al.* 1967), it seems that the critical variable is probably the absence of an obstacle. The Dewsbury and Bernstein suggestion – that barriers impair performance because of some sensorimotor impairment – cannot account for the fact that escape latencies (latencies to respond to shock onset, on trials on which subjects fail to avoid) are, at least in goldfish, as short in forebrainless as in normal fish (e.g. Hainsworth *et al.* 1967; Savage 1969*b*); Kaplan and Aronson (1967) did find, using mouthbreeders, impaired escape behaviour, but this does not affect the general argument – avoidance behaviour may be disrupted when escape behaviour is not, and this indicates that there need not be a gross sensorimotor impairment of the response for disruption of avoidance to occur. Why, then, did the absence of a barrier abolish the difference normally seen between control and forebrainless fish? One suggestion is that the responding seen in the Dewsbury and Bernstein situation may be entirely due to 'the classical conditioning of gross motor activity leading to accidental avoidance' (Flood, Overmier, and Savage 1976, p. 788); that classical contingencies do lead to substantial levels of response in avoidance situations was, it will be recalled, shown in intact goldfish, by both Woodard and Bitterman (1973*a*) and Scobie and Fallon (1974). Flood *et al.* argue, therefore, that as classical conditioning of general movement is not affected by telencephalic ablation, any behaviour in which instrumental contingencies do not play a critical role will not be disrupted by the lesion. It is, of course, not implausible that 'gross motor activity' will result in more responses in an apparatus that allows entirely free unobstructed passage from one compartment to another than in an apparatus with barriers, and it is therefore reasonable to accept the account put forward by Flood *et al.*

The data reviewed above have generated a number of hypotheses, of which three will be discussed in some detail.

The first proposal, due to Aronson (e.g. Aronson 1970), is that the teleost forebrain has a function of general arousal, which consists basically in a facilitation of the activity of lower brain centres. Applied to poor active avoidance performance, the notion is that forebrainless fish find more difficulty in initiating responses, and so in responding prior to shock onset; the application of the hypothesis to appetitive situations is less clear, but Aronson and Kaplan (1968) argue that a deficit in arousal will cause more disruption, the more complex the task.

Relevant to Aronson's account are reports of effects of forebrain ablation on certain behavioural patterns: Shapiro, Schuckman, Sussman, and Tucker (1974) reported that the aggressive gill-cover response of the Siamese fighting fish, elicited by either mirror exposure or a conspecific, occurred less frequently in forebrainless subjects than in controls, although the mean duration of elicited responses did not differ between groups; these authors also reported less spontaneous movement in forebrainless fish from one part of a tank to another. Similarly, Segaar (1961) found that forebrain lesions in sticklebacks reduced the amount of aggressive and courtship behaviour elicited by test male and female subjects contained in glass cylinders. There are also relevant data from brain stimulation studies: Savage (1971) found that forebrain brain stimulation of free-swimming goldfish elicited 'arousal' reactions (rapid extensions of the fins, and tail-flips); Demski and Knigge (1971) found that stimulation of a mediodorsal area of the telencephalon of the bluegill (*Lepomis macrochirus*) elicited nest-building responses.

However, a decline in activity following forebrain lesions is by no means as ubiquitous a finding as is disruption of avoidance responding: Segaar (1965) has reviewed a series of early studies of forebrain ablation in teleosts, the majority of which found no detectable change of any kind in spontaneous behaviour. Segaar, moreover, while finding reductions in aggression and courtship following forebrain lesions in sticklebacks, found in those same fish increases in nest-building activity, a finding that implies shifts in the relative intensities of different motivational systems rather than a general decline in all forms of motivated behaviour (Segaar 1961). Aronson himself (Kaplan and Aronson 1967) found, in forebrainless mouthbreeders, not only no changes in posture or locomotion, but, in their home tanks, the same speed and agility in fleeing a net. Pertinent here are reports of spontaneous activity levels seen in avoidance tasks. Savage (1968*b*) reported that forebrainless goldfish spontaneous activity levels (measured in the start-box of a two-compartment avoidance apparatus, with no CS and the goal-box closed off), were only 1 to 2 per cent of those of normal fish; Overmier and Curnow (1969) reported that forebrainless fish in a sensitization condition, in which only CSs, and no UCSs were delivered, were significantly less active than normals. However, where spontaneous activity is measured in conditions in which shocks are experienced, there do not appear to be any differences between forebrainless and normal fish (e.g. Fujita and Oi 1969; Overmier and Curnow 1969); it appears, from the Overmier and Curnow report, that shocks, whether signalled or not, reduce the level of spontaneous activity of intact fish to that of forebrainless fish, whose activity is unaffected by the presence or absence of shocks.

There are further clear difficulties facing Aronson's proposal. Its application to appetitive tasks is vague, and, as has been seen, there is to date very little evidence that increasing complexity does lead to increasing disruption. Moreover, there are data to suggest that over-responding can be a source of disruption in forebrainless fish in appetitive tasks: Frank *et al.* (1972) found that forebrainless goldfish performing reversals were significantly slower to stop responding to the formerly positive (now negative) stimulus; similarly, Flood and Overmier (1971) showed that forebrainless goldfish extinguished more slowly than unoperated controls following food-reinforced acquisition training in an alley. The Flood and Overmier (1971) study found, in fact, no difference in extinction between telencephalon ablated and olfactory-tract sectioned fish, suggesting a possible source of prolonged extinction simply in the loss of olfaction; but the Frank *et al.* (1972) report also compared forebrainless and olfactory tract sectioned fish, and showed increased reversal perseveration in only the forebrainless animals, suggesting an involvement of forebrain ablation in both findings.

There are data from avoidance tasks that also indicate poor performance in forebrainless subjects even where learning would manifest itself in a reduction of response output: Overmier and Flood (1969), for example, trained goldfish, forebrainless and normals, to swim from a start chamber into a goal chamber containing food. Once all the fish responded rapidly to the lifting of the partition between the chambers, passive avoidance training was begun, entry into the goal chamber now being punished by shocks. Forebrainless animals showed much shorter latencies than controls as a consequence of the shocks—they were, that is, more likely to initiate responses where shocks were contingent on those responses.

Although, therefore, there are data to indicate that forebrain removal sometimes reduces activity in fish, the pattern of results gives very little support to the proposal that the learning deficits observed are due to a reduction in non-specific arousal.

A second proposal (Savage 1968*b*) is that the forebrain is essential for efficient short-term memory function in teleosts. Savage based this hypothesis principally on two findings: first, the severe effect upon forebrainless fish of the introduction of any delay of reinforcement into appetitive situations (e.g. Savage and Swingland 1969), and, second, the beneficial effect, in an avoidance task, of increasing the number of trials per day (Savage 1968*a*). Savage took the deleterious effects of delays as indicating that the forebrainless subjects did not adequately retain the memory of the response, so that the response was not efficiently associated with the (delayed) reward. This account, however, is difficult to reconcile with the subsequent demonstration (Overmier and Savage 1974) of unimpaired trace classical conditioning—a procedure in which, presumably, the memory for the CS must persist for an association to be formed between it and the UCS. The fact that increasing the number of trials a day from 10 or 20 to 30 abolishes the forebrainless subjects' deficit might perhaps be taken to suggest that individual trials set up, in forebrainless fish, rather rapidly decaying traces which can only be consolidated satisfactorily into long-term memory if sufficient trials occur within a relatively short space of time. There are, however, difficulties with this interpretation: the fact that forebrainless fish given 30 trials a day were as efficient as normals seemed (Savage 1968*a*) to be due as much to the performance of the normals deteriorating with increasing numbers of trials as to that of the forebrainless subjects improving; the appetitive tasks showed severe disruption when delays of 5 s were introduced—but the ITI for the avoidance tasks was 1 to 2 min, so that the rates of decay implied by the explanations proposed for the two deficits do not seem comparable; moreover, Flood (1975) showed unimpaired acquisition by forebrainless goldfish of an appetitive discrimination although only ten trials a day were given, and, at the other extreme, Kaplan and Aronson (1967) found impaired shuttle-box avoidance performance in forebrainless mouthbreeders even when 45 trials a day were administered.

An alternative memory deficit account—that forebrain lesions disrupt long-term, but not short-term memory—was advanced by Peeke, Peeke, and Williston (1972) in their discussion of a finding that poses additional problems for Savage's hypothesis. Peeke *et al.* studied the habituation of a predatory reponse (biting at live brine shrimp contained in a clear plastic tube) in forebrainless and control goldfish, the tube being presented for 10 min a day over five days. The rate of habituation within a day—the decline in bites per min over the 10 min period—was identical for forebrainless and control fish, but, whereas the controls showed complete overnight savings over Days 2 to 5, responding, that is, at the same low rate at the start of each day as at the end of the preceding day, the forebrainless subjects showed no overnight savings at all over those days (although they did show some overnight savings from Day 1 to Day 2). Here, then, is evidence that within-day learning is normal in forebrainless subjects, suggesting an intact short-term memory, but that overnight retention is poor, suggesting malfunction of long-term memory. Although this result is intriguing, there is good evidence that forebrainless teleosts do not in fact suffer from any general deficit in long-term memory. Shapiro *et al.* (1974), scoring aggressive displays in Siamese fighting fish, found, after taking into account differing initial levels of response, similar rates of between-day habituation in forebrainless and sham-operated controls. Flood (1975) showed that a post-operatively acquired shape discrimination was retained by both forebrainless and sham-operated goldfish without any loss of accuracy over an eight-day rest period. It does not, then, seem to be the case that there is in forebrainless teleosts any general impairment in either short- or long-term memory.

The final theory initially arose (Overmier and Curnow 1969) from a theoretical approach to the question—why is classical conditioning normal but avoidance conditioning disrupted following forebrain ablation in fish? As classical conditioning is intact, CS onset should arouse, in forebrainless fish, the same motivation (fear) and the same expectancy of shock as in normals; however, it is possible, Overmier and Curnow argued, either that forebrainless fish are not able to respond appropriately in instrumental situations to such classically conditioned internal events, or that they do not find the termination of such conditioned internal states reinforcing. This latter possibility is derived from so-called 'two-process' theories of learning, which propose that one of the events that reinforce responding in avoidance (and other) tasks is the reduction by the response of classically conditioned aversive states (e.g. Gray 1975). Overmier and Starkman (1974) conducted an experiment to determine between the two alternatives suggested by Overmier and Curnow. In Stage 1 of this study, forebrainless and normal goldfish were trained to shuttle to a tone CS to avoid shocks. Once this response was acquired to the criterion of 17 avoidances out of the 20 trials on one day (and forebrainless subjects were slower to achieve criterion, as would be expected), Stage 2 began. In this second stage, the fish were given classical discrimination training in a different tank with two stimuli (red and white light), one of which (CS+) was paired with shock, and the other (CS−) not so paired. In Stage 3, the fish were returned to the shuttle apparatus, and their responses to presentations of the tone (which continued to be paired with shock where no response occurred within 10 s), and to presentations of the CS+ and CS− (neither of which was now paired with shock, and which both terminated either when a response occurred or after 40 s had elapsed) were assessed. The performances of the two groups in Stage 3 were indistinguishable: both normal and forebrainless fish responded rapidly to both the tone and the light CS+, and much more slowly to the CS−. The rapid avoidance response to the (classically conditioned) CS+ was taken to show that brain-damaged fish can indeed respond appropriately to internally generated states (fear, or expectation of shock), and so by default, to support the other alternative proposed by Overmier and Curnow (1969), namely, that forebrainless teleosts do not find the termination of aversive classically conditioned internal states reinforcing.

Applied to avoidance learning this proposal is, then, that forebrainless fish do not find the termination of fear rewarding; this notion can account for poor passive avoidance (Overmier and Flood 1969) on the assumption that punishment takes its effect in intact animals at least partly through the reinforcement, by fear reduction, of responses that are incompatible with the punished response. A precisely similar account can be given of retarded extinction of appetitively motivated responses, given the additional assumption that the omission of reinforcement generates frustration, and that conditioned frustration, like fear, is a motivational state whose reduction is, in normals, but not in forebrainless fish, reinforcing.

Flood *et al.* (1976) have developed this proposal to apply to performance in appetitively motivated tasks by suggesting that 'the telencephalon of the fish is involved in the utilization of changes in conditioned motivational reactions as reinforcers for the acquisition and maintenance of behaviors' (Flood *et al.* 1976, p. 792). Stimuli that are paired with food, then, should not act as 'secondary reinforcers' in forebrainless fish; there are no data directly relevant to this proposal, but Flood *et al.* argue that delayed reinforcement is (at least in mammals) 'critically dependent upon conditioned secondary reinforcing stimuli' which 'function to bridge the temporal delay between the response and the primary reinforcer of food' (p. 793). The drastic effect of delay upon performance in forebrainless fish could, then, be due to their inability to profit by the secondary reinforcers that mediate the delay in normals. The whole question of secondary reinforcement is a difficult and highly contentious theoretical area, but it is reasonable to argue that those learning theorists who hold that delays are mediated by conditioned secondary reinforcers also hold that secondary reinforcers play an important role in chaining together segments of behaviour in any apparatus in which the original initiation of responding occurs at a locus some distance from the eventual site of primary reinforcement, an obvious example of such an apparatus being the runway (or swimway); so that, if forebrainless fish are disrupted in delay situations because of a lack of secondary reinforcement, they should be similarly impaired in swimways—however, as has been reported above, both trials to criterion and asymptotic speed in food-reinforced alleys are unaffected by telencephalic ablation in goldfish (Flood and Overmier 1971).

The hypotheses discussed above each attempted to account for a wide range of consequences induced by forebrain ablation. A theoretically less attractive possibility is, of course, that the telencephalon plays a number of relatively independent roles in various learning situations, so that a deficit seen in, say, avoidance learning, cannot be explained in the same way as some other deficit seen in, for example, appetitively reinforced tasks with delayed reward. This latter possibility is hardly implausible: there are a number of regions in the fish forebrain, clearly discriminable anatomically, so that it is reasonable to expect a variety of functions within the forebrain; although little direct evidence is available, we do know that olfactory input goes to some areas rather than others in the teleost forebrain. It may, then, be best at this stage to put aside the somewhat scanty evidence available on deficits in forebrainless fish in appetitive situations and to consider instead whether the well-documented disruption in avoidance may be accounted for in isolation. One simple suggestion might be that forebrainless teleosts experience very little fear: Savage (1968*b*, p. 136), for example, remarks on the 'peculiar fearlessness' of forebrainless fish. According to this view, forebrainless fish learn to 'expect' shocks as rapidly as normals (and this is confirmed by their performance in classical conditioning situations), but are simply less afraid of them. A difficulty lies, of course, in the fact that classically conditioned fear-inducing stimuli transfer normally to instrumental avoidance situations (Overmier and Starkman 1974); if it is assumed that this across-situation transfer is mediated by some common motivational state (e.g. Rescorla and Solomon 1967), it can hardly be maintained that forebrainless fish experience abnormally low levels of fear. However, classically conditioned CSs may be supposed to have two functions: first, they may induce some appropriate motivational state, and second, they may serve a cue function, in which they induce expectancies of the UCS (Trapold and Overmier 1972). Transfer from the classical to the instrumental procedure may be mediated equally well by the cue properties as by the motivational properties of a CS, so that the Overmier and Starkman result need not be taken as demonstrating normal fear motivation in forebrainless goldfish. This proposal, then, agrees with that of Overmier and Starkman in placing the source of disruption in what may generally be characterized as motivation and enables us to reach two general conclusions concerning forebrain function in teleosts: first, the forebrain does not appear to be involved in classical conditioning, and second, the forebrain is involved in some way with fear. Neither of these conclusions indicates any role for the forebrain in intelligent behaviour, and it might seem that our credulity is now being stretched too far: it is one thing to suggest that the intelligence of fish may not differ substantially from that of mammals, but quite another to suggest that, in addition, the forebrain plays no role in that intelligence. Fortunately, however, this latter suggestion is unjustified: while it is true that there is little evidence of forebrain involvement in intelligence in fish, it is also clear that this probably reflects the fact that few relevant experiments have been conducted. Where experiments using complex tasks have been carried out using brain-damaged fish, deficits have generally emerged (e.g. in reversal learning, and in 'umweg' problems, Frank *et al.* 1972; Warren 1961). But whereas there have been numerous studies of avoidance learning in forebrainless fish, there has been no accumulation of studies using any more complex procedure, and so, no convincing hypothesis of a 'higher' function for the forebrain of teleosts has yet emerged.

Summary and conclusions

This chapter began by emphasizing the diversity of living groups of fish, a diversity so great that extant species are assigned to three different classes. The line of descent leading to terrestrial vertebrates diverged from those leading to all living fish at a very early stage so that in particular teleosts (the only fish for which useful behavioural data relevant to intelligence are available) ceased to share a common ancestry with terrestrial vertebrates some 400 million years ago.

Comparison of teleost forebrain anatomy with that of mammals is complicated by the unique way in what the telencephalon develops in actinopterygians. However in elasmobranchs (and possibly in cyclostomes) the major subdivisions of the forebrain seen in mammals can be discerned, and it may therefore be that there is in these vertebrates a structure homologous to mammalian neocortex.

Although relatively little is yet known of sensory and motor projections to and from fish telencephalon, it is clear that neither the elasmobranch nor the teleost telencephalon is dominated by olfactory projections, which are in fact restricted to local telencephalic areas. There is good evidence for a telencephalic role in vision in sharks, as well as much less substantial evidence for the representation of other senses in shark forebrain, and for projection of visual fibres to teleost telencephalon. It is, moreover, clear that the overall pattern of retinofugal projections in fish is not dissimilar to that found in other vertebrates, these projections being by no means confined to the optic tectum of the midbrain. Virtually nothing is known of the role of fish telencephalon in the control of movement; decerebration does not lead to gross abnormalities of movement in teleosts.

The sections concerned with the intellectual capacities of fish explored a number of hypotheses concerning potential qualitative and quantitative differences between teleosts and other vertebrates; the paucity of data available precludes any attempt at present to make a reasonable assessment of the intelligence of either cyclostomes or elasmobranchs.

An early proposal was that the serial reversal and probability learning tasks might reveal some (unspecified) qualitative difference between fish and rats; this suggestion lost much force through demonstrations that fish may behave in a 'rat-like' way in both tasks under certain conditions, and there remain now only two phenomena observed in such experiments which may discriminate categorically between fish and rats: first, fish have not shown one-trial reversals, and, second, rats have not shown random matching. It was argued that the former finding need not be taken to imply a qualitative species difference in, say, the use of strategies; some evidence that fish do use strategies was found, and it was suggested that the use of strategies by fish might be relatively inefficient due to their insensitivity to the after-effects of reward and non-reward. The demonstration of random matching in fish but not in rats did not appear to point to any clear theoretical interpretation, and was taken to reflect the influence of some as yet unspecified contextual variables.

A second proposed qualitative difference between teleosts and rats is that the former may not possess the mechanisms of selective attention found in the latter. Evidence for this proposal is the failure to demonstrate superiority of IDS over EDS in fish, although a weakness of the experiments concerned is that the designs used differ markedly from those used with rats. Evidence, on the other hand, that fish do show selective attention is provided by demonstrations of two phenomena (overshadowing and blocking) which may be interpreted as a consequence of selective attention; however, alternative non-attentional accounts of these phenomena are available. This second proposal cannot, therefore, be conclusively affirmed or rejected at present.

The most actively investigated hypothesis of qualitative difference is that fish, unlike rats, do not form expectations which anticipate reinforcers, this inability reflecting a more general inability to form associations involving only stimuli (and not responses), an inability, in other words, to show true classical conditioning. We saw, however, that evidence derived from standard classical conditioning procedures (including the effects of omission training) supported the view that fish do, in fact, show classical conditioning (although the failure to observe SPC in fish does provide a difficulty for that conclusion). The notion that fish do not form expectations was originally formulated to account for a series of effects seen when an established reinforcer is either omitted (extinction effects) or reduced (contrast effects). Although the hypothesis provides a generally satisfactory account of such effects, it encounters serious difficulty with the finding that fish do show a simultaneous NCE, an effect which most theorists suppose does involve expectancies of specific amounts of reward.

An alternative account of extinction and contrast effects proposes a quantitative difference between fish and rats in their use of after-effects of reward and non-reward. This account, which supposes that fish are capable of classical conditioning, appeared to provide a good account of the same range of effects accommodated by the absence-of-expectations hypothesis, and had the considerable advantage over that hypothesis that it allowed a satisfactory explanation of the simultaneous NCE. It therefore appears that this account is to be preferred, although a weakness in it is the absence of precision in its predictions. It is argued that adoption of this notion of quantitative difference between fish and rat need not imply a difference in intellectual mechanisms, since the performance of rats comes to resemble that of fish when certain types of reinforcer are used.

The notion that fish might show a quantitative difference from pigeons and rats in stability of attention was assessed, but it appears that there is little evidence from experiments using fish which is directly relevant to that proposal; at best, it provides an incomplete account of the generally poorer performance of fish compared to rats in serial reversal and probability learning tasks, and of the absence of the ORE in fish. In general, then, it seems that whether or not mechanisms of selective attention exist in fish, and whether, if they do, they allow relatively unstable selectivity remain at present very much open questions.

It is a matter for regret that there are currently no reliable data on either within- or between-class differences in intelligence in fish, particularly because the lamprey brain is very much smaller than the teleost brain, which in turn tends to be smaller than that of elasmobranchs, whose brain-size is in some species comparable to that of some birds and mammals. Investigations of the role of the fish télencaphalon in learning have concentrated almost exclusively on the effects of total forebrain ablation. A variety of findings on both appetitively and aversively motivated tasks have been reported, and the most consistent outcome has been a severe disruption in avoidance tasks. It is suggested that the range of deficits observed probably reflects a number of dysfunctions, and that one of these may be a reduction of fear. Classical conditioning appears to be unaffected by forebrain ablation, and there is, indeed, no clear evidence of any specifically 'intellectual' dysfunction following such lesions; this may, however, reflect the fact that few complex learning tasks have been explored in forebrainless fish.

We shall conclude this chapter by referring once again to our survey of intellectual capacity in fish: that survey, while not rejecting outright all proposals for qualitative differences between fish and rats, found no compelling evidence in favour of the proposals. Similarly, while good support was found for one proposed quantitative difference between fish and rats (in reliance on after-effects), that difference was not taken to imply a quantitative difference in intellect. If, then, we adopt as the null hypothesis the notion that there are neither quantitative nor qualitative differences between the intellectual capacities of fish and rats, we come to the perhaps startling conclusion that that hypothesis cannot be rejected at this stage. Whether such an account is plausible, in the light of the considerable divergence of evolutionary history of teleosts and mammals, and of the major differences in their forebrain structure, may perhaps best be considered after the performance of other vertebrate classes has been explored: if a similar interpretation can be applied to all vertebrate classes, it may be reasonable to conclude that the mechanisms possessed existed in some common ancestor, and so, that the capacities necessary for problem-solving were present at a very early stage in vertebrate evolution in more or less their final form. This conclusion may, given the high esteem in which humans hold the human intellect, seem counter-intuitive: it remains the case, however, that it is a conclusion that has not been contradicted by any of the evidence reviewed at this stage.

4. Amphibians

Classification of amphibians

Amphibians were the first vertebrates to gain a measure of independence from an aquatic environment and to spend a substantial portion of their existence on land. Living amphibians are assigned to three orders, the Anura (the frogs and toads, of which there are some 2000 species), the Urodela (the newts and salamanders, with 300 species), and the Gymnophiona (about 80 species of blind burrowing worm-like creatures, also known as apodans, that live in the tropics).

All amphibians derive ultimately from crossopterygian stock, and the earliest tetrapod fossil, dating from the late Devonian period about 350 million years ago, is of an ichthyostegid, an animal that belonged to the subclass known as labyrinthodonts from whom, it is generally believed, modern amphibians ultimately derive. The labyrinthodonts are so called because of the maze-like folds on the enamel of their teeth, which are similar to those of their crossopterygian ancestors.

The earliest anuran fossil dates back to the Triassic (225-196 million years ago), and bears resemblances to fossils of rhachitomes, an order of labyrinthodont amphibians which were abundant throughout the Carboniferous, Permian, and Triassic periods; rhachitomes grew up to 1.5m or more in length, and somewhat resembled crocodiles in appearance. As reptiles derive from another labyrinthodont order, the anthracosaurs, representatives of which co-existed with rhachitomes in the early Carboniferous period, it appears that anurans ceased to share a common lineage with reptiles well over 300 million years ago.

The earliest urodele fossil is from the Jurassic period (195-136 million years ago), and no pre-tertiary fossils of apodans have yet been found; while this may suggest a more recent origin for these groups than for frogs and toads, the gaps in the fossil record make it impossible to specify the relationship between living amphibian groups with any certainty. The many controversies concerning the origins of modern amphibians, which are very well presented in Stahl (1974), do not require further discussion here, as there is general agreement on the points of central interest for our purposes. First, amphibians are descended from crossopterygian fishes; second, urodeles and anurans, the only groups for whom relevant neurological and behavioural data are available, have developed independently for at least 200 million years; and third, the amphibian line that gave rise to reptiles and so ultimately to birds and mammals diverged from the main labyrinthodont stock early in the Carboniferous period, so that the evolution of modern amphibians has proceeded independently of that of other land vertebrates for more than 300 million years. Figure 4.1 summarizes the principal features of amphibian evolution, according to Romer and Parsons's (1977) account.

It is clear that modern amphibians are indeed closer than teleosts to the line of descent of other land vertebrates; whereas we have to go back 400 million years to find an ancestor common to both teleosts and land vertebrates, we need return only some 350 million years to find an ancestor common to both modern amphibians and other living terrestrial vertebrates. But it is, of course, abundantly

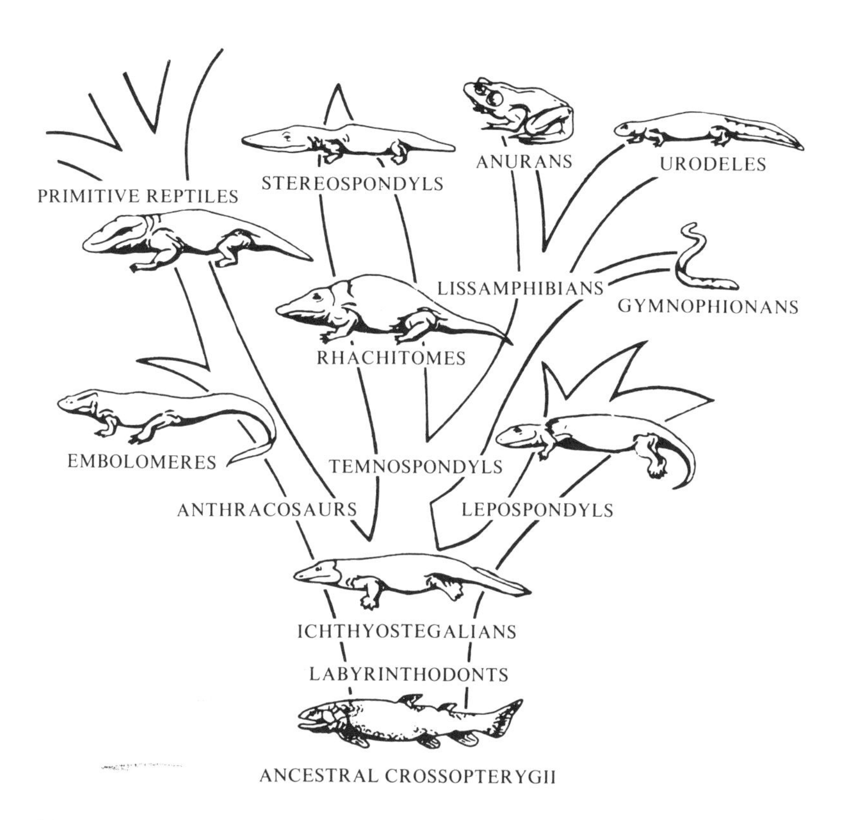

Fig. 4.1. A simplified family tree of amphibians, showing their probable relationships to each other and to reptiles. (From Romer and Parsons 1977.)

clear that we have no more reason in the case of amphibians than we had in the case of teleosts to suppose that we can see in them a primitive stage in the sequence of development of other vertebrate classes. The 'distance' between modern amphibians and their Carboniferous ancestors is emphasized by Romer, who writes: 'A frog is, in many ways, as far removed structurally from the oldest land vertebrates as is a man' (Romer and Parsons 1977, p. 61). The evolutionary history of amphibians does not, then, encourage us to expect to see in modern amphibians simpler or more primitive versions of brain and behaviour in 'more advanced' vertebrates; on the contrary, it would seem to suggest that there may be striking contrasts between them and other vertebrates.

Structure of amphibian forebrain

Despite their long history of independent evolution, forebrain structures in the three living orders of amphibians are grossly similar. Figure 4.2 shows external views of the brains of a frog (*Rana*) and a salamander (*Necturus*, the mudpuppy). In amphibians, as can be seen from the figure, the olfactory bulbs are not clearly

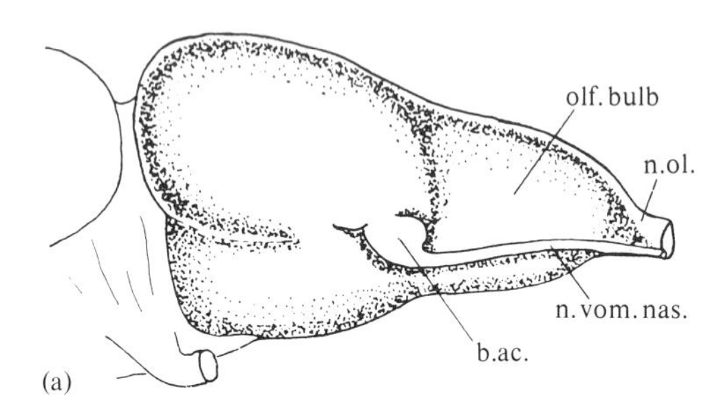

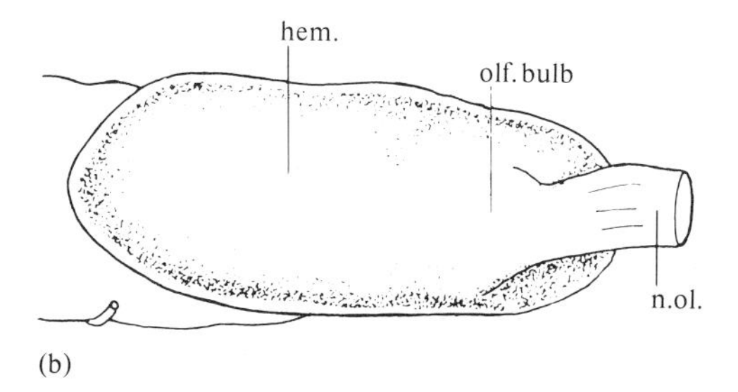

Fig. 4.2. External views of the forebrains of (a) a frog and (b) a salamander. (From Nieuwenhuys 1967.)
Key: b.ac. – accessory olfactory bulb; hem. – cerebral hemisphere; n.ol. – olfactory nerve; n.vom.nas. – vomeronasal nerve; olf.bulb. – olfactory bulb.

separated from the forebrain, but constitute the rostral pole of the hemisphere as well as extending caudally to form part of the lateral (and, in urodeles and apodans, of the medial) hemispheric wall. The bulbar formation in amphibians contains a structure not seen in fish, the accessory bulb. The main olfactory bulb and the accessory bulb are served by nerves which originate in two different types of nasal epithelium. The first type is typically dorsally located, contains the so-called Bowman's glands, and projects to the main bulb; the second type, the vomeronasal epithelium, also known as Jacobson's organ, is typically ventrally located, has no Bowman's glands, and projects to the accessory bulb. Very little is known of the difference, if any, in function of the two types of epithelium; both types are sensitive to chemical stimuli, but it may be that, whereas the olfactory epithelium detects air-borne stimuli, the vomeronasal epithelium detects stimuli immersed in solution (provided, perhaps, by licking with the tongue) and so would in effect be an organ of taste. Support for this proposal is provided by the observation that Bowman's glands are absent in fishes (which do not, of

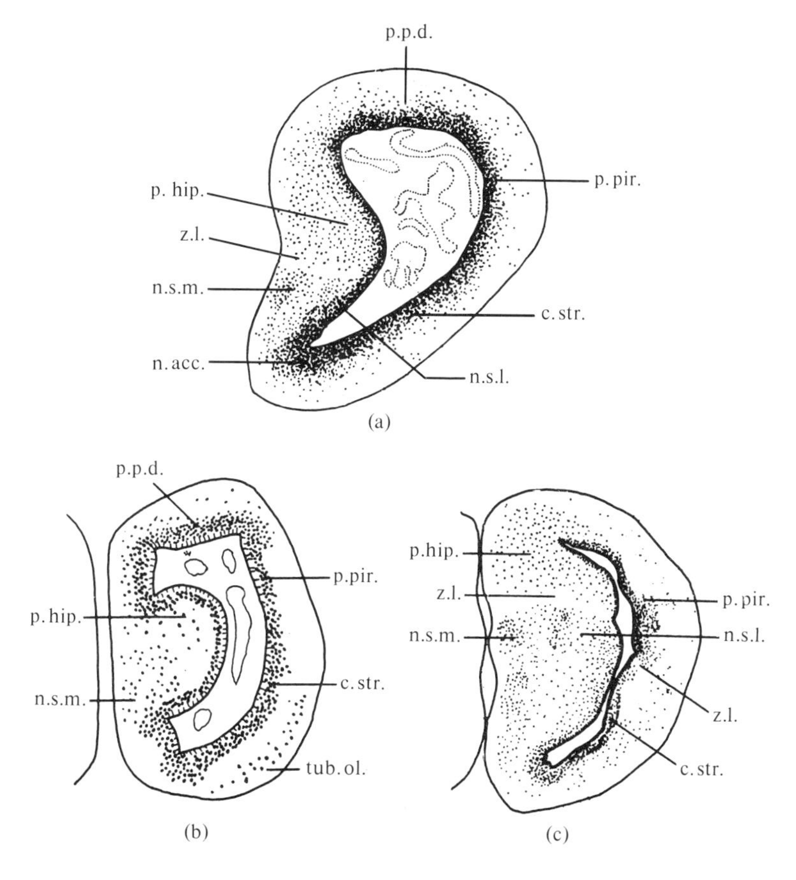

Fig. 4.3. Transverse hemisections of the telencephalon of (a) the urodele *Ambystoma tigrinum*, (b) the apodan *Siphonops annulatus*; (c) the anuran *Rana esculenta*. (From Nieuwenhuys 1967.)

Key: c.str. – corpus striatum; n.acc. – nucleus accumbens; n.s.l. – lateral septal nucleus; n.s.m. – medial septal nucleus; p.hip. – primordium hippocampi; p.p.d. – primordium pallii dorsalis; p.pir. – primordium piriforme; tub.ol. – olfactory tubercle; z.l. – zona limitans.

course, need to detect air-borne stimuli), and it may in fact be that fish do not possess the major vertebrate sense of smell (for a discussion, see Tucker 1971).

Figure 4.3 shows transverse sections through the forebrains of a urodele, an apodan, and an anuran. It can be seen from this figure that the nerve cells tend, particularly in apodans and urodeles, to be concentrated in the periventricular area, and that no superficial layers of cells can be seen. Pallial regions can be distinguished medially from subpallial regions by a zona limitans; laterally, a zona limitans is found only in anurans, so that the lateral demarcation of pallial and subpallial areas in urodeles and apodans is unclear. Nieuwenhuys (1967), along with most authorities, divides the pallial area into three regions, the medially placed primordium hippocampi, the lateral primordium piriforme, and, lying between them, the primordium pallii dorsalis. According to Herrick, 'the homologies of the hippocampal and piriform sectors with those of mammals are clear, as shown by substantially similar nervous connections' (Herrick 1948, p. 55); however, Herrick warns us that the primordium pallii dorsalis is 'of uncertain relationships' (p. 21) and 'that the application of mammalian names to the structures here revealed rarely implies exact homology; these areas are to be regarded as primordia from which the designated mammalian structures have differentiated' (p. 18). Now while it is clear that we must have severe reservations over regarding any structure in a living vertebrate as a primordium for a structure in some other vertebrate, we do share some of Herrick's interest in establishing correspondences, and it should be noted that although some modern work using histochemical techniques (Northcutt 1974) does provide general support for the homologies implied in the nomenclature of the classically recognized subdivisions of both the pallium and subpallium, large areas of uncertainty remain. Northcutt, for example, proposes that the ventral part of the lateral pallium of the bullfrog (*Rana catesbeiana*) may be homologous with the reptilian hypopallium (or dorsal ventricular ridge); Källén (1951*a*), however, concluded from ontogenetic studies of embryological material that no structure could be found in amphibians or fish that corresponded to reptilian hypopallium, that structure being, then, a reptilian innovation.

The subpallial region is divided into a ventrolateral corpus striatum and a ventromedial septal region containing a medial and a lateral septal nucleus; some authors divide the medial septal nucleus in two parts, the most ventral of which is taken to be homologous with the mammalian nucleus of the diagonal band of Broca. There is also a band of cells surrounding the ventral part of the ventricle, and this has been thought by some to be homologous with the nucleus accumbens and by others, with the olfactory tubercle of mammals. Herrick (1948) took the anterior parts of the striatal region to be homologues of the mammalian caudate nucleus, putamen, and globus pallidus, and the posterior part, to be homologous with the amygdala. There is in fact a wide divergence of view on possible homologues for the amphibian amygdala. Nieuwenhuys (1967) and Källén (1951*a*) agree that it is a wholly subpallial structure; Northcutt and Royce (1975), on the other hand, argue that in bullfrogs, the pars lateralis of the amygdala is pallial in origin, the pars medialis being subpallial. To make the matter even more confusing, Northcutt (1974) proposes that the medial (subpallial) part is homologous with the basolateral region of the mammalian amygdaloid complex, whereas Källén (1951*a*) argues that it is precisely that part of the mammalian amygdala that is pallial in origin and that in neither fish nor amphibians is there a structure that corresponds to the basolateral amygdala in mammals. It is, then, with some relief that we recall that our primary purpose is not the establishment of homologies: areas of the brain must be identified so that, for example, lesions may be localized, and homologies implied by the names given may, as has been suggested earlier, be useful in indicating which areas might be of functional interest, but further than that we need not go.

Since the teleost hemispheres form by the process of eversion rather than that of evagination, it is difficult to compare them with the amphibian type: however, there seems little advance in organization of the amphibian brain compared with either the elasmobranch brain or the brain of the crossopterygian coelacanth, which also shows a pallial region and three subpallial regions, characterized as olfactory tubercle, striatum, and septal region (Nieuwenhuys 1967). Moreover, the size of the amphibian brain is no larger than that of teleosts of comparable body-weight; although amphibian data are not plotted by Jerison (1969) (see Fig. 2.2, p. 27), brain- and body-weight data are supplied by Ebbesson and Northcutt (1976) for both the bullfrog and the mudpuppy, and both points fall within the 'lower vertebrates' polygon shown on Fig. 2.3 (p. 28). If, therefore, we accept Jerison's method for comparing animals of different weights, then amphibians have brains that are comparable in relative size to those of teleosts and reptiles, and notably smaller than those of elasmobranchs.

Sensory and motor organization in amphibian forebrain

There are comparatively few modern data on sensorimotor organization in the amphibian telencephalon, but what evidence there is suggests, as was the case for fishes, that the notion that the telencephalon is dominated by olfaction may not be valid. More positive features of the evidence to be reviewed are, first, that it will show, at least for the senses of vision and hearing, a striking similarity upto the diencephalic level between amphibian and mammalian systems, and second, that the concept of progressive encephalization, in particular of visual function, is open to serious doubt.

Olfaction

The best data we have on sensory projection to the amphibian hemispheres are for the olfactory modality, but even here, as Northcutt and Royce (1975) have emphasized, our knowledge is incomplete. This is due in part to the fact that experimental anatomical techniques face two unusual difficulties in amphibians: first, the degenerating particles observed are so fine that it is virtually impossible to discriminate between axonal and terminal degeneration, and so to determine whether axons terminate in or simply pass through a region; second, much degeneration occurs in regions of neuropil that are virtually devoid of cell bodies so that, in order to determine which nuclei (aggregates of nerve cell bodies) receive projections, further contributions to those regions of neuropil must be established.

Studies of projections of secondary olfactory fibres have been carried out in both anurans (Scalia, Halpern, Knapp, and Riss 1968; Scalia 1972; Northcutt and Royce 1975) and urodeles (Royce and Northcutt 1969), and, as these accounts in general agree, a summary of the findings may be given here. The olfactory and accessory olfactory bulbs (the latter of which, it will be recalled, derives its input from the vomeronasal epithelium) send independent projections to the forebrain. The accessory bulb projects, via the ventrolateral olfactory tract, to the amygdala, which in turn sends projections to the hypothalamus. The main output from the olfactory bulb projects via the lateral olfactory tract through the entire extent of the ventrolateral pallium, whose major projection appears to be to the medial pallium. There are also olfactory bulb fibres that project via the medial olfactory tract, to anterior parts of the medial pallium and septal area. The above connections are ipsilateral, and there may also be some contralateral olfactory projections to posterior pallial and striatal regions adjoining the amygdala, and to the amygdala itself. We can see, then, that olfactory input to the amphibian forebrain is indeed fairly widespread—rather more so, indeed, than appeared to be the case for either sharks or teleosts; there are, however, marked similarities in the organization of these projections in amphibians with those in mammals. In particular, the projections of the main and accessory bulbs in mammals are independent, the major destination of the former being piriform cortex and associated cortical regions, and of the latter, the amygdaloid complex.

Vision

The most extensively investigated sense modality in amphibians is, of course, vision. Most of the relevant neurophysiological studies have, however, been concerned with sites outside the telencephalon, and the roles of thalamus and tectum in vision will be briefly reviewed here, as there are important implications of the findings for the notion of encephalization of function. The destinations of the majority of retinal efferents are to be found in the superficial layers of the optic tectum, but there are, nevertheless, retinal projections to a number of sites in the thalamus, to the pretectal region, to the hypothalamus, and to the basal optic nucleus. The great majority of retinal fibres are crossed, although Peyrichoux, Reperant, and Weidner (1978) have shown ipsilateral projections in the edible frog (*Rana esculenta*) having a pattern of distribution similar to that of the contralateral projection. This overall pattern is one that recurs throughout the various classes of vertebrates, including, as we have seen, teleosts (p. 55).

There is evidence for highly complex processing of visual information in anurans, and for specialization of function within different thalamic and tectal regions. Neurophysiological studies (reviewed by Grüsser and Grüsser-Cornehls 1976) show that there are at least six classes of retinal ganglion cells, seven classes of tectal cells, and ten classes of thalamic cells.

Muntz (1962*a*) described blue-sensitive cells in the rostral thalamus of leopard frogs (*Rana pipiens*), and hypothesized that these were involved in the positive phototaxis towards blue light shown by frogs (Muntz 1962*b*); this conclusion has been reinforced by Kicliter (1973), who showed that rostral thalamic lesions disrupted the taxis. The posterior thalamus of toads appears to be involved in the avoidance by these animals of large moving objects: there are cells in this region that are sensitive to relatively large moving visual stimuli (Ewert 1971), and whereas electrical stimulation of the caudal thalamus obtains avoidance responses, lesions of the same region result in toads snapping vigorously at previously avoided objects (Ewert 1970).

Other thalamic cells, possibly in the posterior thalamus, where Ewert (1971) observed units that showed prolonged responses to non-moving stimuli, appear to be involved in the perception of stationary objects: Ingle (1977) has shown that, although frogs were unable, following tectal ablation, either to avoid or to attack moving objects, they were able to avoid barriers of various kinds when jumping. Ingle also concludes from this and other studies that the posterior thalamic units involved in avoidance derive their input from the tectum, so that tectal lesions disrupt avoidance although avoidance is mediated by thalamic units; the absence of effect of tectal lesions on perception of stationary objects suggests that the thalamic cells involved receive direct retinal projections.

Attacks on prey-objects in toads appear to be mediated by tectal units: there are units in this region that have response properties which closely match those of stimuli that elicit prey-catching in intact toads (e.g. they respond optimally to small moving objects, and show rapid habituation); stimulation of the tectum obtains prey-catching responses, and ablation of the tectum abolishes prey-catching (Ewert 1970).

We have, then, seen that there are a number of relatively independent visual systems in anurans, and one gross distinction that may be drawn amongst them is based on sensitivity to movement: those systems that respond to moving stimuli (with attack, if the object is small, or avoidance, if it is large) depend directly or indirectly on retinotectal projections, whereas the systems that respond to stationary objects (blue phototaxis or barrier detection) use direct retinothalamic projections. We shall see in subsequent chapters that parallel distinctions emerge in reptiles, birds, and mammals, and we have already seen that in sharks there is at least some evidence from lesion studies that both the tectum and the telencephalon (which receives thalamic afferents) are involved in vision. These parallels will encourage us to reject the view that vision becomes gradually encephalized, in the sense that 'higher centres' take over functions carried out by 'lower centres' in more primitive forms, and to prefer Ingle's suggestion that 'the anatomical similarities in retinofugal patterns from fish to primate provide the foundation for a basic set of visual functions which have been elaborated but not reconstructed during evolution' (Ingle 1973, p. 1054).

What, then, of telencephalic involvement in vision? Regrettably, very little relevant evidence is available. There are anatomical data to show that the amphibian thalamus does indeed project to the telencephalon (see Kicliter and Ebbesson 1976); the dorsal diencephalon appears to project primarily to the striatum, but also sends projections to pallial regions. Kicliter (1979) reports that one of the thalamic regions believed to receive direct retinal afferents (the rostral posteroventral nucleus) projects to the medial pallium, and that another region (the posterolateral nucleus), which receives efferents from the tectum, projects to the striatum. Since similar independent thalamofugal and tectofugal pathways are found in mammals (and in reptiles and birds) these data provide further support for the parallel noted above between the dual visual systems found in amphibians and those found in other vertebrates. Kicliter (1979) does, on the other hand, point out that the telencephalic destinations of these routes in amphibians, the medial pallium and the striatum, do not appear to be homologous with the regions (both neocortical) to which the corresponding routes project in mammals.

Electrophysiological studies have confirmed that visual information does reach parts of the amphibian telencephalon, but no systematic studies have yet been carried out to show precisely which areas are visually responsive, and which are not. Gruberg and Ambros (1974) observed single-unit responses to visual stimuli in the rostral striatum of leopard frogs but found no retinotopic organisation within the responsive area. Their units were modality specific, showing no responses to tactile or auditory stimuli, had large receptive fields, responded optimally to moving stimuli, and showed rapid habituation. Liege and Galand (1972) found units with similar properties in the telencephalon of edible frogs, but these units were located in the posterior pole of the hemisphere, near the midline (their precise location was not histologically verified). Gruberg and Ambros in fact failed to detect visual units in the hippocampus or septal area in their study, but do not report how extensive their search was; Karamian, Vesselkin, Belekhova, and Zagorulko (1966) found long latency evoked potentials to visual stimuli in the primordium hippocami of two species of frogs, although it is not clear whether these responses were modality specific. These physiological studies allow us to conclude, therefore, that there are indeed visual projections to both striatal and (probably) medial pallial sites in amphibians, but do not help reveal the precise extent of those projections.

Lesion studies of telencephalic function have not been very helpful. Unilateral hemispheric lesions abolish prey-catching responses to stimuli presented to the contralateral visual field, and depress responding to ipsilateral stimuli. However, the superimposition of bilateral lesions of the posterior thalamic-pretectal region restores prey-catching and in fact obtains higher levels of responding than those seen in intact animals (e.g. Ewert 1976). It can be seen, therefore, that although the telencephalon appears to modulate avoidance and prey-catching behaviour, there is no reason to suppose that it participates in the relevant visual analysis. On the other hand, Karamian's (1962) report that, following hemispheric ablation, frogs jump blindly, hitting obstacles, supports the notion that the retinothalamic projection involved in the perception of stationary objects (Ingle 1977) is part of a pathway whose final destination is in the telencephalon.

Hearing

Although there are few data on the telencephalic representation of hearing, the evidence available is sufficient to show marked similarities between the central nervous physiology of hearing in anurans and that in other terrestrial vertebrates (it is generally believed that neither urodeles nor apodans can detect airborne sounds – Campbell and Boord 1974). The auditory portion of the eighth nerve projects to the dorsal medullary nucleus (which may be compared, therefore, to the mammalian cochlear nuclei); the dorsal medullary nucleus in turn projects to the torus semicircularis of the midbrain, both directly and via the contralateral superior olivary nucleus. The torus semicircularis is a complex mass containing a number of nuclear groups, and those regions within it that receive auditory projections may be compared to the mammalian inferior colliculus; the torus semicircularis in turn projects to a region (the posterocentral nucleus) in the posterior thalamus (for a clear summary and detailed references of auditory projections upto the diencephalic level, see Mudry, Constantine-Paton, and Capranica 1977). The organization of the anuran peripheral auditory system shows clear parallels to that of mammals, in terms both of the number of synaptic relay stations between the auditory nerve and the thalamus, and of the gross location of those relay stations.

Finally, there is both anatomical (Kicliter 1979) and physiological (Mudry and Capranica 1980) evidence for an auditory projection from the posterocentral nucleus to the striatum in frogs. The Mudry and Capranica report found evoked potentials to auditory stimuli from both a region in the ventral striatum and a region in the medial pallium, although the anatomical basis for auditory activity in this latter site has not yet been identified. It is of interest to note that Mudry and Capranica used fairly complex sounds, having parameters similar to those of species-specific vocalizations of the animals they studied (bullfrogs); they report that they obtained activity in response to these sounds from sites that did not respond to simpler stimuli, such as clicks and tones. It is not yet clear how close the parallel is between telencephalic organization of hearing in anurans and that in other terrestrial vertebrates; we do not yet know, for example, whether there is in frogs a pallial site that receives direct projections from the thalamic region that is homologous with the medial geniculate body of mammals. What is clear is that the anuran telencephalon does receive auditory information from the thalamus and that is, perhaps, the finding most relevant to our interests in function rather than homology.

Skin senses

There are two reports (Liege and Galand 1972; Vesselkin, Agayan, and Nomokonova 1971) of electrophysiological responses in the hemispheres to cutaneous stimulation. Vesselkin *et al.* stimulated the sciatic nerve directly and observed bilateral evoked responses in the primordium hippocampi, but it is not clear whether the responses were modality specific. Liege and Galand found four single units in the posteromedial telencephalon (somewhat dorsal to their visual units) that appeared to be modality specific, being sensitive to cutaneous, but not visual or auditory stimulation. Apart from these reports, all that can be added on the subject of cutaneous projections is that there are demonstrations of pathways to the telencephalon that could carry such information; there are, for example, direct projections to the telencephalon from not only the thalamus, but also the mesencephalon, and indeed from areas posterior to the mesencephalon (see Kicliter and Ebbesson 1976). Kicliter (1979), moreover, reports that both the striatum and medial pallium do receive projections from thalamic regions that are believed to receive somatosensory afferents; however, the difficulty of establishing functional pathways by anatomical techniques alone is particular marked in the amphibian diencephalon because of the lack of well-defined nuclear groups, and confirmation by physiological studies will be needed before the telencephalic destinations of somatosensory information can be known with confidence.

Motor control

Although a descending pathway from the striatum to the tegmentum has been demonstrated in leopard frogs (Halpern 1972) and in tiger salamanders, *Ambystoma tigrinum* (in which a projection to rostral spinal regions is also seen – Kokoros and Northcutt 1977), there is no direct evidence that the amphibian hemispheres contribute to the organization or control of movement. Abbie and Adey (1950), for example, failed to elicit movements in response to forebrain stimulation (of pallial and striatal regions) in various species of frogs and toads; movements were elicited by tectal stimulation. Similarly, Hoffmann and Lico (1972*a*,*b*) found that although low-intensity brain stimulation did elicit gross body movements in anaesthetized toads (*Bufo paracnemis*), stimulation of the septal area, amygdala or hippocampal region obtained only inconsistent movements, with a high threshold possibly indicating that successful stimulations required current spread to midbrain sites. Hoffman and Lico report a number of autonomic effects of forebrain stimulation (e.g. changes in blood pressure, heart-rate, and pupil size), but these effects too were inconsistent and showed no very clear mapping.

A further finding of Abbie and Adey (1950) was that there were no obvious motor deficits following abolition of forebrain influence by coronal section through the diencephalon. These findings do suggest that there may not be in amphibians any hemispheric area comparable to the motor cortex in mammals; further evidencc in support of this conclusion is provided by Halpern's (1972) report that pallial regions did not project beyond the diencephalon. The dorsal pallium in particular (sometimes regarded as primordium of neocortex) appeared to project only to neighbouring regions, mainly to the primordium hippocampi and dorsal septal area.

Conclusions

Before summarizing this section, it should be emphasized that any conclusions must, given the dearth of hard facts, remain tentative. We have seen evidence which suggests, in the case of motor control, that there may be a contrast between amphibian and mammalian organization, and evidence which suggests that there may be more in common between amphibian and mammalian sensory systems than was previously believed. The possible similarities should not be exaggerated: it remains possible that non-olfactory sensory representation is in fact minimal and possible, indeed, that olfaction does 'dominate' the forebrain. The known facts still allow a number of interpretations; for our purposes, it may be most reasonable to conclude that it is at least possible that integration of sensory information from a number of modalities occurs within the telencephalon, and that there may also be residual capacity for complex information processing in the forebrain. We clearly cannot determine the probable intelligence of amphibians from what is known of the sensorimotor organization of their hemispheres: we must look instead to behavioural data.

Habituation in amphibians

In this and the following sections on learning phenomena in amphibians the results to be discussed will be drawn exclusively from experiments using anurans and urodeles, as there do not appear to be any relevant data available on learning in apodans.

Habituation has been observed in a number of response systems in anurans. Kimble and Ray (1965) reported that the wiping reflex elicited by a tactile stimulus to the dorsal skin of leopard frogs declined over a period of 12 days, with 100 presentations of the stimulus each day; a similar result was obtained by Kuczka (1965) using toads (*Bufo bufo*). Habituation of the optokinetic response (eye and body movements elicited by moving stripes) has been demonstrated in toads (*Bufo bufo*) by Eikmanns (1955). Glanzman and Schmidt (1979) have reported habituation of the nictitating membrane reflex, elicited by electric shocks to the periorbital region of bullfrogs. Finally, orienting movements elicited by dummy prey habituate in both leopard frogs and toads (*Bufo bufo*) (e.g. Ewert and Ingle 1971). As we shall refer to Ewert's work on a number of occasions, the technique he uses will be given in some detail here. His subjects (toads) are placed in a glass cylinder set at the centre of a black disc, around the periphery of which are a number of holes. A white worm-like dummy prey object appears at a hole placed at an angle from the toad's midline, and this elicits an orienting movement designed to bring the toad's body facing the dummy; however, as soon as an orienting movement occurs, the prey is moved again so requiring a further response to centre the prey. By varying the distance apart of the holes on the disc, the visual angle subtended by the prey object can be varied. The procedure is maintained until no orienting response occurs for two minutes, at which point the stimulus series ends. The strength of the orienting response is assessed by counting the number of orienting movements in successive stimulus series: as habituation proceeds, that number declines.

Habituation in a urodele has been investigated by Goodman and Weinberger (1973), who used the mudpuppy (*Necturus maculosus*). These authors showed that either a moving shadow or a vibratory stimulus inhibited gill beating, tail waving, and heart beating, and observed a decline in the amount of inhibition in these systems as a consequence of repeated presentations of the eliciting stimulus. Moore and Welch (1940) showed that larval tiger salamanders rapidly learned not to strike at earthworms contained in a glass tube.

There appears to be much in common between the properties of habituation in amphibians and other vertebrates, and Glanzman and Schmidt (1979) have in fact demonstrated in their studies of the bullfrog nictitating membrane reflex all nine of the characteristics of habituation detailed by Thompson and Spencer (1966). Goodman and Weinberger (1973) found that the rate of habituation varied according to the response system measured, the heart-beat inhibition effect waning more rapidly than the gill-beat suppression. Spontaneous recovery of an habituated response has been observed by Glanzman and Schmidt (1979), Goodman and Weinberger (1973), and Ewert (1967); Ewert, for example, showed that a fairly rapid recovery from a stimulus series occurred within ten minutes, and that the strength of the orienting response reached 65-70 per cent of its original level after 100 minutes, although some degree of retention of habituation was also found over a 24 hour interval.

Dishabituation of an habituated response has been reported by Glanzman and Schmidt (1979), using an intense UCS, and by Goodman and Weinberger (1973), using a vibratory stimulus once habituation had been established to a shadow stimulus. Dishabituation has also been obtained by varying some features of the eliciting stimulus (thus indicating that habituation is stimulus-specific, and not the consequence of receptor or effector fatigue); dishabituation of the toad's prey-catching response has, for example, been obtained by altering the speed, size, or colour of the stimulus (Eikmanns 1955) and by presenting the prey object either to the opposite eye or to an alternative locus in the habituated eye (Ewert and Ingle 1971).

Finally, there is evidence that the repeated presentation of a stimulus may result in both decremental affects (habituation) and facilitating effects (sensitization). Glanzman and Schmidt (1979) report that the use of very intense stimuli frequently obtained sensitization of the nictitating membrane response. Kimble and Ray (1965) showed that, whereas repeated tactile stimulation at exactly the same point on a frog's skin obtained habituation, repeated stimulation of points in one general area, but not identical points, resulted in gradual increase in strength of the reflex obtained; this finding appears to explain Franzisket's (1963) apparently paradoxical failure to demonstrate habituation in either spinal or normal frogs. Ewert and Ingle (1971) found that, in toads, previous habituation to dummy prey presented in one part of the visual field resulted in slower habituation (stronger orienting response) to the same stimulus presented at a locus nearer to the midline of the visual field. Ewert and Ingle also found, in frogs, that habituation to a dummy prey stimulus presented laterally to one eye resulted in a marked preference in favour of that eye between two prey stimuli simultaneously presented in a more frontal position, each displaced either side of the midline at the same angle. Two physiological manipulations provide direct evidence of a process of sensitization at the site of the repeated presentation of a stimulus:

Ingle (quoted by Thompson, Groves, Teyler, and Roemer 1973, p.246) has observed that alcohol may weaken habituation and unmask sensitization on the first test trial in a prey-catching study; similarly, Ewert (1967) has shown that toads with unilateral forebrain lesions show more prey-catching responses in the second of a series of habituation tests than on the first.

This survey of habituation in amphibians has, therefore, found a series of parallels between phenomena seen in amphibians and those seen in other vertebrates, and provides no suggestion that the mechanisms involved differ.

Classical conditioning in amphibians

There have been relatively few reports of the use of classical conditioning procedures with amphibian subjects, and as most of these were poorly controlled, only three will be discussed here.

Two studies have investigated conditioning of the nictitating membrane. In the first of these, Goldstein, Spies, and Sepinwall (1964) suspended leopard frogs in a harness of rubber bands, and used a light touch on the nostril as CS, and a touch on the cornea, just sufficient to obtain lid closure, as UCS. They found satisfactory conditioning using either 0.5 s or 2 s ISI; acquisition was more rapid with the longer ISI, although it should be noted that this may have been an artefact since there was more time available (2 s as opposed to 0.5 s) in which the 2 s ISI subjects could make an 'anticipatory' CR. The 2 s ISI group appeared to reach asymptote (CRs on about two-thirds of all trials) within 75 trials, and Goldstein *et al.* note that this rate of acquisition is comparable to those reported using nictitating membrane or eyelid closure CRs in a number of mammalian species. Goldstein *et al.* also ran two control groups, one of which received CS alone trials, the other (sensitization) group receiving US alone trials with, 2 s after every fifth UCS, a test CS alone presentation. There was a negligible level of responding in each of these groups, and although no 'random' control group was run, the sensitization group does show that unpaired CS presentations against a background of UCS presentations did not obtain CRs.

A subsequent report by Yaremko, Boice, and Thompson (1969) cast doubt on the Goldstein *et al.* 1964) conclusions. Yaremko *et al.* were concerned that a touch to the nostril could, if sufficiently intense, elicit by itself a nictitating membrane closure, and that this suggested the possibility of sensitization of this response by the training procedure. In their study, pairings of a touch to the nostril region with touch to the cornea did obtain 'CRs' on about two-thirds of trials, but the level of CRs showed no change from the beginning to end of training, and a second group, given as CS touch to the cephalic region, which was markedly less reflexogenic, showed a constant very low level of CRs. They conclude that learning did not occur in their study, and suggest that the Goldstein *et al.* findings were a result of sensitization of the initially reflexogenic nostril region.

Although it is clear that Yaremko *et al.* failed to establish conditioning with their procedures, their conclusions do not explain the contrasting performance of the sensitization controls and that of the paired groups in the earlier report: it is possible that had Goldstein *et al.* equated the number of CS alone presentations in the sensitization group with the number of trials given to the paired groups sensitization might have been observed, but this does not seem very likely. It is more plausible that the Yaremko *et al.* failure was due to some procedural difference between their study and the Goldstein *et al.* report; there were, in fact, a number of important procedural differences—for example, Yaremko *et al.*, instead of using a harness, strapped their frogs to a board, and Goldstein *et al.*, before starting training, screened their subjects, rejecting frogs that consistently responded to the CS alone.

Yaremko *et al.* ran a further experiment using toads (*Bufo americanus*), with as CS, light touch on the cephalic region, and, as UCS, touch on the cornea. Three groups were used, one of which received classical training—each CS followed, after a 2 s ISI, by the UCS, the second group of which, avoidance training—each CR having the consequence that no UCS was delivered, and the third of which received a series of explicitly unpaired CS and UCS presentations, in random sequence, having the same ITI (minimum 30 s) as in the other groups. The unpaired group showed a very low level of responding, but both the classical and the avoidance groups showed steady increases in probability of CR—upto 66 per cent—across the five days of training. This study, then, provides good evidence that conditioning of the nictitating membrane response can be established in toads; if we accept the Goldstein *et al.* findings, this conclusion may be extended to frogs also. As we saw in our discussion of classical conditioning in fish, it makes little sense to attempt to compare rates of acquisition between species, although it is interesting that, given the large range of factors that potentially influence those rates, they appear nevertheless to be comparable in amphibians (or at least in anurans) and mammals. It need hardly be added that, given the lack of studies on amphibians, we simply do not know whether some procedural variation might not dramatically alter the rate of acquisition in species of this class.

A final question arising from the Goldstein *et al.* and Yaremko *et al.* studies is whether they have demonstrated classical or instrumental conditioning. It has been argued previously (p. 11) that, to settle this question, omission training procedures must be used. Yaremko *et al.* in fact used precisely such a procedure (avoidance training) with their toads, but unfortunately the outcome does not resolve the issue. In their study, the rates of acquisition of the classical and avoidance groups were indistinguishable: this could be taken to support a classical interpretation, since introducing an explicit instrumental contingency—total avoidance of the noxious tactile stimulus—did not improve performance; on the other hand, if the conditioning was wholly classical, then one might anticipate more rapid learning in the classical group, as this group experienced more CS–UCS pairings than the avoidance group (and this is indeed the result obtained with conditioning of both the human eyelid and rabbit nictitating membrane, Gormezano 1965). Had there, then, been a marked difference between the classical and avoidance groups there would have been good support—depending on the direction of the difference—for either a classical or an instrumental interpretation; as the results fell out, however, either position can be comfortably maintained.

The third study to be considered in this section concerns an interesting observation, despite lack of controls, made by Van Bergeijk (1967), who found that bullfrogs which were fed every weekday at 4 p.m. tended to gather together near the feeding place at the appropriate time on five days of the week. Van Bergeijk found that the frogs could be retrained to go to a new place at a new time, and that both acquisition (which needed about 5-7 trials) and extinction (3-4 trials) were relatively rapid. The findings suggest that frogs can use time-related cues (from either an 'internal clock' or the position of the sun) as CSs, although what the nature of the other CSs involved might be is far from clear. As the frogs were observed in a group, it is not possible to say how many subjects learned and how many merely followed the learners; nor, of course, is it possible to decide whether the learning was obtained by the classical contingency (pairing of CSs with food reinforcers) or by some instrumental contingency (the CR having the consequence, for example, of reducing the delay prior to reinforcement).

Instrumental conditioning in amphibians

The great majority of experiments on instrumental learning in amphibians have used aversive stimuli as the source of motivation, but before reviewing them, three studies that show that positive reinforcers can be effective will be discussed. Moore and Welch (1940) trained larval salamanders (*Ambystoma paroticum*) to rise to the water surface in response to either movement of the experimenter's hand, or an intermittent flashing light. Neither of these stimuli normally elicits approach, and subjects were trained to rise by the technique of lowering a worm on a string, and drawing the subjects to the surface before allowing them to eat it. The larvae, given two trials a day, apparently learned this habit in from four to eight trials. Hershkowitz and Samuel (1973) made use of the fact that both larval and adult crested newts (*Triturus cristatus*) snap spontaneously at small black objects to train a discrimination between a black circle and a black triangle pasted to the outside of the tank: on each trial the stimuli were exposed (simultaneously), and a snap at the correct stimulus obtained a piece of worm dropped into the tank, near the subject's head. Snaps to the incorrect stimulus simply terminated the trial without reinforcement. Subjects, trained against their initial preference, improved from 40 per cent correct over trials 1 to 10 to 90 per cent correct after 170 trials, although whether the discrimination was based on shape or position is not known, as the stimuli were shown in the same place on each trial. Hershkowitz and Samuel showed that the habit was retained without loss over metamorphosis although they were generally unsuccessful in their attempts to train naive adult newts. Finally, Schmajuk, Segura, and Reboreda (1980) have shown that dehydrated toads (*Bufo arenarum*) will traverse runways for the reward of water (taken up through the skin), and that the same reward will motivate acquisition and reversal in a Y-maze.

A number of authors have examined choice behaviour of anurans and urodeles in mazes, the source of reinforcement being escape from some aversive stimulus, such as bright light, or electric shock, present in the start-box (e.g. Yerkes 1903; Munn 1940; Roberts, Heckel, and Wiggins 1962). The reports are generally unsystematic and poorly controlled, but do agree that amphibians are capable of maze learning, and results obtained by Fankhauser, Vernon, Frank, and Slack (1955) suggest that salamanders, at least, may be comparatively efficient learners. Fankhauser *et al.* (1955) tested larval newts (*Triturus viridescens*) in a Y-maze, training half to turn left, half to turn right. The larvae were prodded up the stem on the maze until a choice was made: if correct, 90 seconds of rest was allowed, and if wrong, the prodding continued, a bright light came on, and the larva was sucked up into a pipette, and replaced in the stem. The larvae acquired the habit, to the stiff criterion of ten consecutive correct choices, in, on average, 39 trials (with only nine errors).

McGill (1960) reported, on the other hand, a total failure of leopard frogs to learn to escape from shock onset in one compartment of a shuttle-box by jumping to the other compartment in the box; in fact, McGill's subjects took progressively longer to escape from shock onset, until all his subjects died during the experiment. Subsequent research indicates, however, that the problem lies, not in the intellectual equipment (or lack of it) of the frog so much as in its inability to initiate responses. Boice (1970) compared four species of anurans in a one-way active avoidance task, in which subjects were placed in one end (either black or white) of a box, divided from the other end by a removeable partition: when the partition was removed, the animal had to jump to the other side of the box within 4 seconds in order to avoid a shock which, in the absence of an escape response, lasted for 10 seconds. Two of the species were classified as passive—that is, as showing little activity in their normal habitat—and these were leopard frogs and spadefoot toads (*Scaphiopus hammondi*), the other two species, green frogs (*Rana clamitans*) and Woodhouse's toad (*Bufo woodhousei*) being relatively active. Control animals were yoked to individuals of the same species and received inescapable shocks of the same duration as the subject in the avoidance condition, independent of location in the apparatus. Neither of the passive species showed any avoidance responses, or any change in escape latency over the course of training; however, the Woodhouse toads showed upto 50 per cent avoidance responses, and the green frogs, although achieving only about 10 per cent responses, were significantly superior to their controls. This study cautions us once again that formally similar tasks may pose very different problems, quite unrelated to their intellectual demands, to different species. Evidence to show the importance of 'contextual variables' within a species is provided by Miller and Berk (1977), who varied shock levels in an avoidance task, and showed that the performance of both larval and adult African clawed toads (*Xenopus laevis*) varied with shock intensity, the adults being apparently more sensitive to shocks than the larvae.

Despite the demonstrable importance of contextual variables, no comparative study using amphibians has employed the technique of systematic variation so that, despite the widespread view (e.g. Thorpe 1963) that amphibians are of inferior learning ability, there is no convincing evidence for such a conclusion. This scepticism may be further justified by quoting a study, by Brower, Brower, and Westcott (1960), to show that toads, like teleosts, are capable of one-trial learning: these authors found that captured toads (*Bufo terrestris*) would attack a bumblebee dangled on a string when first presented with one, but not on subsequent occasions, although persisting in attacks on edible prey (dragon-flies). Control toads, given bumblebees whose stings had been removed, persisted in attacking these stingless targets.

There are, from the studies reviewed above, sufficient control data to allow a confident conclusion that amphibians, or at least some amphibians, are capable of associative learning in (apparently) instrumental tasks. There have, however, been very few efforts to establish whether that learning has proceeded through the detection of instrumental as opposed to classical contingencies: it is possible that, in all those studies, subjects demonstrated only approach to (or withdrawal from) stimuli that were classically associated with appetitive (or aversive) reinforcers. We have already discussed Yaremko *et al.*'s (1969) inconclusive study, and only one other report is relevant here. Miller and Berk (1977) trained their *Xenopus* in an unusual trapezoid tank in which there was a shock intensity gradient so that shock was more intense at the narrow end of the tank; there was a smooth white insert at the broad end. Subjects were placed in the narrow end in squads of six, and left in the tank for 24 hours: the number of subjects in the white end was scored every two hours through that period, to provide a measure of avoidance of the black end (this unconventional technique was used for the very good reason that it provided comparable rates of acquisition in both larval and adult *Xenopus*; this was a crucial requirement for the central purpose of the study, which was to compare retention while larval, while adult, and over the period of metamorphosis). Control subjects were run with a barrier between the two ends of the box so that they spent comparable amounts of time in the narrow and wide ends as did yoked experimental groups, but could not detect the consequences of moving from one end to the other. When tested with the barrier removed, the control subjects showed significant avoidance of the narrow end, but to a lesser degree than the corresponding experimental group. That the controls avoided the narrow end provides some evidence that stimulus-reinforcer (classical) associations were formed (although, as there was a gradient in each half of the box, they may have learned to swim away from cues at the narrow end); that their performance was inferior to that of the experimentals suggests, in turn, that response-reinforcer (instrumental) associations were formed in the latter group (although, again, an alternative account might be that the control subjects showed a performance decrement brought about by the change in stimulus conditions occasioned by the removal of the barrier).

We may conclude then that amphibians, or, at least, some anurans and urodeles, are capable of associative learning, and that, although the evidence available is far from extensive, there is no reason to suppose that they cannot detect both stimulus-reinforcer and response-reinforcer contingencies. Amphibians have been shown to be efficient learners in some tasks, and, as we might expect, there is evidence that contextual variables have a marked effect on acquisition rates. There is currently no evidence that suggests differences in intelligence between different groups of amphibians, and, although no efforts have been made to train amphibians on anything but relatively simple tasks, no direct comparative evidence that amphibians are either inferior or superior in intelligence to any other group of vertebrates.

Forebrain lesions and learning in amphibians

Burnett (1912) concluded that 'the decerebrate frog is incapable of forming even the simplest associations' (Burnett 1912, p. 87). Unfortunately, his experimental data, like those of other early investigators that came to similar conclusions, are open to other interpretations. Burnett, for example, trained his frogs (three decerebrates, three controls) in a T-maze in which subjects were shocked for incorrect choices, and encouraged to choose, if too lethargic, by drops of ether onto their backs from a pipette; the 'reward' was simply to escape from the apparatus. The normals learned in about 20 trials, but the decerebrates showed no learning over 100 training trials. However, Burnett reports that the decerebrates showed less spontaneous movement than normals, and took much longer than normals to start any attempt to escape (hence the introduction of the ether). Not only that, but the decerebrates, despite being able to orient to and snap at food objects, in fact ate less than the normals, so that all three gradually lost weight and died. Clearly, there are serious doubts over the motivational condition of the operated animals; we could as well suppose that reduced motivation caused the failure to learn as that it was due to reduced intelligence—and, in fact, similar criticisms apply to a number of other early studies (e.g. Blankenagel 1931; Diebschlag 1934).

There do not appear to have been any modern reports concerned specifically with amphibian forebrain involvement in learning, but there are reports which show that associations can be formed without the hemispheres and, indeed, without the entire brain. Horn and Horn (1969) experimented on clawed toads (*Xenopus mullerei*) whose spinal cord had been sectioned. Their subjects were strapped to a vertical board, with one hindlimb free; the experimenters waited until the free limb had been in the same position for 5 minutes, and then delivered a UCS (a pinch or an electric shock) to the foot, which elicited withdrawal, followed by slow return, of the foot. Horn and Horn showed that if the UCS was re-applied at fixed points on the return path of the limb (e.g. 4 mm above the original resting position), then the limb gradually took longer to return to the position at which the UCS was delivered. Controls, given comparable numbers of shocks irrespective of limb position, showed no change in speed of return to the resting position. A similar phenomenon was observed by Farel and Buerger (1972) in spinal bullfrogs: in their experiment, which was based on a design developed by Horridge (1962) for use with cockroaches, a hindlimb was placed in a bowl of saline so that a flexion response would remove it from the bowl; shock was then delivered whenever the limb touched the saline, and the current circuit was completed through the limb of a yoked control frog (see Fig. 4.4) which, therefore, received shocks of the same intensity and duration as the experimental subject, independent of limb position. Farel and Buerger not only obtained conditioning, as shown by highly significant differences in the amount of time spent in contact with the saline solution between experimental and yoked control subjects, but

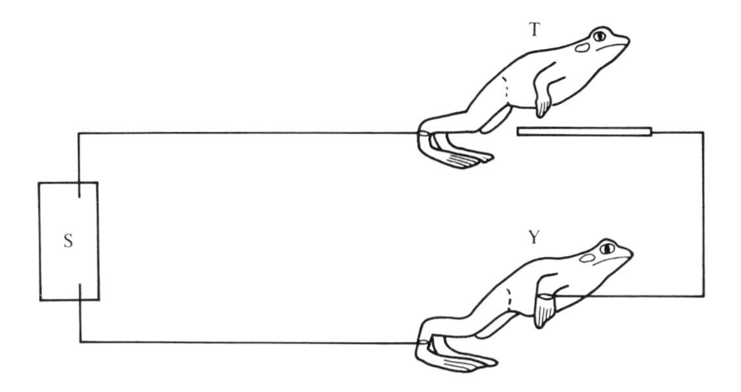

Fig. 4.4. Diagrammatic representation of the training procedure used by Farel and Buerger (1972). (From Peters and Wirth 1976.)
Key: S – stimulator; T – trained animal; Y – yoked animal.

also found that learning proceeded very rapidly – the spinal frogs in fact achieved stable performance within about 15 seconds. Their study included one further control condition which showed that, if the sciatic nerve was cut (preventing central nervous involvement), no learning occurred.

In case the efficiency of learning within the anuran spinal cord should tempt one to suppose that in 'lower' animals conditioning may be an independent, local phenomenon not subject, as in 'higher' animals, to some overall co-ordinating intelligence (the notion that learning too may be gradually 'encephalized'), it should be pointed out here that conditioning has been obtained in the isolated spinal cord of a number of mammals including, perhaps most significantly, man (Ince, Brucker, and Alba 1978).

One final study which may not be entirely relevant, but has considerable intrinsic interest, is that of Fankhauser *et al.* (1955). These authors manipulated the number of brain cells in newts (*Triturus viridescens*) by subjecting fertilized eggs to a temperature of 36 °C for 10 minutes. This technique increases the normal chromosome number (from diploid to triploid), which has the consequence that, although overall body size remains unchanged, the number of cells (of all types) in the body decreases, by about one third, their individual size being correspondingly increased. Fankhauser *et al.* then compared performance of diploid and triploid subjects (full-grown larvae) in the Y-maze described in the previous section, finding that, whereas the mean score of the normals reaching criterion was 39 trials (with 9 errors), the triploid subjects were significantly slower (125 trials, with 44 errors). Now this outcome may provide evidence for 'mass action' (see Lashley 1950) in those regions of the amphibian nervous system concerned with learning, but there are, of course, numerous other interpretations: the decline in cell numbers in the sensory systems may have impaired the triploid subjects' perception of the maze, or the depletion of cells in the motivational systems may have altered motivational state (not to mention possible consequences of changes in non-nervous tissue).

There is clearly little to be said of forebrain involvement in amphibian learning: anurans can learn without the participation of the brain (and so can mammals), but what role the brain, and in particular the forebrain or any region within the forebrain, plays in the intellectual functioning of intact amphibians is simply not known.

Summary and conclusions

Amphibians evolved from crossopterygian fishes, a group quite different from that which gave rise to the teleosts. The line of descent leading to modern amphibians diverged from that leading to the reptiles at an early stage in amphibian evolution so that, in order to find a common ancestor for amphibians and reptiles, birds, and mammals, we must go back at least 300 million years.

The amphibian telencephalon shows a concentration of nerve-cell bodies in the periventricular region, and is not very well differentiated. However, the pallial and subpallial regions are conventionally divided into regions which parallel the major telencephalic divisions seen in mammals and there may, therefore, be an area homologous with neocortex in amphibians.

Although there do seem to be widespread olfactory projections in the amphibian forebrain, there is evidence that other senses (vision, hearing, and the skin senses) are represented there also. There are, moreover, striking parallels between the organization, at least up to the diencephalic level, of the visual and auditory pathways in amphibians and those in mammals. In particular, there appear to be two major routes for visual information, and similarities between the properties of those routes in amphibians and mammals cast doubt on the notion that functions carried out in the midbrain of 'lower' vertebrates are taken over by the forebrain (encephalized) in 'higher' vertebrates.

There is reasonably good evidence that the mechanisms of habituation in amphibians are comparable to those in other vertebrates. However, although there are reports which show that amphibians are capable of association-formation, the available data are insufficient to decide whether either classical or instrumental associations (or both) are to be found in amphibians. The paucity of data on association-formation (and virtual absence of data on 'complex' tasks) make it impossible to provide any comparative assessment of the intellectual capacities of amphibians.

Finally, although it is interesting to note that spinal frogs are capable of association-formation, it is unfortunate that we have as yet no reliable information on the contribution of the hemispheres to intelligence in amphibians.

5. Reptiles

Classification of reptiles

Although the fossil record does not allow a confident account of the transition from amphibian to reptilian forms, it appears that the first true reptiles emerged during the first half of the Carboniferous period (345-281 M a ago). These early reptiles, the cotylosaurs, or 'stem reptiles', are believed to be descended from anthracosaur amphibians, and to be the ancestors of all reptilian groups. A critical step in the transition from amphibians to reptiles was the development of an egg which could be laid on land, and in which the embryo could grow to a size which would allow it, on emerging from the egg, to survive outside an aquatic environment; the egg that evolved was, as with modern reptiles and birds, encased in a hard shell covering within which there were three other membranes, the innermost of which is known as the amnion. The amnion is preserved throughout reptiles, birds, and mammals which are, therefore, collectively known as amniotes.

The early reptiles showed a rapid radiation into a wide variety of groups, soon displacing amphibians as the dominant land vertebrates, and the Mesozoic era (225-66 M a ago) is commonly referred to as the Age of Reptiles. It was during the Mesozoic era that the huge dinosaurs evolved, only to disappear abruptly and completely by the end of the Cretaceous. Just four orders, Chelonia, Rhynchocephalia, Squamata, and Crocodilia, have survived whatever catastrophic changes occurred towards the end of Cretaceous (what these changes might have been—climatic, or increased competition from birds or mammals—remains a matter of debate).

The order Rhynchocephalia possesses only one living species, the tuatara (*Sphenodon punctatus*), found in a few islands belonging to New Zealand. Fossil forms very similar to the modern genus have been found dating back to the early Mesozoic, so that, in that sense, *Sphenodon* can be considered a primitive reptile.

The *Squamata* are the most successful of living reptilian groups, and three suborders are recognized: Sauria (Lacertilia, or lizards), of which there are some 3000 species; Amphisbaenia (Annulata, or worm-lizards), of which there are about 130 species; and Serpentes (Ophidia, or snakes, which evolved from lizards), of some 3000 species. Both rhynchocephalians and squamates are representatives of the lepidosaur subclass, although their evolution has proceeded independently from the late Permian period.

The order Chelonia (turtles* and tortoises), which embraces some 220 species, appears to have diverged from other reptilian groups at a very early stage in the

* In Britain, the semi-aquatic forms are known as terrapins; however to minimize discrepancies between this text and the original research papers, most of which are of American origin, the term 'turtle' will be used to refer to both aquatic and semi-aquatic chelonians.

reptilian radiation and is included in the same subclass (Anapsida) as the cotylosaurs. Modern chelonians have in fact shown very little advance in most regards over ancestral forms of the Triassic (225-196 million years ago).

Finally, the order Crocodilia (crocodiles and alligators), with 21 living species, is the sole surviving representative of the ruling reptiles (archosaurs) from which evolved the two great groups of dinosaurs, the saurischians (e.g. *Brontosaurus* and *Diplodocus*) and the ornithischians (e.g. *Stegosaurus* and *Triceratops*). Although fossils of the order have been traced back to the Triassic, modern crocodilians are in many respects anatomically degenerate, and far removed from typical early reptilian forms. Crocodilians have many features in common with birds, and this reflects the fact that birds too are descended from archosaurs—indeed, as we shall see (in Chapter 6), one modern view is that birds are in effect living dinosaurs.

The lines of descent from cotylosaurs to the ancestors of chelonians, to lepidosaurs and to archosaurs diverged early in the Permian period, so that no extant reptilian order has shared a common ancestry with any other order for at least 225 million years. Moreover, those three lines of descent in turn diverged from that which gave rise to a group known as pelycosaurs, from which ultimately the mammals derived, during the Carboniferous period—about 280 million years ago. Figure 5.1 summarizes the relationships between the major reptilian groups, and between reptiles, mammals, and birds; it can be seen from this figure that, as Romer and Parsons emphasize, although mammals are descended from reptiles 'their relationship to the existing reptilian orders is thus extremely remote' (Romer and Parsons 1977, p. 75).

The conclusion to be drawn from this section is, by now, a familiar one: we do not see in living reptiles forms that are in any sense primitive versions or direct ancestors of mammals. Whatever may emerge of their brains and intelligence, we need not assume that we see in either a stage through which mammals passed in their evolution.

Structure of reptilian forebrain

There is in reptiles a clear distinction between the olfactory bulbs and the cerebral hemispheres; in some reptiles, the bulbs are separated from the hemispheres to which they are attached by stalks (or peduncles) but in others, the bulbs are directly attached to the rostral poles of the hemispheres. Accessory bulbs are also found in all except the crocodilian species, adults of which, as one might expect, lack Jacobson's organ. The sizes of the bulbs and accessory bulbs show considerable variation within reptile groups: although the main bulbs in chelonians, for example, are relatively large, species of that order have a relatively primitive vomeronasal epithelium, and have small accessory bulbs. Similarly, most squamates have well-developed main and accessory bulbs, but some lizards (e.g. chameleons, some iguanids) are microsmotic, and have minute main and accessory bulbs. Figure 5.2 shows external views of the brains of representatives of the

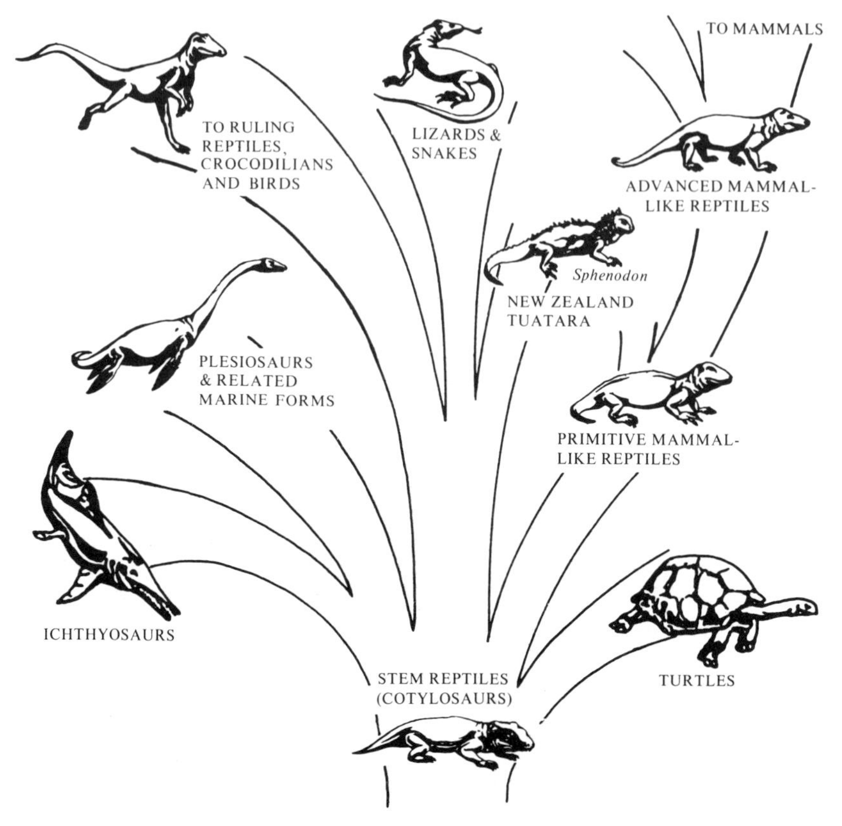

Fig. 5.1. Diagrammatic representation of the relations between the major reptilian groups, and between reptiles, mammals, and birds. (After Romer and Parsons 1977.)

three main extant reptilian orders, *Emys* (a chelonian), *Lacerta* (a squamate), and *Alligator* (a crocodilian).

Although there are marked differences between the forebrains of individual reptile groups, 'nevertheless a common pattern can be traced throughout the different orders' (Kappers, Huber, and Crosby 1960, p. 1313). This common pattern can indeed be discerned also in the amphibian forebrain despite a number of important changes. One of the most striking of these changes can be seen in the pallial division of the forebrain which is, as in amphibians, conventionally divided into three longitudinal regions: a medial archicortical or hippocampal region, a dorsal or general cortical region, and a lateral paleocortical or piriform region. These regions can be seen in the transverse section of a gecko's forebrain shown in Fig. 5.3. The change from the amphibian condition consists in the fact that, whereas in amphibians the cell bodies of the pallial region are concentrated in the periventricular region, in reptiles the cells have migrated towards the brain surface, forming a distinct lamina separated from the ventricles by white matter (axonal and dendritic fibres); this lamina is regarded by most authors as forming a true cortex. Reptilian cortex is thin, about 0.5 mm thick as opposed to the 1-3 mm thickness of mammalian neocortex, and shows in most regions only one prominent layer of cells (Ebner 1976), although in the tegu lizard (*Tupinambis nigropunctatus*) four different layers can be discerned throughout the cortical mantle (Ebbesson and Voneida 1969).

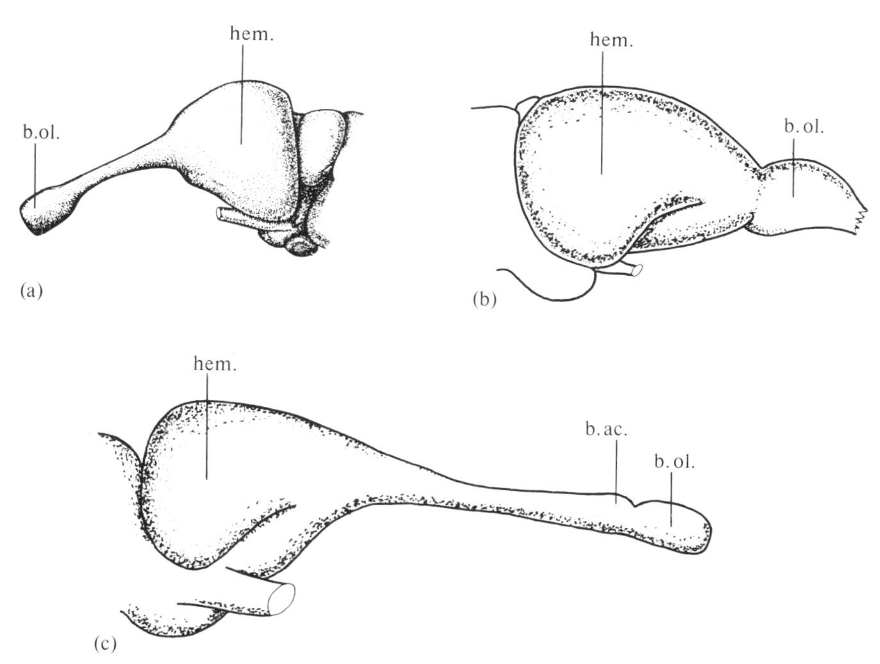

Fig. 5.2. External views of the forebrains of representatives of the three major extant reptilian orders. (a) *Alligator* (a crocodilian). (From Romer and Parsons 1977.) (b) *Emys* (a chelonian). (From Nieuwenhuys 1967.) (c) *Lacerta* (a squamate). (From Nieuwenhuys 1967.).

Key: b.ac. – accessory olfactory bulb; b.ol. – olfactory bulb; hem. – cerebral hemispheres.

In some reptilian species the three longitudinal pallial zones are not clearly distinguishable from one another on architectonic grounds (Goldby and Gamble 1957), whereas in others more than three zones can be distinguished (Ebbesson and Voneida 1969). In the rostral pole of the hemisphere there is a structure, not always clearly distinguishable from the underlying striatum with which it is continuous, known as the 'pallial thickening', the status of which is uncertain. While the traditional view has been that this is an area of fusion of the three cortical zones (e.g. Kruger 1969), Ebbesson and Voneida (1969) regard it as a subcortical nucleus extending into the lateral pallial region and Lohman and Van Woerden-Verkley (1976) regard it as a thickening of the lateral zone exclusively;

these contrasting views reflect in part the fact that as yet little is known of the functional significance of the area.

There is general agreement that the efferent connections of the medial cortical zone in reptiles resemble those of mammalian hippocampus (e.g. Voneida and Ebbesson 1969; Lohman and Van Woerden-Verkley 1976), and further support for the comparison of reptilian and mammalian hippocampal structures is provided

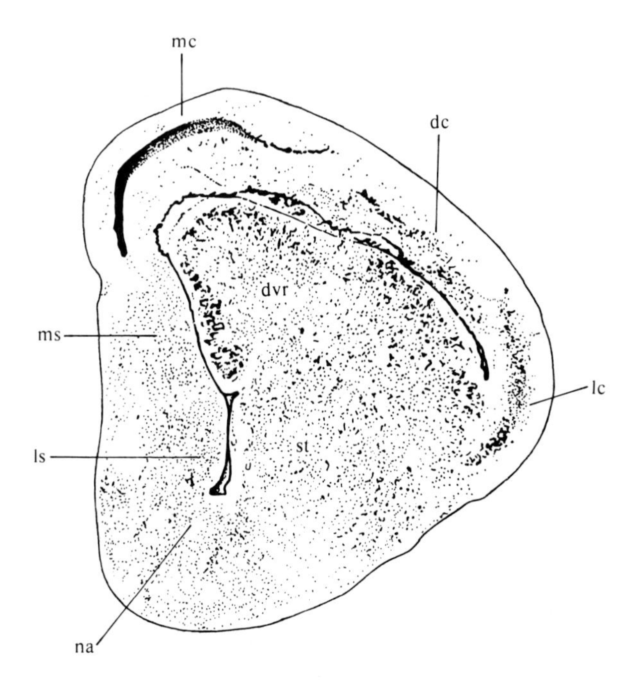

Fig. 5.3. Transverse hemisection through the telencephalon of the lizard *Gekko gecko*. (From Northcutt 1978.)

Key: dc – dorsal cortex; dvr – dorsal ventricular ridge; lc – lateral cortex; ls – lateral septal nucleus; mc – medial cortex; ms – medial septal nucleus; na – nucleus accumbens; st – striatum.

by histochemical analysis of enzyme concentrations (Baker-Cohen 1969). Lacey (1978), however, analysing the detailed neuronal architecture of the hippocampal region of the fence lizard (*Sceloporus undulatus*) concludes that the subdivisions of reptilian hippocampus do not appear to parallel those of the mammalian structure.

There is also agreement, based primarily on the distribution of secondary olfactory fibres (which will be discussed in the following section), that lateral pallium is comparable to mammalian piriform cortex. However, the extent to which reptilian dorsal cortex parallels mammalian neocortex is uncertain: there is evidence that the dorsal cortex should be divided into at least two longitudinal zones, and that the more medial zone has connections that resemble those of mammalian hippocampus, rather than neocortex (Voneida and Ebbesson 1969; Lohman and Van Woerden-Verkley 1976). On the other hand, there are numerous catecholaminergic terminals in the dorsal cortex (and, to a lesser extent, in the hippocampal cortex) of the painted turtle (*Chrysemys picta*) (Parent and Poitras 1974) and evidence that the cells or origin of these terminals lie in a nucleus, the nucleus dorsolateralis tegmenti, which lies near the junction of mid- and hindbrain; such a system closely resembles the dorsal noradrenergic bundle in mammals, which projects from the locus coeruleus in the pons to neocortex and hippocampus. In the section that follows we shall see that although there are thalamic sensory afferent projections to dorsal cortex in some reptiles (turtles – Hall and Ebner 1970*a*) there are not in others (lizards – Lohman and Van Woerden-Verkley 1978), and that there does not appear to be any cortical 'motor area' comparable to mammalian motor cortex. The parallels between reptilian dorsal cortex and mammalian neocortex remain, therefore, tenuous.

The subpallial division of the telencephalon also stands in marked contrast to that of the amphibian forebrain, through an increased differentiation of nuclear areas in general, and a considerable expansion of the region ventral and lateral to the ventricle in particular. The major divisions are the now familiar ones – a medial septal area, containing a medial and a lateral septal nucleus, and a ventrolateral strio-amygdaloid complex, with the posteriorly located amygdalar region not always clearly distinguishable from the striatal area (in some reptiles, the amygdala takes the form of a roughly circular cell-free zone capped by a curved cellular layer, and is known as the nucleus sphaericus). There is also a 'cuff' of cells surrounding the olfactory peduncle, the anterior olfactory nucleus, a nucleus accumbens surrounding the ventricle ventral to the septal region, and, near the ventral surface, an olfactory tubercle which merges caudomedially with another superficial layer of cells, the nucleus of the diagonal band of Broca.

The rostral striatum shows a marked dorsal enlargement, giving the lateral ventricles a pronounced curve, and this particular area is known by a variety of names, including the dorsal ventricular ridge (DVR), the hypopallium, and the neostriatum. Of these the first, DVR, appears to enjoy most support at present (perhaps because it is a neutral term), and we shall adopt it in this survey. There has been a long-standing controversy over the origin of the DVR, with the balance of opinion shifting recently to favour Källén's (1951*b*) opinion that it is not of subpallial, but of pallial, origin; one reason for regarding the DVR as pallial is that in some species it appears that the paleocortical layer continues unbroken around the lateral edge of the ventricle into the underlying DVR (Fig. 5.4). Parallels with mammalian neocortex have been found recently in both the efferent (Hoogland 1977; Voneida and Sligar 1979) and afferent connections of the DVR: there are, as we shall see in the following section, direct thalamic sensory projections to reptilian DVR.

The ventral striatum, sometimes known as the paleostriatum, lies below the DVR and in the main, anterior to the amygdalar (or archistriatal) region. There is evidence from both histochemical (Baker-Cohen 1968; Juorio 1969; Parent and Olivier 1970) studies and from anatomical investigations of efferent pathways (Hoogland 1977; Voneida and Sligar 1979; Brauth and Kitt 1980) for parallels between the reptilian ventral striatal region and the mammalian corpus striatum—consisting of the caudate nucleus, putamen, and globus pallidus. All of these reports show marked contrasts between dorsal and ventral striatum and emphasize that, whether or not the DVR is regarded as pallial, it has very little in common with the striatal regions of mammals.

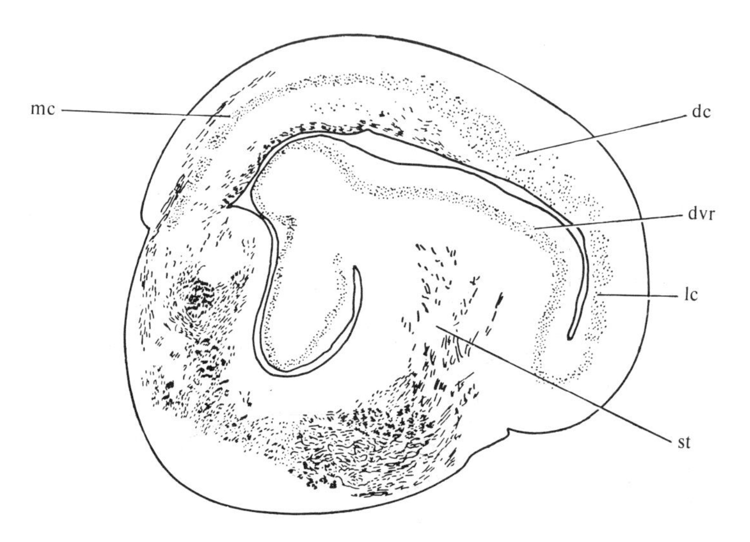

Fig. 5.4. Transverse hemisection through the telencephalon of the tuatara, showing continuity between lateral cortex and dorsal ventricular ridge. (From Kappers, Huber, and Crosby 1960.) Key as for Fig. 5.3.

There are a number of parallels, besides the similarity of topographical location, between the amygdalar region of the reptilian brain and that of mammals; both for example, receive a major afferent input from the lateral olfactory tract and have a major efferent pathway, the stria terminalis, which projects to the hypothalamus. Of direct relevance to our theoretical interests is the fact that there appear to be similarities in behavioural function of the regions in the two classes: Keating, Kormann, and Horel (1970) have shown, using spectacled caimans (*Caiman sclerops*), that direct electrical stimulation via implanted electrodes in the amygdala of free-moving animals elicits flight responses, and that lesions in the amygdaloid complex markedly reduce both attack and retreat responses to noxious or mildly noxious stimuli (such as taps with a 'bottle brush'). It should be noted that these latter effects were not due to any motor deficits—attack and retreat responses could be elicited, but required much more vigorous stimulation in amygdalectomized subjects. Tarr (1977), investigated the effects of amygdalar lesions on the social behaviour of Western fence lizards (*Sceloporus occidentalis*), testing his animals in established social groups. Tarr found that, following operation, his subjects were less spontaneously active, and showed 'a total lack of responsiveness to visual social signals' (Tarr 1977, p. 1157). Amygdalectomized subjects would neither challenge previously subordinate members of their group, nor submit to challenges of dominant members—they simply ignored their fellow group-members altogether, although they showed no visual (or motor) impairments in non-social tests. Control subjects, having lesions in the DVR, showed drops in spontaneous activity, but did respond appropriately to social signals when they 'noticed' them: the implication is, therefore, that the absence of social responding was not a secondary consequence of the decline in activity. It is well-established that in mammals, amygdalar lesions severely disrupt social behaviour and diminish aggressive responses (e.g. Dicks, Myers, and Kling 1969; Goddard 1964), and these results clearly encourage the view that there is a 'general homology' (Goldby and Gamble 1957, p. 347) of the posterior part of the reptilian striatal area with the mammalian amygdalar complex. Further support for this view is provided by Parent (1971) who, studying the distribution of cholinesterase in *Chrysemys picta*, found not only a gross similarity between the reptilian and the mammalian amygdaloid complex, but evidence also for a distinction in the turtle between a basolateral region (rich in cholinesterase) and a corticomedial region. However, it is not yet clear how close the homology in fact is; Goldby and Gamble, for example, point out that there are differences in the internal nuclear organization within the amygdaloid region of different groups of reptiles. Evidence on the ontogeny of the amygdaloid complex is also (as we found with amphibians) confusing: Källén (1951*a*) argues that the reptilian complex consists of both striatal components (homologous with mammalian corticomedial amygdala) and a pallial component (the posterior part of the dorsal ventricular ridge), but that the pallial component is not in fact homologous with the pallial (basolateral) part of the mammalian amygdaloid complex, which has in fact no non-mammalian counterpart.

Sensory and motor organization in reptilian forebrain

The following sections will provide a brief survey of the afferent and efferent connections of the reptilian telencephalon, along with relevant physiological data. It will be seen that much more is known of reptilian forebrain organization than was the case with either fish or amphibians, and one dominant theme will be the similarity in at least gross organization at the telencephalic level of olfactory, visual, and auditory input in reptiles and mammals.

Olfaction

Experimental anatomical investigations of the reptilian olfactory system began with the pioneering studies of Goldby (1937) and Gamble (1952, 1956). Recent reports (Scalia, Halpern, and Riss 1969; Halpern 1976) have in general confirmed the earlier conclusions, so that there is now good agreement on basic olfactory projections.

The secondary fibres from the main and accessory bulbs project via two tracts, the medial and lateral olfactory tracts. Fibres of these tracts terminate in the olfactory tubercle, in the anterior olfactory nucleus, and throughout the lateral pallium; the projections are generally bilateral, with most fibres terminating ipsilaterally. Fibres of the lateral tract also run to the amygdaloid area, although according to Scalia *et al.* (1969) these fibres are, in *Caiman sclerops* (which does not possess an accessory bulb) superficial and do not penetrate the amygdala itself. Halpern (1976) presents evidence that in snakes of the genus *Thamnophis* the projection of the accessory bulb is entirely independent of that of the main bulb, its principal projection area being the central core of the amygdala, known in this group as the nucleus sphaericus.

Scalia *et al.* (1969) have emphasized that there is a close correspondence between the pattern of distribution of secondary olfactory fibres in reptiles with that found in mammals. In particular, these fibres terminate in a restricted cortical area (lateral or piriform cortex), do not project directly to either the hippocampus or the septal area, and show independent destinations for main and accessory bulb fibres.

Vision

The direct retinal projections of reptiles have been investigated using modern experimental anatomical techniques in lizards (e.g. Butler and Northcutt 1971*a*, 1978), snakes (e.g. Northcutt and Butler 1974; Halpern and Frumin 1973; Repérant and Rio 1976), turtles (e.g. Knapp and Kang 1968*a*,*b*; Hall and Ebner 1970*b*), alligators (Burns and Goodman 1967), and the tuatara (Northcutt, Braford, and Landreth 1975). These reports show a wide measure of agreement, so that a confident account of the major destinations of retinal fibres in reptiles may be given.

The great majority of retinal fibres terminate in the superficial layers of the tectum, in a number of nuclei of the pretectal region, and in at least two sites in the thalamus. There is also a projection to the tegmentum, to the nucleus opticus tegmenti, otherwise known as the nucleus of the basal optic root (Reiner and Karten 1978), and in at least some species, a small projection to the contralateral hypothalamus (e.g. Reperant and Rio 1976). All of these sites receive their major input from the contralateral eye, but smaller ipsilateral projections have also been reported, most commonly to thalamic sites; Reperant and Rio (1976) observe that although the contralateral projection pattern appears relatively constant throughout reptiles, there appears to be considerable variability in the ipsilateral projections, which seem to be more extensive in squamates than in other reptilian orders.

The two major thalamic destinations are the dorsal and ventral parts of the lateral geniculate nucleus, or the neuropil associated with those two parts. Hall (1972) discusses the question whether these two sites should in fact be compared to the similarly named regions of the mammalian thalamus, and provides the following parallels: first, lesions of the optic tectum (in red-eared turtles, *Pseudemys scripta*) and of the superior colliculus (in hedgehogs, *Paraechinus hypomelas*) lead to degeneration in the ventral, but not in the dorsal, part of the ipsilateral lateral geniculate nucleus in both species, and second, in both species, the dorsal part projects to the cortex (in hedgehogs, to the striate cortex, and in turtles, to a region of the dorsal or general cortex), showing some degree of topographical organization.

The evidence produced this far shows a close correspondence between reptilian and mammalian organization: given that the thalamic sites are comparable, then the entire basic pattern—including projections to pretectal, tectal, tegmental, and even hypothalamic sites—parallels that reported in comparable mammalian studies (see, for example, Hall and Ebner 1970*b*).

There is, moreover, a further important similarity between mammals and reptiles, and this concerns a route from the tectum (or superior colliculus) to the telencephalon. Tectal lesions in both groups lead to thalamic degeneration in not only the ventral lateral geniculate nucleus but also, in reptiles, in the nucleus rotundus (Hall and Ebner 1970*b*; Butler and Northcutt 1971 and in mammals, in the lateral posterior nucleus (known as the pulvinar in primates) (e.g. Hall and Ebner 1970*b*). The reptilian nucleus rotundus in turn projects directly to a rostrolateral region of the anterior dorsal ventricular ridge (Hall 1972; Pritz 1975; Lohman and Van Woerden-Verkley 1978), while the mammalian lateral posterior nucleus projects to an area of neocortex adjoining the striate cortex, the peristriate cortex.

There are, then, two major routes in both reptiles and mammals from the retina to the telencephalon: one, tectofugal, from the retina via the tectum (superior colliculus) and the thalamus (nucleus rotundus or lateral posterior nucleus) to either dorsal ventricular ridge or peristriate cortex; the other, thalamofugal, from the retina via the dorsal lateral geniculate nucleus of the thalamus to visual cortex. The projection of the thalamofugal system to cortex in reptiles must be qualified by noting the fact that in lizards and snakes the dorsal lateral geniculate projects, not to dorsal cortex, but to a region in the anterior DVR close to but distinct from the area to which the nucleus rotundus projects (Lohman and Van Woerden-Verkley 1978). If the dorsal ventricular ridge is considered a pallial structure, then the parallel with mammalian organization is very close; if, on the other hand, it is regarded as subpallial, then one contrast between the classes is that, whereas the lateral posterior nucleus projects to cortex, the nucleus rotundus projects to subpallium and another that, with the exception of turtles, the dorsal lateral geniculate of reptiles also projects to a subpallial site.

There are, in view of the wealth of anatomical studies, surprisingly few physiological investigations of the reptilian visual system; moreover, all of those investigations have used chelonians. Robbins (1972) has shown, by single-unit recordings, that tectal cells in *Pseudemys scripta* code colour information and reports preliminary data suggesting that the receptive fields of these cells are organized into spectrally opponent areas consisting of circular centres and annular surrounds. As one would expect from the anatomical data, tectal lesions do not cause total blindness in reptiles. Optokinetic responses can still be obtained after large bilateral tectal lesions (e.g. Bass 1977; Hertzler 1972) although a decline in the rate of responding has been observed in some studies (e.g. Hayes and Hertzler 1967). Similarly, using a 'visual cliff' apparatus (Walk and Gibson 1961) in which turtles had to choose onto which of two sides to descend from a narrow runway, where the (checkered) floor of one (shallow) side was 3.5 in (8.5 cm) below the runway, the other, similarly patterned floor on the deep side being 18.5 in (44.5 cm) down, Hertzler (1972) found that, although choice latencies were increased by tectal lesions, the preference for the shallow side was not affected.

However, tectal lesions do cause severe visual disturbances. Bass, Pritz, and Northcutt (1973) have shown that bilateral lesions in the side-necked turtle (*Podocnemis unifilis*) abolished the ability to relearn a preoperatively acquired pattern discrimination. Bass *et al.* also report other visual peculiarities following tectal lesions: the turtles frequently showed misdirected head and neck movements when feeding, and failed to respond to threat in any part of the visual field. Bass (1977) further reports that turtles with tectal lesions cannot directly approach a salient food-containing stimulus, but instead move along the side of their tanks and appear to be guided entirely by tactile and other non-visual cues; the absence of orientation to visual cues following tectal lesions is similar to that found in hamsters following collicular lesions (Schneider 1969), and reminiscent of the deficits reported in responses to visual stimuli in frogs with tectal lesions.

At the thalamic level, there is evidence that light flashes obtain single-unit responses in both the lateral geniculate complex and the nucleus rotundus, that these responses are specific to photic stimuli, and that the rotundus responses are abolished by tectal ablation (Belekhova and Kosareva 1971). There is also a report (Reiner and Powers 1978) on the effect of rotundal lesions, showing that such lesions severely impair retention and reacquisition of both intensity and pattern discriminations by painted turtles; the precise nature of their deficit requires further investigation: Reiner and Powers report that, following lesions, their animals were initially lethargic and, interestingly, did not bite accurately at food from the magazine. A subsequent report (Reiner and Powers 1980) suggests that more extensive lesions, involving both the nucleus rotundus and the lateral geniculate nucleus, have a more severe effect than lesions of the rotundus alone; this report found that two turtles were wholly unable to relearn a preoperatively acquired brightness-discrimination following such extensive lesions.

There is little information on the role played by the reptilian telencephalon in vision. Studies using gross electrodes have reported visually evoked responses in the dorsal cortex of turtles (Orrego 1961; Belekhova and Kosareva 1971); on the other hand, Gusel'nikov, Morenkov, and Pivovarov (1971) found that visually responsive single units in turtle dorsal cortex showed relatively long latencies and rapid habituation and so suggested that the cells did not receive a direct thalamic projection from the lateral geniculate. Bass *et al.* (1973) failed to find any effect of either dorsal cortex lesions or combined lesions of dorsal cortex and dorsal ventricular ridge, on post-operative retention of a pattern discrimination—the same discrimination that was abolished by tectal lesions. Hertzler and Hayes (1967) found that, following dorsal cortex lesions, visual cliff choices were unaffected although the choices were made more slowly. The same pattern of results, it will be recalled, follows tectal lesions (Hertzler 1972): combined cortical and tectal lesions did disrupt choice performance, so that the subjects no longer discriminated between the two sides, although now their speed of choice was unaffected. Reiner and Powers (1980) report that basal telencephalic damage involving the lateral forebrain bundle severely retarded reacquisition of visual discriminations in two turtles (one performing a brightness, the other, a pattern, discrimination); these lesions would be expected to disrupt both the tectofugal and the thalamofugal routes to the telencephalon and provide some evidence that one or both of the telencephalic destinations of these pathways does indeed play an important role in visual perception.

There are, then, in reptiles as in mammals two major pathways from the retina to the telencephalon. There is evidence that damage to the tectofugal pathway at either the tectal or rotundal level has a profound effect upon vision, and indications that at least part of the deficit concerns orientation to visual stimuli; there is, however, very little evidence at present for involvement of the thalamofugal pathway in visual performance.

Hearing

Although virtually nothing is known of the auditory capacities of reptiles, recent neuroanatomical studies using reptiles and crocodilians have shown that the basic organization of the reptilian auditory system is not dissimilar from that of mammals.

Primary auditory fibres, whose cell bodies lie in the intraotic ganglion, project to two medullary nuclei, the nucleus angularis and the nucleus magnocellularis, which form part of a medullary nuclear complex known as the auditory tubercle (De Fina and Webster 1974; Foster and Hall 1978). Fibres from the nucleus angularis project to four sites: ipsilaterally, to the superior olive and contralaterally, to the nucleus of the trapezoid body, the nucleus of the lateral lemniscus, and the central nucleus of the torus semicircularis (Foster and Hall 1978). This latter nucleus in turn projects to a region of dorsal thalamus known in lizards as the nucleus medialis, and in crocodilians, as the nucleus reuniens. Finally, thalamic auditory projections ascend to the telencephalon, terminating principally in a medial region of the dorsal ventricular ridge, as well as in a more central area of the DVR and a more ventral striatal region (Foster and Hall 1978; Pritz 1974).

There are few physiological studies to support the anatomical findings, but single-unit investigations have confirmed the existence of cells sensitive to auditory stimuli in both the torus semicircularis (Manley 1971, who found a degree of tonotopic organization within the torus) and the DVR (Weisbach & Schwartzkopff, 1967).

The organization revealed in these reports appears broadly similar in lizards (Foster and Hall 1978) and crococilians (Pritz 1974); we saw in the case of amphibians a similarity between their auditory organization up to the thalamic level with that of mammals, and the same comparability exists in reptiles also. However, in reptiles the comparison may be taken a step further, as the telencephalic projections of the thalamic auditory nucleus are better known; these too show a parallel. The mammalian auditory thalamic nucleus, the medial geniculate body, projects both to auditory neocortex and to restricted striatal regions (Ebner 1969), and in the same way the reptilian auditory thalamic nucleus projects to two distinct regions, one striatal and the other (the dorsal ventricular ridge) which may be pallial in origin.

Foster and Hall (1978) draw attention to parallels between reptiles and mammals but also enter a caveat that is worth quoting here: 'in spite of these overall similarities, the iguana clearly lacks many of the nuclear subdivisions along the auditory pathway that are found in groups, such as birds and mammals, which apparently possess more sophisticated auditory capacities' (Foster and Hall 1978, p. 827).

Skin senses

Very little is known of the processing of somatic sensory input by the reptilian forebrain. There are electrophysiological studies, using the evoked potential method, which show that somatic input reaches the pallial regions: Kruger and Berkowitz (1960), for example, report responses from widespread regions of both dorsal and hippocampal (but not piriform) cortex to stimulation of the contralateral sciatic nerve in *Alligator mississippiensis,* and Orrego (1961) reports evoked responses in a rather more restricted region of dorsal cortex (anterior to the region from which he obtained visual responses) of *Pseudemys scripta* to stimulation of ascending spinal cord pathways at the cervical level of the cord. Moore and Tschirgi (1962) recorded both single-unit responses and gross evoked potentials from the general cortex of *Alligator mississippiensis*, finding no evidence of modality specificity (using somatic, auditory, and visual stimuli) in either the single unit or evoked potential data. The responses they obtained had relatively long latencies, and showed marked lability, features which, along with the lack of modality-specificity, argue that the responses were not the product of a direct thalamic sensory projection system to the cortex. Moore and Tschirgi (1962) conclude: 'reptilian general cortex does not exhibit those properties commonly associated with the primary sensory cortical areas of the mammalian brain' (Moore & Tschirgi, 1962, p. 206).

There are data also to show that cells in the thalamus process somatic input: Belekhova and Kosareva (1971) report single unit responses in turtles to skin stimulation in cells located in a ventral zone of the thalamus composed of a number of heterogeneous structures lying between the nucleus rotundus and the hypothalamus. The great majority of the units detected were modality-specific, in that they showed no response to visual stimulation.

The Belekhova and Kosarova (1971) report takes on particular significance in the light of anatomical studies, using caimans, which indicate that the nucleus medialis, which occupies a position medial and somewhat caudal to the nucleus rotundus, receives direct spinal projections and in turn projects to the central part of the anterior DVR (Pritz and Northcutt 1980). If physiological studies confirm the existence of somatosensory cells in the dorsal ventricular ridge, then the parallels between the functions of that structure and the mammalian neocortex will be further strengthened.

Our information is, then, extremely sparse. Somatic input reaches cortical regions, but it seems likely that it does so via polysynaptic pathways; there is probably also a direct somatic projection from the thalamus to the DVR.

Motor control

It appears that the reptilian hemispheres play very little direct role in the control of movement. Extensive hemispheric ablation, although it may reduce the amount of spontaneous movement 'does not lead to inability to perform any specific movement seen in the normal animal under ordinary laboratory conditions' (Goldby and Gamble 1957, pp. 408-9). Similarly, direct electrical stimulation of pallial and striatal areas, although it may elicit a variety of responses, such as alerting, flight, turning movements or even retching (Schapiro 1968), does not elicit 'forced' movements, or show any indication of topographical representation of the musculature within the hemispheres (Goldby and Gamble 1957; Distel 1978; Sugerman and Demski 1978).

These physiological findings are reinforced by anatomical studies, which have so far failed to show any telencephalic efferents which run directly to the spinal cord (and which might, therefore, be considered comparable to the corticospinal tract of mammals) (e.g. Lohman and Van Woerden-Verkley 1976; Ulinski 1978; Cruce 1975). It should be added that there are, in non-telencephalic areas of the brain, striking parallels between reptiles and mammals as regards sites giving rise to fibres that descend to the spinal cord (Ten Donkelaar and De Boer-Van Huizen 1978).

This negative conclusion parallels that reached for amphibians; while it indicates the absence of a development found in mammals, it does have the implication, of course, that there may be that much more forebrain tissue available for cognitive functioning in the forebrain of cold-blooded vertebrates.

Conclusions

Although many questions remain to be answered, the data from those senses for which a reasonable amount of information is available (the senses of smell, vision, and hearing) all agree in showing marked similarities in reptiles and mammals, one major remaining contrast being the apparent absence of any close pallial involvement in motor control in reptiles. It is clear that, as in mammals, the telencephalic representations of the various senses are separate and not, as some authors (e.g. Moore and Tschirgi 1962) have supposed, overlaid in an amalgam such that the reptilian 'higher centres' could not distinguish between stimuli from different modalities. Moore and Tschirgi, for example, concluded that 'the stereotyped behavior of these creatures in response to so many diverse stimuli may in fact be the outward expression of the stereotyped input which their higher centres receive' (Moore and Tschirgi 1962, p. 208). The modern studies indicate that no such stereotypy need be anticipated from the telencephalic organization as currently understood. Similarly, the reptilian telencephalon does not appear to be dominated by any one sense, and although forebrain mapping studies are clearly far from complete as yet, there is no reason to suppose that the amount of tissue devoted to sensory analysis (and perhaps to that extent less available for purely cognitive processes) is particularly small (or large) in reptiles.

Habituation in reptiles

The great majority of experiments on learning in reptiles have used chelonians or squamates (excluding amphisbaenians); there have been a few studies of crocodilians, one using the tuatara, but none using amphisbaenians. For a comprehensive review of learning studies in reptiles, with a very different theoretical stance from that adopted here, see Burghardt 1977.

Reports of habituation appear to date to be confined to squamates and chelonians, but within these groups, a considerable variety of response systems have been investigated. A number of studies have concerned responses to stimuli that are potentially dangerous. Smith (1964) reports anecdotal evidence that various British lizards and snakes can be 'tamed': slow worms (*Anguis fragilis*), for example, struggle violently when first caught by man, but within a relatively short time, maybe only a few minutes, may be handled easily. Boycott and Guillery (1962) found that red-eared turtles showed gradually declining latencies to eat food introduced into an initially novel tank. Withdrawal of the head into the carapace by turtles has been shown to habituate when elicited either by a looming shadow (Hayes and Saiff 1967) or by mechanical taps on the carapace (Humphrey 1933; Farris and Breuning 1977); habituation of head withdrawal to a shadow stimulus has also been observed in garter snakes (*Thamnophis radix*) (Fuenzalida, Ulrich, and Ichikawa 1975).

Responses that appear to be investigatory in function have also been studied. Chiszar, Carter, Knight, Simonsen, and Taylor (1976) introduced a variety of snakes (garter snakes, four species of rattlesnakes, *Crotalus*, and Eastern hognose snakes, *Heterodon platyrhinus*) to a novel open field for five minutes a day, for five days, and scored the amount of gross locomotion and the number of tongue-flicks in each minute of the time spent in the apparatus. Marked reductions in both measures were seen within each daily period of confinement. Czaplicki (1975) similarly observed a decline in the rate of tongue-flick responses to cotton swabs emitting odours from prey extracts, the swabs being presented 1 to 2 centimetres from the snout of diamond-backed water snakes (*Natrix rhombifera*); Czaplicki also obtained a corresponding decline in the number of overt attack responses on the swabs. Gubernick and Wright (1979) studied habituation of the dewlap response in the American chameleon (*Anolis carolinensis*) when shown a live house sparrow (*Passer domesticus*) in a transparent cylinder. The dewlap response, which consists of distension of a 'throat fan' located beneath the throat, occurs in a number of social displays and showed a substantial decrement over the course of 40 30-s exposures to the sparrow, the ITIs used varying from 5 to 120 seconds.

As a final example, Hayes, Hertzler, and Hogberg (1968), using red-eared turtles, and Gubernick and Wright (1979), using the American chameleon, studied the optokinetic response (following movements of the head and eyes) elicited by placing their subjects within a rotating drum of vertical black and white stripes, and obtained decrements in that response also.

It appears, then, that habituation is a widespread property of reptilian response systems; there have, however, been few analytical studies and some of the studies quoted above lack even elementary controls. Reported rates of habituation vary, as one might expect, and there is some evidence that procedural details are of importance. Hayes *et al.* (1968), for example, found that, whereas the optokinetic response declined from day to day when the rotating drum exposed 19 one-inch stripes, the same response increased across days when the drum exposed only four stripes of similar width, widely spaced apart. Ireland, Hayes, and Laddin (1969) found a positive correlation within a group of red-eared turtles between the amplitude of the first response (head withdrawal to a looming shadow stimulus) and the number of responses to a criterion of habituation (three successive trials without any response), indicating a systematic effect of individual differences in responsiveness on trials-to-habituation scores. Gubernick and Wright (1979) showed that whereas large chameleons showed habituation of the dewlap response using ITIs of 5, 30, or 120 seconds, smaller (younger) anoles habituated at only the short (5-second) ITI. Such observations underline the fact that comparisons of rates of habituation within reptilian species or between reptilian and non-reptilian species would face formidable difficulties; as yet, no effort to make any such comparisons has been made.

Very few further analytical data of interest are available. Some investigators (Chiszar *et al.* 1976, Experiment 2; Hayes *et al.* 1968, Experiment 2; Hayes and Saiff 1967) find overnight retention of habituation, others (Fuenzalida *et al.* 1975; Hayes *et al.* 1968, Experiment 3; Gubernick and Wright 1979) do not. As one of the former studies (Hayes and Saiff 1967) finds good retention of habituation over a ten-day period, there seems no reason to assume any basic lack of stability in reptilian retention of habituation. There is also evidence from a number of studies for stimulus specificity: Czaplicki (1975) found that water snakes, having habituated attack and tongue-flick responding to minnow extracts, responded readily to goldfish extracts; similarly, both Hayes *et al.* (1968) and Gubernick and Wright (1979) found that the optokinetic response, habituated to rotation in one direction, was readily obtained by rotation in the opposite direction; finally, Hayes and Saiff (1967) showed that, after habituation of the head-withdrawal response to a looming shadow stimulus, approach of a human still elicited the response.

There is evidence that dishabituation may be obtained; the earliest account is by Humphrey (1933), who reported (with few details) that, following habituation of a turtle's head withdrawal response to taps with a hammer, a few taps with a mallet were sufficient to restore the response to the hammer. The best recent evidence is provided by Hayes *et al.* (1968), who showed that following one hour's continuous rotation of a striped drum in one direction, one minute of either rotation in the opposite direction or no movement at all obtained, in the following minute, increases relative to that seen in the final minute of the original one-hour period in the rate of optokinetic response to the restoration of the drum rotation in the original direction.

The reports of dishabituation, along with those indicating stimulus specificity, are sufficient to show that genuine habituation, as opposed to fatigue of either the sensory or motor apparatus, is obtained in reptiles.

There has been only one report in reptiles of sensitization on initial trials preceding eventual habituation, and that report gives no details: Gubernick and Wright (1979) simply state (p. 128) that 'Although not shown here, only the dewlap response showed an initial increment in responding above Trial 1 response levels'. There are two reports, both using turtles, that describe sensitization effects, but in each case, the phenomenon described appears significantly different from that commonly observed in other vertebrates. Hayes and Ireland (1972), for example, studied the optomotor response of turtles placed (unrestrained) at the centre of a rotating striped drum (this response differs from the optokinetic response in that the subjects move their entire bodies, rather than just their head and eyes, in response to movements of the stripes). The optomotor response increased steadily over five minutes of exposure, indicating a process of sensitization; however, no extended periods of rotation were attempted, so that there is no evidence that habituation would have been obtained eventually (and Hayes and Ireland assume in fact that habituation will not be obtained using the optomotor response). The second report, by Hayes *et al.* (1968), which has already been described, while it does appear to show a sensitization of sorts, concerns a time interval—24 hours—very different from that conventionally associated with sensitization in organisms other than reptiles. The dearth of reports of sensitization may be curious, but it currently seems likely that this is due to the lack of details in the reports available, which rarely give trial-by-trial scores for the initial presentations of the eliciting stimulus. We may conclude, therefore, that habituation does occur in reptiles, and that, although the data are insufficient to support any confident statement, the process appears grossly similar to that found in other vertebrate species.

Before leaving this section, we shall consider one further report of a phenomenon which, while demonstrating neither habituation nor sensitization (at least in any conventional sense) does have much in common with those phenomena. Burghardt and Hess (1966) gave three groups of newly hatched snapping turtles (*Chelydra serpentina*) one type of food (horsemeat, fish, or worm, one type used exclusively for each group) for 12 days, and then gave the hatchlings a choice between all three food types. The turtles showed a strong preference for the type of food with which they had become familiar. Burghardt and Hess point out the parallel between this phenomenon and that of imprinting, much studied in newly hatched nidifugous birds (Thorpe 1963). What is of interest here is the demonstration that, as in conventional studies of habituation, the mere repeated presentation of a UCS is sufficient for the establishment of a discrimination between that UCS and others, mediated, presumably, by some learned internal representation of the UCS.

Classical conditioning in reptiles

There have been very few studies of classical conditioning in reptiles, and only one of those (Loop 1976) has used an appetitive CS. Loop's subjects were four Bengal monitor lizards (*Varanus bengalensis*) run in a chamber containing in one wall two translucent response keys both placed somewhat to the right of an automated feeder that dispensed defrosted mouse pups. The lizards were exposed to an autoshaping procedure in which food rewards were preceded by a CS (flickering light behind a response key—the left key for two subjects, the right for the other two) of 15 seconds duration. All four subjects formed the CR of responding to the appropriate response key when lit (and not otherwise). The CRs, which were, of course, without any effect on the arrival of the food rewards, were 'open-mouthed bites or pecks directed at the CS' (Loop 1976, p. 575) and Loop concludes, appropriately enough, that the Bengal monitor 'will direct biting-like responses to a stimulus light which simply predicts the presentation of food' (*idem*, p. 575). It is not possible to assess the efficiency with which the habit was acquired, as the same subjects had previously failed to autoshape in a different apparatus.

Three reports which used aversive UCSs will be discussed here; each used a different type of UCS. Davidson and Richardson (1970), using collared lizards (*Crotaphytus collaris*), paired a 10-s sound and light CS with a 1-s UCS (a train of shocks delivered to the hind-limbs), and recorded changes in respiratory rate, pulse rate, and leg movements. CRs of all three kinds were obtained, and the rate at which these emerged varied with the response measured. Respiratory CRs, for example, reached the criterion level (more than 80 per cent CRs) within 20 pairings, whereas the leg CR did not achieve criterion until 120 pairings had been delivered.

Farris and Breuning (1977) paired light onset with hammer-tap UCSs and established head-withdrawal CRs in red-eared turtles; they went on to explore the effects of habituation to the UCS, presented alone, on the previously established CR, finding that extinction (decrement in CRs over a series of CS alone trials) proceeded much more rapidly following UCS habituation. This same finding has been reported in rats by Rescorla (1973) and provides interesting evidence for the involvement of common mechanisms in the establishment of CRs. Rescorla, for example, argues that the result indicates that the CRs seen are the consequence of an association between internal representations of the CS and of the UCS (a stimulus-stimulus association) rather than of an association involving some representation of the response (a stimulus-response association): whereas habituation of the UCS might be expected to alter an internal representation of the UCS, it should not affect any such representation of the UCR—which is still as readily elicited by other previously effective UCSs, barring only that UCS.

Although the results described above seem generally reliable, it should be pointed out that none of the studies reviewed has controlled satisfactorily for the possibility of either sensitization or pseudoconditioning; Farris and Breuning (1977) did give their subjects random sequences of CSs and UCSs (ten of each) before pairing trials began, but did not run a group that continued to receive random presentations comparable in number to those given to the successfully conditioned subjects. Moreover, none of the studies ran any omission-training control groups, so that there is no direct evidence that the CRs obtained were classical rather than instrumental in origin; the only indirect evidence appears to be Loop's observation of the resemblance between the autoshaped CR and the UCR, and the Farris and Breuning (1977) account of the effects of UCS habituation after conditioning.

The final study (Czaplicki, Porter, and Wilcoxon 1975) used a procedure somewhat different from the designs that we have considered previously. It is well-established that, in certain mammals, aversions to specific foods may be conditioned by following ingestion of those foods by systemic poisons, even if there is a considerable delay between ingestion and the poison administration (see Rozin and Kalat 1971, for a review). Czaplicki *et al.* explored the possibility that similar long-delay associations might be formed in garter snakes (*Thamnophis sirtalis*). Food (pieces of minnows or of worms) was tested for acceptability by holding it for a maximum of 30 seconds about 2 cm from the snout of a snake, and scoring the latency of capture by the snake. A number of tests were carried out using minnows (which had been their exclusive diet for six weeks prior to the experiment), and then pieces of worm were introduced for the first time (and shown to be readily captured). Thirty minutes later, experimental snakes were injected with a poison (lithium chloride), while controls received injections of isotonic saline. Subsequent tests showed that experimental subjects were very reluctant to attack pieces of minnow that had been dipped into a solution of worm extract, while remaining willing to attack untreated minnows rapidly. Control subjects showed rapid responses to both treated and untreated minnows. It is, of course, possible that the experimental subjects were simply avoiding all novel smells as a consequence of having been poisoned: a second experiment found that, following similar treatment to that in the first experiment, experimental snakes would refuse minnows dipped in worm extract, but not minnows dipped in extracts prepared from either salamanders or guppies. Unfortunately, this second experiment is hardly conclusive, as no evidence is provided to show that either the salamander or the guppy extract provided salient olfactory cues – they may, that is, not have been detected, and so not classified as novel. However, this may seem an unlikely possibility and, if so, it may be assumed that specific conditioned aversions may be established in garter snakes. Such an assumption argues again for a parallel between reptilian and mammalian conditioning mechanisms and, incidentally, demonstrates one-trial learning in a reptile, with a 30 minute CS–US interval.

We may conclude, then, that although few data are available some interesting parallels have emerged between reptilian and mammalian studies, suggesting that common mechanisms are involved. On the other hand, controls have generally been less than rigorous and there is, in particular, very little to show that classical rather than instrumental learning has been obtained.

Instrumental conditioning in reptiles

Representatives of all four living orders of reptiles have been successfully trained by instrumental procedures, using a wide range of reinforcers. Some experimenters have reported considerable difficulty in training their animals, but have generally concluded that the problem has lain in providing effective motivation rather than in any deficit in capacity to form associations: Davidson (1966), for example, reports that alligators failed to master a simple discrimination for food reward, despite 18 weeks of continuous food and water deprivation – but, using the same apparatus, demonstrated efficient learning where correct choices enabled the alligators to escape from excessive heat.

Procedures using appetitive reinforcers

The most popular subjects in these studies have been lizards and turtles, and the most common apparatuses, the runway and the T-maze. Burghardt (1977) lists many such studies, in which the reinforcers used have almost invariably been either food reward or (in many lizard studies) access to a warm (about 40 °C) area. Other techniques and other reinforcers have, however, been successfully used: Bitterman (1964) describes an apparatus for use with turtles in which the response is a panel press, and Van Sommers (1963) reports training red-eared turtles to lever press (under water) for a brief period of access to air.

Other species have also been trained. Northcutt and Heath (1973), in what appears to be the only report so far of learning in the tuatara, trained two individuals of that species in a T-maze, against their initial side preference, where the reward was simply access to the home pen. The tuatara's performance was, incidentally, reasonably efficient (they achieved a level of 80 per cent correct choices within 20–25 trials), and closely comparable to that found earlier in the same apparatus using caimans (Northcutt and Heath 1971). Kleinginna (1970) trained indigo snakes (*Drymarchon corais*) to press a response key (of the kind used in pigeon operant chambers) for water rewards, and Gossette and Hombach (1969) obtained a modified lever-press in alligators and crocodiles for food reward.

There is nothing in any of the reports to suggest that, once adequate motivation is established, acquisition is anything but efficient. Rates of acquisition do, of course, vary and there is evidence to show that a substantial amount of that variation may be due to 'contextual variables': Krekorian, Vance, and Richardson (1968), for example, ran desert iguanas (*Dipsosaurus dorsalis*) in a simple maze, the reward being access to a goal box in which the temperature was ten degrees higher than that in the maze, and showed that efficiency, in terms of errors, increased with rises in the temperature at which the iguanas were run.

Procedures using aversive reinforcers

Aversive stimuli have been used successfully to motivate reptilian subjects in escape, active avoidance and punishment tasks. Most studies have used lizards, but some reports on performance of crocodilians and snakes are also available.

Williams has carried out a series of experiments (e.g. Williams 1968) using caimans in a T-maze, the floor of which is permanently electrified throughout each trial, escape to a water-filled pan being available by choice of the correct side of the maze; performance in this task, as we shall see in the section on reversal learning, can be extremely efficient. Davidson (1966) has also trained crocodilians (*Alligator mississippiensis*) to escape in a simple discrimination apparatus (a modified Grice box), the aversive stimulation in this case being provided by heat lamps, which were sufficiently intense to raise the floor temperature to 105 °F. Two final examples of escape learning using other aversive stimuli will be briefly described: Kemp (1969) trained desert iguanas to press a black disc in one wall of a chamber to lower the temperature in the chamber, which otherwise constantly grew hotter. The lizards performed in such a way as to maintain a relatively narrow band of temperature in the chamber, initiating responding as soon as their body temperatures reached 42-43 °C, and stopping responding when their temperatures fell to about 36-39 °C. Crawford and Holmes (1966) trained rat snakes (*Elaphe obsoleta*) to move from one chamber of a two-chamber box to the other in order to escape floor vibrations; learning was demonstrated by a progressive decline in response latencies across the series of trials.

A number of authors have reported difficulty in establishing active avoidance, as opposed to escape, learning in reptiles (e.g. Powell and Mantor 1969; Crawford and Holmes 1966; Bicknell and Richardson 1973); there are, however, positive instances and there is evidence that procedural variables—including in particular the nature of the aversive UCS—affect performance (e.g. Williams 1967; Yori 1978).

One technique that has obtained efficient avoidance performance using a shock UCS is that described by Granda, Matsumiya, and Stirling (1965), whose subjects were red-eared turtles. In their study, the onset of the CS (a small patch of light some 7 cm from the turtles's left eye) preceded the shock onset by 5 seconds and any CR (head withdrawal) in that interval avoided the shock. Their two subjects achieved criterion (85 per cent avoidance responses over three consecutive 20-trial sessions) within 150-200 trials, and were very slow to extinguish the response. Granda *et al.* also showed that the latency of the avoidance response varied systematically with the brightness of the CS—further evidence of the importance of contextual variables in avoidance tasks. Yori (1978), using the tegu lizard *Tupinambis teguixen*, demonstrated avoidance learning in a two-compartment box in which the UCS was a rapid rise in the start-box temperature (induced by a battery of six infrared heat lamps over the chamber) and the required response, running from the start chamber to the goal chamber; Yori showed that, although avoidance responding was obtained with a 15-s CS–UCS interval, responding was considerably more efficient with a 30-s interval. This in turn suggests that one of the problems facing reptiles in these tasks may be concerned with response initiation—we saw a clear demonstration of a comparable effect in amphibians (Boice 1970). Bicknell and Richardson (1973) have suggested that characteristics of the 'safe' area may be important in avoidance learning, a suggestion that arose from their failure to obtain consistent avoidance performance within 250 trials in either collared lizards or desert iguanas using an apparatus which required subjects to run from a white to a black compartment to avoid a shock UCS. Efficient performance was, however, obtained in collared lizards when the apparatus was modified so that the goal-box was fitted with a tunnel-like structure, although the iguanas still performed poorly. The authors conclude: 'It may be that the tunnel represents a more "natural" escape route for collared lizards, which live among rocks and crevices but not for desert iguanas, which live among sand and bushes' (Bicknell and Richardson 1973, p. 1062). However, a subsequent experiment (Richardson and Julian 1974) using a similar apparatus with some procedural changes (including a reduction in the shock intensity from 20–100 mA to 0.25–1 mA) produced good avoidance acquisition in desert iguanas (80–220 trials to reach a criterion of 80 per cent correct over three consecutive ten-trial sessions) and elicited a quite different rationale: Richardson and Julian state that 'The desert iguanas used in the present study typically escape into burrows' and conclude 'It might be reasonable to speculate that the dark pseudo-tunnel used in the present study acted as a releaser for the familiar natural burrow escape response of these lizards' (Richardson and Julian 1974, p. 39). We may conclude that procedural variables are probably of major importance in reptilian avoidance learning, but that appeals to 'natural' modes of escape may yield somewhat imprecise predictions.

In conclusion, although there are clearly problems in establishing active avoidance learning in reptiles, there is no reason to suppose that these reflect defects in intelligence, and ample evidence that contextual variables are of crucial importance.

There do not appear to be any comparable problems using passive avoidance, or punishment procedures; two reports (Andrews 1915, Boycott and Guillery 1962) show that turtles (of various species) can be trained not to eat in the presence of specific auditory, olfactory, or visual discriminative stimuli, and two reports, one (Mrosovsky 1964) using turtles (*Pseudemys ornata*) and the other (Vance, Richardson, and Goodrich 1965), collared lizards, that initial brightness preferences can be reversed by punishing choice of the preferred stimulus. Rates of acquisition in these tasks are of restricted interest, as they are clearly dependent to an indefinite extent on such factors as the discriminability of the stimuli, and the strength of the initial preferences. A final report (Burghardt 1969) demonstrates one-trial learning in 12-day old garter snakes (*Thamnophis sirtalis*) which, after one experience of attacking a slug given them in their home cages refused, 24 hours later, to attack a second slug although continuing to attack worms. It is not, incidentally, clear why these snakes so dislike slugs, which are in fact readily eaten by other snakes of the same genus.

Conclusions

This section has reviewed studies using instrumental procedures in reptiles, and has generally found, as in the review of studies using classical procedures, that learning can be demonstrated using these techniques, that it is generally efficient, and that variations in efficiency may reasonably be ascribed to contextual factors. However, our interest in classical and instrumental techniques is not limited to whether or not efficient learning is obtained, but also—and more importantly—to whether the studies demonstrate the formation in reptiles of stimulus-stimulus (classical) and response-stimulus (instrumental) associations. Unfortunately, none of the studies reviewed in either section can be taken as conclusive evidence for the formation of one, as opposed to the other, type of association: no study has compared classical with omission training, and there is no very compelling internal evidence of any other kind in the reports surveyed relevant to the classical/instrumental dichotomy. On the other hand there is, of course, no evidence that either type of association is *not* formed, and no reason to suppose that reptiles differ from other vertebrate classes in the formation of these associations.

Complex learning in reptiles

This section will be devoted almost exclusively to the successive proposals made by Bitterman concerning possible qualitative differences between reptiles and other vertebrate classes. Most of the tasks to which reference will be made were introduced in the corresponding section of Chapter 3, and familiarity with the nature of the tasks will be assumed here; similarly, the rival theoretical accounts of Bitterman and Mackintosh will not be repeated in this section, and the implications of the results for their views will not be spelled out in detail. As our starting-point we take Bitterman's classification (see Table 3.1, p. 70) of vertebrates as either 'fish-like' or 'rat-like' in performance on serial reversal and probability tasks. We shall then review evidence on Bitterman's more recent proposals concerning the existence of expectations and selective attention in non-mammals. It will be noticed that, although a variety of species are used in the experiments discussed, Bitterman's own studies have used turtles exclusively.

Serial reversal

Bitterman (1965*a*) initially supposed that, although turtles showed improvement on spatial reversal tasks, they did not on visual problems. However, it now appears that this contrast was due to some misleading preliminary results (Bitterman 1965*b*) of a study subsequently reported by Holmes and Bitterman (1966). This latter study explored the performance of painted turtles in either spatial discrimination tasks (where both response keys were yellow, and either the left or the right key was consistently rewarded) or visual tasks (in which discrimination was between either red and green or blue and yellow keys). Two groups of subjects performed the spatial tasks, one group being reversed daily, the other, every four days; both groups showed considerable improvement, both in reducing the number of errors in the early trials of each day (indicating, perhaps, overnight 'forgetting') and in reducing errors over the second 10 trials of each day's 20 trials (indicating an improvement not attributable to forgetting). Three groups were run in colour reversals, one group being reversed after meeting a criterion of 17 out of 20 correct choices in one day, another, after two days training, and the third, every four days. All three groups showed improvement across the series, although this was least marked in the four-day group, which was in fact the condition whose early results were reported in Bitterman (1965*b*). The criterion and two-day reversal groups both showed substantial improvement, and their later reversal scores were well below those in original acquisition; if an increase in overnight forgetting, due to proactive interference, was the only factor concerned in serial reversal improvement, performance would not, of course, improve over that seen in acquisition, and this finding therefore indicates that the improvement seen involved some factor other than increased forgetting.

Two further reports (Gossette and Hombach 1969; Williams 1968) agree in showing substantial improvement across a series of spatial reversals, using crocodilians. Gossette and Hombach trained alligators and crocodiles (*Crocodilus americanus*) in a food-rewarded lever-pressing task, finding superior performance—which they attributed to possible differences in motivational level—in the crocodiles. Williams, who used spectacled caimans in a T-maze escape task, reversed his subjects as soon as they met a criterion (either 6 or 12 successive correct choices), whether this occurred in mid-session or not, and found marked improvements in subjects run to the criterion of 12 correct, but not in the group run to the weaker criterion. However, this latter finding may have been due to a floor effect, as the number of errors in this group was very low throughout, and one member of the group in fact performed all 21 problems (original acquisition and 20 reversals) with a total of 36 errors, making only one error on 13 of the problems. This is an interesting finding which emphasizes the potential efficiency of learning in reptiles, and provides evidence, more convincing than that obtained in goldfish (p. 78), for the adoption of strategies like 'win-stay, lose-shift' by reptiles. It may be added that, although one-trial reversals have been reported in rats (e.g. Theios 1965; Pubols 1957), they are by no means the invariable result of serial reversal training in that species (e.g. North 1950*a*, *b*).

There is, finally, evidence that serial reversal improvement may also be obtained in squamates: Kirkish, Fobes, and Richardson (1979), for example, trained banded geckos (*Coleonyx variegatus*) in a spatial discrimination using a Y-maze and found striking improvement across a series of eight reversals. There were five subjects, and the group mean acquisition score was over 40 errors; the later reversals showed mean error scores as low as 3, strongly suggesting (although individual data were not provided) that at least some one-trial reversals were obtained.

The evidence is, then, that reptiles, like teleosts and mammals, do show improvement across a series of reversals, and that the improvement is not due solely to the occurrence of overnight forgetting. Although comparatively few studies have been carried out, the Williams (1968) and Kirkish *et al.* (1979) reports are sufficient to show that reptilian performance can be extremely efficient.

Probability learning

Bitterman's (1965*a*) classification proposed that turtles, like mammals, maximize on spatial probability tasks, but, like fish, match on visual tasks. It will be recalled from the corresponding section (in Chapter 3) on fish, that Bitterman's early proposals had to be modified in the light of the fact that at least in certain conditions, rats and other mammals do not maximize but follow other systematic strategies (e.g. reward-following—choice of stimulus most recently rewarded), some of which give the appearance of matching. We concluded there that random matching—matching, that is, which appears in individuals and does not appear to be the consequence of any systematic strategy—has not yet been obtained in mammals and it is, therefore, only the contrast between systematic strategies—maximizing or otherwise—and random matching which is of theoretical interest here.

The two available reports on probability learning in reptiles agree that, where the discrimination is based on spatial cues, then either maximizing or reward-following, or a mixture of both, is obtained; one of the reports (Kirk and Bitterman 1965) used painted turtles in a food-rewarding key-nosing task, and the other (Williams and Albiniak 1972), caimans in the Williams (1968) shock-escape T-maze.

One report (Kirk and Bitterman 1965) describes the outcome of probability discrimination training where visual (colour) cues are relevant. These authors found that their turtles matched reward ratios of both 70:30 and 50:50, and that their performance did indicate that random matching had been obtained. The authors found 'no evidence of reward-following or of any relation at all between choice on any trial and the events of the immediately preceding trial' (Kirk and Bitterman 1965, p. 1485). We are, unfortunately, not given any details of the ways in which sequential dependencies were assessed, nor are we given any analysis of the type of errors that occurred; it would, for example, be of interest to know whether errors showed, as Mackintosh, Lord, and Little (1971) might expect, a tendency to occur on a preferred side key.

Kirk and Bitterman conclude that 'random probability-matching is a precortical phenomenon which tends to be suppressed by cortical development, more effectively in the spatial modality than in the visual' (Kirk and Bitterman 1965, p. 1484). There are, limiting ourselves to the behavioural data, a number of objections to this proposal. First, given the paucity of data available, it seems premature to conclude either that systematic strategies will rarely, if ever, be found in visual probability tasks using reptiles or that random matching will not occur in spatial tasks, and premature to conclude even that random matching is more likely to occur in visual than in spatial tasks using reptiles, particularly as there has been so far only one report on visual probability discrimination learning, using only one species; Bitterman's own work (with goldfish) shows that while one apparatus may obtain random matching, another, using the same modality, may obtain maximizing (cf. Behrend and Bitterman 1966; Woodard and Bitterman 1973*b*). Second, the distinction between spatial and visual tasks appears to be entirely arbitrary, a response to a pattern of results that appeals to no other theoretical underpinning. It is, for example, not clear how animals solve spatial problems—to what extent, say, visual or somesthetic cues may be involved—but, unless the suggestion is that olfactory cues are particularly relevant to spatial tasks, there seems no reason why emerging cortex should be better equipped to process spatial as opposed to purely visual information.

The performance of turtles on spatial probability learning tasks resembles (superficially, at least) that of mammals; the only available report on visual probability learning tasks claims that turtles show random matching, a phenomenon seen under some conditions in fish but not yet obtained in mammals. This report does not seem in itself to provide sufficient grounds for accepting a qualitative distinction between turtles and mammals, particularly since there is no convincing account available of the nature of that distinction.

Effects of reward shifts

This section will assess evidence relevant to the proposal (Pert and Gonzalez 1974) that reptiles (like fish) do not form 'expectancies'. Such expectancies, or anticipational processes are, according to Bitterman and his colleagues (See Chapter 3) responsible (in mammals) for the inverse relationship between reward size and resistance to extinction, for the spaced-trial PREE, and for the successive NCE, and we shall ask whether these phenomena are indeed absent in reptiles. We should preface this discussion by noting that our earlier survey of classical conditioning is relevant to this issue, since the inability to form expectancies may be taken (p. 86) to reflect an inability to form stimulus-reinforcer (classical) associations: regrettably, that survey was not able to provide convincing evidence that reptiles can (or can not) form such associations.

There have been two reports relevant to the first phenomenon listed above: Pert and Bitterman (1970) ran two groups of painted turtles in a white runway for a reward of either one or five pieces of fish, finding that the large reward group ran faster in acquisition and in extinction (although this latter finding was only marginally significant). Pert and Gonzalez (1974) found a clear superiority in resistance to extinction in turtles that had received large rewards in a key-nosing apparatus. To date, then, there are no reports of an inverse relationship between reward size and extinction in reptiles, whose performance therefore appears closer to that typically reported in fish than to that of mammals.

Two reports (Murillo, Diercks, and Capaldi 1961; Wise and Gallagher 1964) agree that a PREE is obtained in red-eared turtles, using a brief (30-s) ITI in a food-rewarded task. The question of major theoretical interest, however, is whether a spaced (24-h ITI) PREE is obtained, and it is, therefore, unfortunate that the available data are inconclusive. Two reports (Eskin and Bitterman 1961, using 'sliders'—immature *Pseudemys scripta*; Graf 1972, using desert iguanas) found no difference in rates of extinction following either continuous or partial reinforcement in runways; one report (Pert and Bitterman 1970, using painted turtles) found an 'inverse' PREE—faster extinction, that is, following partial as opposed to continuous reinforcement, and one report (Gonzalez and Bitterman 1962), again with *Chrysemys picta*, found a conventional PREE. Although it would clearly be preferable that many more data were available, the most parsimonious account at present would seem to be that reptiles—or, at least, turtles—may show a conventional spaced-trial PREE, but that the range of conditions in which this is found may be somewhat restricted. Pert and Bitterman (1970) in fact claim that Gonzalez and Bitterman (1962) did *not* obtain a spaced-trial PREE, but the claim seems odd in the light of the conclusions reached in the original report, where Gonzalez and Bitterman (p. 111) write that 'while the Partial Group ran more slowly at the outset [of extinction], the Consistent Group ran more slowly towards the end. This new divergence of the curves reflects a differential resistance to extinction, greater in the Partial Group than in the Consistent, which began to be manifested only in the last 10 days of extinction'. It should be added that one reason for the slow emergence of the superiority of the Partial Group is that they ran more slowly in the acquisition phase and so, at the start of extinction.

The last of the three phenomena attributed by Bitterman to the operation of anticipational processes is the successive NCE. Two reports (Pert and Bitterman 1970, using spaced trials; Pert and Gonazlez 1974, using massed trials) agree that when painted turtles are shifted from a large to a small reward, their response levels (speed of running a runway or rate of nosing a key) gradually decline to those observed in groups maintained on small reward throughout, but not below those levels: there is, in other words, no sign of a successive NCE in either experiment. Performance in this task appears, as was the case with the relation between reward size and extinction, to parallel that usually found with goldfish: running speeds decline, but do not fall below those of unshifted controls.

Over the three tasks, only one result—the spaced-trial PREE observed by Gonzalez and Bitterman (1962)—directly contradicts the proposal that there may be no expectancy mechanism in reptiles; the exception does, however, seem serious, and raises the question whether an account based on Mackintosh's view that all three phenomena are in fact mediated by after-effects might be more plausible. It might, then, be proposed that reptiles are less dependent on after-effects than are mammals: two difficulties face this notion. First, the imprecision of this account would allow it to accommodate almost any pattern of results—thus, while it is not overthrown by the demonstration that, perhaps only rarely, a conventional PREE is obtained, it clearly does not predict under what circumstances such an effect should be obtained. Second, this difficulty is not alleviated, as it was in the case of fish, by direct evidence that after-effects are not easily conditioned in reptiles; there have, for example, been no studies of single alternation in reptiles.

Finally, we shall consider reports of two further types of contrast effect in reptiles, the simultaneous NCE and a phenomenon not previously discussed, positive behavioural contrast.

Pert and Gonzalez (1974) describe an experiment which showed that, where turtles formed a discrimination between a stimulus associated with low reward and a stimulus associated with high reward (each stimulus being separately presented on 10 of each day's 20 trials) the rate of response to the negative stimulus was depressed below that maintained by subjects for whom both stimuli were associated with low reward: in other words, a simultaneous NCE was obtained It will be recalled from Chapter 3 that it was argued there that the occurrence of a simultaneous NCE in fact provides good evidence of 'expectations' in animals, and that alternative accounts put forward by Bitterman and Gonzalez did not seem convincing.

Positive behavioural contrast is a phenomenon that occurs in tasks in which a discrimination is formed between two separately presented stimuli, one of which (the positive stimulus) is associated with reward, and the other of which (the negative stimulus) is associated with extinction. It has been found, with both avian and mammalian subjects that, after the formation of such a discrimination, rates of response to the positive stimulus are higher than rates obtained to two stimuli each of which is associated with the identical reward. In other words, the introduction of the discrimination potentiates responding to a stimulus whose association with reward is unchanged. Pert and Gonzalez (1974) have demonstrated a similar phenomenon in turtles and it is, therefore, reasonable to conclude that whatever mechanism is responsible for the effect in birds and mammals is also present in reptiles. Unfortunately, there is as yet no unanimous opinion on the causation of behavioural contrast, and until a confident account is available, no very safe inferences concerning mechanisms in turtles can be made. What is, of course, clear is, what we have often seen before, that if further research with, say, birds or mammals does resolve the question, there will be important implications for accounts of mechanisms of intelligence in species of other vertebrate classes. One interpretation of behavioural contrast, for example, supposes that the phenomenon is a consequence of classically conditioned (autoshaped) 'extra' responses being added to the current rate of response as a result of the increased signal-value of the positive stimulus (see, for example, Gamzu and Schwartz 1973). If this 'autoshaping' theory is correct, then we might well infer that classical conditioning does indeed occur in reptiles. An alternative theory (e.g. Scull and Macphail 1976) supposes that the introduction of non-reward results in the occurrence of frustration, that frustration is conditioned to background cues, common to both positive and negative trials, and that conditioned frustration energises responding to the positive stimulus. If this view is eventually vindicated, we might conclude that frustration occurs in reptiles and this in turn would provide further evidence for the formation of 'expectancies' in reptiles since, according to standard accounts of the theory (e.g. Amsel 1962), frustration is generated by non-reward only in a context in which reward is anticipated.

Mechanisms of attention in reptiles

There has been no systematic series of experiments on attention in reptiles, and only one study requires some comment here. Graf and Tighe (1971) trained painted turtles in a food-rewarded Y-maze having translucent end walls which could show stimuli that varied in hue and brightness. Groups of turtles were trained with hue (or brightness) relevant, and brightness (or hue) irrelevant, and then shifted to a condition in which either the same dimension (hue or brightness) was relevant, but the values of the stimuli had been reversed (the reversal shift condition) or the alternative dimension was now relevant, the previously relevant condition now being irrelevant (the extradimensional shift condition). The results showed that extradimensional shift performance enjoyed an overall superiority to reversal shift performance and this, along with some other complex additional internal evidence, suggested that the turtles had solved individual stimulus-pairs independently, rather than 'switching in' the analyser relevant to all the stimulus-pairs encountered in a given stage of the experiment. This interpretation might in turn suggest that turtles may—as Bitterman has suggested is the case in fish—have no mechanisms of selective attention. However, precisely the same outcome is obtained when this particular experimental design is used with rats (Tighe and Frey 1972) or four-year-old (but not ten-year-old) children (Tighe and Tighe 1967), so that the result does not suggest any major distinction between turtles and any other vertebrate. There is currently no evidence that turtles (or any other reptiles) either do or do not differ from any other non-human vertebrate on a task that has been supposed to demonstrate the operation of selective attention.

Concluding comments

We have been able to find many fewer studies on complex learning in reptiles than was the case in the corresponding survey in fish, and so can feel even less confidence about forming generalizations based on them. It will be apparent that the data available do not allow comparisons to be drawn between the various orders of reptiles, but it may be of interest to note that there do not appear to be major differences in relative brain size amongst lizards, snakes, chelonians, and crocodilians (Platel 1979). There is considerable overlap between all these groups, the largest relative brain size being found amongst the lizards, and the smallest, amongst the snakes (both squamates); none of the reptilian data overlap the comparable avian and mammalian data plotted by Jerison (see Fig. 2.3, p. 28), and Platel's findings do not conflict with the earlier conclusion that, on average, reptilian brains are about ten times smaller than those of birds and mammals of the same weight.

There have, of course, been other studies which have not been reviewed here because, for example, they have not seemed to offer data amenable to convincing theoretical analyses. But it is interesting, and perhaps reassuring, that at least some of these studies have concluded that there may be very little difference between the reptilian and mammalian intellects. Yerkes (1901), for example, studied the performance of a speckled turtle (*Clemmys guttata*) in two multi-choice mazes, with access to a nest as a reward. Yerkes began with what is, perhaps, the conventional attitude towards reptilian intelligence: 'Turtles certainly appear to be very stupid—so much so, indeed, that one would not expect much in the way of intelligent actions' (Yerkes 1901, p. 520). However, Yerkes' turtle did not conform to those expectations: 'It learned with surprising quickness to make the proper turns and to take the shortest path' (*idem*, p. 522). Tinklepaugh (1932) was also surprised when he tested a common wood turtle (*Clemmys insculpta*) in a food-rewarded multi-unit T-maze. The correct sequence of choices in this maze was: R, L, L, R, R, L—the turtle made four errors on Trial 1, three on Trial 2, two each on Trials 3 and 4, and none on the final three trials. Tinklepaugh, like Yerkes, surprised by the results, wrote: 'In my estimation, the learning of the turtle equalled the expected accomplishment of a rat in the same maze under ordinary experimental conditions' (p. 205), and concluded that '. . . the physical sluggishness and awkwardness of the turtle may have earned for him an undeserved reputation for stupidity' (p. 206). Tinklepaugh's conclusion seems just as reasonable now as it did almost 50 years ago.

Forebrain lesions and learning in reptiles

There have been to date very few investigations of reptilian forebrain involvement in learning, and there is currently no evidence (excepting the studies by Reiner and Powers 1980, whose positive results presumably reflect perceptual disturbances) that any forebrain lesion affects learning. Morlock (1972), in what appears to be the only modern study, ablated the dorsal cortex in painted turtles and found no interference with the acquisition of a position discrimination. Goldby and Gamble (1957) similarly report observations of Diebschlag (1938) to the effect that dorsal hemispheric lesions (not histologically confirmed) in lizards did not disrupt food-reinforced visual associations. It is clear that in our present state of almost entire ignorance, nothing useful can be said of the relation of reptilian forebrain structures to intelligent behaviour.

Summary and conclusions

The four extant orders of reptiles are far removed from those stem reptiles that were common ancestors of both mammals and modern reptiles; living reptiles, therefore, cannot be viewed as steps on the line of descent to mammals, a line which diverged from that leading to the extant reptilian groups some 280 million years ago.

The organization of the reptilian brain, which shows no increase in size over that of amphibians and teleosts, differs in important respects from that of the amphibian brain: the reptilian pallium, for example, shows a true cortex and there is also a region, known as the dorsal ventricular ridge, which, although lying ventral to the lateral ventricle, appears to be pallial in origin and has a number of functional parallels with mammalian neocortex. The reptilian telencephalon is in no sense dominated by olfaction, and it is clear that vision, hearing, and probably the somatic senses also gain independent representation in pallial areas (including the DVR). There is good anatomical evidence for the homology between the reptilian striatum and the mammalian basal ganglia, as well as for the parallel between reptilian and mammalian amygdala; this latter parallel is supported also by the observation that amygdalar lesions in both groups lead to tameness. There is no good evidence, however, for pallial involvement in motor control in reptiles.

Investigations of reptilian visual pathways have shown the existence of two independent routes, thalamofugal and tectofugal, from the retina to the telencephalon. These pathways appear to parallel those found in mammals and cast doubt on the notion that visual functions are progressively 'encephalized' in 'higher' vertebrates.

There have been many fewer behavioural studies of reptiles than of teleosts. It is clear that habituation is found in reptiles, and that both classical and instrumental training procedures do obtain association-formation in reptiles. However, the data are inadequate to show convincingly that the characteristics of habituation in reptiles are the same as those of habituation in other groups, or whether the associations formed are genuinely classical or instrumental.

The survey of complex learning suffered from lack of adequate data, and found few phenomena which might possibly reflect qualitative differences between reptiles and mammals. Those phenomena may be summarized rapidly: first, random probability matching has never been reported in mammals, but has been obtained in one study (using visual cues) in turtles; second, an inverse relation between reward size and resistance to extinction in runway tasks has frequently been found in rats but has not yet been obtained in studies using turtles. There is, then, no evidence for any reptile but the turtle in support of the view that there may be qualitative differences between reptiles and mammals, and, even for turtles, very little support indeed. It has been argued above that the absence of random matching in spatial as opposed to visual tasks (if accepted) makes very little theoretical sense and, mainly in Chapter 3, that the extinction (and successive NCE) data can be interpreted as reflecting quantitative differences in after-effects rather than qualitative differences in anticipatory mechanisms.

There is no substantial evidence either for quantitative intellectual differences between reptiles and mammals. It was argued in Chapter 3 that differences in after-effects may well be taken as reflecting the operation of contextual variables; it has also been shown that alligators are capable of one-trial reversals, an outcome which not only emphasizes the potential efficiency of reptilian learning but also provides good support for the notion that reptiles can use strategies that enable choices to be made relatively independent of the reinforcement history of the stimuli concerned.

There are no data available that allow us either to compare the intellectual capacities of different reptilian groups, or to assess the contribution of the various telencephalic regions to those capacities.

6. Birds

Classification of birds

Reptiles (of different groups) were the immediate ancestors of the final two classes of vertebrates to be considered, the birds and the mammals. In both of these classes endothermy, the capacity to maintain body temperature relatively independent of environmental temperature, has evolved and allowed the invasion of ecological niches too cold for other, ectothermic, vertebrates. The most striking innovation in birds, however, is the exploitation of the air, a niche previously unoccupied by vertebrates; although some reptiles were capable of gliding through the air, powered flight was unknown in vertebrates prior to the evolution of birds. Unfortunately, the details of the steps in the series of adaptations required for successful conquest of the air remain obscure, primarily because of the relative dearth of fossils of immediate ancestors of birds.

The earliest fossils which are universally accepted as avian are of *Archaeopteryx*, a bird about the size of a crow, with a long tail and three claws on the 'fingers' of its wings. *Archaeopteryx*, which probably glided rather than flew, lived in the late Jurassic period, about 150 M a ago, and most authorities agree that no plausible fossil ancestor to *Archaeopteryx* is known more recent than approximately 70 M a prior to *Archaeopteryx*. One possible ancestor is known as *Euparkeria*, a group of small bipedal reptiles belonging to the thecodont order of Archosaurs, and *Euparkeria* fossils are dated as being from the early Triassic, about 220 M a ago. The 70 M a gap in the fossil record is generally regarded as a frustrating mystery, although a recent theory argues that in fact very much closer ancestors are to be found amongst dinosaur fossils (Bakker 1975). This case is supported by evidence suggesting that some dinosaurs were in fact endotherms, and that one indisputably dinosaurian fossil, *Deinonychus*, appears so similar to *Archaeopteryx* that some close relative of *Deinonychus* must have been ancestral to *Archaeopteryx*. If this conclusion (which is, to say the least, controversial—see Charig 1976) is accepted, it may be reasonable to go further and adopt Bakker's suggestion that there is now insufficient reason to separate dinosaurs and birds into different classes, and that an acceptable alternative would be to undertake a major reclassification of reptiles, birds, and mammals, in which birds would be classified along with fossil dinosaur species and so be living dinosaurs. Intriguing as these speculations are, we need not go further into them at this point: there is no dispute that birds evolved from archosaurian reptiles, and that their closest living reptilian relatives are the crocodilians, and so, no doubt that to find a common ancestor with mammals, we need to go back to the beginning of the reptilian line, some 300 M a ago.

The Cretacean period (135-66 M a ago) saw a marked radiation of birds, there now being 35 species known from that period (Brodkorb 1971). Brodkorb argues that, as genera that showed marked differences from *Archaeopteryx* are found in the lower Cretaceous, a maximum of 25 M a after *Archaeopteryx* lived, it is unlikely that *Archaeopteryx* is the ancestor of all living birds (and so, not necessarily the ancestor of *any* living bird). By the end of the Cretaceous, representatives of seven modern orders of aquatic birds had appeared, and by that stage at least the aquatic bird fauna was of 'distinctly modern appearance' (Brodkorb 1971, p. 40). It must be presumed that there also existed land birds at that time, but none of these have so far been found fossilized.

The first, Tertiary, period of the Cenozoic era saw further dramatic radiation of avian forms so that, by the end of the Miocene epoch (5.5 M a ago) representatives of all non-passerine as well as most, if not all, passerine families had appeared. However, none of the actual species that existed in the Cenozoic exists today, and it appears that all modern species have evolved within the last one million years.

Modern birds are classified into some 8600 species (making them, by this yardstick, by far the most successful group of terrestrial vertebrates), arranged in families within 28 orders (Storer 1971). Avian taxonomists conventionally list these orders, and the families within the orders, in such a way as to reflect a ranking of the birds from least evolved, and so most primitive, to most evolved (e.g. Storer 1971). Ratites, which are large, flightless birds like the ostrich (*Struthio*), the emu (*Dromiceius*), and the rhea (*Rhea*) are generally placed low in the list (although it is now known that these birds, like all modern flightless birds, evolved from ancestors that did fly), and the passerines, or song-birds, at the apex (see Fig. 6.1). It is not easy to spell out the criteria for the ordering of these lists, although it is clear that complexity of anatomical organization is of prime importance: one of the reasons for which song-birds are at the apex is that the muscles of the syrinx – the organ used to produce bird song – are most highly developed in passerines. Whether a ranking of this sort should be expected to correlate with a ranking according to intelligence is doubtful, particularly since, as Romer and Parsons point out, 'several orders of water birds and oceanic types are customarily placed at the beginning of the series, although there is little evidence that they are actually primitive' (Romer and Parsons 1977, p. 76). The fact remains, of course, that passerines are popularly believed to be the 'highest' birds, and the most intelligent; this is not only a popular view – taxonomists too appear to subscribe to it, and not only that, but to use (presumed) intelligence as an aid to ordering the list. According to Storer (1971), for example, most European taxonomists place the crow family (Corvidae) and their relatives (including bowerbirds, Ptilonorhynchidae, and birds of paradise, Paradisaeidae) at the top of the list – on the grounds that crows have relatively great mental capacity, that bowerbirds have very complex behaviour, and that birds of paradise have extremely complex plumage and displays. The European ranking does not, then, expect crows to be the most intelligent birds because they are the most evolved – on the contrary, the crows are regarded as the most evolved precisely because they

Fig. 6.1. Representative members of the major orders of birds (excluding ratites). The birds are arranged (from left to right, and from top to bottom) in ascending order of (hypothetical) degree of evolution. (From Romer and Parsons 1977.)

are (presumed to be) the most intelligent. Virtually all American authorities, on the other hand, place the 'nine-primaried' passerines (e.g. New World blackbirds, of the Icteridae family; New World sparrows – Emberizidae; New World warblers – Parulidae) at the top on the grounds that 'they are believed to be undergoing a rapid and extensive adaptive radiation' (Storer 1971, p. 7). We may conclude from this discussion that the methods of ranking used appear in fact to have little direct relevance to the intellectual capacities of the various groups of birds.

We shall have cause to refer to these lists again, when contrasting the abilities of various avian species. It may be helpful at this stage, however, to add two facts of general relevance: first, pigeons, the birds that have been used in far more behavioural experiments than any other avian species, belong to an order (Columbiformes) that is ranked about two-thirds of the way up the list; and second, passerines are by far the most successful group of birds, about 60 per cent of the existing species belonging to the group.

Structure of avian forebrain

Olfactory bulbs

There is considerable variation in size and shape of the telencephalon of the various orders of birds, but the same major divisions are found in all birds, and those divisions show many affinities with those seen in reptiles. The olfactory bulbs, which are in some birds (e.g. the house sparrow) fused into a single structure, are invariably closely attached to the rostral poles of the hemisphere; there are no accessory bulbs in birds. Although the olfactory bulbs are generally rather small, there is a wide variation in their size, from less than 10 per cent of the size of the hemispheres (in passerines) to more than 30 per cent (in the kiwi, *Apteryx*); Pearson (1972) shows that, in general, tree-living species have relatively smaller bulbs than species living on water, and suggests that the size of the bulb is related, as one would expect, to the importance of smell. Figure 6.2 shows the external appearance of a number of avian brains.

Immediately adjacent to the bulb, in the rostral ventrolateral wall of the hemisphere, is a small cortex prepiriformis which is caudally continuous with the anterior olfactory nucleus. Each of these structures is small in relation to the overall size of the hemisphere, reflecting, presumably, the restricted role of olfaction in birds.

Neostriatum

The five major constituents of the avian telencephalon are structures which, in the terminology most widely used today (e.g. Karten and Hodos 1967) are labelled as 'striatum', and these are the paleostriatum, the archistriatum, the neostriatum, the ectostriatum and the hyperstriatum; Fig. 6.3, which shows cross-sections of the pigeon telencephalon, indicates the location of these and other structures. Although most early workers believed that at least the first four of these regions

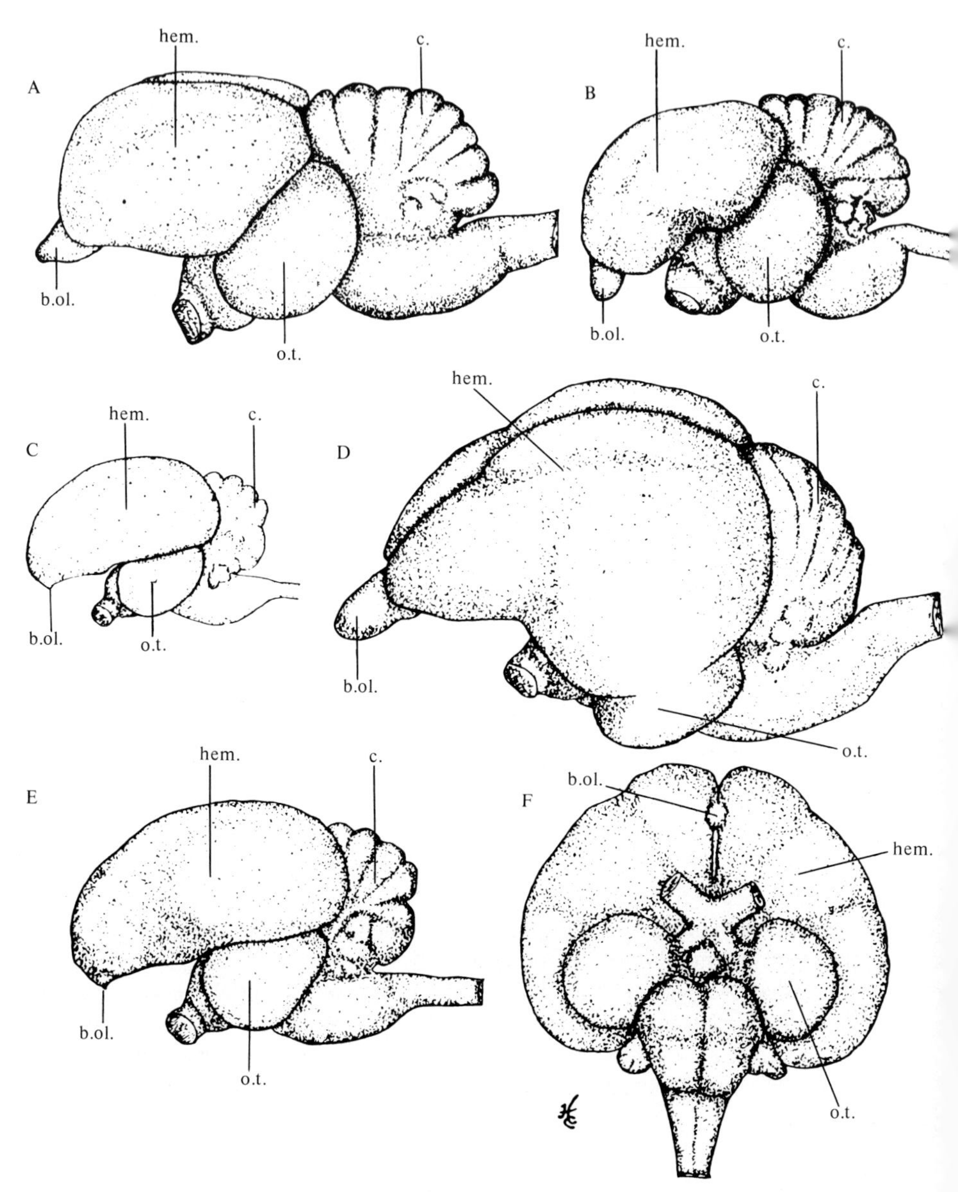

Fig. 6.2. (A–D) Dorsolateral views of avian brains. (A) chicken; (B) pigeon; (C) house sparrow; (D) duck. (E) Lateral view of sparrow brain. (F) Ventral view of sparrow brain. (From Kappers, Huber, and Crosby 1960.)

Key: b.ol. – olfactory bulb; c. – cerebellum; hem. – cerebral hemisphere; o.t. – optic tectum.

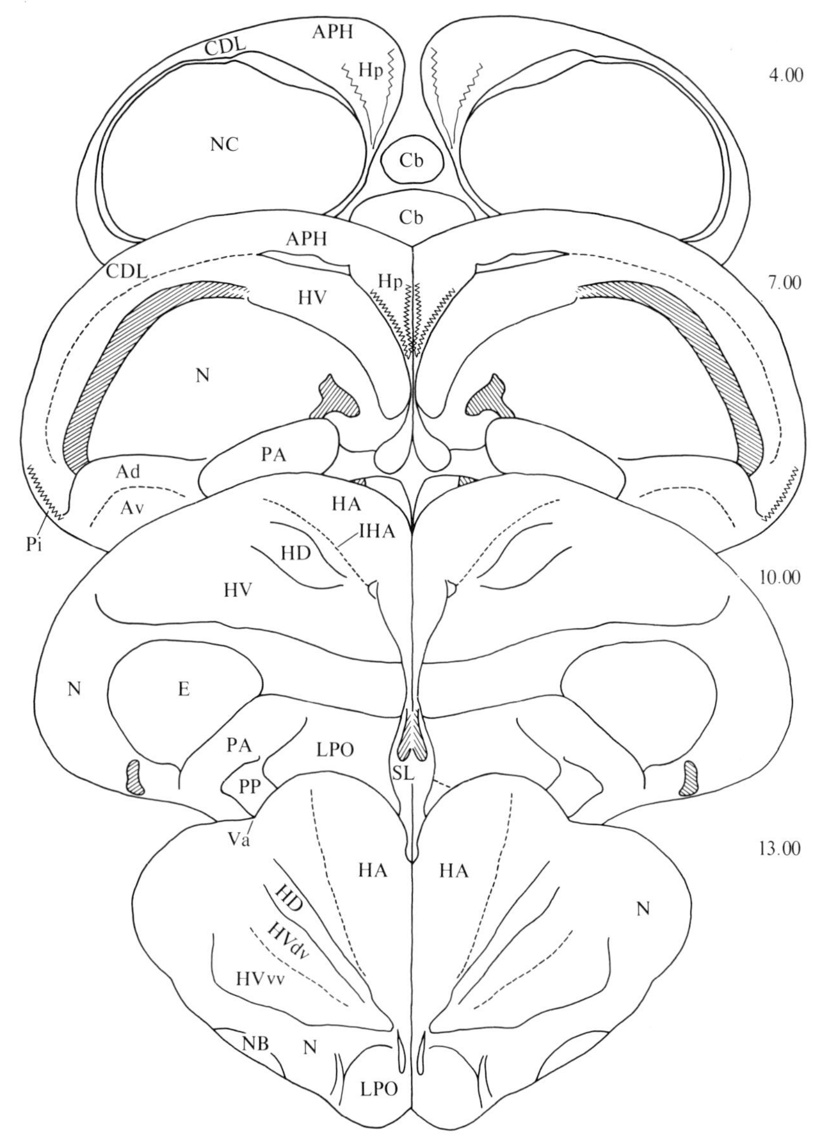

Fig. 6.3. Transverse sections through the pigeon telencephalon; the figures at the side of each drawing indicate the location of the section (in millimetres anterior to the interaural zero, using the vertical plane of the Karten and Hodos atlas). (After Karten and Hodos 1967.)

Key: Ad – archistriatum dorsale; APH – parahippocampal area; Av – archistriatum ventrale; Cb – cerebellum; CDL – area corticoidea dorsolateralis; E – ectostriatum; HA – hyperstriatum accessorium; HD – hyperstriatum dorsale; Hp – hippocampus; HV – hyperstriatum ventrale; HVdv – hyperstriatum ventrale dorso-ventrale; HVvv – Hyperstriatum ventrale ventro-ventrale; IHA – intercalated nucleus of the hyperstriatum accessorium; LPO – lobus parolfactorius; N – neostriatum; NB – nucleus basalis; NC – neostriatum caudale; PA – paleostriatum augmentatum; Pi – cortex piriformis; PP – paleostriatum primitivum; SL – lateral septal nucleus; Va – vallecula.

were striatal, and homologous with parts of the mammalian basal ganglia, more recent embryological studies (Källén 1962; Jones and Levi-Montalcini 1958) have concluded that only the paleostriatum derives from the basal or striatal part of the embryonic telencephalon and that all the other avian 'striatal' structures derive from the dorsal part, and so are in fact of pallial origin. These conclusions are, as we shall see, supported by anatomical and physiological demonstrations of similarities between certain avian 'striatal' structures and mammalian cortical structures so that we shall adopt this more recent view of the proper distinction between pallial and subpallial regions in birds, while retaining the traditional nomenclature.

It can be seen from Fig. 6.3 that the largest single cell mass in the forebrain is the neostriatum which has in fact traditionally been divided into three regions, the neostriatum frontale, intermedium, and caudale. There is evidence that the caudal neostriatum is involved in hearing, but little is known of the functions of the two anterior parts of the neostriatum. The ectostriatum lies laterally and ventrally in the anterior neostriatum, separated from underlying structures by a fibrous layer, the dorsal medullary lamina. Histochemical analysis of enzyme concentrations in the ectostriatum (Baker-Cohen 1968) show a marked similarity between the region and an area within the reptilian dorsal ventricular ridge, and further support for such a parallel will be provided by evidence of a direct projection from the thalamic nucleus rotundus to ectostriatum.

Hyperstriatum

Overlying the neostriatum, and separated from it by the lamina hyperstriatica, is the hyperstriatum ventrale which, along with the ectostriatum and neostriatum, forms the avian equivalent of the reptilian dorsal ventricular ridge. Little is known of the functions of the hyperstriatum ventrale which, like the neostriatum, appears to possess primarily (but not exclusively) 'intrinsic' connections to and from other telencephalic regions.

Another lamina, the lamina frontalis superior, divides the hyperstriatum ventrale from the other two major constituents of the hyperstriatal complex, the hyperstriatum dorsale and hyperstriatum accessorium. The latter two regions, which are divided by the lamina frontalis suprema, form the so-called 'Wulst', an elevation of the hemispheres, more marked in some species than in others, which lies medial to a longitudinal groove, the vallecula. The functions of the Wulst are currently a subject of active speculation and research, much of which concerns the role of a posterior part of the region (the nucleus intercalatus hyperstriati accessorii – IHA) in vision, details of which will be considered in a later section. Karten, Hodos, Nauta, and Revzin (1973) have shown that the efferent connections of the anterior Wulst in the owl (*Speotyto cunicularia*) distribute to areas within the brainstem and spinal cord that show general agreement with the distribution of components of the mammalian pyramidal tract, so that there may be a parallel between anterior Wulst and mammalian motor cortex; they have also shown that the ventromedial Wulst, a region adjacent to the avian hippocampal complex, receives afferents from the dorsomedial anterior thalamic complex.

Although the afferent connections of this thalamic region are not yet known, the anterior thalamic complex of mammals projects to the cingulate gyrus, part of the so-called 'limbic cortex', so that the ventromedial Wulst may be comparable to mammalian limbic cortex (there are, however, no behavioural data relevant to this suggestion yet available).

Källén's (1962) embryological studies led him to suggest that the hyperstriatum accessorium is homologous with reptilian cortex, but that all other hyperstriatal regions are homologues of the reptilian dorsal ventricular ridge. Whether Källén's suggestion is accepted or not, it is clear that two major contrasts have been seen between the reptilian and avian pallial regions: first, there has been an increase in the differentiation of the areas, and second, there is no longer a clear cortex—the hyperstriatum accessorium, although it may be homologous with reptilian cortex, has not a layered but a nuclear structure. There is in fact in birds a thin layer of cells which can usually be detected over most of the dorsal and lateral surface of the hemispheres (terminating ventrolaterally at the small piriform cortex), and this layer is generally regarded as cortical or, more cautiously, as 'corticoid'. According to Pearson (1972), there may be a correlation between the size of the corticoid components and that of the olfactory bulb in the various orders of birds; however, little is known of the connections or function of the corticoid layer, which seems at best a vestigial structure.

Archistriatum

The archistriatum occupies a ventrolateral position in the caudal third of the hemisphere, below the neostriatum. This region has been traditionally seen as homologous with the mammalian amygdalar complex, although it is clear that, given the complex organization of both regions, no confident homologies for the nuclei within the regions may be made. Both Källén (1962) and Jones and Levi-Montalcini (1958) agree that the avian archistriatum is of pallial origin. Källén, however, does not see the avian archistriatum as homologous with the pallial (basolateral) regions of the mammalian amygdala, and so, not homologous with any part of the mammalian amygdala, and Källén in fact holds that part of the avian paleostriatal complex is homologous with striatal (corticomedial) parts of the mammalian amygdala. Other data do, however, support the traditional view to some degree. Parent (1971), for example, has shown that the lateral regions of the pigeon archistriatum are, like the basolateral region of the rat amygdala, rich in cholinesterase, the medial regions in both species being less rich. There is also evidence (Phillips 1964) that archistriatal lesions in birds have a 'taming' effect, similar to that found in both reptiles and mammals following amygdalar lesions; this finding will be discussed in more detail when the effects of archistriatal lesions on learning tasks are reviewed. Zeier and Karten (1971) made a detailed study of the nuclear organization and connections of the pigeon archistriatum, making the following observations. First, the avian archistriatal complex may be divided into four main regions—an archistriatum anterior, intermedium, posterior, and mediale, there being within each region from four to eight discrete nuclei.

Second, the efferent connections of the latter two (posterior and mediale) regions do resemble those of mammalian amygdala, principally in showing a wide distribution in the hypothalamus. Third, the efferent connections of the archistriatum anterior and intermedium (which form about two-thirds of the complex) resemble those of motor cortical regions in mammals, descending in particular to regions of the hindbrain and spinal cord. Zeier and Karten suggest, therefore, that the posteromedial regions of the avian archistriatum be regarded as 'limbic' archistriatum, comparable to the mammalian amygdala, and that the anterior archistriatal regions be seen as somatomotor, comparable to regions within mammalian neocortex.

Hippocampal complex

The final pallial region to be considered is the hippocampal complex, which consists of the medially located hippocampus and an adjoining region, lying dorsolateral to the hippocampus, the parahippocampal area, whose boundaries, medially, with the hippocampus and laterally, with the hyperstriatum accessorium, are not clearly defined. Anatomical studies have shown that the avian hippocampus, like the mammalian hippocampus, projects extensively to the septal area and nucleus of the diagonal band, showing some degree of topographical organization (Krayniak and Siegel 1978*a*). Similarly, both the efferent (Krayniak and Siegel 1978*a*) and afferent (Benowitz and Karten 1976) connections of the parahippocampal area resemble those of the mammalian subicular cortex, part of the hippocampal complex lying between the hippocampus proper and the neocortex. There are, of course, differences between birds and mammals in many aspects of these connections, and the most striking of these is that there does not appear to be a direct pathway from the avian hippocampal complex to the hypothalamus, and so, no homologue of the postcommissural fornix of mammals (Benowitz and Karten 1976; Krayniak and Siegel 1978*a*).

Subpallial structures

The principal subpallial structures are the anterolateral nucleus basalis, the lobus parolfactorius, which merges with the paleostriatal complex, and the medially located septal area. The nucleus basalis is of peculiar interest in that it receives a direct sensory projection from the principal sensory nucleus of the trigeminal ganglion; until relatively recently, it was believed that no (non-olfactory) sensory input in mammals reached the telencephalon without an intervening thalamic relay, and this led Cohen and Karten (1974) to suggest that the nucleus basalis might be regarded, despite its topographical location, as a diencephalic rather than a telencephalic structure. Norgren (1974) has in fact shown that gustatory afferents run direct from a region in the pons of rats to an area, the substantia innominata, of the basal telencephalon, so that sensory projections having no thalamic relays may not be unique to birds. We shall, however, see that there may be in birds more than one such pathway, none of which is gustatory.

The lobus parolfactorius, which receives secondary olfactory fibres from the olfactory bulb (Rieke and Wenzel 1978), projects to the lateral preoptic area and anterior hypothalamus and may therefore be comparable to part of the mammalian limbic system (Karten and Dubbeldam 1973). The adjoining paleostriatal complex contains three structures, the outermost being the paleostriatum augmentatum, which envelops the paleostriatum primitivum within which is the nucleus intrapeduncularis. There is now general agreement that the paleostriatum augmentatum should be compared with the mammalian caudato-putamen complex (Karten and Dubbeldam 1973; Brauth, Ferguson, and Kitt 1978), although Källén (1962) argues that part of the paleostriatum augmentatum is homologous with the striatal (corticomedial) part of the mammalian amygdalar complex. The grounds for the parallel are embryological (Källén 1962), histochemical and hodological. The avian paleostriatum augmentatum, like the mammalian caudato-putamen complex, is rich in cholinesterase and dopamine, and receives its major afferents from overlying pallial regions (Karten and Dubbeldam 1973); the sources of these afferents appear to be the corticoid area and the temporal-parietal-occipital region, both regions of the lateral surface, not clearly distinguishable from the adjoining neostriatum (Brauth *et al.* 1978). Efferents of the paleostriatum augmentatum in turn project topographically onto the underlying paleostriatum primitivum and nucleus intrapeduncularis, those structures being the source of the ansa lenticularis, which projects to nuclei of the diencephalon and mesencephalon; this pattern of inputs and outputs supports Källén's (1962) proposal that the internal part of the avian paleostriatal complex is the homologue of the mammalian globus pallidus. It has been suggested (Parent and Olivier 1970) that the paleostriatum primitivum may be homologous with the external segment of the globus pallidus, the nucleus intrapeduncularis, which is rich in cholinesterase, being the homologue of the internal segment; Brauth *et al.* (1978), however, argue that the nucleus intrapeduncularis may not correspond to any part of the globus pallidus, because, first, there is no resemblance between the cytoarchitecture of the two regions, and, second, the nucleus intrapeduncularis has reciprocal interconnections with the temporal-parietal-occipital region, no such comparable connections being known for the globus pallidus.

The olfactory tubercle lies on the ventral surface of the hemispheres, below the lobus parolfactorius, merging caudomedially with the nucleus accumbens which, like the nucleus of the diagonal band, forms part of the ventral septal area. The efferents of the dorsal septal area, which contains both a lateral and a medial septal nucleus, include projections to the lateral hypothalamus, the dorsomedial thalamus, and the ventral midbrain (Krayniak and Siegel 1978*b*); these connections show a close resemblance to those of the mammalian septal nuclei. Efferents of the ventral septal region include projections to sites in both lateral and medial hypothalamus, the dorsomedial thalamus, the rostral midbrain, the hippocampus and parahippocampal area, and the lateral habenular nucleus. These connections too resemble those of the corresponding mammalian area, although the absence of a projection to either the medial habenular nucleus or the amygdala indicates a less close parallel than that of the dorsal area (Krayniak and Siegel 1978*b*).

Sensory and motor organization in avian forebrain

Olfaction

Recent anatomical (Rieke and Wenzel 1978) and electrophysiological (Macadar, Rausch, Wenzel, and Hutchison 1980) work has shown that the olfactory bulbs of the pigeon project to both ipsilateral and contralateral telencephalic regions. The electrophysiological and anatomical data agree in showing direct ipsilateral projections to the prepiriform cortex, the hyperstriatum ventrale, and the lobus parolfactorius. The contralateral projection goes via the anterior commissure to terminate primarily in the paleostriatum primitivum and, to a lesser extent, in the paleostriatum augmentatum and the caudal lobus parolfactorius. Rieke and Wenzel (1978) also report electrophysiological observations which indicate that, whereas stimulation of the contralateral bulb activates neurons in the paleostriatum primitivum, stimulation of the ipsilateral bulb inhibits those neurons. Rieke and Wenzel assume that this latter effect is mediated by a relay in the hyperstriatum ventrale (although it should be noted that neither Karten and Dubbeldam (1973) nor Brauth *et al.* (1978) find a projection from the ventral hyperstriatum to the paleostriatum primitivum).

There are some intriguing contrasts between this pattern of projections and that reported for reptiles. It may have been anticipated that, assuming a lesser importance of olfaction in birds, the avian olfactory projections would be somewhat restricted, so that the small cortical olfactory projection (to the prepiriform cortex) and the absence of any projection to the amygdalar complex may not be surprising (particularly as there are no accessory bulbs in birds). This reduction is compensated, however, by direct projections to the hyperstriatum ventrale, an area that presumably corresponds to part of the reptilian anterior dorsal ventricular ridge (which does not receive direct olfactory input) and to the paleostriatum primitivum. This latter projection is of particular interest in view of the evidence supporting a homology between paleostriatum primitivum and globus pallidus (which in mammals receives no direct olfactory input). Rieke and Wenzel (1978) make the speculative suggestion that, if the paleostriatum primitivum is indeed involved in motor activities, its olfactory input may be concerned with the use of olfactory information in homing or navigation.

Vision

The avian visual system has been the subject of intensive research in recent years, and a number of excellent reviews of this work are available (e.g. Hodos 1976; Donovan 1978).

Retinal efferents project to a variety of destinations, and the overall pattern of these projections resembles that already described in amphibians and reptiles. The principal destination of the majority of ganglion cell fibres is the contralateral optic tectum. There are two major thalamic areas that receive retinal input and these are, first, the pars ventralis of the lateral geniculate nucleus, and, second, a group of nuclei in the anterior dorsal thalamus (including the nuclei lateralis anterior, dorsolateralis anterior, dorsolateralis anterior pars magnocellularis, and suprarotundus) now known collectively as the principal optic nucleus of the thalamus. Direct retinal projections run also to the pretectal complex (the area pretectalis and the nucleus lentiformis mesencephali), to the nucleus suprachiasmaticus hypothalami, and to the nucleus ectomammillaris (the accessory optic

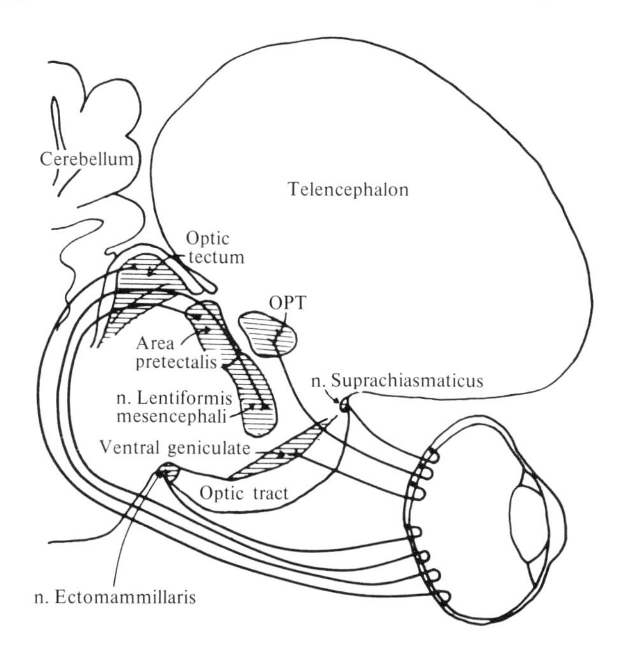

Fig. 6.4. Diagrammatic representation of the terminations of the optic tract in birds. Fibre pathways are indicated by solid lines; cell groups are indicated by hatching. OPT–principal optic nucleus of the thalamus. (From Hodos 1976).

nucleus). Most authors (e.g. Meier, Mihailović, and Cuénod 1974) agree that the retinal projections are entirely crossed, although Repérant (1973) does report some uncrossed fibres to thalamic nuclei in the pigeon. Figure 6.4 summarizes the retinal projections in birds.

There are in birds (and, as we have seen previously, in reptiles and mammals) two independent pathways for visual information to the telencephalon. One of these pathways, known as the thalamofugal pathway, runs from the retina to the principal optic nucleus of the thalamus and from there bilaterally to a medial, dorsal telencephalic region within the Wulst known as 'visual Wulst'. This region consists primarily of the nucleus intercalatus hyperstriati accessorii (IHA), a granule cell layer lying between the dorsal and the accessory hyperstriatum, although some thalamic fibres terminate in the dorsal hyperstriatum itself. The second pathway, the tectofugal pathway, runs from the retina to the tectum, and from there to the ipsilateral nucleus rotundus of the thalamus, from which it projects to the ipsilateral ectostriatum in the telencephalon. Figure 6.5 summarizes these two routes to the telencephalon. The thalamofugal pathway invites comparison with the geniculostriate system in mammals, whereas the tectofugal pathway resembles the mammalian projection via the superior colliculus to the nucleus

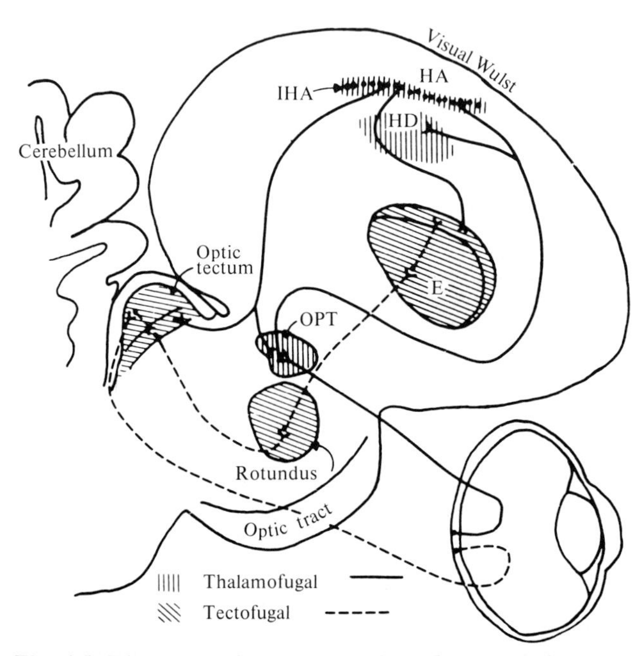

Fig. 6.5. Diagrammatic representation of the tectofugal and thalamofugal visual pathways in birds. Fibre pathways are indicated by continuous and broken lines. Cell groups are indicated by hatching. Abbreviations as for Figs. 6.3 and 6.4. (From Hodos 1976).

lateralis posterior (or pulvinar) of the thalamus, and finally to cortical regions adjacent to the striate cortex (Hall 1972).

A further point of similarity between the avian and the mammalian systems is that, although the two pathways ascend independently, there is ample scope for their interaction, primarily through descending projections from the Wulst. There are, for example, direct projections from the Wulst to both the optic tectum itself and to a periectostriatal belt adjacent to the central core of the ectostriatum (Karten *et al.* 1973); although the nucleus rotundus projects exclusively to the ectostriatal core, the core in turn is believed to project onto the periectostriatal belt (Karten and Hodos 1970). The visual Wulst also appears to modulate its own input, via a projection onto the principal optic nucleus of the thalamus. These three Wulst projections resemble mammalian striate interactions with the superior colliculus (McIlwain and Fields 1971), the peristriate cortex (Wilson 1978*b*), and the lateral geniculate nucleus (Singer 1977).

Electrophysiological studies of the two pathways reveal marked differences between them in the types of stimulation to which they are selectively responsive. The tectal input to the rotundus arises primarily from the deeper layers of the tectum, and cells in those layers have relatively large receptive fields and are generally movement sensitive (for a detailed summary of response properties of cells in the pigeon visual system, see Donovan's 1978 review). The great majority of cells in the rotundus have, in turn, even larger receptive fields (120-180°) and are responsive to movement anywhere within those fields. Finally, ectostriatal units are also 'wide-field' movement detectors, although they show more selectivity for a given direction of movement than is the case with rotundal units (Kimberly, Holden, and Bamborough 1971). The movement-sensitivity of cells in the tectofugal system is reminiscent of that reported for tectal units in amphibians (p. 121) and mammals (Goldberg and Robinson 1978).

A major difference between cells in the tectofugal and thalamofugal pathways is that the latter (cells in the principal optic nucleus of the thalamus and in the visual Wulst) have generally smaller receptive fields (usually less than 10°). Different nuclear groups within the thalamic visual complex show somewhat different response properties, and generally prefer moving stimuli, although showing responses to stationary stimuli also. Pettigrew and Konishi (1976), recording from neurons in the principal optic nucleus of the barn owl (*Tyto alba*), found precise retinotopic organization, both sustained and transient responses to stimulation, concentrically organized receptive fields with either on or off centres, and responses to stimulation of the contralateral eye only; although ganglion cells from both eyes project to the cat lateral geniculate, individual cells are driven by one eye only so that there is close agreement between the properties of cells in the avian principal optic nucleus and those in the mammalian lateral geniculate nucleus. Pettigrew and Konishi in fact assert that neurons in the principal optic nucleus 'were indistinguishable from the neurons of lateral geniculate nucleus which provide input to the visual cortex of the cat' (p. 677).

Striking results indicating similarity between visual Wulst cells and mammalian striate cortex cells were obtained by Revzin (1969), who found that cells in the visual Wulst of pigeons were arranged in columns vertical to the surface, the cells in a given column being selectively responsive to bar stimuli of a particular orientation; this organization of course closely resembles that described in the striate cortex of cats by Hubel and Wiesel (1962). Further evidence for similarity in processing between avian hyperstriatal and mammalian striate cortex neurons is provided by Pettigrew and Konishi (1976), who found in barn owls units that were selective not only for orientation but also for binocular disparity. Similar disparity units have been observed in cat brain by Barlow, Blakemore, and Pettigrew (1967), and are assumed to be involved in stereoscopic vision; it is therefore of interest that stereopsis has been demonstrated in a bird (the American kestrel, *Falco sparverius*) by Fox, Lehmkuhle, and Bush (1977) and that this is the only non-mammalian vertebrate in which any such unequivocal evidence has been produced to date.

Investigations of response properties of cells in the thalamofugal system show, then, not only that they differ from those of the tectofugal system but that they have much in common with the properties of units in corresponding stations of the mammalian geniculostriate pathway.

The contrast between the large receptive fields characteristic of the tectofugal system and the smaller fields of thalamofugal system has suggested that whereas the latter system seems well adapted to pattern vision, the former system may have the spatially imprecise role of alerting the organism to movement somewhere in the visual field. Revzin (1970), for example, suggests that 'the functions of the tecto-telencephalic system are to direct or to redirect the attention of the animal from its current object to other significant events in the environment' (p. 203). However, we shall see that the results obtained to date from lesion studies do not support any such simple dual function account of the systems.

A series of experiments by Hodos and his colleagues tested the effects of lesions at various sites in the tecto- and thalamofugal pathways on post-operative retention of an intensity (black-white) and three pattern discriminations (horizontal versus vertical bar; inverted versus normal triangle; inverted versus normal triangle, the shapes composed of small dots instead of lines). These experiments found substantial impairments caused by lesions at all levels of the tectofugal system (of the tectum – Hodos and Karten 1974; the rotundus – Hodos and Karten 1966; the ectostriatum – Hodos and Karten 1970); the impairments were severe for all the problems tested, although reacquisition was generally possible after many more trials had been administered than were necessary in original training.

Hodos and Karten (1974) noted that, after large tectal lesions, their pigeons seemed disoriented in the experimental chamber, and would feel their way along the walls of the chamber with their beaks or the side of their bodies, and point out that this behaviour is similar to that reported by Schneider (1969) in hamsters (*Mesocricetus auratus*) having collicular lesions. It might be thought, then, that the discrimination deficits are a consequence of this disorientation. However, this seems unlikely, following a report by Jarvis (1974) who tested pigeons with lesions of either the rotundus or the tectum in a discrimination task for which the importance of orientation was minimized by constricting the movement of the pigeons, and having the two discriminanda close together; despite the minimal dependence on orientation, Jarvis found substantial impairment in both the intensity and pattern problems. Jarvis also ran two localization tasks, one of which required pigeons to locate and respond to a lit key in one of four corners of the chamber and found that, as expected, rotundal and tectal lesions disrupted performance; birds which were poor on the discrimination tasks tended to be poor on the localization tasks, and vice versa.

It appears, then, that lesions in the tectofugal system do disrupt orientation but that they also disrupt visual discrimination of pattern and intensity, and this is a result that conflicts with data obtained in mammals (e.g. Schneider 1969).

The severity of impairment caused by tectofugal system damage suggests that the thalamofugal system may in fact play a minor role in these tasks, and Hodos, Karten, and Bonbright (1973) have shown that this is indeed the case: lesions of either the principal optic nucleus of the thalamus or the intercalated nucleus of hyperstriatum accessorium caused minimal impairment of intensity and pattern discrimination, although some role for the thalamofugal system was indicated by the tendency for lesions involving both the rotundus and the principal optic nucleus to cause more impairment than lesions involving the rotundus alone.

Hodos has suggested that thalamofugal lesions may have been ineffective because the intensity and pattern tasks used were relatively coarse discriminations, incapable of detecting small decrements in sensitivity. In two papers, Hodos reports the effects of principal optic nucleus (Hodos and Bonbright 1974) and visual Wulst (Pasternak and Hodos 1977) lesions on pre-operatively established incremental brightness thresholds, using a sophisticated psychophysical technique. In each case, thresholds were raised as a result of the lesions; following principal optic nucleus lesions, the elevations in thresholds were somewhat larger than those seen after Wulst lesions, and the elevations were permanent, the elevations produced by lesions in the Wulst being in all but the most severe cases temporary. It is of interest that Hodos and Bonbright (1974) showed that rotundal lesions also caused threshold elevations, but that these were, unlike those caused by principal optic nucleus lesions, temporary.

There is, then, evidence of severe (but generally temporary) impairment of visual localization, and of brightness and pattern discriminations, following lesions in the tectofugal pathway; lesions of the thalamofugal pathway do not appear to cause any gross visual disturbances, but do cause elevations (permanent in the case of principal optic nucleus lesions) in sensitivity to brightness differences. The parallel between avian visual Wulst and mammalian striate cortex, so powerfully supported by anatomical and electrophysiological data, must be treated with some reserve until the absence of pattern discrimination deficits following Wulst lesions is better understood.

One further possible route for visual information to the telencephalon has been reported by Bradley and Horn (1978), who find evidence, based on retrograde transport of horseradish peroxidase, for a direct projection (bilaterally) form the optic tectum to the hyperstriatum ventrale in chicks. Brown and Horn (1978) have recorded evoked responses in the ventral hyperstriatum of chicks finding that those vary (unlike those of the overlying intercalated nucleus of the hyperstriatum accessorium) with visual experience (dark-reared chicks showing fewer responses than visually experienced chicks). The significance of the variation in response with experience may be associated with a role for the hyperstriatum ventrale in imprinting: Bateson, Horn, and McCabe (1978), for example, have shown that bilateral lesions of the medial hyperstriatum ventrale abolish imprinting in newly hatched chicks.

The demonstration of a visual pathway to the telencaphalon that does not synapse in the thalamus is of particular interest in the light of evidence, that we shall consider in later sections, of similar pathways for both auditory (Delius, Runge, and Oeckinghaus 1979) and somesthetic (Cohen and Karten 1974) information. Such 'by-passing' of the thalamus is found in mammals for gustatory input alone (see p. 176) and if this is indeed a widespread property of avian sensory projection systems, a striking contrast will have emerged between birds and mammals.

Hearing

We saw that, for both amphibians and reptiles, the anatomical organization of the auditory system appeared grossly similar to that of mammals up to the thalamic level and that in reptiles there was a close parallel in telencephalic projections also. It is therefore not surprising that in birds too the auditory system clearly resembes that of mammals.

Primary auditory fibres from the cochlea terminate in two medullary nuclei, the nucleus magnocellularis and the nucleus angularis; this projection is topographically organized, and the pattern of projections suggests that the nucleus magnocellularis and part of the nucleus angularis correspond to the mammalian ventral cochlear nucleus, the remainder of the nucleus angularis corresponding to the mammalian dorsal cochlear nucleus (Campbell and Boord 1974).

The nucleus magnocellularis projects bilaterally to the nucleus laminaris which in turn projects to both the superior olive and the nucleus mesencephalicus lateralis dorsalis; Campbell and Boord suggest that the nucleus laminaris is the homologue of the mammalian medial superior olive, pointing out that birds such as owls, which rely on acoustic localization of prey, have a particularly large nucleus laminaris.

Nucleus angularis and nucleus magnocellularis project to the superior olive, the nucleus of the lateral lemniscus, and the nucleus mesencephalicus lateralis dorsalis, this latter nucleus being taken to be the homologue of the mammalian inferior colliculus. Fibres from the nucleus mesencephalicus lateralis dorsalis run to the nucleus ovoidalis of the thalamus, and thence to a region of the caudal medial neostriatum known as Field L. This region is, according to Karten (1968), the only telencephalic region that receives a direct auditory thalamic projection and this would suggest a difference between birds and reptiles (and mammals), which have both pallial and striatal auditory projection areas (see p. 148). Fig. 6.6 summarizes the organization discussed here, and includes certain details that have not been described in the text.

A final observation on the anatomy of the system is that in songbirds there appear to be links between auditory areas of the telencephalon and areas concerned with song-production. Kelley and Nottebohm (1979) have shown that Field L in the canary (*Serinus canarius*) projects to the borders of a caudal part of the ventral hyperstriatum that is (as we shall see in the section on motor control) concerned in the control of song, and that two areas adjacent to Field L project directly into the caudal hyperstriatum ventrale and to the borders of an area of the archistriatum known as the nucleus robustus which, too, is involved in song-production. These anatomical observations gain support from Zaretsky's (1978) observation that auditory stimuli obtain responses from single cells in the caudal hyperstriatum ventrale of canaries and zebra finches (*Peophila guttata*).

There have been many more physiological studies on the avian auditory system than on the reptilian system and these point to two basic conclusions. First, tonotopic organization is preserved through the relays of the system up to the telencephalic level (Zaretsky and Konishi 1976) and, second, some cells in the system respond selectively to complex properties of sound stimulation. Knudsen and Konishi (1978), for example, recording from cells in the nucleus mesencephalicus lateralis dorsalis of barn owls, found units that responded to sounds originating in a well-defined area of space, virtually independent of the nature and intensity of the sounds; direction sensitivity of this kind is, of course, only possible through the combination of information from both ears. Leppelsack and Vogt (1974), recording from cells in Field L of the starling (*Sturnus vulgaris*), found that only two-thirds of the units in that region responded to 'simple' sounds (pure tones or white noise), the remainder requiring more complex stimuli. A subsequent report (Leppelsack 1978) showed that 'natural' stimuli (parts of the starling's song) were effective complex stimuli, and that most Field L neurons were selective in their responses to different parts of the starling song. Both tonotopic organisation and complex feature extraction are important properties of the mammalian auditory system (e.g. Evans and Whitfield 1964; Wollberg and Newman 1972), and these physiological investigations support the conclusions indicated by the anatomical work, namely that there is good agreement between the principal features of organization of the auditory system in both mammals and birds.

Before leaving this subject, mention should be made of the observation by a number of authors (e.g. Delius *et al.* 1979) of short-latency responses to auditory stimuli from a region of frontal telencephalon in or near the nucleus basalis. Delius *et al.* report that the latency of these responses is about half that of responses recorded from Field L sites, and present evidence to counter the proposal (e.g. Karten 1968) that the responses are mediated by the known projection from the sensory nucleus of the trigeminal nerve to the nucleus basalis and are therefore somesthetic rather than auditory. If these are indeed auditory responses, then their latency suggests that they are mediated not through a pathway having a thalamic relay, but through some more direct projection from a site in the medulla. At present, no anatomical support for such a projection is available, and until some plausible pathway is detected, the proposal should perhaps be regarded with some reserve.

Skin senses

Very little is known of the anatomy of ascending pathways from the spinal cord in birds. Fibres do ascend from the dorsal roots to the cuneate and gracile nuclei

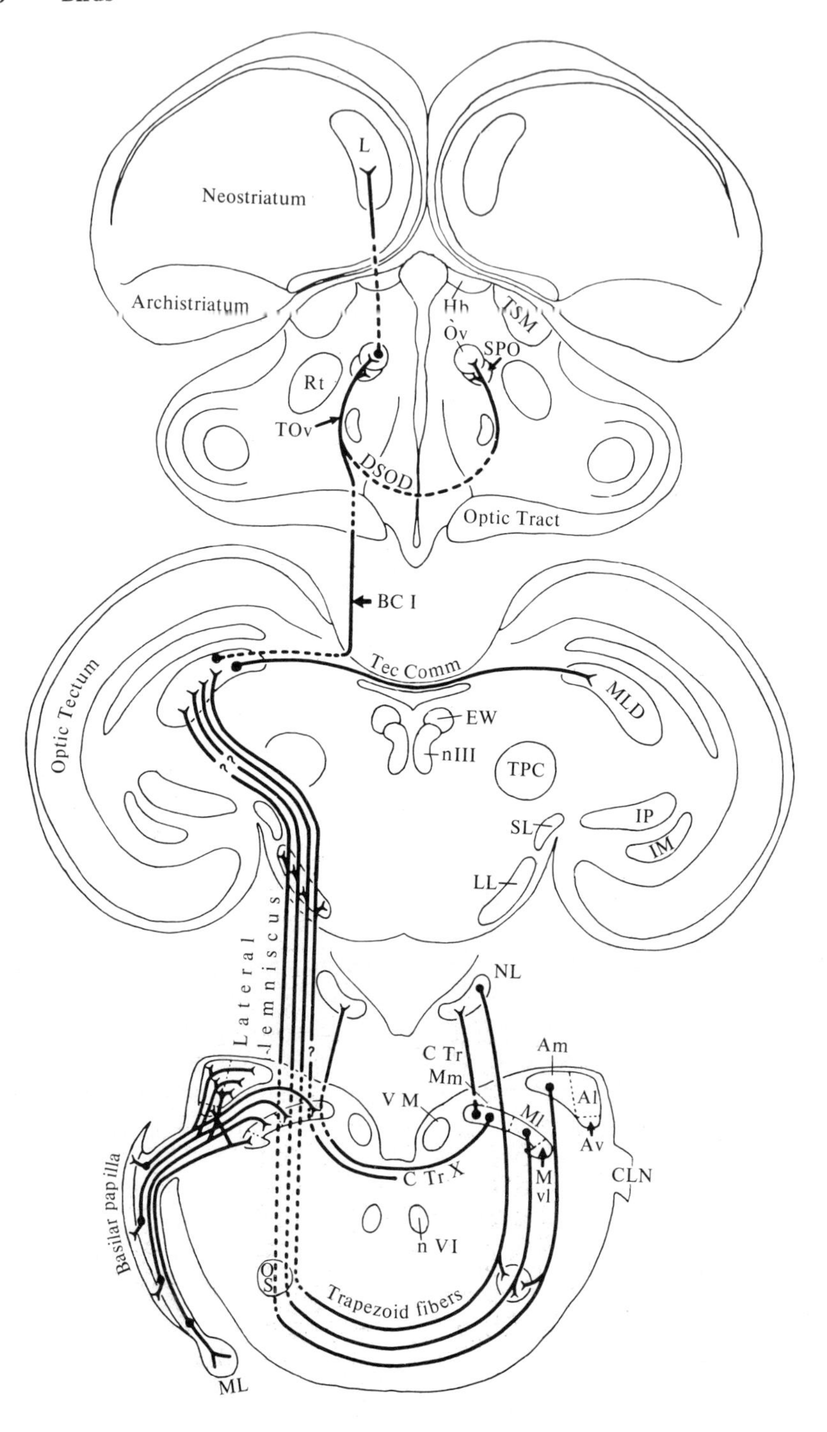
L
Neostriatum
Archistriatum
Hb
TSM
Ov
SPO
Rt
TOv
DSOD
Optic Tract
BC I
Tec Comm
Optic Tectum
MLD
EW
nIII
TPC
SL
IP
IM
LL
Lateral lemniscus
NL
C Tr
Mm
Am
Al
V M
Ml
Av
C Tr X
M vl
CLN
Basilar papilla
n VI
OS
Trapezoid fibers
ML

in the medulla, and are topographically arranged, but the ascending projections of these medullary nuclei have not yet been investigated (Cohen and Karten 1974). There are also direct spinothalamic projections which appear to terminate in two thalamic nuclei, the nucleus dorsolateralis posterior and the nucleus superficialis parvocellularis, located dorsal to the nucleus rotundus (Karten and Revzin 1966). Although there is evidence (Powell and Cowan 1961) that these nuclei do project to the telencephalon, there is as yet no anatomical information on which areas in the telencephalon receive the projections.

Delius and Bennetto (1972) have explored the diencephalon and telencephalon of the pigeon, seeking evoked potentials to stimulation of either the body surface or peripheral sensory nerves, and have found one diencephalic and two telencephalic sites responsive to stimulation (showing maximal responses to contralateral stimulation). The diencephalic site appeared to coincide with a small area, dorsal to the rotundus, centred on the nucleus superficialis parvocellularis. One telencephalic site lay in the anterior Wulst, in the hyperstriatum intercalatus superior, and the other, in the medial part of the caudal neostriatum, somewhat anterior and dorsal to an area from which maximal auditory responsiveness was obtained. Delius and Bennetto provide evidence for modality specificity of the telencephalic sites, and for some degree of topographical organization.

A further source of sensory input to the telencephalon, one to which reference has been made previously, is the direct projection from the principal sensory nucleus of the trigeminal nerve to the nucleus basalis. Cells in both these structures are primarily responsive to stimulation of the beak and mouth region, the receptive fields of cells in the principal sensory nucleus being, however, much smaller than those of the nucleus basalis (Zeigler 1974). Zeigler suggests that these structures form part of a feeding behaviour system, whose efferent limb is formed by a projection from the nucleus basalis to the dorsolateral archistriatum and thence, via the occipitomesencephalic tract, direct to the midbrain and spinal

Fig. 6.6. Diagram of the ascending auditory conduction pathways from the peripheral receptor (basilar papilla) to the telencephalon of the pigeon. Pathways for which the anatomical evidence is equivocal are indicated by question marks. (From Boord 1969.)

Key: Al – nucleus angularis pars lateralis; Am – nucleus angularis pars medialis; Av – nucleus angularis pars ventralis; BCI – brachium of inferior colliculus; CLN – cochlear and lagenar nerves; Ctr – uncrossed cochlear tract; CtrX – crossed cochlear tract; DSOD – dorsal supraoptic decussation; EW – Edinger–Westphal nucleus; Hb – habenula; IM – nucleus isthmi pars magnocellularis; IP – nucleus isthmi pars parvocellularis; L – Field L of Rose; LL – nucleus of lateral lemniscus; Ml – nucleus magnocellularis pars lateralis; Mm – nucleus magnocellularis pars medialis; Mvl – nucleus magnocellularis pars ventrolateralis; ML – macula lagenae; MLD – nucleus mesencephali lateralis dorsalis; NL – nucleus laminaris; Ov – nucleus ovoidalis; OS – superior olive; Rt – nucleus rotundus; SPO – nucleus semilunaris parovoidalis; Tec Comm – tectal commissure; TOv – tractus ovoidalis; TPC – nucleus tegmenti pedunculopontinus pars compacta; TSM – tractus septomesencephalicus; nIII – oculomotor nucleus; nVI – abducens nucleus VM – medial vestibular nucleus.

cord; Zeigler and his colleagues have shown that lesions placed in any of these areas disrupt feeding, while leaving drinking intact.

It has already been pointed out (p. 176) that the nucleus basalis is an area of exceptional interest in receiving a projection that has not relayed in the thalamus, and mention has been made of Cohen and Karten's (1974) suggestion that perhaps the nucleus basalis should be regarded as a diencephalic structure. Before leaving the topic, it may be of interest to quote Zeigler's specific proposal for the sensory role played by quinto-frontal structures in feeding: 'A comparison of the sensory control and time course of mandibulation with the receptive field characteristics and adaptive properties of PrV [principal sensory nucleus of the trigeminal nerve] and NB [nucleus basalis] units suggests that these nuclei contain neurons that signal the presence of a kernel of grain at the beak tip, provide complementary information about its static position and movement within the mouth, and monitor the extent of beak displacement' (Zeigler 1974, p. 122).

Motor control

There is anatomical evidence for the existence of two tracts, arising in disparate telencephalic regions, that descend direct to the spinal cord and project also to sites in the brainstem. One of these, the septomesencephalic tract, arises in the anterior Wulst (Karten *et al.* 1973; Zecha 1962) and the other, the occipito-mesencephalic tract, in the anterior archistriatum (Zeier and Karten 1971; Zecha 1962), and each tract may, therefore, be considered comparable to some component of the corticospinal projections of mammals, and might be expected to have some role in the control of movement. A further telencephalic region which could be expected to be involved in movement control is the paleostriatal complex, given the grounds (p. 177) for supposing that it may be homologous with the mammalian basal ganglia, which form an important part of the mammalian extrapyramidal motor system; this latter expectation gains support from Rieke's (1980) report that unilateral destruction of cell bodies in the paleostriatum by the neurotoxin kainic acid results in disturbances of posture and of movement (including rapid rotation and 'involuntary' movements) similar to those reported in mammals after comparable damage to the basal ganglia (caudate nucleus, putamen and globus pallidus). However, the results of experiments in which the hemispheres have been totally removed (e.g. Phillips 1964) show that gross bodily movement appears to be unaffected by the operation. Although decerebrate birds show little or no spontaneous movement, they can fly, avoid obstacles, and land without difficulty; decerebrate mallards (*Anas platyrhynchos*) are also 'able to body-shake, head-shake, tail-wag, wing-leg-tail stretch, head-scratch, wing-flap, and preen in apparently normal fashion' (Phillips 1964, p. 141). The lesion data, then, suggest that the telencephalon does not play a major role in movement control and the results of experiments using brain stimulation point to a similar conclusion; although some investigators (e.g. Bremer, Dow, and Morruzzi 1939) have reported movements in response to forebrain stimulation, there is no evidence for any extensive or topographical representation of the body musculature in the hemispheres (Putkonen 1967; Phillips and Youngren 1971).

The efferent limbs of two specific behavioural systems do, nevertheless, appear to originate in telencephalic regions, and in each case the archistriatum, site of origin of one of the telencephalo-spinal tracts, is involved. The first of these is the feeding behaviour system, described by Zeigler (1974), which conveys sensory information about the beak region via the nucleus basalis to the archistriatum: lesions of the archistriatum interfere with feeding in both pigeons (Zeigler, Green, and Karten 1969) and ducks (Phillips 1964), although the disruption is not as severe as that seen following basalis lesions (Zeigler *et al.* 1969). Phillips (1964) produces evidence that this disruption may have been due to interference with motor components by showing that stimulation of the archistriatum or occipito-mesencephalic tract produced 'rapid movement of the bill and often of the head and neck, movements that looked very much like searching and gabbling (feeding movements)' (Phillips 1964, p. 149).

The second system is that for the production of song in songbirds. Two telencephalic regions have been shown to be critical for the production of song and these are, first, a region in the caudal part of the ventral hyperstriatum and, second, a large-celled round nucleus in the somatomotor part of the archistriatum, the nucleus robustus. Unilateral lesions in either of these sites cause severe disruption of song-production in canaries (Nottebohm, Stokes, and Leonard 1976) and it will be recalled that each of these areas receives projections from Field L or adjacent regions of the neostriatum, Field L being the telencephalic projection region of the thalamic auditory nucleus, the nucleus ovoidalis. It is also of interest that the nucleus robustus has not yet been described in the pigeon or in any other non-songbird species, although it has been observed in other passerines (Nottebohm *et al.* 1976).

The effects of hemispheric lesions on song production appear to be more severe following left than right hemispheric lesions (Nottebohm *et al.* 1976), and this observation invites comparison with the well-known phenomenon of left hemisphere dominance in most humans. The parallel, however, does not appear to be close. Two major differences are, first, that the avian unilateral dominance extends throughout the song-production system, so that the left part of the syrinx (the song-production organ, located at the base of the trachea) produces 90 per cent or more of the components of bird songs (Nottebohm *et al.* 1976), and, second, that there is no evidence of avian cerebral dominance in any other skill: in humans, of course, cerebral dominance for speech is clearly a central affair—there is no suggestion that the dominance may be the result of asymmetrical importance of the two halves of the larynx in voice-production—and left-hemispheric dominance is associated (to some degree—see p. 314) with the phenomenon of right-handedness. Two minor differences are that the avian left-hemispheric dominance is effected by an ipsilateral tract so that the corresponding 'skill' is on the left side of the bird's body, and that unilateral dominance of song-production by the left syrinx is much more marked in birds, as to date, from studies on 91 individuals of five different songbird species, only two have shown reverse dominance (Seller 1979), whereas some 7 per cent of the human population do not show left cerebral dominance (p. 314).

Conclusions

The data reviewed in this section lead to conclusions that are in good general agreement with those reached for reptiles. There are, for example, clear parallels between avian and mammalian organization of ascending sensory pathways, and the various senses appear to gain independent representation in the hemispheres. Although there is in birds anatomical support for tracts that might correspond to components of the corticospinal tract in mammals, physiological data suggest that, as in reptiles, the forebrain does not play a major role in movement control.

There are, on the other hand, some interesting contrasts between birds and reptiles, the most interesting of which is the possibility that information from three of the senses (touch, hearing, and vision) may reach the telencephalon from sites caudal to the diencephalon without an intervening thalamic relay. Further contrasts concern secondary olfactory projections to the avian hemisphere, which appear to differ significantly from those found in reptiles, and are surprisingly extensive, and a general increase in differentiation, seen particularly in the archistriatal complex, in avian as opposed to reptilian forebrain.

The bird forebrain, then, receives the inputs necessary to allow sensory integration, but does not seem to be totally dominated by sensory input—there appears to be ample 'space' for mechanisms devoted to intellectual operations.

Habituation in birds

There are in the literature a vast number of reports on the behaviour of birds, both in the field and in the laboratory. The reasons for the popularity of birds are not hard to find. Ethologists are particularly attracted by the fact that, being diurnal, they are easy to observe, and by their complex social organization, involving as it does a variety of relatively stereotyped displays and calls. Psychologists find useful the fact that birds have good vision, and are also attracted by more mundane considerations, such as cheapness and hardiness, in their choice of the pigeon as the second most popular animal (next to the rat) for laboratory studies. The great majority of these studies have not been comparative in nature, in that they have not been designed to explore contrasts between one species and another. Ethologists have been interested in the organization of behaviour in one species, studied for its own sake, but given also the assumption that principles of that organization may emerge which have a more general application. Psychologists have commonly used pigeons on the simple (if tacit) assumption that the rules controlling their learning certainly will have extremely wide application to other species, including species from other classes. Psychologists have, that is, used the pigeon rather as they have used the rat, assuming that each species will reflect in its learning the operation of basic laws of learning that have a near-universal application.

This is not the appropriate place at which to consider whether the psychologists' assumptions are justified or not; the point to be made here is that it would not be possible – or even desirable – to attempt a comprehensive survey of learning studies using birds. The numerous excellent surveys of animal learning currently available refer to studies using pigeons as easily as to studies using rats – as though, indeed, the species were interchangeable – and the reader is referred to those surveys for accounts of the performance of pigeons in what might be called standard experimental paradigms. The sections that follow here will point to selected instances of habituation, classical conditioning and instrumental conditioning in birds, and concentrate on potential contrasts between avian and mammalian performance; the instances quoted are of course chosen somewhat arbitrarily, and represent the tip of an iceberg of published work.

Reports are available of habituation in a wide variety of avian response systems, including post-rotational nystagmus (head movements in blindfolded pigeons – Fearing 1926), the mobbing response to a predator model (Hinde 1970), fear responses to model predators (Melzack 1961), aggressive responses to recordings of conspecific songs (Petrinovich and Peeke 1973), the cardiac component of the orienting response to a novel light stimulus (Cohen and MacDonald 1971), and the tonic immobility response (animal hypnosis), induced by manual restraint (Nash and Gallup 1976). The rates of decrement vary according to both the response measured and the stimuli used: Melzack (1961), for example, found that whereas the orienting response of mallard ducklings to an overhead model of a hawk showed little if any habituation, overt fear responses to the models showed marked habituation; Rouse (1905) found that respiratory responses of pigeons habituated to 'meaningless' sounds, even to pistol shots, but remained responsive to 'significant' sounds, such as calls and wing flaps of other birds.

Hinde has carried out a systematic series of studies on habituation of calls elicited in chaffinches (*Fringilla coelebs*) as a component of the mobbing response to a stuffed owl (see Hinde 1970, for a review), and this work has shown that many characteristics of habituation in other groups are observed in birds. For example, the rate of calling to a stuffed owl, while it eventually habituates, shows initially a gradual increase, reaching a peak after two to three minutes' exposure (sensitization). Spontaneous recovery of the response is seen: Hinde measured the strength of calling after a number of stimulation-free intervals following 30 minutes' exposure to the owl, and found a fairly rapid recovery of responsiveness (to 56 per cent of the initial level within 24 hours). The response decrement is to a considerable extent stimulus specific: the mobbing response may also be elicited by a toy dog, and after 24 hours' continuous exposure to either the stuffed owl or a toy dog, chaffinches show greater responsiveness, re-tested 24 hours later, to the object to which they have not previously been exposed.

Finally, it appears that dishabituation can be obtained in birds, although data on the phenomenon are surprisingly sparse: Nash and Gallup (1976) report that the tonic immobility response may, following habituation, be abruptly restored by novel external stimuli..

From this brief survey, it seems reasonable to conclude that the processes involved in habituation in birds are comparable to those involved in other groups of vertebrates, and to accept the parsimonious assumption that there are neither qualitative nor quantitative differences in habituation between birds and other vertebrates.

It may be recalled that in the corresponding section on reptiles, a phenomenon resembling imprinting was introduced on the grounds that it represented learning following presentations of a UCS not explicitly paired with any other stimulus. The topic of imprinting in birds has generated a large literature, and there are a number of rival explanations of the phenomenon, some of which assume that imprinting can be accounted for in terms of laws of learning applicable in other paradigms, and others, that imprinting proceeds according to its own particular rules (for a review, see Bateson 1966). This debate will not be discussed here, for reasons that have been advanced earlier (p. 5): if imprinting is a special form of learning, then it may not be relevant to our interest in mechanisms of general intelligence; if it is in fact a product of general learning mechanisms, then what we derive from it—and we shall have other instances pointing to a similar conclusion—is that birds may show stable learning following a single exposure or trial in certain paradigms. This may also be the place at which to reiterate that navigation and homing in birds will also be excluded as they seem even more likely than imprinting to be 'special' systems, the mechanisms of which may not be available to birds in learning other types of problem.

Classical conditioning in birds

Classical conditioning techniques have been used successfully in birds, with both aversive and appetitive reinforcers. Aversive UCSs have included electric shocks (e.g. Davis and Coates 1978) and drug-induced illness (e.g. Wilcoxon, Dragoin, and Kral 1971), and CRs measured have included the nictitating membrane response (Davis and Coates 1978), changes in heart rate and respiration (Cohen and Durkovic 1966), and poison aversions (Wilcoxon *et al.* 1971). The most widely used appetitive UCS has been food, although other reinforcers, such as heat (Wasserman 1973) and water (Jenkins and Moore 1973) have also served successfully. The most common CR used with an appetitive reinforcer has been pecking a key illuminated immediately prior to food delivery, although, again, other responses, such as approach to a stimulus correlated with food (e.g. Hearst and Franklin 1977) and general movement at onset of a CS associated with food (Longo, Klempay, and Bitterman 1964) have also been established as CRs.

It is apparent from these reports that conditioning may be rapid (e.g. after one trial, in the Wilcoxon *et al.* 1971, report) and that the effects obtained are not due to either sensitization or pseudoconditioning: a number of authors (e.g. Wasserman, Hunter, Gutowski, and Bader 1975) have run a truly random control condition to show that the contingency between the CS and the UCS is indeed necessary for the establishment of CRs. It will be recalled that in this procedure, advocated by Rescorla (1967), subjects receive the same number of CSs and UCSs as those in the experimental (paired) condition, but there is no systematic relationship between the delivery of the stimuli.

The best-known of the classical training procedures is that in which the illumination of a key is followed by the delivery of food and the subject (usually a pigeon) responds by pecking the key, although this does not, of course, have any effect on food delivery. This procedure, first described by Brown and Jenkins (1968) is known as autoshaping and is of major theoretical importance, partly because it shows that a skeletal response may be conditioned by a classical technique, and partly because it has provided strong evidence in favour of a stimulus-substitution theory of classical conditioning; for example, if the UCS used is water rather than food, then the pecks that the bird directs at the key closely resemble the movements involved in the consummatory drinking response and are of longer duration than those elicited by a food UCS, which in turn resemble the rapid pecks made to grain (Jenkins and Moore 1973). However, what is of central relevance here is that experiments using this technique have provided convincing demonstrations that the responses are in fact classical rather than instrumental in origin. Williams and Williams (1969) arranged a negative contingency between pigeons' responding to the lit key and food delivery (omission training) and showed that, despite the adverse consequences of key-pecks, responding was maintained indefinitely by this procedure. Although there have been attempts to provide alternative, instrumental, accounts of the results of omission training (e.g. Wessels 1974), the great majority of investigations agree in concluding that it is indeed the (classical) contingency between the stimulus and the reinforcer that is primarily responsible for the responses seen (e.g. Peden, Browne, and Hearst 1977).

Classical conditioning is, then, an effective procedure for establishing learning in birds, and we may safely conclude that 'genuine' classical conditioning can be obtained in birds—a conclusion that has been much less certain for all the other non-mammalian classes.

Instrumental conditioning in birds

Procedures using appetitive reinforcers

There is, of course, a huge literature concerning the use of instrumental procedures with birds and although key-pecking by pigeons is the response most commonly studied, a large variety of different species and apparatuses have been tested. Thorpe (1963) gives a good impression of the range of both species and techniques, describing, for example, experiments using mazes and puzzle-boxes, and tests requiring complicated motor skills, such as string manipulation. What this flexibility reflects at least in part is the increased sensorimotor capacities of birds compared with cold-blooded vertebrates, but it does not necessarily indicate any increase in general intellectual capacity. Our interest in this section, given that efficient performance is obtained in instrumental tasks, is to ascertain whether the responses seen are 'genuine' instrumental responses, and this is a question that has not been extensively discussed until recently.

It was argued in the preceding section that the results of omission training procedures indicated that autoshaped key-pecks in pigeons are classical and not instrumental in origin. These results, however, might be taken to have larger consequences, since in effect they show that the pigeons's key-peck is not sensitive to its consequences—pecks, that is, although 'punished' by non-delivery of food, are not withheld. Some authors (e.g. Moore 1973) have argued from such results that we should consider the possibility that the key-pecks elicited in conventional 'instrumental' paradigms using pigeons may in fact be of classical origin. Moore elaborates his case with ingenuity, and cites a variety of studies showing failures to modify features of the pigeon's key-peck (e.g. duration or force) by arranging differential consequences for different types of peck, as well as findings which suggest that other 'instrumentally' conditioned responses may also be classical in origin. In general, Moore argues that classical conditioning is a primitive form of learning which is supplemented by instrumental learning as intelligence evolves, and that although instrumental conditioning no doubt plays some role in birds, it plays a greater role in mammals. Our first step in discussing this proposal will be to see whether there is indeed proof that instrumental conditioning occurs at all in birds, bearing in mind that our earlier discussion, in Chapter 1, led us to expect that such proof will not easily be obtained.

There have been a number of reports of failures to maintain responses established by autoshaping procedures on the introduction of omission training: Woodruff and Williams (1976), for example, established key-pecking to a light using water injected directly into the mouth as a reinforcer, but found that responding ceased when key-pecks resulted in omission of the reinforcer. Such results suggest that at least some avian responses may be instrumental in origin, but leave open the possibility that the most common response, key-pecking for food, is wholly classical. There is, however, evidence that even this restricted proposal may not be valid. Schwartz and Williams (1972) compared the responding of pigeons to a stimulus that regularly preceded food provided no response was emitted with that to a yoked stimulus that was presented the same number of times, and was followed by a reinforcer the same number of times, but with no relation to responding in this case. They found that response rate was lower to the first stimulus than to the second. That is, where responses prevented the delivery of food, response rate was lower than in the case where responses had no consequence but were elicited by a stimulus having an identical relationship to food delivery. The same result has been established by Wasserman *et al.* (1975) using chicks and a heat reinforcer. There are, then, studies which show that key-pecking CRs, for water, food, or heat rewards, are reduced by omission procedures, as would be expected if instrumental contingencies were effective; on the other hand, such CRs do affect the stimulus-complex and it has been pointed out (p. 12) that, this being so, a classical interpretation of the reduction in performance is possible.

We shall conclude this section by observing that there is one report (Rudolph and Van Houten 1977), which shows that key-pecking by pigeons for food reward may be maintained in total darkness; this is a result which is difficult to interpret as an instance of classical conditioning, since it is hard to conceive of an invisible CS eliciting approach and acting as a substitute UCS. We shall return to Moore's case for a contrast between birds and mammals in the relative importance of instrumental conditioning after introducing the data available from studies using aversive reinforcers.

Procedures using aversive reinforcers

There are reports of the use of both active and passive avoidance (punishment) procedures. Two interesting examples of the latter are, first, a report by Cherkin (1969) who punished chicks' spontaneous pecks at a glass bead by coating the bead with an aversive-tasting solution and, second, a study by Matthews, McHugh, and Carr (1974), who punished pigeons' key-pecks for food with electric shocks. Cherkin's study is of interest in that he established reliable avoidance in one trial. The Matthews *et al.* report contained evidence that the contingency between responding and the aversive UCS was detected by the pigeons: responding was more suppressed during a stimulus in which responses were shocked than during a yoked stimulus in which shocks occurred at the same rate but were now independent of responding; while this suggests, of course, that at least some of the suppression observed was instrumental in origin, it must be conceded that it is not proof since, as we have seen above, key-pecking is a response which is correlated with a change in the stimulus-complex. In the present case, it may have been that responding was less suppressed in the presence of the yoked stimulus because stimuli seen when close to the key were not (as they were in the case of the 'master' stimulus) more likely to be shocked than those seen at a distance from the key.

Experiments on active avoidance appear to be confined to pigeons, using electric shock UCSs. An early report (Graf and Bitterman 1963) used gross body movement as the target response and established responding using a Sidman or unsignalled avoidance procedure, in which shocks occur at regular intervals (every 20 seconds in this case), unless a response occurs. Each response postpones the next shock (again, for 20 seconds in the Graf and Bitterman study). Graf and Bitterman ran two yoked control birds, who received shocks at the same time as the experimental birds but whose responses were without effect: the control birds did not maintain responding, and this experiment provides good evidence for instrumental learning in pigeons. There was no manipulandum for the master birds to approach, so that the CRs in this study did not systematically affect the stimulus-complex; it is therefore reasonable to suppose that the master and yoked subjects experienced closely comparable CS–UCS pairings, and that the difference in responding seen reflects the influence of the instrumental contingency.

Subsequent experiments have shown that active avoidance can be established in a shuttle-box (Macphail 1968) and using a treadle-press response (Smith and Keller 1970). These reports agree in showing that avoidance responding may be rapidly established in pigeons and support the view that the difficulty experienced in training pigeons to peck keys to avoid shocks (e.g. Hoffmann and Fleshler 1959; Rachlin and Hineline 1967) is due, not to pigeons' difficulty in detecting the contingency between a response and shock-omission, but to incompatibility between the target response (key-pecking) and some innate pattern of response (e.g. flight, withdrawal) elicited in fear-inducing conditions.

Conclusions

In this and the preceding section we have seen that efficient learning has been demonstrated in birds using both classical and instrumental procedures, with either appetitive or aversive UCSs.

There is convincing evidence, largely from results of studies using omission procedures and appetitive UCSs, that birds detect classical, stimulus–reinforcer, contingencies. It is, however, more difficult to demonstrate instrumental learning, and we should now consider Moore's (1973) suggestion that instrumental conditioning may play a smaller role in birds than in mammals. We have seen that imposing response–reinforcer contingencies does, as one would expect if instrumental learning did occur, influence responding; it is possible, for example, to eliminate key-pecking for a water reward using an omission procedure. However, alternative, 'classical' interpretations of such effects are possible, and we found only two examples of learned performance which seem particularly resistant to an instrumental interpretation, namely maintained key-pecking for food in the dark, and effective avoidance with a response that does not produce systematic changes in the stimulus complex. Does the difficulty in finding unequivocal examples of instrumental conditioning in birds suggest that it is a relatively rare phenomenon in the class, and, in particular, rarer in birds than in mammals? Such a conclusion is at present premature, since the difficulty in proving instrumental conditioning in birds appears to reflect the general difficulty in obtaining such proof for any species, a difficulty foreshadowed in Chapter 1. The very difficulty of proving the effectiveness of instrumental contingencies has meant that few convincing examples are available for any species (see Mackintosh and Dickinson 1979). It may still be possible to argue that instrumental learning does not occur in any (non-human) vertebrate, but there does not currently seem to be any reason to suppose that it is better established in mammals than in birds. It may be added that there appear to be at least some responses in mammals (e.g. the nictitating membrane response of rabbits, *Oryctolagus cuniculus*—Coleman 1975) which are not sensitive to instrumental contingencies so that, to prove a radical difference between birds and mammals, it would seem to be necessary to prove that none of a series of avian responses showed instrumental learning. There is little evidence to support such a broad conclusion at present, and it is therefore reasonable to conclude that there are no significant differences between birds and mammals in performance of 'simple' learning tasks.

Complex learning in birds

The topics discussed under the early subheadings of this section were originally introduced in Chapter 3, and details of the general procedures used, as well as of conflicting theoretical interpretations, are to be found there. Topics discussed under the later subheadings tend to be less strictly comparative in nature; they report miscellaneous abilities and disabilities of birds in a variety of tasks, which appear to throw light on the general capacities of the class, without explicitly contrasting them with the capacities of species from any other class.

Serial reversal learning

Progressive improvement over a series of reversals has been observed in a variety of birds from different orders, in both spatial and visual problems (e.g. Bullock and Bitterman 1962*a*; Gossette, Gossette, and Riddell 1966). Bitterman therefore concludes that birds are 'rat-like' in serial reversal performance, and that the major source of improvement is, in birds as in rats, due to an increase in 'forgetting' as the series progresses (e.g. Gonzalez, Behrend, and Bitterman 1967; Bitterman 1969).

Mackintosh (1969*a*, *b*), while agreeing that there is no qualitative difference between birds and rats in serial performance, argues that there is a quantitative difference, birds, 'more particularly Columbiformes and other lower birds' (Mackintosh 1969*b*, p. 176) being less efficient than rats. This difference, he believes, reflects a difference in stability of attention which is, along with forgetting, an important determinant of serial reversal performance. The restriction to 'lower birds' by Mackintosh is a consequence of demonstrations (e.g. Gossette *et al.* 1966) that birds from different orders show rather different levels of efficiency in reversal, with some passerines, such as magpies and mynah birds, showing considerably more improvement than chicks or quail (both Galliformes). These inter-species differences will be discussed further in the appropriate section, and we shall restrict ourselves here to a discussion of the performance of pigeons and doves, to see whether for these species, Mackintosh's case can be sustained.

Prima facie evidence for the existence of a quantitative difference between rats and pigeons is provided by a survey (Mackintosh 1969*a*) of reports from various laboratories on serial reversal performance in rats and pigeons. This survey showed that marked improvement across a reversal series is more reliably obtained in rats than in pigeons and that one-trial reversals, which have not uncommonly been obtained with rats, had not been reported using pigeons. Now this might, of course, reflect an 'intellectual' difference, or variation in some contextual variable. That contextual variables are, as one would expect, of importance, is indicated by the fact that reports of improvement in pigeons vary considerably: Reid (1958), for example, trained pigeons at 95 per cent of their free-feeding weights, and found no serial reversal improvement; Gossette and Hood (1968) explored the effects of varying deprivation level and incentive magnitude on serial reversals of a spatial discrimination by pigeons in a non-automated 'formboard' apparatus (see p. 222) and found (in a study published too late for inclusion in Mackintosh's 1969*a* review) highly efficient learning for high incentives, which clearly involved a substantial number of one-trial reversals. There can, then, be no guarantee that the generally lower level of performance of pigeons is not due to inappropriate values of contextual variables.

As was discussed previously in Chapter 3 (p. 75), Mackintosh has suggested, after presenting data on rats, pigeons, and fish, that 'since initial learning scores are similar for the three groups, it seems reasonable to compare absolute levels of performance' (p. 144). However, it will be recalled that Gonzalez, Berger, and Bitterman (1966*a*—see Fig. 3.10) presented data to show differences in reversal performance within one species (pigeons), obtained by variations in the apparatus which resulted in comparable acquisition scores. There are, then, grounds for remaining sceptical about there being a reliable quantitative difference in the intellectual mechanisms involved in reversal performance in pigeons and rats.

The second, more analytical, aspect of Mackintosh's case concerns the proposal that serial reversal improvement, in both pigeons and rats, reflects in part increased attention to the relevant dimension. Schade and Bitterman (1966) cast doubt on this proposal by running an experiment in which, on different days, either the position or the colour of the stimulus was relevant, the other dimension being irrelevant: a series of reversals of the values of the stimuli on both dimensions was carried out, and pigeons showed improvement across the series for both dimensions. The authors argue that as improvement is occurring concurrently on both dimensions, the improvement cannot be due to increasing attention to the relevant dimension since, according to attention theory, any increase in attention to one dimension should be gained only at the expense of attention to other dimensions. This conclusion is weakened, however, by Mackintosh and Little's (1969*a*) demonstration that a series of reversals involving alternating relevant dimensions did not obtain as efficient performance as that found following a series of reversals in a single dimension—an effect specific to the dimension that was relevant throughout was, then, found in addition to any improvement seen following reversal training per se. Given this evidence that, during the course of reversal training, pigeons do learn something specific to the relevant dimension, we now turn to evidence that this dimension-specific learning is weaker in birds than in rats.

Mackintosh (1969*a*) reports experiments in which rats and doves (*Streptopelia risoria*) were given a series of reversals with one dimension (brightness or position) relevant, and then shifted to a discrimination in which the previously irrelevant dimension was now relevant. At the end of reversal training, rats were performing more efficiently than were the doves: however, the rats' non-reversal shift performance was poorer than that of the doves (Fig. 6.7). Mackintosh interprets these results as showing that, during reversal performance, the attention of the rats was more firmly fixed on the relevant dimension than was that of the doves so that, when that dimension became irrelevant, the rats perseverated longer on the now irrelevant dimension, and so were inferior to the doves. This is an interesting result and Mackintosh's interpretation is convincing: it is, however, the only finding from the serial reversal task that provides direct support for the stability of attention hypothesis.

One finding that may weaken Mackintosh's case is his own demonstration (Mackintosh and Cauty 1971) of poorer performance in pigeons than rats in a serial reversal task in which position was the relevant dimension, and no salient irrelevant stimuli were present (both response keys, for example, were white on

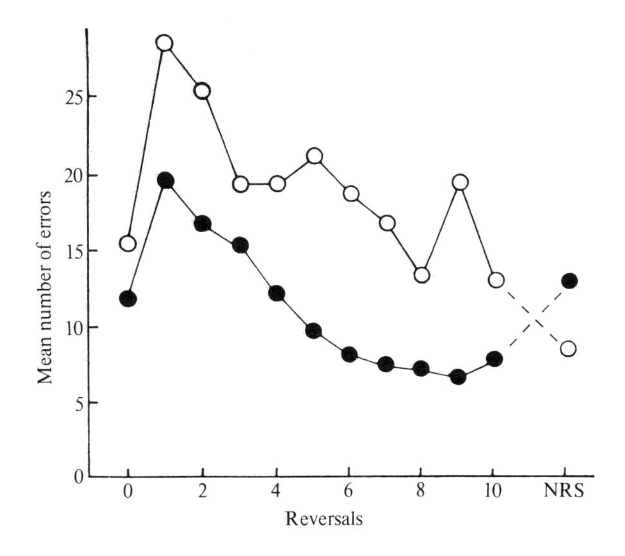

Fig. 6.7. Performance of rats (filled circles) and doves (open circles) on a series of reversals and on a final non-reversal shift (NRS). (From Mackintosh 1969*a*).

all trials). Although this result may be taken as evidence for a possible quantitative difference between rats and pigeons, it does not suggest that the difference lies in stability of attention, since the effect of any such difference should be minimal where no distracting stimuli are used. This in turn indicates that some other factor besides forgetting and selective attention plays an important role in serial reversal—so that evidence for quantitative differences in performance between pigeons and rats does not necessarily provide support for the notion that the species differ in stability of attention.

In our earlier discussion of serial reversal in fish, it was suggested that an important factor in improvement might be the development of strategies which allow rapid behavioural adjustment, relatively independent of the reinforcement history of the stimuli involved. As was reported in that discussion (p. 77), there is direct evidence for the use of 'win-stay, lose-shift' strategies in blue jays, evidence which will be introduced in the section on learning sets in this chapter. Mackintosh and Little (1969*a*), however, have suggested that pigeons may not be capable of adopting strategies of this kind (it will be recalled that Mackintosh has suggested that pigeons' serial reversal performance might be inferior to that of 'higher' birds, such as corvids). They support their case by reporting experiments which showed that pigeons found it difficult, if not impossible, to master a complex 'delayed double-conditional' task whose solution appeared to pose the same formal demands as those required for mastery of 'win-stay, lose-shift' strategies. There are, however, data from the serial reversal task itself to suggest that pigeons can indeed adopt such strategies. First, as we have seen, pigeons do under some circumstances achieve one-trial reversals (Gossette and Hood 1968), and this provides good *prima facie* evidence for the use of strategies. Second, Williams (1971) has shown that pigeons' serial reversal performance is improved by decreases in the inter-trial interval. One account of Williams' finding is that the birds retain the outcome of the immediately preceding trial, and use that information to guide choice on the current trial; this account is supported by a subsequent analytical study (Williams 1976) that provides good evidence that pigeons, after experience on serial reversals, do base their choices on the outcome of the preceding trial, and that retention of that information decays rapidly over a period of approximately 30 to 60 seconds. Finally, Miller Hansen, and Thomas (1972) report that pigeons given extensive serial reversal training (of a successive colour discrimination) showed superior performance on (a single) reversal of a subsequent line-orientation discrimination, compared to controls given non-discriminative pre-training alone; it is important to note that there were no differences between the groups on acquisition of the line-orientation discrimination, so that the results cannot be simply ascribed to such 'general' factors as practice, habituation to the apparatus, and so on. On balance, then, the evidence does support the view that pigeons, like blue-jays, do adopt strategies during the course of serial reversal training.

It may, then, be agreed that, in both pigeons and rats, proactive interference, stability of attention to the relevant dimension and the adoption of strategies all play a role in serial reversal improvement. Although most reports show pigeons performing at a lower level than that commonly seen in rats, it is clear that contextual variables play a large role in determining pigeons' performance, and not clear how those variables take their effect. We may incline, therefore, to agree with the conclusion expressed by Miller *et al.* (1972), that theoretical controversies over the nature of differences in levels of serial reversal performance will remain premature until a much greater range of contextual variables has been explored.

Probability learning

Bitterman's studies of probability learning in pigeons (Bullock and Bitterman 1962*b*; Graf, Bullock, and Bitterman 1964) showed that whereas the birds tended to maximize on spatial problems, they generally matched on visual problems, and so were more 'fish-like' than 'rat-like' in visual tasks. However, Bitterman's own studies indicated that pigeons do not always match on visual tasks: Graf *et al.* (1964) ran a 70:30 colour discrimination under five different conditions, finding matching in four of these conditions, but maximizing in a fifth (a condition in which a centre key was used to set up each trial, and in which errors were followed by 'guidance' trials, in which only the correct stimulus was presented). Subsequent studies (Shimp 1966; Mackintosh *et al.* 1971) have, moreover, found maximizing in two of the conditions in which Graf *et al.* found matching (on a successive visual problem, using a central set-up key and a correction procedure; and in the conventional simultaneous problem, without a centre key, and using a guidance procedure), so that the only (weak) claim that may now be made is that pigeons may be less likely to maximize in visual than in spatial problems. Bitterman, moreover, provides no evidence, such as data from individual birds or detailed analyses of error patterns, to show that truly 'random' matching—the phenomenon that has not been observed in mammals (p. 80)—ever occurs in pigeons, so that there is no compelling reason to suppose, from the data available, that there is a qualitative difference between birds and mammals reflected in probability learning performance.

Mackintosh (1969*a*) does, however, argue that there are quantitative differences between rats and pigeons in probability learning tasks, and that these reflect differences in stability of attention; he claims that neither rats nor pigeons literally maximize, and that the errors made by both groups generally reflect deviations of attention, in the face of inconsistent reinforcement, from the relevant dimension. Mackintosh's case is that, because attention in birds is less stable than in rats, more such errors occur.

To support the claim that pigeons when 'maximizing' do make more errors than rats, Mackintosh (1969*a*) quotes published studies to show that, when trained on a 70:30 spatial problem, rats generally reach an asymptote of more than 95 per cent choices of the majority stimulus, noting that the asymptote reached by pigeons in the Graf *et al.* (1964) study was only 82 per cent. Mackintosh also describes an experiment of his own which compared the performance of rats and chicks in the same apparatus (a Grice-box) in 200 trials of 75:25 problems with either spatial or brightness cues relevant. The performance of rats and chicks was essentially indistinguishable on the brightness problem (about 90 per cent choice of the majority stimulus), but chicks (at about 80 per cent majority choice) were notably inferior to rats on the position problem (where rats again performed at the 90 per cent level). Mackintosh accounts for the relatively good chick performance on the brightness problem by arguing that visual cues tend to be more dominant in birds, so that attention would be less likely to stray on such problems.

Further support is provided by evidence that in rats, pigeons, and fish, errors tend to be controlled by an irrelevant dimension: for example, 88 per cent of all errors made by rats in the Mackintosh (1969*a*) spatial 75:25 problem were to the preferred (black) value of the irrelevant (brightness) dimension. Mackintosh argues, therefore, that subjects are not simply from time to time choosing the minority stimulus, but that they are switching their attention intermittently to an irrelevant dimension, and choosing the stimulus on that dimension having most approach strength. Where birds are trained with a salient dimension relevant (e.g. brightness for chicks) then their performance may match that of rats: however, where there is a salient irrelevant dimension, then birds are more likely to be distracted than are rats, since training does not lead to the stability of attention found in rats.

Although Mackintosh presents a persuasive case for the view that in both rats and birds deviations from maximizing are to be explained in terms of attentional processes, two problems remain for the proposal that there are quantitative differences in attention between the two groups. The first problem is that one report (Mackintosh *et al.* 1971) finds performance of pigeons on 70:30 problems using either spatial or visual (colour) cues reaching a level of better than 90 per cent by the termination of training. In the spatial discrimination the stimuli from the irrelevant dimension (red and green) were present but irrelevant, so that in at least one case, a salient, irrelevant dimension was presumably available, and yet performance achieved a level very little below that typically found in rats. In other words, in at least some cases, the performance of birds is so close to that of rats that any quantitative difference appears trivial; precisely which variations in conditions yield good performance in pigeons remains, of course, obscure at present.

A second weakness in Mackintosh's case is that there is as yet no direct evidence that errors in probability learning by birds are caused by deviations in attention. It could be, for example, that there is a weak tendency to select the minority stimulus and that this summates with the tendency to choose a preferred value on the irrelevant dimension to produce an error. In this case, fluctuations in the tendency to select the minority stimulus could influence overall error scores, independent of any fluctuations in attention, and yet yield the systematic error patterns detected by Mackintosh. There is evidence that, in rats, manipulations designed to strengthen or weaken attention to the relevant analyser (pretraining on a 100:0 problem on either the relevant or an irrelevant dimension) do raise or lower asymptotic performance on a 70:30 problem (Mackintosh and Holgate 1967), and this may be taken to show that variations in strength of attention in rats do play a significant role in probability learning, but we have no comparable data on any other species.

We conclude, then, that since there is no evidence that random-matching occurs in birds, there is no reason to suppose that probability learning performance shows any qualitative difference between birds and mammals. Although birds (or, at least, some birds) may perform at levels lower than those achieved by rats, there is good reason to suppose that their performance can be improved by changes in contextual variables, and little direct evidence that the differences reflect differences in stability of attention.

Effects of reward shifts

As there appears to be general agreement that birds respond in similar ways as rats to reward shifts, this section will be brief, and will confine itself to a summary of the evidence leading to that conclusion, with a minimum of theoretical analysis.

Pigeons, like rats, show a spaced-trial (24 h ITI) PREE (Roberts, Bullock, and Bitterman 1963, using a runway), and at a shorter (6 s) ITI, show response patterning where reward and non-reward trials alternate (Roberts *et al.* 1963, using a key-pecking apparatus). Brownlee and Bitterman (1968), while not giving details, report that in extinction after either large or small rewards, and over a series of acquisitions and extinction, pigeons 'performed in much the same way as rats' (Brownlee and Bitterman 1968, p. 345). These data, then, provide good evidence that Bitterman's 'contrast' mechanism exists in pigeons (see p. 85), and so, that pigeons learn to anticipate food rewards. There is no published report to date of a successive NCE in pigeons, but there are reports of a simultaneous NCE in pigeons in both a discrete-trial (Brownlee and Bitterman 1968) and free-operant (Gonzalez and Champlin 1974) discriminations.

There is therefore good reason to suppose that the mechanisms engaged by shifts in reward are comparable in mammals and birds (or at least in rats and pigeons), there being no evidence to date of either qualitative or quantitative differences between the species.

Mechanisms of attention

The discussion in this section will be directed at the assessment of two propositions: first, that birds have no mechanisms of selective attention (Couvillon *et al.* 1976), and second, that the stability of analysers in birds is less than that in rats (Mackintosh 1969*a*).

The Couvillon *et al.* (1976) proposal arises from three failures to find any advantage of intra-dimensional over extra-dimensional shift learning in pigeons. Two of these failures involved the use of experienced pigeons, and in one of those studies, the stimuli were presented successively rather than simultaneously as in the conventional mammalian studies. It has already been argued (p. 96) that although such results may be embarrassing for attention theory, they do not necessarily indicate any difference in attention between birds and mammals; moreover it is not unlikely that minor modifications to attention theory could accommodate difficulties in obtaining IDS superiority in experienced subjects or in successive-presentation designs. The third failure reported by Couvillon *et al.* is of considerably more importance, since in it they used naive pigeons, and a simultaneous choice design: the two dimensions concerned were colour and line orientation, and, although IDS birds did make fewer errors than EDS subjects, the difference fell short of statistical significance. The problem with this result is that, in an experiment of almost identical design, Mackintosh and Little (1969*b*) did find a significant superiority of IDS over EDS subjects. Table 6.1 compares the error scores obtained in the shift phase of the two studies, and it can be seen that there is a close similarity between them. This similarity suggests in turn that the Couvillon *et al.* report may have failed to find a significant effect, not because there was no real effect to be found, but because there was too much variance in their data. This possibility gains plausibility from the fact that only 12 subjects were used in their study, so that each of the error scores shown in Table 6.1 from their study represents the mean of only three subjects (the Mackintosh and Little study used 16 birds). Unless further negative reports appear, the most plausible account is that, when designs comparable to those used with mammals are employed, there is in pigeons a superiority of IDS over EDS learning.

That attention theory is indeed applicable to birds is further supported by demonstrations in birds of four further phenomena. Two of these, overshadowing and blocking, have previously been described (p. 98); both phenomena can be obtained in pigeons (Miles and Jenkins 1973; Mackintosh and Honig 1970) but, as was pointed out in Chapter 3, although an attentional analysis of the phenomena is available, an alternative analysis which does not rely on selective attention is available (Rescorla and Wagner 1972). Two further phenomena, transfer along a

Table 6.1. *Comparison of the results of Mackintosh and Little* (ML) *with those of Couvillon, Tennant, and Bitterman* (CTB)

		Mean errors to criterion	
	Condition	ML	CTB
IDS	Colour–Colour	13.5	13.0
EDS	Orientation–Colour	30.5	21.3
IDS	Orientation–Orientation	54.0	43.0
EDS	Colour–Orientation	67.0	58.3

continuum and latent inhibition, have not been described in previous chapters (and, indeed, have not been reported in any cold-blooded vertebrate), and each provides good support for attention theory.

Transfer along a continuum, or the easy-to-hard effect, consists of the demonstration of superior learning of a difficult discrimination following pretraining in an easy discrimination using stimuli from the same dimension. For example, Williams (1968) gave pigeons 200 rewarded trials on a difficult simultaneous size discrimination, comparing their terminal performance with that of birds given 50 trials on an easy size discrimination, followed by 150 trials on the difficult discrimination, and found superior performance in the latter group. Attention theory accounts for this effect by supposing that the pretraining on the easy discrimination allows the relevant (size) analyser to be switched in firmly so that it remains switched in despite the errors, possibly perceptual in origin, that occur in the difficult discrimination. Conversely, the birds trained throughout on the difficult discrimination do not, as a consequence of persistent errors, learn to attend consistently to the relevant dimension, and so generate errors both through perceptual failures and through shifts of attention. Further support for this interpretation and against, say, accounts of transfer in terms of stimulus generalization, comes from demonstrations by both Williams (1968) and Mackintosh and Little (1970, using colour as the relevant dimension) that reversing the relative values of the stimuli in the shift from the easy to the hard discrimination still obtains the effect: that is, even if the positive stimulus in the easy phase is, say, the larger of the two stimuli, and the positive stimulus in the difficult phase is the smaller, positive transfer is found.

Latent inhibition (Lubow 1973) is the term used to refer to an effect commonly found in rats, in which non-reinforced pre-exposures to a stimulus significantly retard subsequent associative learning involving that stimulus. This phenomenon, which may be akin to habituation, can be plausibly interpreted as indicating that selective attention to a stimulus wanes if the stimulus does not predict any significant event. Tranberg and Rilling (1978) exposed one group of pigeons to 500 presentations of a key-light stimulus and showed that subsequent autoshaping to that stimulus was significantly retarded when compared to that of pigeons given comparable pre-exposure to the experimental chamber, but no non-reinforced light presentations.

This survey has then found good evidence for the proposal that pigeons do possess mechanisms for selective attention, or at least for the proposal that the phenomena which are taken by some theorists to demonstrate selective attention are found in pigeons as well as in rats. We shall now turn, therefore, to the question whether attention is less stable in birds. Some relevant evidence has previously been considered, in the sections on serial reversal and probability learning. There is also some support from the phenomena already considered in this section: the fact that Couvillon *et al.* (1976) could not replicate Mackintosh and Little's (1969*b*) finding of significant IDS superiority in pigeons might suggest that the phenomenon is less reliable in pigeons than in rats; similarly, although Tranberg and Rilling (1978) found a latent inhibition effect in pigeons, three previous reports (Cohen and MacDonald 1971; Mackintosh 1973; Wasserman and Molina 1975) failed to find a significant effect of non-reinforced presentations of a stimulus (although in all of these reports, the trend was towards a latent inhibition effect).

The final phenomenon to be discussed in this section, the overtraining reversal effect (ORE, see p. 95), is considered by Mackintosh to provide evidence for a lower stability of analysers in birds as opposed to mammals. In a survey of this topic, Mackintosh (Sutherland and Mackintosh, 1971) discusses a number of studies (e.g. Brookshire, Warren, and Ball 1961; Schade and Bitterman 1965) using birds which obtained either no effect of overtraining on reversal, or a reverse ORE (poorer reversal performance following overtraining on the original discrimination). Mackintosh argues that these early failures were probably due to the use of easy discriminations, having a salient relevant dimension. In support of this proposal, Mackintosh points out that the ORE is not obtained in rats in easy discriminations (Mackintosh 1969*c*) and that, where birds are given difficult discriminations, then no reverse ORE is obtained, and in some cases at least, a positive ORE is found (e.g. Williams 1967; Mackintosh 1965).

Now although an ORE can be found in birds, it is not easily obtained, and is small when it is found; moreover, a reverse ORE is very rarely reported in rats, and only on very easy spatial discriminations. Accordingly, Mackintosh proposes that differences in reversal following overtraining between birds and rats do reflect real differences, but that these differences are quantitative, and not qualitative in nature. Specifically, Mackintosh proposes (Sutherland and Mackintosh 1971) that changes in analyser strength as a consequence of rewards and non-rewards proceed, relative to changes in response strength, more rapidly in birds than in rats, so that, first, overtraining in birds will tend to increase response strength rather than (as in rats) analyser strength and, second, the introduction of non-reward at the outset of reversal following OT will tend to switch out the relevant analyser before response strength has had time to reverse, as it does in overtrained rats. Mackintosh goes on to cite evidence from an experiment (Mackintosh 1965) which undertook a detailed analysis of response patterns during reversals by over-trained and non-overtrained rats and chicks; the outcome of these analyses supported the view that where overtraining did benefit reversal (in rats or chicks), analyser strength had been selectively strengthened so that response strengths reversed before analysers were switched out, and that where overtraining retarded reversal (in chicks), then overtraining had not increased attention to the relevant dimension, while it had increased response strength. Now although these analyses give good support to an explanation of the ORE in terms of attention theory, they do nevertheless pose problems for the proposal of interest here, namely, that birds' analyser strengths are relatively more labile than their response strengths. The problem lies, as we have found before when considering proposals of quantitative species differences, in the imprecision of predictions to be derived. It is not the case, for example, that an ORE is predicted whenever a difficult discrimination is involved, but neither is it the case that the ORE will never occur: the proposal is, that the ORE will sometimes occur in difficult discriminations, and sometimes not, there being no way of saying which discriminations are of appropriate difficulty. It is, in other words, not possible to predict, from the salience of the relevant dimension, whether an ORE will be obtained in birds, and so, reasonable to suppose that quite other factors are involved. There is, moreover, evidence that failures to obtain the ORE do not invariably reflect similar difficulties: for example, whereas Mackintosh (1965) found that over-training chicks on an (easy) brightness discrimination led to increased position habits in reversal training, Matyniak and Stettner (1970) found that overtrained bob white quail (*Colinus virginianus*) in a (difficult) pattern discrimination showed no increase in position responding, although they were, like the overtrained chicks, slower to reverse than non-overtrained subjects.

The conclusion reached here, then, is that although there is good evidence from both rat and chick studies to support an attentional account of the ORE, there is little evidence for the proposal that the reduced likelihood of obtaining the ORE in birds is due to differences in stability of attention. It seems equally likely that some contextual variables might be critical, and it may be relevant to note here both that two contextual variables, reward size and task difficulty, have been identified by Mackintosh (1969*c*) as being critical for the occurrence of ORE in rats, and that the ORE has never yet been obtained in rats or any other species in an automated apparatus (Mackintosh, personal communication).

Learning-set formation

Harlow (1949) showed that monkeys given a series of discrimination problems having the same method of solution show a progressive improvement over the series, and come eventually to solve new problems after only one trial. The impressive interproblem learning shown is known as learning-set formation: the animals appear to be learning to learn. Since it is difficult (but not impossible—see Reese 1964) to interpret learning sets in terms of associations between particular stimuli and responses, many psychologists have regarded learning-set formation as evidence for higher-order learning or rule-learning, and comparative studies of learning-set formation in mammals have been carried out in the belief that such higher-order learning might discriminate between species of differing intellectual ability. Encouragement for such a view was provided by a number of early studies which suggested that optimal learning-set performance was found in primates, the performances of other mammals being ranked in a way that agreed roughly with 'intuitive' estimates of their intelligence. The mammalian studies will be discussed in a later section, and our task here is to contrast the performance of birds with that seen in primates, and to ask whether any observed differences encourage the view that birds are either qualitatively or quantitatively inferior to primates in this task.

In our previous discussions of serial reversal learning, we introduced the notion of 'strategies', contrasting them, implicitly or explicitly, with the incremental processes conventionally assumed to be engaged by reward and non-reward. Little attention was given to the precise nature of the strategies involved, apart from observing that the increases in the rapidity of adjustment to the changes in reward value involved in reversals could be accounted for in part by supposing that a 'win–stay, lose–shift' strategy was operating. There are more specific accounts available of the similarly rapid adjustment to novel reward conditions seen in learning-set formation, and we shall now set the background for the comparison of birds and primates by considering two of the most influential of these accounts.

Restle (1958) distinguished between three types of cues present in learning set tasks, which he labelled Type A, Type B, and Type C cues. Type C cues are irrelevant to the solution of any problem (cues such as general features of the apparatus) and the neutralizing or adapting out of such cues could provide some positive transfer across a series of problems. Type B cues are relevant to the solution of a particular problem—the colour, shape, and texture, for example, of one of the objects concerned, and building up appropriate associative strengths to Type B cues would solve a given problem, but would not show positive transfer to other problems. Type A cues are valid for all problems, and are abstract, contingent cues, analogous to conditional cues, which may be conceived of as being like, for example, 'the object which was correct on the previous trial'.

Restle's analysis distinguishes between the cue-value of reward, especially on Trial 1 of a problem, and its reinforcing value, pointing to the important role played by reward (or non-reward) in forming the Type A cue, That performance comes to be independent of the reinforcing value of the reward has been emphasized by experiments in which the correct Trial 1 object covered, not a reward, but a marble; Trial 2 performance in such experiments by experienced monkeys is actually better with the marble than with the reward on Trial 1 (Riopelle, Francisco, and Ades 1954). Restle provides a mathematical analysis of learning-set acquisition data, showing that acquisition can be accounted for by supposing gradual incremental changes in associative strengths of the three types of cues, culminating in the domination of type A cues. What Restle's analysis shows, then, is that a relatively traditional account of learning-set formation can be provided, given the assumption that we regard as a single cue two pieces of information: first, the object chosen on the preceding trial and, second, the outcome of that trial.

A second analysis (Levine 1959) argues that performance in learning-set acquisition may be more systematic than is envisaged in Restle's account, and supposes in fact that animals adopt (and reject) various strategies or hypotheses as they master the task, only one hypothesis being engaged on any given trial. Examples of such hypotheses, are, for example, select (or avoid) the position rewarded (or not rewarded) on the preceding trial (win-stay, lose-shift with respect to position), follow the most preferred stimulus, irrespective of reward and non-reward (stimulus preference) and select (or avoid) the object rewarded (or not rewarded) on the preceding trial (win-stay, lose-shift with respect to the object— the only appropriate strategy for an object-discrimination learning set). Levine (1959) has provided an elegant mathematical model which allows assessment of the use of eight alternative strategies in learning-set formation; the model analyses performance over the first three trials of a problem, and assumes that only one strategy will be used over those three trials. Since Levine's method of strategy analysis has been applied to data from birds, we shall be able to compare the results of that analysis with results obtained from primates, and there does not appear to be any need to decide whether Levine's account is in general to be preferred to Restle's.

Common to both Restle's (1958) and Levine's (1959) accounts is the notion that choice on a given trial is determined, in a learning-set experienced subject, by the events of the immediately preceding trial, so that no animal incapable of retention over the inter-trial interval of the information provided by that one trial (the object chosen, and the outcome of the trial) could be said to be controlled by type A cues or to be adopting a win-stay, lose-shift strategy. This view of learning-set performance, while well-adapted to mathematical analysis of data, may be too restrictive if applied to strategies in general: it has been argued (p. 78) that fish, who are, it seems, relatively insensitive to the after effects of reward and non-reward, might engage a win-stay, lose-shift strategy, but would require a number of rewards and non-rewards for its activation. This present discussion has shown that Restle's analysis allows an incremental view of learning-set formation, and has introduced Levine's method of strategy analysis which will be valuable in the comparison of birds' and primates' performance. We shall, however, continue to use the word 'strategy' to refer to those methods of problem-solution that emancipate animals from control by the incrementally changed associative strengths of the specific stimuli present in a given problem, whether or not such strategies rely wholly on information retained from the immediately preceding trial.

Learning-set formation has been studied in a variety of avian species, including non-passerine species, and contrasts between avian species will be discussed in a later section. In this section we shall concentrate on a series of reports by Kamil and his colleagues which describe the performance of northern blue jays; we shall see that these reports have provided analyses of learning-set performance in blue jays which allow detailed comparisons of their performance with that of rhesus monkeys (*Macaca mulatta*).

Kamil's studies have all concerned object-discriminations, in which a new pair of junk objects (e.g. toys, household items, statuettes, etc.) are used for each problem, and response to one of these objects is consistently associated with reward. In an early report, Hunter and Kamil (1971) showed that performance, assessed by percentage correct choices in Trial 2 of each discrimination, improved over 700 problems to 75 per cent correct; this figure, while below the 90 per cent and better reported for rhesus monkeys (e.g. Harlow 1949) falls well within the mammalian range, and is close to the levels reported for squirrel monkeys (*Saimiri sciureus*) and marmosets (*Callithrix*) (Miles 1957; Miles and Meyer 1956). However, the absolute level of performance tells us little of the manner in which it is achieved, and Hunter and Kamil went on to apply Levine's hypothesis-testing model to their data. Application of the model confirmed that the blue jays had adopted a 'win-stay, lose-shift with respect to object' strategy; this, although of course, expected, is not a necessary consequence of a 75 per cent Trial 2 accuracy, since that level could be achieved in a number of other ways—for example, by a bird that adopted a 'win-stay' strategy, but chose at random following a non-reward. A further interesting result of the analysis was that, particularly in the early stages of learning-set formation, the birds showed frequent following of a 'win-stay lose-shift with respect to position' strategy, as though they had grasped the 'conditional' nature of the appropriate cue, and had retained the outcome of the trial, but either had lost object information or allowed their attention to shift to an irrelevant dimension. Finally, Hunter and Kamil (1971) derived a statistic, the percentage of variance explained, introduced by Levine as a measure of the internal consistency of his model: in essence, this statistic reflects the accuracy with which the use of the various hypotheses can be predicted for one set of trial outcomes by the results of the application of the analysis to another set of trial outcomes. The value obtained for the percentage of variance explained in blue jays was 85 per cent; this agrees well with those reported by Levine (1965) for rhesus monkeys (from 67 to 94 per cent), which in turn argues that the performance of blue jays is as systematic as is that of monkeys.

Kamil, Lougee, and Shulman (1973) report further similarities between the performance of blue jays and rhesus monkeys: first, they found that retention of a learning-set was virtually perfect over a five-month interval indicating long-term retention of the appropriate strategy (or long-term adaptation of Type B and C cues): Second, Kamil *et al.* found that, if a delay was introduced between Trials 2 and 3 of a problem, considerable intra-problem forgetting occurred; this latter effect was more marked on problems where Trial 1 choice had not been rewarded than where Trial 1 had been rewarded. This shows that short-term retention is, as expected by both Restle and Levine, critical in learning-set formation; Kamil *et al.* go on to produce evidence to show that the relative lack of effect in problems where Trial 1 had been rewarded reflected simply the fact that the birds choose initially the most preferred or attractive object, and, following a delay after Trial 2, re-selected that object, not because of any memory for its association with reward, but because, having lost such associations, choice was based again on attractiveness. The above two phenomena have also been reported in rhesus monkeys (Bessemer and Stollnitz 1971). A third parallel reported by Kamil *et al.* (1973) is that, if one of the pair of objects seen on Trial 1 is replaced on Trial 2, that replacement is without effect if the object replaced was not the one chosen on Trial 1; a similar dependence of Trial 2 performance on the presence of the object selected on Trial 1 was reported by Lockhart, Parks and Davenport (1963), using pigtail macaques (*Macaca nemestrina*). This phenomenon indicates that neither blue jays nor monkeys learn about the object not chosen on Trial 1 – they attend solely to the object chosen.

Kamil and Mauldin (1975) explored intra-problem retention during the course of learning-set acquisition by naive blue jays. In this experiment, a delay (of from 1 to 5 minutes) was introduced between Trial 3 and Trial 4 of each problem, and a progressive decline in retention over the delay interval was observed as performance improved: once again, a similar finding has been reported in rhesus monkeys by Deets, Harlow, and Blomquist (1970) who showed that increasing inter-trial interval had a more disruptive effect on performance in learning-set experienced subjects. Kamil and Mauldin interpret this result as showing that, in the early stages of acquisition 'problems are probably solved in a gradual, incremental way which is based upon changes in the response tendencies or habits elicited by the stimulus objects themselves' (p. 130), and that object-discrimination learning-set acquisition, 'especially in terms of high levels of performance on Trial 2 of new problems, reflects the acquisition of the conditional discrimination such that the subject's choice behaviour on these early trials is primarily under the control of memory traces for recent trial events' (p. 130). This interpretation accords well with Restle's (1958) view that Type B cues will gradually exert less control over choice as Type A cues gain associative strength.

Finally, support for the view that blue jays are capable of forming a generalized win-stay lose-shift strategy is provided by the demonstration by Kamil *et al.* (1977) of substantial positive transfer from serial reversal training (using one pair of objects) to object discrimination learning-set formation. Similar positive transfer has been reported also in chimpanzees (*Pan troglodytes*) (Schusterman 1964) and in macaque monkeys (Schrier 1974), although Warren (1966) found no transfer from serial reversal to learning-set formation in cats. The finding has two implications for blue jays: first, that in serial reversal training, they adopt a strategy (win-stay, lose-shift) which is sufficiently general to transfer to problems using novel stimuli and, second, that the strategy is indeed relevant to learning-set formation in blue jays.

The work from Kamil's laboratory provides, then, a series of impressive parallels between the performance of blue jays and primates in learning-set formation, and more than justifies the claim that 'it appears that the differences between macaque monkeys and blue jays in object discrimination learning-set behaviour are more quantitative than qualitative' (Kamil *et al.* 1973, p. 403). That there are quantitative differences in performance cannot be denied, but it is not clear that these in turn reflect differences in intellectual capacity. There is good evidence that manipulation of contextual variables affects the performance of rhesus monkeys (Devine 1970), and evidence too from Kamil *et al.* (1973) that an increased tendency towards perseverating with preferred stimuli may contribute to a lowering of asymptotic performance in blue jays compared to rhesus monkeys. It is possible, therefore, that the blue jays' performance could be improved and that one technique for improvement may be the use of stimuli that do not elicit such powerful preferences, so that the strongest conclusion that emerges from this review is that there is no reason to suppose that the superior performance of (some) primates reflects the possession of any learning mechanism not available to blue jays.

Sameness–difference concept

In this and the following two sections, we shall consider three types of concept formation which, although no doubt involving very different modes of solution, do have in common that they involve discrimination learning tasks in which the positive stimulus is identified, not by any simple sensory attribute, but by its conforming to some abstract rule which generalizes to stimuli of differing sensory qualities. If we adopt Restle's (1958) analysis of learning-set formation, then we are looking here at further instances of 'Type A' cues, an observation made by Restle himself (1958) with specific reference to the concept we shall discuss in this first section.

Acquisition of the sameness–difference concept is typically studied by discrimination tasks in which three stimuli (two of which are identical) are shown on each trial, and the subject is required to respond to either one of the two identical stimuli (in a matching task) or the different stimulus (in an oddity task). The sensory qualities of the 'same' and 'different' stimuli are varied from trial to trial so that only the relationship between the stimuli is relevant to the solution of the problem. However, where a limited stimulus ensemble is used, at least three modes of solution are available: subjects may indeed respond on the basis of an abstract concept such as identity or oddity–in which case they should show positive transfer to novel sets of stimuli. Alternatively, subjects could respond on the basis of the configurational relationship between the stimuli. This latter mode of solution is not particularly unlikely in the standard testing design, in which the subject is asked to respond to the left or right stimulus, but not to the centre stimulus, on which the odd stimulus is never displayed; subjects could, where stimuli were either red or green, learn to match by choosing the left-hand key when shown either red-red-green or green-green-red, and the right-hand key when shown red-green-green or green-red-red. A third solution is to solve the problem as a conditional discrimination—to (say) choose red, if the centre stimulus is red, and green if the centre is green. These latter two modes of solution would be expected to show comparatively little transfer to other stimuli.

Although there has been little support for the notion that matching or oddity tasks might be solved by configurational learning, there have been a number of reports supporting the view that these tasks are simply examples of conditional discriminations (e.g. Cumming and Berryman 1961; Berryman, Cumming, Cohen, and Johnson 1965; Carter and Eckerman 1975). Berryman *et al.* (1965) trained pigeons in an oddity task in which the stimuli might be red, green, or blue. When the birds had mastered the task, a yellow stimulus was substituted for the blue stimulus, and transfer to yellow, as a standard (centre key) or comparison (side key) stimulus, was analysed. The birds' performance in the transfer phase was poor, and best described by assuming that, in the acquisition phase, three rules (one for each standard stimulus) of the form: 'if see green, approach red or blue' had been learned, and that, when yellow was the standard stimulus, it was 'coded' as red. Carter and Eckerman (1975) showed that although a conventional colour matching task was mastered more easily than a 'symbolic' matching task in which horizontal or vertical lines served as conditional stimuli to indicate which colour of stimulus to choose on the side keys, the difficulty of the symbolic task was due in large part to the difficulty in achieving a successive discrimination between the two line stimuli.

However, while there are reports which indicate that pigeons may not always use a concept of identity or oddity when it would be appropriate but not necessary for problem solution, they do not, of course, indicate that pigeons cannot form such concepts, and there is in fact good evidence that they can. Carter himself (Carter and Werner 1978) has shown that pigeons learn a simultaneous colour matching problem much more rapidly than a 'symbolic' matching task (in which the conditional or standard stimulus is a colour that differs from each of the comparison stimulus colours); this design eliminates differential discrimination difficulty between the two tasks and gives results that are not easily explained except in terms of concept formation. Further convincing evidence is provided by Zentall and Hogan (1974, 1975, 1976) who have trained groups of pigeons on either a matching or an oddity task using stimuli varying along one dimension (e.g. colour) and then transferred the birds to stimuli differing along a new orthogonal dimension (e.g. brightness or shape), either shifting the nature of task (from matching to oddity or vice versa) or leaving the task unchanged. The results of these studies consistently show better transfer performance in the birds for whom the nature of the task remained unchanged than in shifted birds. Neither the configurational nor the conditional accounts of matching and oddity can provide any explanation of the outcome, which provides direct evidence for the proposition that pigeons find more in common between (say) a colour matching task and a brightness matching task than between a colour oddity task and a brightness matching task. Carter and Eckerman (e.g. 1975) object that appropriate controls are required to show that positive transfer in the non-shifted group has occurred, and suggest that the major effect is one of negative transfer in the shifted groups. Now although it would be interesting to know whether in fact positive transfer is obtained, the overall force of the objection is weak: the concept-formation account can easily explain negative transfer in the shifted groups, and assume that weak positive transfer in non-shifted groups is due to generalization decrement occasioned by the use of novel stimuli (but see Zentall and Hogan 1978), whereas neither the configurational nor the conditional accounts can provide any explanation at all for differential transfer effects obtained by the different training procedures.

Our conclusion is, then, that pigeons can use concepts of sameness and difference; however, transfer of these concepts from one problem to another has been unimpressive in most studies, and certainly not at the level reported for rhesus monkeys by Moon and Harlow (1955), who found, in an oddity task using objects, performance on Trial 1 of a novel pair of objects rising to almost 90 per cent correct after about 130 problems. The difference, may, of course, be due to procedural factors: Zentall and Hogan (1978) have in fact shown somewhat more substantial transfer effects to novel stimuli in pigeons, following training in which there were interspersed among conventional trials 'negative instances'. By negative instances Zentall and Hogan mean trials in which no correct response is possible (in which, for matching subjects, neither comparison stimulus was the same as the standard, or, for oddity subjects, all three stimuli were identical, and in which responding was punished by delay of the next trial).

A further demonstration of the role of contextual variables in this area has been provided by Zentall, Hogan, Edwards, and Hearst (1980), who explored oddity learning in pigeons when a variable number (from 2 to 24) of alternative (matching and incorrect) stimuli were exposed along with the single odd stimulus. Zentall *et al.* found that performance was very much more efficient as the number of alternatives increased, and that, although performance with 2 alternatives did benefit from prior training with a large number of alternatives, there was, nevertheless, a large drop in performance as the number was reduced (e.g. from over 90 per cent selection of the odd stimulus when five matching stimuli were presented to approximately 70 per cent when only two matching stimuli were presented). Although no clear account is available of the importance of increasing the number of alternatives, Zentall *et al.* present evidence to show that the perceptual salience of the odd stimulus is enhanced where more matching stimuli are shown. They also show that increasing the number of alternatives does not enhance the attractiveness (in the absence of differential training) of the odd stimulus, and quote earlier findings of improved oddity learning with more alternatives in both chimpanzees (Nissen and McCulloch 1937) and canaries (Pastore 1954). Birds' acquisition of oddity is, then, like that of chimpanzees, affected to an unexpected degree by apparently perceptual factors, and birds are capable of better concept formation than that reported using the conventional training procedures. Moreover, the monkey studies have used objects as stimuli, and it is known that three-dimensional stimuli are generally more effective than two dimensional stimuli for monkeys (Meyer, Treichler, and Meyer 1965). The learning-set experiments using corvids, reviewed above, showed rather impressive performance, and in these studies, objects were used as discriminanda: it may be that birds' matching-oddity performance would be similarly impressive if objects were used.

Number concepts

There has been a long history of experiments on number concepts in animals, and many successful reports may be criticized on the grounds that a solution based on attributes other than number was available (Honigmann, 1942). The most convincing successful reports, all of which use birds, are from Koehler's laboratory in Freiburg, and these are reviewed and criticized by Honigmann (1942), Thorpe (1963), and Wesley (1961). The studies use two basic experimental techniques which, according to Koehler (1950) demonstrate the existence in birds of two prelinguistic abilities which are an essential prerequisite to the development in man of counting; Koehler holds that birds do not actually count, but 'do think unnamed numbers' (Koehler 1950, p. 42). Koehler's first ability is that of 'being able to compare groups of units presented *simultaneously* side by side by seeing numbers of those units only, excluding all other clues' (p. 42). This ability was demonstrated (in a raven and a parrot) by showing that the subjects could choose, from a collection of five boxes having lids with two, three, four, five, and six spots on them, that box whose lid had the same number of spots as spots (arranged in various arrays) on a 'key' card.

The implications of results using the simultaneous technique depend upon theoretical interpretations of the birds' performance. Koehler (1950) argues that birds do not count, because birds do not use words, and do not *name* objects or places. The suggestion appears to be that detection of identity in number of sets of units in varying spatial arrays is in some way a direct or perceptual process. But whether any such process could detect number without a 'counting' process is far from clear; it is not easy to imagine how a machine could be programmed to respond differentially to different numbers of items without actually counting the items. Perhaps it is most useful to suggest that a distinction could be made between counting devices that are relatively specific and those that are general; specific devices, tied to particular perceptual systems, might generate differential outputs according to numbers perceived, but their counting ability would not be available for information derived from any other channel. This notion is similar (if not identical) to the concept of 'subitizing', introduced by Kaufman, Lord, Reese, and Volkmann (1949) to account for the rapid perception by humans of a number of items when shown a few shapes using brief tachistoscopic exposure.

Koehler's second ability is 'to remember numbers of incidents following each other and thus to keep in mind *numbers presented successively in time,* independent of rhythm or any other clue which might be helpful' (Koehler 1950, p. 43). For example, pigeons were trained to eat a fixed number of peas delivered via a shute, punishment being delivered if the birds took more than the specified number. The interval between delivery of the peas was varied, so that temporal cues were not relevant. In another, similar, task, birds were trained to take a certain number of baits from a larger number of boxes (some of which contained no reward). A difficulty with these problems is that there may have been a physical correlate – quantity of food in the crop – that correlated with the number of rewards taken: evidence against such an interpretation comes from a famous observation made by Schiemann, one of Koehler's collaborators, and described by Thorpe (1963), Honigmann (1942), and Koehler (1950). A jackdaw, trained to take five baits, was presented with six boxes, there being one bait in the first box, two in the second, one in the third, none in the fourth, one in the fifth, and none in the sixth. The bird, in fact, took only four baits before retiring to its cage; but, after a brief pause, the jackdaw then returned to the row of boxes, bowed its head once before the first box, twice before the second, once before the third, and then continued along the row until it obtained the fifth bait, whereupon it left the sixth box untouched and 'went home with an air of finality' (Koehler 1950, p. 43). This remarkable report may, perhaps, be too slim a foundation on which to build an entire theory of counting in birds (see Wesley 1961, for a highly critical account of these European experiments). However, if taken at its face value, it suggests something very close indeed to counting as we understand it in humans, and rather distant from any direct perceptual apprehension of number.

Koehler (1950) reports that both simultaneous and successive techniques agree on the same limit for number-discrimination, and that this limit varies from five to six in pigeons, to six in jackdaws (*Corvus monedula*), and seven in ravens (*C. corax*) and parrots (Psittaciformes). There have been relatively few studies of counting in mammals (excluding humans), but the results of these studies (which will be reviewed in Chapter 7) are generally less impressive than those reported by Koehler. Hayes and Nissen (1971), for example, who raised Viki, a chimpanzee, at home, report (p. 77) that 'Viki's numerical abilities . . . fell short of the achievement of counting in people and Koehler's birds'. Birds, then, are as good as, if not better than, any other non-human animals at tasks which appear to require the ability to count; it must, of course, be acknowledged that the relevance of these findings to an account of general intellectual capacity will not become clear until a well-specified theory of counting in non-humans is developed.

Visual concepts

Complex visual concept formation has been successfully demonstrated in a number of studies using pigeons, required to discriminate between photographs according to whether they did or did not show instances of the relevant concept. A variety of relevant concepts have been used, including people, trees, bodies of water, a specific person, natural (as opposed to man-made) objects and letters of the alphabet (Herrnstein and Loveland 1964; Herrnstein, Loveland, and Cable 1976; Lubow 1974; Morgan, Fitch, Holman, and Lea 1976), and there is general agreement that pigeons' acquisition rates are rapid in these tasks. In all of these problems, the birds have shown the ability to pick out (novel) pictures illustrating a given object (e.g. a tree), despite extensive variations in size, colour, and shape, in both positive and negative pictures. Herrnstein *et al.* (1976), for example, found that pigeons discriminated between photographs containing trees (of various types, at various distances, with or without leaves) and photographs showing, say, a stick of celery or a vine climbing a wall. It is clear that such discriminations cannot be solved in terms of any one sensory attribute, and most unlikely that they can be solved in terms of the values of a few sensory attributes, and it is this that gives emphasis to the complexity displayed by the birds.

The difficulty in characterizing further what it is that the birds do when forming visual concepts is that psychologists do not currently possess any adequate theory of object recognition. In place of attempting (in vain) to sketch out a possible theory, it is more appropriate here to point out that whether or not these demonstrations are more properly regarded as requiring explanations in terms of theories of perception rather than theories of intelligence, they do point to similarities between avian and mammalian (or at least, human) performance. It may be added here that, although the pigeons performing in the studies described may have detected some single feature common to all positive displays, Lea and Harrison (1978) have shown that no such assumption need be made. Their study demonstrated acquisition by pigeons of a 'polymorphous' concept, in which positive displays contained (at least) two instances of a set of three 'positive' stimuli and not more than one instance of a set of three 'negative' stimuli, with three stimuli (on a single display) presented each trial. None of the positive stimuli in itself was either necessary or sufficient to guarantee reward, and Lea and Harrison (1978) discuss arguments favouring the view that such polymorphous concepts are typical of the concepts used by humans. We do not, then, know how visual concepts are acquired but, if complex learning processes are involved, then these processes appear to be available and efficient in pigeons.

Memory

Memory in non-human animals has received very little attention compared to the enormous effort devoted to the analyses of learning, and there are insufficient data on this subject from fish, amphibians, or reptiles to have warranted the introduction of this topic in previous chapters. The survey here will concentrate on studies relevant to short-term memory, and this is because, apart from Bitterman's early proposal, discussed in Chapter 3, that fish might not show proactive interference, there have been no theoretically interesting proposals for differences in long-term memory between groups of vertebrates. It will be sufficient, therefore, to note simply that in pigeons, as in mammals, at least some memories are stable over long periods (e.g. Hoffman 1965), and that long-term memories are subject, again as in mammals, to interference effects (Behrend, Powers, and Bitterman 1970) which may be a principal cause of forgetting.

A discussion of investigations of short-term memory in birds may be somewhat unexpected in a survey of studies supposedly related to general intelligence and it, may, therefore, be appropriate to outline first the reasons for their introduction.

One rather general consideration relating short-term memory to intelligence is that it seems reasonable to assume that certain types of problem-solving require the manipulation of more than one piece of information simultaneously, so that limitations of the capacity of any short-term store in which such items might be held would be relevant to restrictions on types of problems that are soluble; Bryant and Trabasso (1971), for example, have shown that failures of young children on certain reasoning tasks may be attributed to shortcomings in their memory, rather than to logical deficiences.

There are, however, more specific grounds for our interest in short-term retention. We have seen in a number of places that differences between groups of animals in the use of after-effects of recent events might account for differences in performance in various learning tasks. Now it would be extremely difficult to compare directly rates of decay between species: it would, for example, clearly be necessary to be able to compare the initial strengths of traces, before attributing a difference to differential rates of decay, and such comparisons would run into formidable (and familiar) problems. However, we can ask a related question – whether some animals and, perhaps, not others, are able to counteract the decay of information, and to maintain the strength of traces across long time intervals.

We have, then, introduced two notions: first, that there may be a store with a limited capacity in which recent information may be held, and, second, that there might be ways in which traces of recent events could be made to resist decay. These are, of course, familiar ideas in theories of human memory, and we shall introduce here a brief account of one model of human short-term memory, our goal then being to establish whether a comparable model can be applied to birds.

Broadbent (1958) was among the first to articulate clearly the notion that there are in human memory two distinct stores: a long-term store (LTM) of unlimited capacity, susceptible to interference but not to decay, and a short-term store (STM), of limited capacity, susceptible to decay, subject to the proviso that items in STM can be maintained indefinitely by a process of rehearsal. Transfer from STM to LTM is, in this model, facilitated by rehearsal of an item in STM.

Support for the dichotomous model of memory comes from a number of sources. First, humans show a finite limit to the number of items for which they show perfect immediate recall (the immediate memory span, or digit span); at one time, this limit (about seven items) was believed to give an indication of STM capacity, although it is now held that immediate recall involves not only items held in STM, but a contribution from LTM as well, and STM capacity is believed to be approximately 3–4 items (Crowder 1976).

Further support for the model is provided by analysis of serial position effects. Where a human is asked to recall a list of items and the length of that list exceeds the immediate memory span, then he will show better recall for the initial and final few items of the list than for items in the middle of the list. It has been argued that the enhanced recall for the early items of the lists—'the primacy effect'—reflects a contribution from LTM, this a consequence of the fact that the early items in the list enjoy superior opportunities for rehearsal (and so for transfer to LTM) before the capacity of STM is exceeded as the list grows. The enhanced recall for the final items on the list—the 'recency effect'—has been taken to reflect a contribution from STM, and to represent those items currently in STM, which have not yet decayed. These interpretations are supported by experiments which selectively affect either the primacy or the recency effect: for example, Glanzer and Cunitz (1966) have shown that the recency but not the primacy effect can be abolished by interposing a delay between list-presentation and recall, and filling that interval with a distracting activity that is presumed to prevent rehearsal. In the absence of rehearsal, the items in STM decay, and this accounts for the loss of the recency effect. Similarly, the primacy (but not the recency) effect may be reduced by speeding up the presentation rate. As the primacy effect is believed to reflect improved transfer to LTM of early items of a list, due to increased rehearsal, a technique, such as rapid presentation, which should reduce the importance of rehearsal, should reduce primacy. We shall, then, survey studies of recent memory in birds with a view to establishing whether anything comparable to an immediate memory span is observed, and whether either primacy or recency effects occur; the results obtained from birds will be interpreted in the light of the notions outlined above.

Very few experiments on short-term retention by birds of a list of items have been carried out, but the few relevant studies (Shimp and Moffitt 1974; Shimp 1976; Macphail 1980) agree that in pigeons the probability of retention of an item in a list is an orderly function of time since presentation: in those studies, if all stimuli were exposed for the same duration, then in no case was more than one item retained at anything approaching the 100 per cent level, and, where lists of more than one item were used, there was no evidence of any sudden deterioration following a given number of items. Macphail (1980), for example, exposed lists of up to five items in an avian analogue of a recognition paradigm, finding, where each item was displayed for one second, an orderly decline in recognition with position from the end of the list, the first item in the five-item list not being recognized above chance level. To the extent that these studies fail to find any abrupt decline in performance with increasing list length, they fail to find an avian analogue of immediate memory span. This failure in turn points to the possibility that there may not be in birds a discrete limited capacity store in which recent information is held.

A further important feature of these studies of list-learning in pigeons is that there was no sign in any of them of a 'primacy' effect—of a superiority in retention for the first over the second item in a list. The primacy effect is, as we saw above, commonly taken as evidence for a rehearsal process in STM and if this is an accurate account of primacy, then apparently pigeons do not rehearse in the list-learning experiments reported to date.

Further evidence which casts doubt on the proposal that pigeons rehearse information is provided by Roberts (1972), who used a delayed-matching-to-sample (DMTS) design in which the standard (centre key) stimulus is exposed for a fixed period prior to a delay, after which the side keys, showing the comparison stimuli, one of which matches the standard, are illuminated. Roberts explored the effect of increasing the duration of the standard stimulus, prior to the delay, finding that the longer the duration, the better was matching performance. However, when the standard stimulus was exposed for the same amount of time, but broken into discrete presentations, each separated by six seconds, performance was poorer than when the stimulus was presented continuously. Roberts contrasts the deterioration caused by 'spaced' presentations of a stimulus in pigeons with the beneficial effects of spaced presentation found in man (e.g. Melton 1970; Tversky and Sherman 1975); one account (e.g. Rundus 1971) of the beneficial effects of spacing in humans is that it encourages more rehearsal than does massed presentation, so that if this account is correct, we are again led to conclude that pigeons do not rehearse in these tasks.

We have found so far no very compelling reason to suppose either that there is in pigeons a limited capacity STM or that pigeons can maintain traces (and perhaps facilitate their transfer to an LTM) by a process of rehearsal. Before considering the broader implications of the findings, we shall now consider a model which has been developed to account for some of the effects obtained by investigations of pigeons' memories for recent events.

Roberts and Grant (1976) have proposed a model of short-term retention in pigeons in which stimuli set up traces which decay spontaneously over time, and which are independent of one another. They produce evidence to show that the initial strength of a trace can be varied by, for example, increasing its exposure duration, but in all their studies, traces appear to decay continuously—they too, then, find no evidence of active maintenance of traces. The independence of traces carries the important theoretical implication that traces do not compete either for space in a limited store, or for access-time to some active process of maintenance. Evidence for trace-independence is found in the demonstration that, where the standard stimulus in the DMTS task is preceded by a distractor stimulus, then performance deteriorates if the distractor stimulus appears as one of the comparison stimuli, but not otherwise. The interference effect can be interpreted as indicating response competition between co-existing traces of the standard and distractor stimuli; the absence of effect where the distractor is not re-presented indicates that the trace of the standard stimulus is not in any way degraded by the co-existence of the distractor trace. Similar evidence for independence of traces is provided by Macphail (1980) who found, first, that recognition for the last two items of a three-item list was no worse than that for the two items of a two-item list, and, second, that increasing the duration of the first item of a two-item list improved recognition for that item but did not affect recognition for the second item. None of these results suggests that STM in pigeons is subject to capacity limitations: traces, it seems, are set up and decay, independent of the number or strengths of other co-existing traces.

The Roberts and Grant model of recent memory in pigeons is very different from the model of human STM described earlier. It would, however, be premature to deduce a radical difference between avian and human memory at this stage, for a number of reasons. In the first place, of course, the avian studies are still in their infancy, and theoretical accounts very much incomplete: it is difficult, for example, to see how pigeons retain any memories over a long period, if all traces decay (as appears to be the case in the current Roberts and Grant model). A more general reason for proceeding cautiously on decision about differences between avian and human memory is that we have contrasted pigeons' memory for visual stimuli with humans' memory for verbal material (since evidence for the human STM/LTM dichotomy derives almost exclusively from verbal learning studies).

There are difficulties in attempting to study non-verbal learning in humans, since we tend to give verbal labels to supposedly non-verbal material, even where 'nonsense' patterns are used. There is, however, good reason to suppose that there may be important differences between the way in which people remember visual stimuli which they have not verbalized, and verbal information. Phillips and Christie (1977), for example, tested recognition for items from lists of up to eight visual patterns (not readily verbalized), finding no primacy effect, and a recency effect for the final item only (all the items preceding the final item being recognised at the same level of accuracy). They found no beneficial effect of slowing the presentation rate of the items, suggesting that transfer to LTM was not facilitated by rehearsal, but evidence that the final item was maintained by rehearsal, since the recency effect was abolished by requiring subjects to perform mental arithmetic in the delay interval. Phillips and Christie suggest that there is a dichotomy between STM and LTM in visual human memory, but that visual STM has a capacity of only one pattern, and that rehearsal does not appear to contribute to the strength of traces in LTM.

Other studies of human visual memory have produced results which, while not in accord with predictions which might be derived from the 'verbal STM' model, disagree with each other, and with the Phillips and Christie account. Some authors (e.g. Shaffer and Shiffrin 1972; Weaver and Stanning 1978) find no serial position effects and others (Potter and Levy 1969; Hines 1975), while finding (as did Phillips and Christie) a final item recency effect, find an inverse primacy effect, the first item showing poorest retention. There is general agreement that increasing stimulus duration does improve retention (e.g. Potter and Levy 1969; Shaffer and Shiffrin 1972; Hines 1975; Weaver and Stanning 1978), but Shaffer and Shiffrin (1972) find that slowing the presentation rate impairs performance and so disagree with both Phillips and Christie (1977) and Weaver and Stenning (1978).

The variety of findings on human visual memory have generated a variety of interpretations, some authors (e.g. Shaffer and Shiffrin 1972) arguing that there is no rehearsal of visual material, and some (Weaver and Stanning 1978), that there is no visual STM, all items entering directly into LTM. The confusion no doubt arises partly from such factors as the extent to which the material was in fact verbalized, and the amount of masking (backward or forward) that may have occurred with particular methods of presentation. For present purposes, it is sufficient to conclude that absence of evidence for rehearsal or for a separate STM in pigeons need not indicate a radical difference between pigeon and human retention of visual information. This in turn, however, should not obscure the fact that there may nevertheless be a radical distinction to be found in the organization of memory: it may be that the dichotomous model of STM and LTM is indeed applicable to verbal material alone, in which case it may be that animals that do not talk do not possess discrete memory stores. This is a point to which we shall return in Chapter 8, when we ask whether non-human mammals show evidence for either rehearsal or STM as distinct from LTM. If we were to decide that mammals but not birds did possess a mechanism for the active maintenance of traces then that, of course, should be of benefit to them in those tasks, such as serial reversal and learning-set formation, to which after effects of recent events are particularly relevant.

Interspecies comparisons in birds

Comparisons among birds from different groups are of interest for two main reasons. First, there is a widespread belief that birds of 'higher' or more evolved orders are of superior intelligence to those of lower orders. Second, there are a large number of measurements of relative brain size available (for a summary, see Pearson 1972), so that, if relative brain size is a predictor of intelligence, then differences amongst birds should agree with these measures. The following discussion of relative brain size in birds should, of course, be viewed in light of the fact (see Fig. 2.3, p. 28) that in general the avian brain is of comparable weight to that of a mammal of similar body weight and approximately ten times as large as that of a reptile, amphibian, or bony fish of similar body weight.

The most extensive avian brain size data are provided by Portmann (1946, 1947), and his findings, along with data from other workers, are summarized in Pearson (1972). Portmann derived a superficially complicated method of estimating the relative degree of encephalization of a species by taking the ratio of the weight of the cerebral hemispheres in that species to that of the 'brainstem rest' (diencephalon, midbrain excluding the optic lobes, and hindbrain, excluding the cerebellum) in a gallinaceous species of comparable body weight. The rationale for this procedure is that the size of the brainstem rest provides an estimate of the relative amount of brain tissue that is devoted to purely vegetative and somatic function. Since the brainstem rest size is larger, for a given body weight, in some species than in others, some of even this part of the brain may be devoted to 'higher' functions, and the best estimate of the minimal volume required for somatic functions is provided by the volume in gallinaceous species, which have consistently lower stem rest weights for a given body weight than do other species (Portmann 1946). However, the formula used for calculating the value of the brainstem rest in gallinaceous species of various body weights is in fact the equation of simple allometry ($E = kP^a$, where E is in this case the estimated gallinaceous brainstem rest weight—see p. 29), and in the case of the stem rest of gallinaceous birds, $a = 0.52$ (Portmann 1946). Now this system in effect produces exactly the same ranking of species as would be obtained by simply applying the equation of simple allometry having $a = 0.52$ to the raw hemisphere data (although the absolute values of the indices differ, by a constant amount, from those that would be obtained in that way). As the weight of the hemispheres, of the stem rest and that of the whole brain are highly correlated (Portmann 1946), the data provided by Portmann differ very little from those that would have been obtained by calculating the value of k (the cephalization index, see p. 30), assuming $a = 0.52$, for conventional whole brain- and body-weight data. Since Portmann (1946) and Jerison (1973) agree that the mean value of the exponent applicable to whole-brain data from a wide cross-section of avian species is 0.56, the implication is that heavier birds will obtain relatively higher rankings (larger indices) using Portmann's techniques than had an exponent of either 0.56 or 0.66 (the exponent favoured by Jerison for all vertebrate data) been used. With these reservations in mind, we may consider the data provided by Portmann.

A survey of hemisphere indices allows two conclusions of interest. First, the indices do show a general increase from 'lower' to 'higher' orders: the index for a chicken, for example, is 3.27, that for a pigeon, 4.0, and that for a magpie (*Pica pica*) is 15.81. Second, there are nevertheless large variations in the index for some orders, the range for psittaciform species being from 7.40 to 28.02, and, for passerines, from 4.28 to 18.70. Variations between species from the same family appears to be much less marked, and it is of interest that, within passerines, the corvid family, with a mean index of 14.99, contains the only passerine species with indices greater than nine; there is, then, considerable overlap in these indices between orders, and this may be emphasized by quoting the hemisphere index of the penguin (*Spheniscus demersus*) which, at 9.31, is higher than that of all non-corvid passerine species. The penguin is, of course, widely supposed to be a primitive bird (see Fig. 6.1, p. 170).

There are no published index data for some of the species to be discussed in this section, and perhaps the central findings to bear in mind as we consider behavioural studies are that both the conventional ranking of orders and the hemisphere indices agree in expecting chickens to be inferior to pigeons and pigeons to be inferior to passerines, of which corvids should be the most intelligent.

Serial reversal

An extensive series of comparative serial reversal studies using various species of birds has been reported by Gossette and his colleagues, using a 'formboard' apparatus in which the birds have to displace one of two small wooden discs to gain access to foodwells; most of these studies involved spatial discriminations, in which the discriminanda were both white and the reward was consistently in one position, but there is a report also concerning reversals of a brightness discrimination, in which one of the discriminanda was black, the other white. Before discussing Gossette's findings, it should be pointed out that rather small numbers of subjects were run in these experiments and that, except where pigeons or chickens were used, no more than three members of any species was run in any of these studies. It may well be doubted whether samples as small as this can give a reliable indication of the capacities of a species.

Gossette, Gossette, and Riddell (1966) compared chickens, bob-white quail, parrots (*Amazona ochrocephala*), and red-billed blue magpies (*Urocissa occipitalis*) in a series of 29 spatial-discrimination reversals, finding fewer errors in the magpies and parrots than in the chickens and quail, there being no significant differences between magpies and parrots or between chickens and quail. A report on greater hill mynahs (*Gracula religiosa*) found very good performance after 19 reversals (at which point the study terminated) (Gossette, Gossette, and Inman 1966); the mynahs' performance over the final block of reversals (about three errors per reversal) seemed to be superior to that reported by Gossesse, Gossette, and Riddell (1966) for parrots and magpies, which both averaged about five errors for reversal over reversals 14–19. Two reports on pigeons (Gossette and Cohen 1966; Gossette and Hood 1968) find error rates which appear directly comparable with those reported for the magpies and parrots; the Gossette and Hood report, which explored different levels of drive and incentive, appears indeed to show better performance in pigeons—including a number of one-error reversals—under the optimal conditions (low drive with high incentive). These results agree with expectations derived from the ranking of orders and of cerebral indices in finding that parrots and passerines are superior to chickens and quail (both gallinaceous species); however, the pigeon appears to be performing well above expectation, and the mynah, which belongs to a passerine family (the Sturnidae—starlings—with hemisphere indices well below those of corvids) appears to have outperformed the magpie, a corvid.

The significance of these findings becomes rather more confused when we consider a report of reversals of brightness discriminations (Gossette 1967), which shows, across 20 reversals, that magpies were somewhat superior to mynahs, which were considerably superior to parrots, parrots in turn being superior to chicks. The striking feature of this report is that parrots were now much worse than magpies and mynahs: the total error scores for three magpies and three mynahs ranged from 172–361 errors, whereas the two parrots had scores of 440 and 446 errors.

If the ordering of species in serial reversal performance can be changed by altering the relevant dimension, it seems clear that serial reversal in itself cannot give a reliable measure of general intelligence. There is good evidence that other procedural variables (e.g. the amount of training for reversal: Gonzalez, Berger, and Bitterman 1966; level of drive and incentive: Gossette and Hood 1968) also affect asymptotic performance, and this is not unexpected. It should be added that there is also a report (Bacon, Warren, and Schein 1962) of very efficient serial reversal performance in chickens of a pattern discrimination, having multiple relevant cues: these birds, given 25 reversals, averaged only three to four errors per reversal by the end of the series, so that the lowly position found by Gossette for chickens in both spatial and brightness discrimination reversals can indeed be improved by altering the task demands.

Gossette has also proposed that the reversal index (RI), recommended by Rajalakshmi and Jeeves (1965), may have advantages over serial reversal performance as a method for studying species differences. This index is obtained by dividing the number of errors (or trials) on the first reversal by the number recorded in original acquisition: since procedural variables are held constant in acquisition and reversal, it may be that this index will reflect the capacity to cope with the intellectual demand common to all reversal tasks. One advantage of this index over serial reversal performance is that, of course, it is less time-consuming and so allows more species to be tested; another important advantage according to Gossette and Hood (1967) is that it is, unlike acquisition or serial reversal performance, unaffected by changes in incentive or drive (at least in pigeons). The results obtained are, however, not encouraging: Gossette and Gossette (1967) report RIs obtained from five mammalian and nine avian species, using a spatial discrimination. These show that although the lowest mean index (0.65)—indicating most efficient reversal—is found in the capuchin monkey (*Cebus*), the squirrel monkey has an index (2.83) which is above that for both pigeons (1.78) and bob-white quail (2.75), that there is considerable overlap of RIs derived from various species of widely differing orders of birds, and that the mean index for the pigeon is lower than that for either the mynah or the common crow (*Corvus brachyrhynchos*). A further, more serious, difficulty is provided by the fact that whereas in this study the index for pigeons (1.78) is lower than that for chickens (4.16), a subsequent study by one of Gossette's students found, using reversal of a colour discrimination, a significantly lower RI in chickens than in pigeons (Levine 1974). It therefore appears that RI is not a useful measure of intelligence: a similar conclusion is reached by Warren (1967*a*), who analysed reversal data from a number of studies using cats and kittens, finding large and generally unsystematic variations in RI across studies, with, in particular, the same cats showing very different indices in different types of apparatus.

Learning-set formation

The only extensive series of studies of learning sets in birds is that reported by Kamil and his colleagues, using blue jays. There are, however, reports of studies using several other species, including birds from non-passerine orders.

One experiment using chickens (Plotnik and Tallarico 1966) found rather striking learning-set formation over a series of 50 object discrimination problems; Trial 2 performance over the last five problems was at the 70 per cent correct level, and so comparable with that reported for blue jays. Unfortunately, however there are methodological weaknesses in this study, the major difficulty being that only 50 objects were used for the 50 problems, so that, since two objects were used in each problem, some re-use of the same objects, with the possibility of direct transfer across problems, must have occurred. A further reason for lack of confidence in the finding is Kamil and Hunter's (1970) report that analysis of Plotnik and Tallarico's original data shows that although Trial 2 performance over the last block of problems was 70 per cent correct, Trial 3 performance at the same stage was only 58 per cent correct. It seems, then, that we should reserve judgement on the ability of chickens to form learning sets.

There are two reports of learning-set formation in pigeons, each of which used coloured photographs of objects as discriminanda. Zeigler (1961) reported marginal improvement (from 62 to 72 per cent correct over the last eight trials of 24-trial problems) across a series of 120 problems. Unfortunately, Zeigler used a remarkable criterion for choice: both stimuli were illuminated at the beginning of a trial, and the birds were judged to have chosen that stimulus to which the first response was made *after a 40-second interval had elapsed.* No other rèport on learning sets, using bird or mammal, has used a comparable technique, and it could, of course, be that requirement which produced so little improvement.

A second report (Wilson 1978*a*) using pigeons attempted to compare their performance directly with that of corvids (a crow, *Corvus corone*, and two rooks, *Corvus frugilegus*), using an identical apparatus for all birds. Wilson's first experiment found that although pigeons were better than corvids in early problems, they were markedly poorer by the end of the series of 800 problems; whereas the corvids showed a steady improvement across the series, the pigeons if anything deteriorated. Analysis of their performance showed that, over the last 100 problems, pigeons made 76 per cent choices of the stimulus on their preferred side, and, accordingly, a correction technique, designed to minimize position habits, was used in a new series of problems. Pigeons now did show some improvement across the series, although performance was unimpressive, reaching a level of about 55 per cent correct Trial 2 choices for novel stimuli at the end of training, the corvid performance lying somewhere between 60 and 70 per cent correct. These results agree with Zeigler's in showing that pigeons can show inter-problem transfer, but show that even in an apparatus in which original acquisition is superior in pigeons, corvids show superior learning-set formation.

The crow and rooks in Wilson's study did not appear to reach the same level of performance as that achieved by Kamil's blue jays (which achieved 73 per cent correct Trial 2 performance); this finding echoes an earlier result of Hunter and Kamil (1975), who used three common crows in a series of (three-dimensional) object discriminations, and found that after 300 problems performance appeared to have reached asymptote, Trial 2 choices being approximately 60 per cent correct. Hunter and Kamil, contrasting this performance with that of blue jays, argue that procedural factors may have been responsible (the crows tended to react emotionally to handling and seemed particularly wary of novel objects); on the other hand, they also point out that, using Levine's hypothesis testing model, the crows did not choose at random, but performed systematically, and had large scores for both position and stimulus preference strategies.

Finally, another study using a passerine species (the greater hill mynah, a member of the starling family), found highly efficient performance (one bird showing 78 per cent correct on Trial 2 over problems 901–1000) in an object-discrimination task, a level only marginally lower than that found in blue jays using the same apparatus (Kamil and Hunter 1970).

In summary, if we reserve judgement on the Plotnik and Tallarico (1966) report, then the optimal learning-set performance to date from birds has been found in passerines—in mynahs and blue jays. Crows tested in identical conditions performed rather worse than either mynahs or blue jays; in an apparatus using two-dimensional stimuli crows showed an absolute level of performance that was again lower than that obtained from mynahs and blue jays using three-dimensional objects, a difference which might, of course, be attributable to the very different conditions of testing. Pigeons tested using two-dimensional stimuli performed rather more poorly than crows, despite superior performance in early problems. The data are clearly inadequate for firm conclusions at present: it would be interesting for example, to know the capacities of pigeons tested using objects rather than pictures as stimuli (or the performance of mynahs or blue jays, with pictures). The best evidence for a real species-difference comes from Wilson's (1978*a*) demonstration of inferior learning-set formation by pigeons compared with crows in an apparatus in which the pigeons were initially the better learners. It should, however, be pointed out that there is no guarantee that some procedural factor may not have grown in importance as training proceeded (Hunter and Kamil 1975, for example, reported that the emotionality of their crows in response to handling increased as the experiments progressed). Moreover, our earlier survey of serial reversal learning did produce evidence from a number of sources to suggest that pigeons are capable of adopting win–stay, lose–shift strategies.

Brain mechanisms and learning in birds

It was argued in Chapter 2 that our main goal in analysing the effects of brain damage is to gain insights into the structure of intelligence. Now, although we have seen in the preceding sections very little reason to suppose any dramatic differences between the intellectual capacities of birds and those of mammals, it may nevertheless of course be the case that, given differences in brain organization, there might be quite different behavioural breakdowns induced by forebrain lesions. In that case, analysis of effects of brain-damage in birds might well cast new light on intelligence in mammals—that is, given similarity of intellect and differences in brain between birds and mammals, implications of breakdowns induced by brain damage in either group could be equally relevant to analysis of performance of intact animals from both groups.

Given that analysis of avian brain function may have broad relevance, it is encouraging that there is good evidence for disruption of performance in learning tasks following localized forebrain lesions in birds. However, even in the case of birds, the amount of work reported to date is extremely small compared to that available on mammals, and insufficient for even tentative hypotheses concerning functions of any but two areas, the archistriatum and the hyperstriatal complex.

Archistriatal lesions

It will be recalled (p. 176) that Zeier and Karten (1971) proposed a major division of the avian archistriatal complex into an anterior (somatomotor) division, which might be comparable to parts of mammalian neocortex, and a posteromedial (limbic) division, comparable to mammalian amygdala. Support for this latter proposal is gained from the fact that Phillips' (1964) experiment on mallard ducks found that the critical region for taming effects (assessed by various measures, including the ease with which they could be caught) appeared to lie in the medial archistriatal area. Phillips (1968) also found modification of emotional behaviour in peach-faced lovebirds (*Agapornis roseicolis*) with lesions in the medial archistriatum. Lovebirds tend to mob dangerous or novel objects, and emit calls as part of this response; medial archistriatal lesions markedly reduced the number of calls to novel objects, and increased the speed with which the birds began to eat food in the presence of a novel object. Where no novel object was present, archistriatal birds were somewhat slower to eat than controls, indicating (at least) that their appetite for food had not been increased by the lesions.

Tameness and docility following lesions of the 'limbic' medial archistriatum have also been reported in pigeons (Zeier 1971; Dafters 1975, 1976); Zeier (1971) further reports that laterally placed lesions of posterior archistriatum (part of the 'limbic' division) have a taming effect, whereas more anteriorly placed lateral lesions have a tendency to increase fearfulness. This latter finding agrees with a report by Maser, Klara, and Gallup (1973) of increased tonic immobility responses in chickens following anterolateral archistriatal lesions: there is good evidence that this response is potentiated by increased levels of fear (Gallup 1974).

Lesions within the archistriatal complex have been shown to have significant effects on a number of learning tasks and one of our concerns when considering these effects will be to decide whether they are related to the somatomotor or emotional functions of the area in question, or whether some other 'intellectual' function for the area is implied.

Experiments using aversive shocks as a reinforcer have been reported by Cohen (1975) and Dafters (1975, 1976), both using pigeons. Cohen found that extensive posteromedial archistriatal lesions impaired the acquisition of a classically conditioned heart-rate change to a light CS which signalled a shock to the foot. Where less than 80 per cent of the posteromedial region was destroyed, conditioning proceeded normally, and extensive damage to the anterior and intermediate archistriatum had no effect. Dafters (1976) similarly found a greatly reduced effect of a stimulus signalling shock in pigeons with medial archistriatal lesions: in this case, conditioning was assessed by the conditioned suppression technique, in which pigeons responding for food delivered at irregular intervals show a decline in response rate in the presence of a stimulus that signals inescapable shock. Pigeons with medial archistriatal lesions showed no suppression to a tone signalling shock, in contrast to sham-operated controls, which showed marked suppression. Dafters (1975) also reported that pigeons with similar lesions were severely impaired on both acquisition and retention of treadle-press avoidance responses, using both discrete-trial and free-operant avoidance training techniques.

These studies can most plausibly be interpreted as demonstrating a reduction in fear in birds with lesions of the limbic archistriatum: Dafters (1975) provides good evidence against an alternative explanation in terms of sensory change—for example, increased pain thresholds—by showing that the escape latencies of operated birds in the treadle-press task, once a shock had been delivered, were no higher in archistriatal than in control birds.

There are, however, effects of limbic archistriatal lesions on food-motivated tasks in which no aversive UCSs are delivered, and it is not obvious that these can be interpreted simply in terms of fear reduction. Zeier (1968, 1969) investigated the effects of archistriatal lesions in pigeons on a variety of operant schedules; these experiments were carried out before the distinction between somatomotor and limbic archistriatum had been established, and the lesions concerned invaded both archistriatal divisions, so that, once the distinction had emerged, no clear interpretation was possible. Accordingly, Zeier (1971) ran birds with lesions of either the limbic or the somatomotor archistriatum, as well as controls which were either unoperated or had neostriatal lesions, in two operant schedules that had shown effects of lesions in his earlier studies. The first of these, a variable-interval (VI) schedule, delivers food for responses after intervals of unpredictable duration, and normally obtains a relatively steady rate of response throughout a session; variations in the subjects' rate of response have relatively little effect on the number of rewards obtained, so that optimal performance is a low rate of response. The second schedule, a differential reinforcement for low response rate (DRL) schedule, delivers rewards for responses emitted after a minimum interval of no responding, that interval remaining constant throughout a session. Optimal performance here is again a low rate of response, with responses spaced to exceed the specified interval; however, in this case, increases in rate have a severe effect on food delivery, and this provides an important contrast with the VI schedule in which increases in response rate have very little effect and, if anything, may marginally increase the number of rewards obtained. Limbic archistriatal lesions reduced response rate on the VI schedule, but increased rate on the DRL schedule: somatomotor lesions had precisely the opposite effects, increasing VI rate and decreasing DRL rate. Neostriatal lesions were without effect, and it is of interest that a further group which were unoperated but subjected to increased food deprivation showed an increase in VI responding and, although showing very little increase in DRL rate, obtained fewer rewards on that schedule: as this pattern differs from those seen in the two experimental groups, neither the somatomotor nor the limbic lesion effects can be attributed simply to changes in overall motivational level.

In interpreting his results, Zeier drew a distinction between the automatic or monotonous pecking generated by the VI schedule, and the pattern obtained by the DRL schedule, in which birds turn away from the key after non-rewarded pecks, and subsequently show a 'much more complex' approach response: Zeier proposed that the two archistriatal divisions interact in determining response rate, so that 'monotonous' pecking is inhibited by the somatomotor, but facilitated by the limbic archistriatum and 'approach' responses are inhibited by the limbic, but facilitated by the somatomotor archistriatum. Of course, until many more data are available, no confident account will be forthcoming, and it is not clear in any case how the distinction between 'monotonous' and 'approach' responses should be understood. This in turn, however, may emphasize that behaviour theory has as yet provided no clear account of the processes involved in DRL (as opposed to, say, VI) responding: it may be that future physiological investigations may help to clarify the contrasts between control of responding in these (and other) operant schedules.

In summary, there is good evidence that 'limbic' archistriatal lesions reduce fear, and that performance in avoidance learning tasks is affected accordingly. The effects of archistriatal lesions on food motivated responding have been little explored, but are probably complex: at present it seems likely that these effects may be independent of the effects of the lesions on fear.

Hyperstriatal lesions

A number of authors have reported effect of lesions within the hyperstriatal complex (hyperstriatum ventrale, dorsale, and accessorium along with their associated intercalated nuclei) on learning. Most of the studies cited in this section use lesions that are large, and invade several subdivisions of the complex; at this stage therefore, very little can be said of which region or regions are responsible for the effects observed.

The lesion-induced deficit of most theoretical interest was first reported by Stettner and Schultz (1967), who used bob-white quail having extensive lesions of the hyperstriatal complex and adjacent regions (including the hippocampus). The birds were tested on acquisition of a simultaneous pattern discrimination (vertical versus horizontal stripes), and on a series of reversals of the discrimination. Stettner and Schultz found that, although acquisition of the discrimination was unaffected by the lesions, reversal performance was severely disrupted, the disruption taking the form of perseverative responding to the former positive stimulus at the outset of each reversal. Now the disruption of reversal, but not of acquisition, suggests that the lesion had produced a breakdown in behaviour that is of interest to the psychologist: good acquisition indicates that no gross sensory, motivational or motor deficits have occurred, and poor reversal suggests that reversal involves some mechanism that is not essential to acquisition. If it is possible to clarify further the nature of that mechanism, then a contribution may be made to understanding the processes of acquisition and reversal in intact animals. It should be pointed out that the studies to be discussed here generally concern either one or a few reversals, and have not been directed at analysing the phenomenon of serial reversal improvement: our concern is with the mechanisms engaged by the first, or first few, experiences of reversal of reinforcement contingencies.

One possible interpretation (Macphail 1975*a*) of the deficit, which is suggested by the perseverative responding found by Stettner and Schultz (1967) in hyperstriatal birds, is that the subjects are less capable than normals of response inhibition. This notion has the advantage that it can account for a number of otherwise puzzling findings in the literature. Although there is good evidence (see p. 183) from performance in simultaneous discriminations that birds with hyperstriatal lesions do not suffer from gross perceptual deficits, there are reports from studies using a go no-go successive visual discrimination technique of deficits in acquisition in birds having hyperstriatal lesions (e.g. Zeigler 1963; Pritz, Mead, and Northcutt 1970; Macphail 1971). In simultaneous discriminations, animals are required to choose between two simultaneously presented stimuli, so that difficulty in inhibiting responses need not cause disruption unless, as in reversal learning, a powerful preference for the negative stimulus exists. In successive discrimination, on the other hand, only one stimulus is seen at any one time, and the birds' task in the presence of the negative stimulus is precisely to withhold responses; Macphail (1971) showed that the disruption obtained by hyperstriatal lesions in a successive red-green discrimination was indeed confined to responding in the presence of the negative stimulus, as would be expected if inhibition of responding had been impaired.

Further support for the response inhibition interpretation of hyperstriatal function is provided by the Reynolds and Limpo (1965) report of disruption of fixed-interval but not fixed-ratio performance by hyperstriatal lesions in pigeons: the fixed-interval schedule delivers reward for the first response emitted after a fixed time interval has elapsed, and establishes, not surprisingly, a pattern of responding in which the birds pause after each reward; in fixed-ratio schedules, reward is delivered as soon as a fixed number of responses has occurred, and birds in this schedule typically respond very rapidly. Clearly, the disruption of the typical fixed-interval pattern by hyperstriatal lesions can be interpreted as being a consequence of the loss of ability to withhold responses.

Macphail (1976*a*) investigated the effects of hyperstriatal lesions in pigeons that performed three serial reversals each day of a red-green discrimination. An earlier report (Macphail 1970) using intact birds, had shown that in this task pigeons, rather surprisingly, require fewer trials to reach criterion in the second of each day's reversals, as opposed to the first; this appeared surprising since, due to overnight 'forgetting', birds had no strong preference for either stimulus at the start of the first of each day's reversals, whereas they did, of course, have a strong preference for the incorrect stimulus at the outset of the second reversals. An interpretation of this effect, which is somewhat too complex to expand fully here, suggested that the increased second-reversal efficiency reflected the fact that in the second, but not the first, reversal of each day, birds generated inhibition to the negative stimulus. Hyperstriatal lesions in the 1976 study selectively impaired performance on the second of each day's three reversals, and this result both strengthens the response inhibition account of hyperstriatal function and supports the interpretation proposed for the performance of intact birds in this task.

Finally, Benowitz (1972) found that chicks with extensive lesions of the dorsomedial telencephalon involving both hyperstriatal and hippocampal structures showed deficits in retention of a one-trial punishment task (in which the UCS was an aversive taste solution): this result extends the generality of the response inhibition account to tasks in which an aversive reinforcer is employed.

In summary, the response inhibition account is supported by evidence drawn from three species of birds—pigeons, quail, and chicks—in a variety of tasks, including serial reversals, successive discriminations and punishment procedures. However, there are difficulties for the account, and these have led to the examination of alternative hypotheses.

One problem for any simple response inhibition hypothesis was first noted by Reynolds and Limpo (1965), who pointed out that their hyperstriatal pigeons occasionally had long pauses in the middle of fixed-interval components. Macphail has similarly noted (1975*b*, 1976*a*,*b*) that hyperstriatal pigeons abandon responding altogether during reversal performance more frequently than controls, and although this is, perhaps, a consequence of the increased number of errors made by lesioned birds, it is hardly consonant with an inability to withhold responses. A second difficulty is that, although successive go no-go discrimination performance is disrupted by the lesions, the disruption is very much less striking than that seen in reversal performance; Macphail (1975*a*) has suggested that in fact go no-go discrimination is disrupted in difficult discriminations only, a conclusion that is, again, not necessarily anticipated by the response inhibition hypothesis.

An alternative account of the lesion-induced effects is that hyperstriatal birds suffer from a deficit in selective attention; this suggestion has been put forward by Stettner (1974), who analysed reversal data of two hyperstriatal quail, and concluded that the source of major difficulty for the birds lay, not in perseverative responding to the former positive stimulus, but in exaggerated position responding. It is pertinent here to note that Karten *et al.* (1973) have suggested, on the basis of anatomical evidence of interaction between the thalamofugal and tectofugal systems, that the thalamofugal system (of which IHA, lying within the hyperstriatal complex, is an important constitutent) may be involved in selective attention by, perhaps, modulating input through the tectofugal system. There are, however, results which appear to rule out this interpretation, at least as applied to simultaneous discrimination and reversal tasks: Macphail (1976*b*) has shown that hyperstriatal birds, pretrained on a simultaneous position discrimination and then shifted to acquisition and reversal of a colour discrimination, showed no deficit in the shift from position to colour discrimination, and no increase in position habits over normals across a series of colour reversals; this latter finding was also obtained in Macphail (1976*a*). Moreover, Macphail (1976*a*) found severe disruption of serial reversals of a position discrimination in which no salient irrelevant dimension was available. Perseverative errors to the former positive stimulus were markedly increased in the reversal series of these two studies. It seems clear, therefore, that reversal performance may be severely disrupted in the absence of any evidence for exaggerated attention to irrelevant stimuli; Stettner's observation of increased position habits may perhaps be accounted for on the supposition that these were themselves a reaction to the failure to solve the reversal problem, due to the initial perseverative errors.

At present, then, the response inhibition account of hyperstriatal function appears to enjoy most support, although there are some findings that do pose serious problems. It may, of course, be the case that we are not looking at a unitary syndrome—that the reversal deficit is independent of the tendency to abandon responding—and this would not be surprising given the diversity of structures invaded by the lesions: Macphail (1969) in fact found evidence, using lesions essentially confined to hyperstriatum accessorium, of enhanced response inhibition (more rapid extinction, improved punishment performance). It must be hoped that eventually we shall have a clearer account of which sites are critical for the deficits observed, and firmer evidence on which deficits are independent, and which are manifestations of the same underlying disruption.

There is one further learning deficit following hyperstriatal damage, to which reference has previously been made (p. 183), and that is the disruption in imprinting obtained by lesions of the medial hyperstriatum ventrale (Bateson *et al.* 1978). These workers have provided further evidence, using different techniques, for a role of this region in imprinting: Horn, McCabe, and Bateson (1979) have shown, using autoradiographic techniques, that metabolic activity in the hyperstriatum ventrale (and not in other forebrain areas) of two-day-old chicks increases as a consequence of exposure to a relatively novel stimulus in an imprinting paradigm; the increase seen was relative to the activity recorded following exposure to the same stimulus in chicks that had had more previous experience of the stimulus and therefore, presumably, had less to learn about it. McCabe, Horn, and Bateson (1979) have been able to modify chicks' preferences for flashing lights of various frequencies by stimulation of the hyperstriatum ventrale at a particular frequency (which becomes the preferred frequency). It seems most likely that these effects are quite independent of those (e.g. reversal deficits) previously discussed, and if it emerges that lesions of this region affect imprinting but not learning in other paradigms, then support will accrue for the view that imprinting is indeed a special 'type' of learning which may proceed according to rules that differ from those governing other types.

It should, however, be added that there are findings in the literature which appear to conflict with those described above: Salzen, Parker, and Williamson (1975) report that imprinting in chicks is not affected by large medial lesions of either the anterior or posterior telencephalon (involving extensive damage to the hyperstriatal complex) but is prevented by relatively small lesions of the lateral telencephalon, which involved parts of the neostriatum, archistriatum, and paleostriatum; Collias (1980) finds no imprinting in totally decerebrate chicks, but normal imprinting in chicks with only the paleostriatum and lobus parolfactorius intact, following extensive telencephalic ablation. It is of course to be hoped that these reports can be reconciled by future investigations of what is an important topic.

Involvement of other brain regions in learning

Although, as was stated earlier, only the hyperstriatal and archistriatal regions have attracted detailed attention, there are scattered reports on the effects of lesions elsewhere on learning, and the results will be summarized briefly here.

Lesions of the septal region have no effect on tonic immobility (Maser *et al.* 1973), on docility (Phillips 1964), on classical heart rate conditioning (Cohen and Goff 1978), or on food-reinforced responding on a variable-interval schedule (Zeier 1968); Zeier (1968), however, reports that septal-lesioned pigeons show increased fear—a result that contrasts with the Maser *et al.* finding.

There are two reports which hint at a possible involvement of the hippocampal complex in processing of spatial information—a possibility which, in mammals, has received considerable attention, as we shall see in Chapter 8 (p. 323). First, Nott (1980) found that pigeons with hippocampal lesions are impaired on reversals of a spatial (position) but not of a visual (pattern) discrimination. Second, Krushinskaya, in an experiment summarized by Vinogradova (1978) found that cedar birds (*Nucifraga caryocataces*), which hide cedar nuts in stores, forget the location of those stores following hippocampal lesions if they are removed from their experimental cages (in which the stores have been made) for more than three hours.

Wesp and Goodman (1978) report that lesions of the parolfactory lobe result in impaired fixed-interval schedule responding, and in increased fear and aggression; Cohen and Goff (1978), on the other hand, found no effects of parolfactory lobe lesions on heart-rate conditioning. Wesp and Goodman also report a temporary fall in fixed interval response rates in pigeons having paleostriatal lesions: Zeier (1968) found that such lesions marginally (but not significantly) increased response rates on a variable interval schedule.

Finally, there are many reports of absence of any effect of neostriatal lesions in a variety of tasks, including successive visual discriminations (Pritz *et al.* 1970; Zeigler 1963), operant schedule performance (Zeier 1969; Wesp and Goodman 1978), and serial reversals of a position discrimination (Macphail 1975*b*). One positive report is that of Krasnegor (cited by Stettner 1974), who found temporary elevation of thresholds in auditory intensity and frequency discriminations following lesions of Field L in the caudal neostriatum; as Field L is the avian telencephalic auditory projection region (see p. 184), these effects presumably reflect sensory rather than intellectual deficits.

Relevant data have also been provided by the use of a technique that has so far received little discussion, that of brain stimulation and, in particular, assessment of the reinforcement value of stimulation at various sites. A number of authors (e.g. Macphail 1967; Andrew 1969; Goodman 1970; Zeigler, Hollard, Wild, and Webster 1978) have shown that electrical stimulation at forebrain loci may be rewarding, punishing, ambivalent (i.e. rewarding by one criterion and punishing by another), or neutral, depending on the locus. However, there is still nothing like a comprehensive mapping of reinforcement value of forebrain sites in birds. There is good evidence that stimulation within the paleostriatal complex is rewarding (e.g. Macphail 1967; Hollard and Davison 1978; Zeigler *et al.* 1978), and most authors find the septal area an aversive site (e.g. Macphail 1967; Phillips 1964; Cannon and Salzen 1971 – but see Andrew 1967). There are isolated reports of rewarding sites elsewhere (e.g. in the neostriatum and hyperstriatum – Macphail 1967), but in total the data are insufficient to sustain currently any general hypothesis concerning their overall anatomical organisation.

It is, of course, not entirely clear what would be gained from a full mapping of reinforcement values (apart, perhaps, from the ability to contrast findings with those obtained in other vertebrate classes); this in turn reflects the fact that theoretical interpretation of brain stimulation reward remains controversial. However, some authors have used brain stimulation in birds in an interesting way: Andrew (1969), for example, has tested the widely held view that 'twitters' in young chicks are calls indicating pleasure by adopting the hypothesis that brain-stimulation reward is pleasurable, and asking whether twitters accompany self-stimulation in chicks. Andrew finds that they do not, at least once self-stimulation is well established, and uses this result to support his view that twitters belong to a system of calls that are connected with degree of attention, rather than with pleasure. Zeigler *et al.* (1978) note that at least some of the sites involved in the feeding circuit (described earlier, p. 189) sustain self-stimulation (e.g. the nucleus basalis and the fronto-archistriatal tract) and point out further that, even where sites not associated with this circuit are stimulated, the pecking response obtained resembles, in both topography and duration, that seen in conventional food-rewarded pecking. Such observations invite further exploration of possible links between self-stimulation and feeding.

The reports on reinforcement value of brain stimulation in birds may not, then, contribute much to date that is of comparative interest, or throw much light on the organization of the avian forebrain: on the other hand, some workers have used the technique to ask interesting questions of a more general nature.

Summary and conclusions

Birds are derived from quite different reptilian stock from that which gave rise to mammals, their closest living relatives being the crocodilians. Although the modern orders of birds are commonly ranked according to some notion of evolutionary progressiveness, there is no good reason to suppose that birds of the 'higher' orders should be more intelligent than those of 'lower' orders.

The avian forebrain conforms generally to the pattern seen in reptiles, but shows a reduction in cortication, and a marked increase in the dorsal 'hyperstriatal' areas. The basic organization of the visual and auditory projections to the telencephalon show very clear parallels to the mammalian pattern, but one possibly unique feature of avian sensory pathways is that there may be tactile, visual and auditory projections direct to the telencephalon, (by-passing the thalamus) from sites in the mid- or hindbrain. There are anatomical data to suggest telencephalic involvement in movement, as well as direct physiological evidence for a telencephalic role in two specialized systems, one for feeding, the other (in songbirds) for song-production; this latter system is of particular interest in that it shows lateralization of function. Extensive cerebral damage does not, however, lead to striking deficits in control of other, general, types of motor activity.

The characteristics of habituation in birds are similar to those established in other vertebrates; similarly, both classical and instrumental training procedures, using a variety of reinforcers, are successful in birds. There is very good analytical evidence to show that birds do detect contingencies between stimuli and reinforcers, and although there were few reports of instrumental learning that could not be re-interpreted as instances of classical conditioning, there seemed no reason to distinguish between birds and mammals on that ground. It was concluded that both classical and instrumental conditioning are obtained in birds, and that learning proceeds no less efficiently than in mammals.

The survey of complex learning in birds found little evidence to suggest that there may be qualitative intellectual differences between birds and mammals. Pigeons can (under some circumstances) achieve one-trial reversals and blue jays show both learning-set acquisition and positive transfer from serial reversal training to learning-set performance; these (and other) findings indicate that birds can adopt win–stay lose–shift strategies. There is evidence also that birds do form expectations, that they possess mechanisms of selective attention, and that they appear to be able to form a variety of concepts, including a 'sameness–difference' concept, number concepts, and various complex visual concepts. Analysis of recent memory in pigeons did, however, suggest the possibility that they may not be capable of maintaining traces by a process of rehearsal, and that there may not be in pigeons a dichotomy between a short-term and a long-term memory store. Whether that conclusion, if sustained by further experiments, in turn indicates a substantive difference between birds and non-human mammals or man remains very much an open question.

A number of quantitative differences between birds and certain mammals in levels of performance in various tasks were found. Although these differences might reflect quantitative intellectual differences, in, for example, stability of attention or use of strategies, it also seems possible that they reflect the operation of contextual variables; where a number of different procedures have been tested (in, for example, studies of serial reversal, probability learning and acquisition of the sameness-difference concept), efficiency of avian performance has varied across a wide range, and it is not currently possible to reject the hypothesis that there are neither qualitative nor quantitative differences in intellect between birds and mammals.

The most extensive data available on comparisons between the various orders of birds concern serial reversal performance; the findings do not, however, appear to coincide with rankings based on either relative brain size or the supposed degree of progressiveness of the various orders, and, moreover, show different rankings for different stimulus modalities. There is some evidence that pigeons may be inferior to corvids in learning-set performance, but insufficient data are yet available for a firm conclusion: pigeons have not, for example, apparently been tested with three-dimensional stimuli.

Our knowledge of the role of avian brain structures in intellectual performance shows a considerable advance over that available for fish, amphibians and reptiles, but is still primitive when compared to that for mammals. There is good evidence that the archistriatum is (as in reptiles and mammals) of critical importance in fear-motivated behaviour, but little understanding of the other functions of that structure. Damage to the hyperstriatal complex disrupts reversal learning, but the precise nature of this deficit is not yet clear. There is scattered evidence for involvement of other forebrain regions in learning, but in general it cannot be said that the investigations currently available have made a real contribution to our understanding of the organization of intelligence in birds.

This chapter completes our surveys of brain and intelligence in non-mammalian vertebrates. The preceding three chapters emphasized parallels between non-mammals—fish, amphibians, and reptiles—and mammals; this chapter finds yet more parallels for birds, amongst which may be singled out comparability in brain size—a fact which may make more plausible the notion that the avian and mammalian intellects are also comparable.

7. Mammals

Classification of mammals

Mammals derive ultimately from a group of reptiles, the pelycosaurs, which first appeared in the late Carboniferous period (about 290 Ma ago), and were abundant during the early Permian. From pelycosaurs evolved a further reptilian groups the therapsids, which were the most successful vertebrates of the late Permian and early Triassic; it was from members of the therapsid order that mammals, which first appeared in the late Triassic (some 200 million years ago) evolved. It will be recalled from Chapter 5 that the line of descent leading to mammals diverged from those leading to modern reptiles and to birds at a very early stage in the evolution of reptiles, so that to find a common ancestor, one must go back about 300 million years—almost as far, indeed, as we should need to go to encounter a common ancestor with amphibians.

One subclass of mammals, the Prototheria, appears to have diverged from the main line of descent at a very early stage, possibly even being derived from a different group of therapsids from that which give rise to the other mammalian subclasses. The Prototheria, for which there is a poor fossil record, are now represented only by the monotremes—the platypus (*Ornithorhynchus*) and the spiny anteaters (the echidnas, *Tachyglossus* and *Zaglossus*) found in the Australasian region. These animals, like other mammals, maintain a relatively constant body temperature, have a hairy skin (modified into spines in the case of the anteaters) and suckle their young; however, they also have a number of characteristics in common with their reptilian ancestors, and not found in other mammals, the most striking of which is that they lay eggs. There are, then, grounds for regarding monotremes as the most primitive living mammals.

The second subclass, the Theria, in turn diversified to give rise, before the end of the Cretaceous period (66 Ma ago), to two infraclasses, the Metatheria, or marsupials, and Eutheria, or placentals. The marsupials differ from the placentals in a number of structural characters, but the best-known distinction is that marsupials lay their young in a larval form, after a brief gestation period, and continue nurturing them in a pouch. The marsupials isolated in Australia have radiated into highly successful groups of animals, including, for example, the Tasmanian wolf, *Thylacinus*, the koala bear, *Phascolarctos*, and the wallabies and the kangaroos (the Macropodidae). However, they have not thrived elsewhere, and this appears to be due to their inability to withstand competition from placentals: some marsupials do survive in the Americas, the best-known example being the Virginia opossum *Didelphis virginiana* which is sometimes described as a 'living fossil', since it appears to have changed very little since the Cretaceous period. The opossum appears to resemble the ancestor from which other living marsupial groups have evolved, so that it too, like the monotremes, may be regarded as in that sense a primitive mammal: however, there are no grounds for regarding the living marsupials in general as primitive relative to placentals.

The final mammalian subclass, the placentals, has, since its origin some 70 Ma ago, proved itself by far the most successful mammalian group: there have been identified to date only three (living or extinct) genera of monotremes, all belonging to one order, 127 genera of marsupials (all, again, of the same order) but some 2650 genera of placentals, belonging to 28 orders, of which 16 have living genera (Colbert 1955). The radiation of mammals took place very rapidly so that, from their beginning in the Cretaceous, there had evolved by the end of the Eocene (36 Ma ago), representatives belonging to all modern placental orders. The details and dating of this radiation remain the subjects of considerable debate (e.g. Stahl 1974) and we shall once again follow Romer's account here, noting that contrary opinions do not affect the broad conclusions that we wish to draw. Romer (1966) takes the view that a common ancestor to all eutherians was an insectivore which, although primitive, was sufficiently similar to modern insectivores (e.g. the shrews, moles, and hedgehogs) to be included in the same order. Romer and Parsons (1977) also suggest that the tree-shrews (Tupaiidae), whose phylogenetic status has long been in dispute (Romer himself at one time included them amongst the primates), are probably the most primitive living placentals, in the sense of being closest to the ancestor common to all placentals. Figure 7.1 summarizes the relationships amongst the placental orders, according to Romer (1966).

Man belongs to the primate order, so that his closest relatives are the other living primates; outside the primate order, man's kinship with other placental mammals is mediated through the primitive insectivores from which primates arose, and is, therefore, relatively distant. Evolution within the primate order has naturally been a subject of considerable interest, and although there is still much that is unknown about the extinct ancestors of living primates, and especially of man, the broad relationships between living genera are generally agreed. The primates may be divided into two suborders, the 'lower' of which, the prosimians, consists of the lemurs (including the loris family), and the tarsier (*Tarsius*). Each of these prosimian groups gives us a concrete picture of a stage of development similar to one through which man's ancestors passed. The second primate suborder, the anthropoids, contains animals that are structurally more advanced than the prosimians, showing in particular greater manual dexterity and more developed visual capacities. The anthropoid suborder is itself divided into two major infraorders, the platyrrhines (New World monkeys), and the (structurally) more advanced catarrhines (Old World monkeys and the hominoids). Three families constitute the hominoid superfamily and these are, the hylobatids (gibbons and siamang), the pongids (the great apes) and the hominids (man and his immediate fossil ancestors).

Man's closest living relatives are the great apes (the orangutan, *Pongo pygmaeus*, the gorilla, *Gorilla gorilla*, and the chimpanzee), and it is of interest to know at

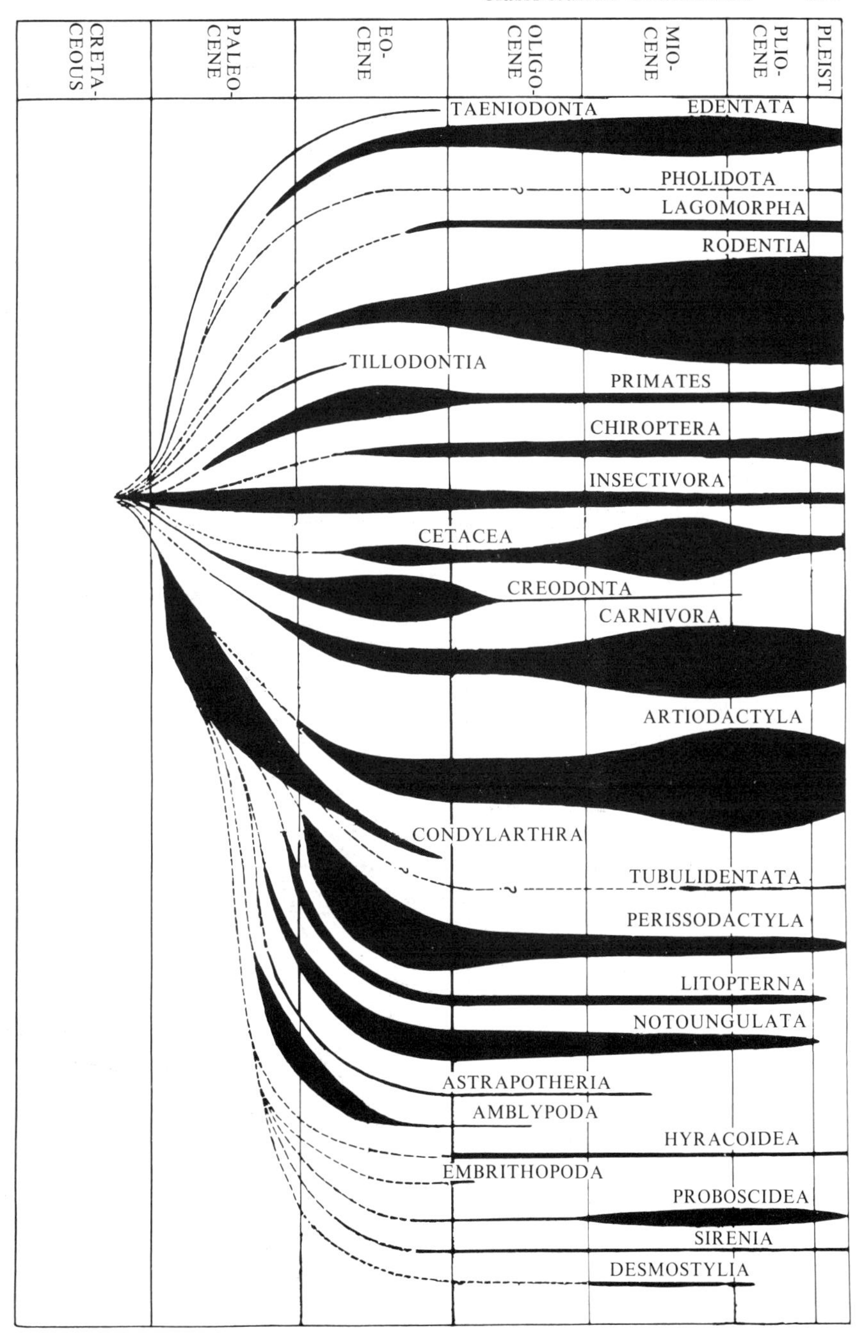

Fig. 7.1. Diagrammatic representation of the relationships between placental mammals, and their chronological distribution. (From Romer 1966).

what stage man's evolution diverged from that of the apes and, of course, through what forms our ape-like ancestors passed on the way to becoming modern man. These are both questions to which no confident answers may be given at present, although new fossil discoveries are rapidly leading to better understanding. There seems to be general agreement that an extinct apelike creature, *Dryopithecus*, which flourished during the late Oligocene and the Miocene epochs, was a common ancestor of both apes and man. Simons (1977) suggests that somewhere between 10 and 15 Ma ago, *Dryopithecus* gave rise to other groups of apes, including *Ramapithecus*, which may be the first true hominid. *Ramapithecus* fossils appear to date from as recently as 8 Ma ago (Tauxe 1979), but there is then a large gap in the record, as fossils of the next apparent ancestor, *Australopithecus*, are less than four Ma ago years old. *Australopithecus* fossils have so far been discovered only in Africa (although both *Dryopithecus* and *Ramapithecus* apparently had a widespread distribution in Africa, Europe, and Asia), and there is evidence (from fossilized footprints) to show that *Australopithecus* did walk upright some 3.3 to 3.8 Ma ago (White 1980).

Evidence for the transition from *Australopithecus* to the first universally accepted fossils of the genus *Homo* (those of *Homo erectus*) is currently scant, and open to a number of interpretations. One recent report (Johanson and White 1979) proposes that the earliest australopithecine fossils should be assigned to the species *A. afarensis*, and that the line of descent from *A. afarensis* divided, about 2.7 Ma ago, into two lines; one of these lines gave rise to an australopithecine species (*A. africanus*), from which a further species, *A. robustus*, evolved some 2 Ma ago; the other line gave rise, some 2 Ma ago, to *Homo habilis* from which *H. erectus* in turn evolved some 1.7 Ma ago. This general view, according to which species of *Homo* and *Australopithecus* were in fact contemporaneous, gains some of its support from evidence (summarized by Pilbeam and Gould 1974) which suggests that (relative) brain size was growing more rapidly in those hominids classified as belonging to the genus *Homo* than in those classified as australopithecines. Johanson and White's account is, as seems almost inevitable in this area, highly controversial (e.g. Leakey and Walker 1980), and alternative accounts are to be found in Walker and Leakey (1978). The earliest *H. sapiens* fossils date from 100 000 years ago, and are of neanderthal man (*H. sapiens neanderthalensis*); neanderthal man had a mean brain-size of approximately 1400 cm^3, which appears to be somewhat larger than that of modern man, although the range of brain size of modern man runs, in non-pathological cases, from less than 1000 to more than 2000 cm^3. Finally, the earliest fossils of modern man, *H. sapiens sapiens*, date back to about 40 000 years ago, and since that time there appear to have been no substantial changes in anatomy (Washburn 1978).

New techniques are constantly being brought to bear on the problem of human evolution, and one which may result in important modifications in the account given above will be mentioned here. The relationships deduced above were derived primarily from evidence of fossils, and from anatomical similarities of living species to fossil forms. However, biochemical similarities of many different kinds in living species may also be explored, and studies of this nature 'consistently suggest a much more recent origin of the man–chimp–gorilla separation than was previously imagined, namely in the range of 4–6 M years ago' (Zihlman, Cronin, Cramer, and Sarich 1978, p. 744). In other words, if the results of these techniques can be reconciled with the fossil record (and that, presumably, will not occur until more ape fossils are discovered), we may come to the conclusion that our kinship with chimpanzees and gorillas is surprisingly close.

We may conclude this section by considering what expectations might be derived, from knowledge of phylogenetic relationships, concerning the relative intelligence of different groups of mammals. On the one hand we should bear in mind that all living mammalian species have been subjected to evolutionary pressures for the same span of time and that, from that point of view, it is just as sensible to see man as a stage in the evolution of the chimpanzee as it is to see the chimpanzee as a step on the road to man. On the other hand, if we assume that man is the most intelligent mammal then, if there are any living forms exhibiting levels of intelligence through which man's ancestors passed, then those forms are most likely to be found in man's closest relatives, the primates. The same principle—closeness of relationship to man—does not lead us to expect (given the data shown in Fig. 7.1) any systematic differences in intelligence amongst other placental orders, since there is little difference between them (with the possible exception of the insectivores) in degree of phylogenetic affinity to man. In particular, Fig. 7.1 shows that there is essentially the same degree of relationship to man to be found in the most commonly used laboratory species (rats, cats, and dogs); this conclusion is, it may be added, confirmed by biochemical evidence (e.g. Washburn 1978). In other words, there are no phylogenetic grounds for supposing that differences in intelligence should be found amongst rats, cats, and dogs, and, moreover, very little reason for expecting different grades of intelligence to be exhibited in other mammalian groups, whether 'primitive' (e.g. monotremes, the Virginia opossum, insectivores) or 'advanced' (e.g. most marsupials, most placental mammals).

Brain size in mammals

This section will survey the data available on relative brain size in mammals with the aim of deciding what ranking of intelligence should be predicted if it is assumed that brain size is an important determinant of intelligence.

There is little evidence concerning the intellectual abilities of the great majority of mammalian species, but it will nevertheless be of interest to provide a brief general review of brain size across a range of mammals including species for whom behavioural data will not be presented, and to reconsider along with behavioural data the brain size measures for those species for whom both types of information have been collected. Von Bonin (1937) and Jerison (1973) have presented analyses of the relationship between brain weight and body weight across a wide range of mammals of different orders and both agree (although using quite independent raw data) that the relationship is well described by the equation of simple allometry, $E = kP^a$ (see p. 29), and that the value of a is approximately 0.66 (Von Bonin found that the best-fitting line had a slope of 0.655, and Jerison, 0.69). In general, then, the volume of the mammalian brain varies regularly with the square of the linear dimensions of the body (see p. 29), and if we take the view that differences between species in the value of k (Dubois's 'index of cephalization') represent variations in the amount of brain available over and above that required for purely bodily functions, then we might expect species with higher values to be more intelligent than those with lower values.

Jerison (1973) reports that the average value of k for a broad sample of living mammals is 0.12, and introduces a new statistic, the encephalization quotient (EQ), which is the ratio of k, the index of cephalization for a given species, to that average mammalian value. This measure has the advantage that it immediately gives a rough indication of the relative encephalization of the species concerned: for example, an EQ of 2.0 indicates that the species has a value of k twice as high as that expected in a mammal of comparable weight having average encephalization, and an EQ of 0.5 indicates a level of encephalization half that of an 'average' mammal. We shall, then, use EQ to compare the encephalization of various species, bearing in mind that a will be taken to be 0.66, and that EQ can be converted to k by simply multiplying by 0.12.

It may be appropriate to begin this survey by considering brain size in monotremes and marsupials. According to brain and body weight data cited by Jerison (1973), EQ for the platypus is 0.74, and, for the echidna, *Tachyglossus aculeatus*, 0.61. Estimates for EQ in the Virginia opossum range from 0.22 (Jerison 1973) to 0.42 (Von Bonin 1937), and Von Bonin's data for two Australian marsupials (*Trichosurus vulpecula*, the brush-tailed possum, and *Onychogalea frenata*, the bridled nail-tailed wallaby) yield EQs of 0.75 and 1.00, respectively. Although most of these values lie below 1.0, and so indicate lower than the mammalian average encephalization, they do show that monotremes and at least some marsupials have relative brain sizes that fall well within the range of placental mammals (as we shall see from data in Table 7.1). While the figures suggest that the Virginia opossum may have the small-sized brain that might have been anticipated from its generally primitive development, they do not indicate any comparable reduction in brain size in monotremes, the most primitive order of mammals. It should be emphasized that all the values reported here clearly belong in the 'mammalian' rather than the 'reptilian' of Jerison's polygons (p. 28); it appears, therefore, that all mammals show a sharp increase in encephalization over reptiles.

Some indication of the range of EQs found in eutherian mammals can be seen in Table 7.1, which shows EQs for a variety of species, drawn from many different orders, and having widely differing body weights. Table 7.1 provides a ranking that agrees with at least some of the widespread intuitions concerning relative intelligence of mammals: man, for example, is at the top, and the dolphin (of unspecified genus), an animal with a high reputation for intelligence, lies second, followed by a group of primates; the dog is above the cat, which just improves on the horse and so on. Now while this may be reassuring it is, of course, one of our central purposes to enquire whether intuitive rankings of intelligence are supported by good evidence, and that we shall do in later sections.

One shortcoming of the method of data presentation employed in Table 7.1 is that it does not enable us to estimate relative brain sizes in species not shown there, since a level of encephalization in a species from one order does not necessarily indicate the level to be expected in other species from that order; one cannot, then, sensibly talk, given only the data shown in Table 7.1, of the level of encephalization in, say, carnivores in general. One attempt to derive some generalizations about differences between some groups of mammalian species has been made by Jerison (1973, Appendix II), who has presented a statistical

Table 7.1. *Encephalization quotients of 21 mammals arranged in descending order. (After Russell 1979)*

1. Man	7.44	8. Marmoset	1.71	15. Cat	1.00
2. Dolphin	5.31	9. Fox	1.59	16. Horse	0.86
3. Chimp	2.49	10. Walrus	1.23	17. Sheep	0.81
4. Squirrel Monkey	2.32	11. Camel	1.17	18. Ox	0.54
5. Rhesus Monkey	2.09	12. Dog	1.17	19. Mouse	0.50
6. Elephant	1.87	13. Squirrel	1.10	20. Rat	0.40
7. Whale	1.76	14. Wild pig	1.01	21. Rabbit	0.40

analysis of EQs reported for various collections of living and fossil mammals. This analysis finds (excluding the fossil data) that the anthropoids (data from 46 species, excluding man, with a mean EQ of 2.08) have significantly higher EQs than carnivores (15 species, mean EQ 1.2), prosimians (20 species, mean EQ 1.11) and ungulates (20 artiodactyl and five perissodactyl species, mean EQ 0.96), and that the latter three groups do not differ from each other. Carnivores, prosimians and ungulates in turn have significantly higher EQs than rodents (37 species, mean EQ 0.65), whose EQs are higher than those of insectivores (33 species, mean EQ 0.44).

These data indicate, then, that the higher primates are in general more 'encephalized' than the lower primates, that the lower primates are, like carnivores and ungulates, about average in their degree of encephalization relative to mammals as a whole, and that rodents and insectivores are below average.

We may conclude this section with the reminder that none of the data reviewed here have any necessary link with intelligence: only behavioural data can show the significance or otherwise of the level of encephalization of a species. What we have from this survey is a possible ranking of species which needs to be validated by behavioural data before we can assume a role for brain size in the determination of intelligence.

Forebrain organization in mammals

Introduction

No attempt will be made here to describe in detail either the anatomy or the physiology of the mammalian forebrain: such information can readily be obtained from any standard textbook of neuroanatomy or physiological psychology. What we shall do is, first, point out similarities (and differences) of the overall organization in mammals to that in non-mammalian vertebrates and then discuss potentially significant differences in organization between various orders of mammals. The discussion will go on to ask whether the neocortex (the structure which most strikingly differentiates between mammals and non-mammals) plays a general role in intelligence, whether any specific neocortical regions serve specific intellectual functions and, finally, whether the functional roles of the various cortical association areas vary across mammals.

The broad principles of telencephalic organization remain constant throughout mammals, despite substantial differences in gross appearance amongst the various orders. The major structure of pallial origin is the neocortex, which is bounded medially by the hippocampal formation (the archicortex, from which it is divided by the hippocampal fissure) and laterally by the piriform cortex (the paleocortex), from which it is separated by the rhinal fissure. The conventional definition of neocortex holds that it is cortex which passes through a six-layered stage in its development; other cortical regions possess fewer layers. The transition from neocortex to paleo- or archicortex is not abrupt, and various schemes for further subdivisions within the pallial mantle have been introduced (for discussions, see Pribram and Kruger 1954; Stephan and Andy 1970). According to one system of nomenclature, the paleo- and archicortex are collectively known as allocortex, and the cortical regions immediately adjoining them, as periallocortex—the peripaleocortex (the insular cortex) and periarchicortex (which includes, for example, the entorhinal, presubicular, and retrosplenial cortex). The cortical regions immediately adjoining the neocortex (sometimes termed isocortex, to avoid implications of evolutionary history) tend to resemble neocortex rather more than allocortex, and are known as juxtallo- or proisocortex; the major juxtallocortical structure is the cingulate cortex which, while not going through a six-layered developmental stage, does closely resemble neocortex in its final appearance.

Although details of sensory projection systems will not in general be presented here, it may be appropriate to point out that the olfactory bulb in mammals does project to the paleocortex and associated regions (to prepiriform and periamygdalar cortex, and to the anterior olfactory nucleus, the olfactory tubercle and the medial amygdaloid nucleus) (Turner and Mishkin, 1978). There will be a limited discussion of hippocampal (archicortical) function in a later section, but nothing further will be said of the role of either the periallocortex or the juxtallocortex, and this is largely because there is at present little evidence to suggest that these regions play any significant role in learning or that there may be inter-species differences in such roles between different mammalian orders.

The neocortex is the dominant type of cortex throughout mammals, and its expansion medially has resulted in the distorted folding of the hippocampal cortex, giving its appearance (in cross-section) of a seahorse; similarly, the lateral expansion of neocortex has forced the paleocortex into a ventral position. Figure 7.2 shows the relative location of the major cortical regions in an echidna. The expansion of neocortex should not be allowed to obscure the basic plan of the

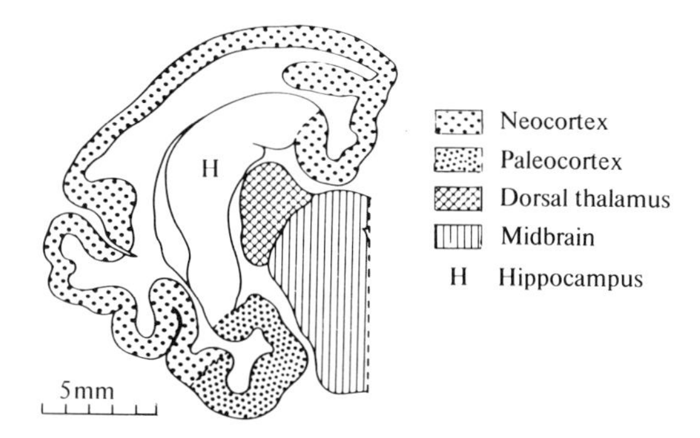

Fig. 7.2. Diagrammatic representation of transverse hemisection from an echidna's telencephalon. Continuity between neocortex and archicortex (hippocampus, which is at this level contiguous ventrally with paleocortex) is clearly seen in this acallosal species. (From Welker and Lende 1980).

telencephalic pallium which is just as it was in reptiles, the reptilian 'general' cortex being bounded medially by hippocampal cortex and laterally by paleocortex. Other structures of pallial origin, such as the basolateral amygdala, also occupy regions which show a general correspondence with those described in non-mammalian vertebrates.

It has been pointed out in previous chapters that the mammalian structures of subpallial origin (including the septal nuclei, the caudate-putamen complex, the globus pallidus and the corticomedial amygdala) also lie in topographically similar positions to those occupied by the corresponding non-mammalian structures. It is therefore reasonable to speak of the basic organization of vertebrate forebrain. On the other hand, it would be unwise to ignore the differences in organization, as if only quantitative differences could prove to be of significance. The similarity of groundplan should not, therefore, obscure certain major differences, of which the following may be singled out: first, a greatly expanded six-layered neocortex is found throughout mammals, whereas only a very limited appearance of layered cortical tissue occurs outside mammals; second, there is a greater degree of telencephalic differentiation, at least in terms of architectonic distinctions, in mammals than in non-mammals, so that there may exist in mammals many areas for which there is no corresponding area in non-mammals; third, there are regions in non-mammals (the dorsal ventricular ridge of reptiles, the hyperstriatum and neostriatum of birds, for example) which may have no corresponding region in mammals.

None of the preceding observations has any necessary implications for intelligence; there is no obvious reason, for example, for expecting animals with multi-layered cortical tissue to be more intelligent than those without (extensive layering is, after all, found in the superior colliculus of non-mammalian vertebrates). The significance of anatomical data depends on the relation of structure to function, and that is a problem about which little is known at present (we cannot, that is, predict the role of a structure from an examination of its gross architectonic appearance).

Before leaving this introduction, it may be as well to emphasize, as has been done in earlier chapters, that the terms paleo-, archi-, and neocortex should no longer be interpreted as indicating the order in which the structures evolved; in particular, the absence of six-layered cortical structures in non-mammals does not necessarily mean that no structure homologous with neocortex is to be found outside mammals. It is legitimate and sensible to enquire whether the possession of neocortex confers intellectual advantage; it is not reasonable to attempt to settle that question without investigating both mammalian intellectual capacity and the relevance of neocortex to that capacity.

Variation in cortical organization within mammals

One gross difference in the anatomical organization of the brain of monotremes and a number of marsupials as compared to that of placentals is that the former two groups do not possess a corpus callosum, the dorsally located bundle of fibres that interconnect homotopic areas of neocortex. There does not, however, appear to be any great significance in this distinction. Some marsupials (e.g. *Macropus*, the kangaroo) *do* have a corpus callosum, and those non-eutherian mammals that do not (e.g. echidnas) have instead anterior commissures that are considerably larger than those found in placentals, and which carry fibres linking neocortical regions (Kappers, Huber, and Crosby 1960). Ebner (1969), moreover, has shown that the pattern of interhemispheric connections established in the Virginia opossum (which has no corpus callosum) and the Pakistani hedgehog *Paraechinus hypomelas* (which does have a corpus callosum) are indistinguishable, so confirming the view that absence of the corpus callosum may be of little functional significance.

A second gross distinction between the brains of mammals is that the brain surface which is, as is well known, highly convoluted in man, is in some species devoid of convolutions (the brains of such species being called 'lissencephalic'). This distinction also seems to have minor significance. Although many of the sulci in the primate brain do indeed mark divisions between areas of differing functional significance (Welker and Campos 1963), the major reason for folding of the brain appears to be a mechanical one (Le Gros Clark 1945). The thickness of the neocortex (which forms the greater part, if not all, of the surface in mammals) varies very little either within or between mammals. For example, the thickness of the human cortex ranges from about 1.5 to 3.5 mm in various areas; the parietal cortex of the mouse is just over 1 mm thick, while that of the elephant is 4 mm thick (Le Gros Clark 1945). Now the shape of the brain is roughly spherical so that if a surface layer of near-constant depth is to maintain itself as a constant proportion of the total volume of the brain, folding will be inevitable: this is, of course, because the volume of the brain will increase according to the cube of the radius of the sphere, whereas that of the surface will increase according to the square of that radius. The implication of the argument that cortical folding is the consequence of maintaining (or increasing) the ratio of neocortex volume to brain volume with increasing brain size is, then, that small-brained species should be lissencephalic, and large-brained species, convoluted (or gyrencephalic). Von Bonin (1941) devised an index of cortical folding (the ratio of the total cortical surface—as if the cortex was laid out flat, unfolded—to the total *exposed* cortical surface) and index of folding data collected by both Von Bonin (1941) and Elias and Schwartz (1969) indicate that, across a wide range of mammals (including marsupials) the index increases fairly regularly with gross brain volume; in particular, man, with a mean index of 2.86 (range 2.25 to 3.07, sample size 20) shows less folding than was found in seven cetaceous species whose indices ranged from 4.0 (Baird's dolphin, *Delphinus bairdii*) to 8.55 (Atlantic pilot whale, *Globicephala melaena*) (Elias and Schwartz 1969). Moreover, although the brains of most of the cetaceans were indeed larger than the human brains, Baird's dolphin has a brain volume (722 cm^3) considerably below that of the human average (1198 cm^3, in the Elias and Schwartz sample). Cortical folding is, then, largely a matter of brain volume; some mammals show in both absolute and relative terms considerably more folding than man, and there is no reason to suppose that the convolutions of the human brain reflect particular functional complexity.

The discussion of cortical folding raises a further question: given that folding could be the inevitable consequence of maintaining a constant ratio of cortical volume to whole brain volume, does this ratio in fact remain constant (as might be suggested by the reasonably regular relationship between brain size and index of folding), or is there an increase in relative neocortical volume in some mammalian groups? Relevant data have been provided by a number of authors, and the various reports agree that there is within different mammalian orders a good correlation between brain size and both neocortical volume (Harman 1957) and neocortical surface (Elias and Schwartz 1969). Figure 7.3 shows data (Harman 1957) for a series of carnivores, and Fig. 7.4 shows comparable data (Passingham 1975, who used data provided by Stephan, Bauchot, and Andy 1970) for primates, including man. There are, however, some differences between orders: Fig. 7.5 shows that rodents show less relative neocorticalization than primates, carnivores, and ungulates (between which there appears to be very little difference), and that the Virginia opossum shows even less than the rodents. In general, then, two

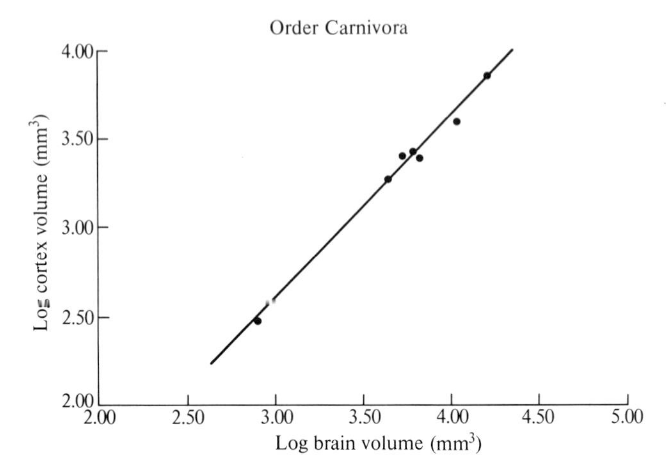

Fig. 7.3. Double logarithmic plot showing the relationship between neocortical volume and brain volume in carnivores. The relationship is rectilinear, and the slope of the line approaches 45°. (From Harman 1957.)

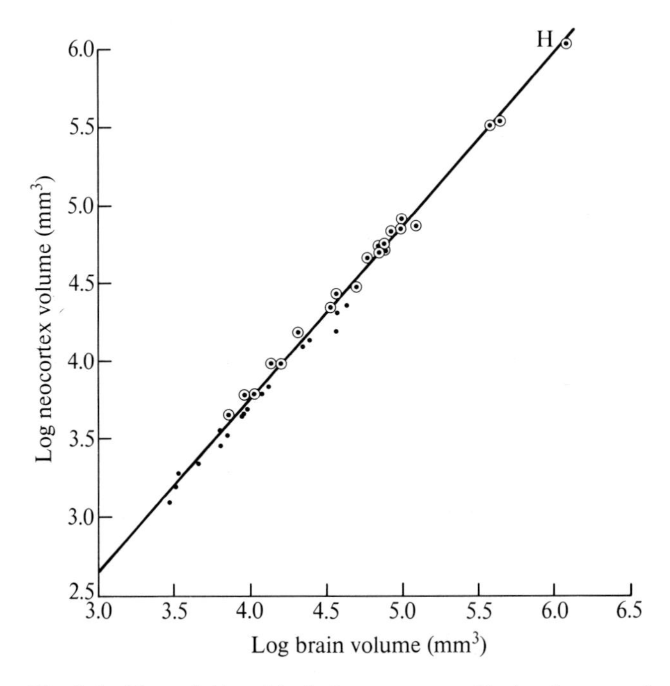

Fig 7.4. The relationship between neocortical volume and brain volume in primates. (From Passingham 1975.)

Key: single dots—prosimians; circled dots—anthropoids; H—*Homo*.

conclusions emerge: first, that man appears to possess no more neocortex than would be expected in a primate of that brain size; and, second, although primates possess relatively more neocortex than some other mammalian orders, they are clearly far from unique in the extent of their neocorticalization.

It may be somewhat unexpected to find that man and primates in general show so little superiority in amount of neocortex; the neocortex is, after all, the neuroanatomical structure that most strikingly differentiates mammals from other vertebrates, and there is a widespread belief that the growth of the neocortex is responsible for the evolution of intelligence. In the light of this view, it should be pointed out that there are other ways of regarding these same data. Stephan and Andy (1969), for example, prefer to compare the size of a structure in a given mammal with the size that would be expected in a 'basal insectivore' of comparable body-weight. Basal insectivores include species from three families, the tenrecs (Tenrecidae), hedgehogs (Erinaceidae), and the shrews (Soricidae), which are regarded as less specialized in comparison with 'progressive' insectivores—species with special adaptations for, say, burrowing or digging, or for searching for food in water. The expected size of a structure in a basal insectivore for a given body weight is calculated for each structure by applying the simple allometry formula to data gathered from basal insectivores of various weights, and extrapolating in the usual way to other body weights. Stephan and Andy label the ratio between the actual size of a structure and the expected size in a basal insectivore the 'progression index', and Fig. 7.6 shows progression indices for a number of brain structures in prosimians and anthropoids. This figure appears to show that the neocortex shows a dramatic progression in size as we go from the insectivores to the 'lower' primates, and from the 'lower' to the 'higher' primates; moreover, this increase in the neocortex far outstrips the progressive size increases seen in other brain structures. How is this interpretation to be reconciled with the data considered previously? The answer comes in two parts. First, the neocortical progression indices reflect primarily differences in gross brain size between basal insectivores, prosimians, and anthropoids (see p. 243), along with minor differences in the ratio of cortical to total brain volume (Fig. 7.4, for example, shows that prosimians do tend to show marginally less neocortex for a given brain volume than do anthropoids). Second, whereas the neocortex maintains a relatively constant percentage of brain volume as brain volume expands

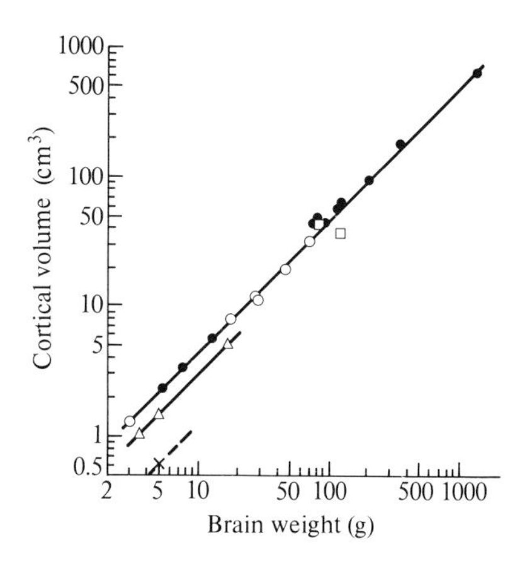

Fig. 7.5. The relationship between neocortical volume and brain volume in primates (solid circles), carnivores (open circles), ungulates (squares), rodents (triangles), and opossum (cross). (From Jerison 1973.)

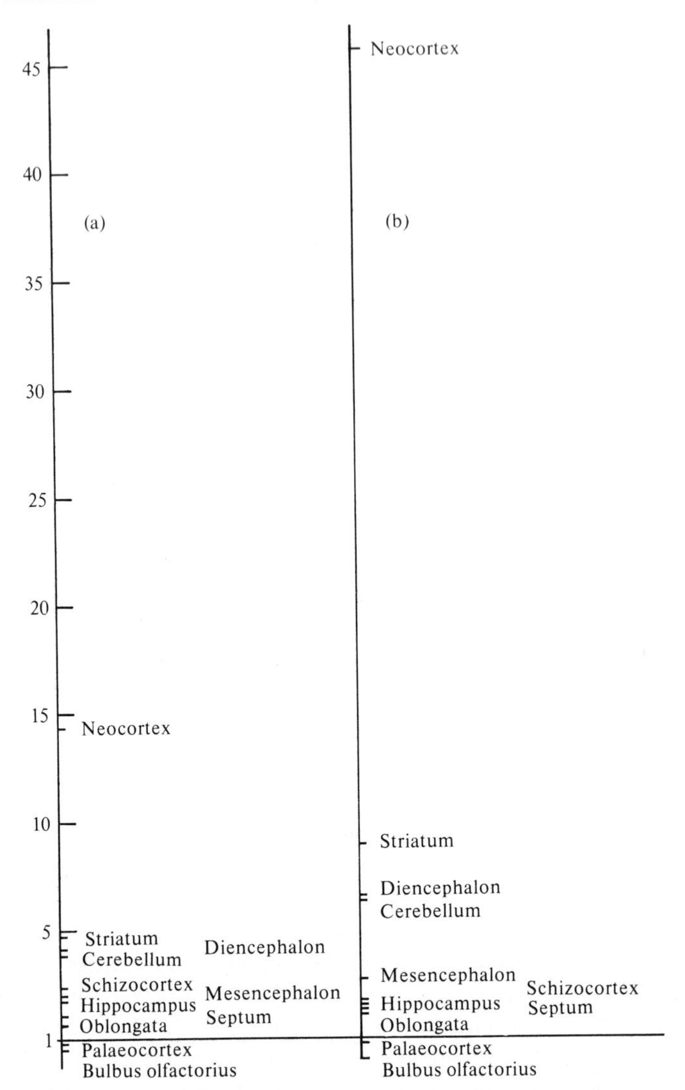

Fig. 7.6. Average progression indices of brain structures in 20 prosimians (a) and 21 anthropoids (b). Reference base (1) is the average progression index of these structures in basal insectivores. (From Stephan and Andy 1969.)

(reflected in the 45° slope of the lines drawn by Jerison on Fig. 7.5), other fore-brain structures, although showing a similarly close relation to whole brain volume, enjoy a gradually decreasing percentage of brain volume as that volume increases. This latter point is illustrated in Fig. 7.7, which shows the relationship between hippocampus and brain volume in various primates: although the relationship is orderly, the slope of the line through the points is less than 1, and, on a double logarithmic plot, this means that the size of the hippocampus becomes progressively less relative to that of the whole brain as the brain size increases. As with the neocortex, however, this is not the whole story—Fig. 7.7 also shows that, in general, prosimians of a given brain weight *do* have relatively more hippocampus than anthropoids of the same brain weight and this is a factor that will also be reflected in the progression indices of Fig. 7.6. In summary, then, knowledge of brain size alone enables us to predict with reasonable accuracy both neocortical

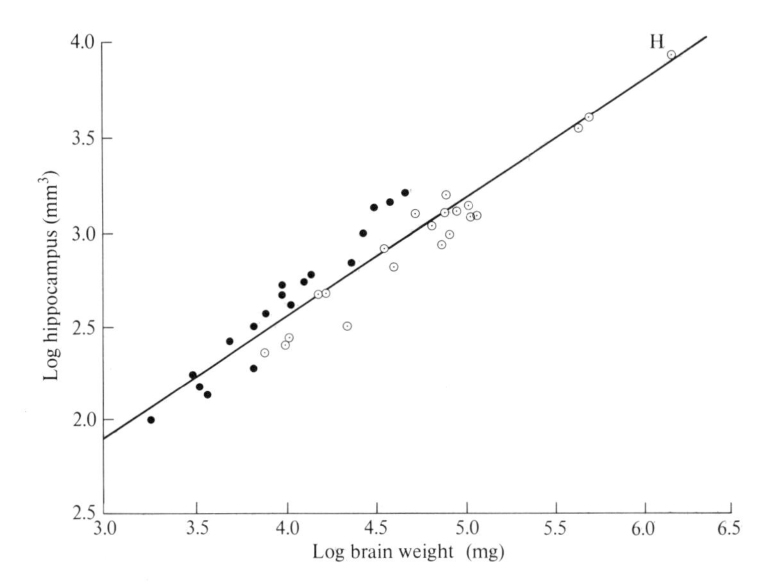

Fig. 7.7. Relationship between hippocampal volume and brain weight in primates. Symbols as for Fig. 7.4. (From Passingham and Ettlinger 1974.)

and hippocampal volumes, and, in fact, this is true in primates for all major brain regions, excluding only the olfactory bulbs (e.g. Sacher 1970). Man does indeed possess very much more neocortex than would be expected in a basal insectivore of comparable *body* weight; but man does not possess very much more neocortex than that expected in a basal insectivore of comparable *brain* weight. Moreover, man possesses just that amount of both neocortex and hippocampus that would be expected in an average primate of comparable brain size. If man has a gross neuroanatomical advantage over other primates, it would seem to lie in the possession of more brain for a given body weight.

A further question that might be raised is whether gross intracortical organization remains the same throughout mammals, or whether there is a progressive relative decrease in the proportion of neocortex devoted to purely bodily (sensorimotor) functions: that is, does the relative amount of 'association' cortex increase as we come closer to man?

This question requires us to digress briefly to ask what it is that we mean by association cortex? In previous references, we have assumed that association cortex includes all cortical areas that are not primarily sensory or motor in nature, and this is, indeed, the key concept involved in the traditional accounts of association cortex. However, certain other assumptions were also made by the early neuro-anatomists. One of these was that association cortex made only intracortical connections—that is, that it neither received nor sent projections from or to areas outside the cortex. A second assumption was that association cortex could be distinguished architectonically from either motor or sensory cortex; whereas motor cortex showed a prominent fifth (pyramidal cell) layer, and sensory cortex showed a prominent fourth (granule cell) layer, association cortex showed generally no particular development of any single layer, and was known as eulaminate.

The notion that association cortex made only intracortical connections had to be revised in the light of the discovery that association cortex did in fact receive projections from the thalamus—for example, the frontal cortex receives projections from the dorsomedial nucleus of the thalamus, and 'visual association cortex' (areas adjoining the primary striate cortex, originally assumed to be involved in elaborating information received from the primary visual area) receives projections from the pulvinar. These findings led to the proposal that cortical areas should be classified according to their extracortical connections, and with the thalamus in particular . Now although the areas traditionally conceived of as association areas had been shown to receive thalamic projections, it was believed that the thalamic nuclei from which these projections originated had themselves no extrathalamic (extrinsic) inputs, and so, again, were primarily neither motor nor sensory in nature. In other words, one of the original presumed properties of association cortex was simply transferred to certain thalamic nuclei, and cortical areas connected with such nuclei were regarded as association areas. However, this distinction also has run into the same difficulty that faced the original account of cortical association areas, namely that many of the thalamic nuclei originally believed to have only intrathalamic (intrinsic) input are now known to have extrinsic input. The most striking example, of course, is the pulvinar, which receives a substantial projection from the superior colliculus—which in turn receives direct retinal projections. A recent survey (Diamond 1979) of work in this area using cats, tree-shrews (*Tupaia glis*) and bush-babies (*Galago senegalensis*) suggests indeed that if areas now known to have prominent connections with sensory projection systems are excluded from association cortex, then there is no association cortex lying between the auditory and visual areas of those mammalian species for which adequate information is available; there is, in other words, no 'auditory' or 'visual' association cortex—all parts of the cortical fields concerned with those modalities are directly connected, through the thalamus, with the corresponding ascending sensory pathways so that the 'primary' fields are much larger than had previously been thought.

It is important to note that Diamond's (1979) development of the notion of an extensive cortical field concerned with a given sensory modality does not deny that there are important subdivisions within a field. One of these subdivisions, which Diamond refers to as a 'core area', in fact corresponds to the traditional primary sensory area, and can indeed be distinguished by its prominent fourth layer. Two other distinguishing features of the core area are, first, that one of the thalamic nuclei concerned with the relevant modality projects only to the core area, and, second, that the topographic organization of the projection is most precise in the core area.

A major implication of this discussion of association cortex is that the only major cortical area that is not prominently associated with either motor or sensory projection systems is the region traditionally known as the frontal association cortex—the area served by the dorsomedial nucleus of the thalamus; does the extent of this frontal area vary systematically between various groups of mammals? Traditionally it has been believed that the frontal association cortex is poorly developed in 'lower' (non-primate) mammals and Kappers, Huber, and Crosby (1960) in fact stated that there was no frontal association cortex visible in insectivores, microchiroptera (bats, excluding Old World fruit bats and 'flying foxes') and many rodents. However, this was a conclusion reached on architectonic evidence alone, and more recent studies have shown that, if receipt of projections from the dorsomedial nucleus is used as a criterion, then quite different conclusions may emerge. Leonard (1969) for example, has shown that, using that criterion, there is a very much more extensive region of frontal cortex in the rat than had been supposed on the basis of architectonic studies. Unfortunately, however, we do not possess detailed studies of thalamocortical projections in more than a few species, and we are forced to continue this section by considering only somewhat unsatisfactory architectonic evidence.

The first series of data to be discussed concern the ratio of eulaminate to total neocortex (Passingham 1973). In general, eulaminate cortex will include all traditional association areas—that is, both the frontal area and the various 'sensory' association areas. If there are differences between groups in either of these categories, then they should be reflected in the ratios. Figure 7.8 shows a plot of eulaminate against total neocortex for a series of primates (four anthropoid genera, *Homo*, *Pan*, *Cercopithecus*, and *Callithrix*, and one prosimian, *Tarsius*), and it can be seen that the amount of eulaminate cortex varies directly with the amount of neocortex; man, in other words, has no more eulaminate cortex than would be expected in any primate of comparable neocortical size. This indirect measure does not, then, suggest that the relative size of frontal cortex varies systematically with phylogenetic status. It may be added that if the dismissal of the traditional 'sensory' association areas is thought too abrupt, it is in any case true that the relative size of those areas does not vary in any obvious way across the primate series.

Passingham (1973) also discusses direct measures (using architectonic criteria) of amounts of frontal cortex in man and chimpanzee and shows that, according to data provided by Brodmann, one of the early champions of architectonics, man and chimp have significantly more frontal neocortex than would be expected in other primates of comparable neocortical extent. On the other hand, subsequent data provided by Rose show much less frontal neocortex in man, and Passingham suggests that confusion has arisen through differences in methods of classifying areas of cortex whose architectonics are transitional between one category and another. Passingham argues that we should at present assume no special increase

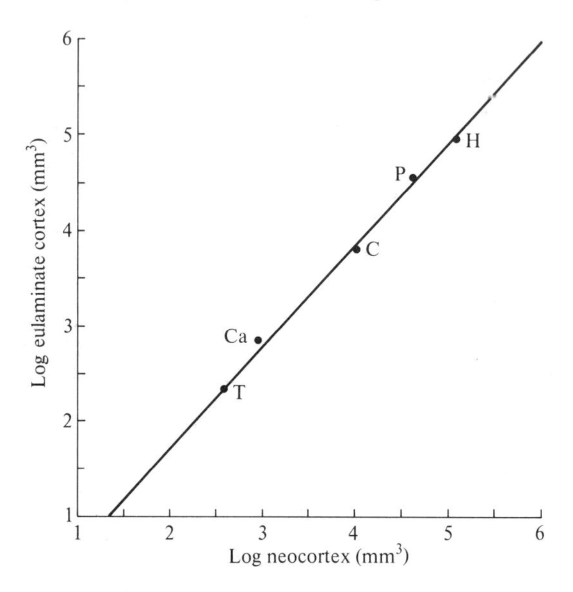

Fig. 7.8. Relationship between volume of eulaminate cortex and total neocortical volume in primates. (From Passingham 1973.)

Key: T–*Tarsius*; Ca–*Callithrix*; C–*Cercopithecus*; P–*Pan*; H–*Homo*.

in frontal neocortex in man; to do otherwise, he argues, would imply (given the relative eulaminate cortex data) that man has relatively less 'sensory' association cortex—and Passingham goes on to provide direct evidence that this is not the case.

Before leaving this somewhat untidy topic, one observation on 'association' cortex in a primitive mammal, the echidna, may be made: Fig. 7.9 shows a drawing of the cortical surface of the spiny anteater, on which are marked sensory and motor areas, established by recording and stimulation techniques (Lende 1969). There is, as can be seen, a large expanse of association cortex, which indeed 'exceeds in proportion the amount of frontal cortex in the highest primate' (Lende 1969, p. 272). Lende adds that 'There are no broad hints of its function from either neuroanatomy or behavioural observations', and this is, of course, unfortunate, although the implications of the organization seen in echidnas are, like the data reviewed earlier in this section, clear enough: if man is indeed the most intelligent vertebrate, there seems little reason to suppose that that eminence was achieved through any special enlargement of either neocortex in general or association cortex in particular.

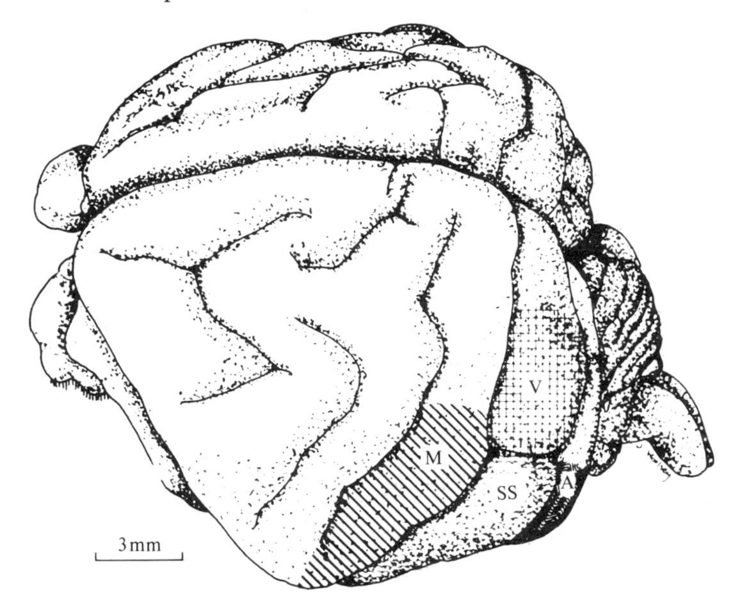

Fig. 7.9. Dorsolateral view of an echidna's left hemisphere, showing motor (M) and sensory (V – visual; A – auditory; SS – somatic) areas. Note caudal location of these areas, and lack of associational cortex between them. (From Lende 1969.)

Neocortex and intelligence

The discussion of the relative development of neocortex and association cortex was based on the notion that the cortex, or parts of it, might bear some special relationship to intellectual functioning. Our conclusion has, however, been that the only statistic that makes substantial distinctions between mammals is relative brain size, so that it may be useful to examine directly the relationship between cortex and intelligence at this point, with two questions in mind. First, does the neocortex (or any subdivision within it) play an important role in intelligent behaviour, and, second, does that role vary across different mammalian species?

A discussion of the role of the neocortex in intelligence must inevitably begin with the pioneering experimental work of Lashley (e.g. 1929). Excellent surveys of this work are available (e.g. Lashley 1950; Zangwill 1961), so that we may concentrate here on the principal features of Lashley's findings and on the interpretation of them. Lashley's early work led him to two major conclusions. The first was that the retention (but not the acquisition) of simple discriminations based on brightness was impaired by damage to the occipital third of the cortex (containing the striate visual projection area), but not by damage to any other cortical region. Within the occipital third of the cortex, the amount of retention loss was correlated with the extent, but not the locus, of the lesion. Lashley's second major conclusion concerned maze habits, and here Lashley found that both acquisition and retention of maze habits were impaired by lesions anywhere within the neocortex, the amount of impairment being correlated with the extent of the lesion but independent of its locus. To account for these findings, Lashley introduced the concepts of mass action and equipotentiality. The principle of mass action proposes that the parts of a given functional area of cortex act in concert so that removal of any of those parts leads to a decline in overall efficiency. The various parts of a functional area are said to be equipotential in so far as efficiency declines as a function of the total amount of the area removed, independent of its locus within the area. The contribution of the various parts of a functional area, that is, appear to equal. It must be emphasized that these notions are not in opposition to the notion of functional localization: they do not state, for example, that all parts of the neocortex are equipotential—the occipital third of the cortex is clearly different from the remainder of the cortex in at least one important functional respect—but those parts of the cortex that are involved in a given function are, according to Lashley's principles, equipotential with regard to that function.

Although our central concern will be the proper interpretation of the maze habit findings, it is important to bring Lashley's findings on brightness discriminations somewhat more up to date. First, Lashley's own subsequent work (Lashley 1935) showed that in fact retention of brightness discriminations was not at all impaired provided any small part of the primary visual thalamocortical projection area was left intact (so that the concept of mass action does not apply to these results). Second, the fact that acquisition of brightness discriminations is not affected by occipital lesions (whereas retention is impaired) is probably due to the use in acquisition by rats having lesions of a different cue (luminous flux as opposed to albedo) from that used by unoperated subjects: where intact rats are compelled to use luminous flux cues in acquisition, subsequent cortical lesions do not impair retention (Bauer and Cooper 1964). Since occipital lesions do *not* differentially affect retention (as opposed to acquisition) of brightness discriminations, we need not suppose any special role for occipital cortex in the storage of brightness 'engrams' (but see Braun, Meyer, and Meyer 1966, for evidence that the occipital cortex may have a role in accessing brightness engrams).

Lashley's maze-learning results are of particular relevance to our main interest because they suggest both that there may be some quite general capacity involved in mastering complex multimodal problems (such as maze-learning) and that that capacity is the province of the entire neocortex. Alternative accounts of Lashley's findings have been advanced, however, and one of these which is of particular importance is that the impairments seen are the consequence of accumulated sensory deficits, and that the correlation of impairment with extent of lesion simply reflects increasing invasion of diverse sensory cortical areas. Lashley's response to this criticism consisted in attempts to show that peripheral blinding led to less serious deficits in maze learning than did lesions of occipital cortex. Such lesions, it was argued, should affect only visual sensory processes, so that any difference between blind and lesioned rats must reflect the participation of occipital cortex in processing of non-visual information; similarly, if occipital cortex lesions have an effect on the performance of blind rats, that effect could not be due to impairment of visual processes.

Tsang (1934, 1936) and Lashley (1943) ran a series of experiments using conventional alley-mazes and showed first, that visual cortical lesions had a considerably more severe effect on both acquisition and retention of maze-learning than did peripheral blinding; second, that visual cortical lesions severely affected both acquisition and retention of maze-learning in already blind rats; and, third, that acquisition of maze-learning in rats blind from birth was severely impaired by visual cortical lesions (this latter result ruling out any explanation in terms of involvement, through visual experience, of the occipital cortex in some kind of spatial co-ordinate system).

There are, however, other results that do not confirm Lashley's expectations. Pickett (1952), for example, found no effect of posterior cortical lesions (involving about 20 per cent of the entire cortical area) on retention of complex elevated maze performance by blind rats; anterior cortical lesions of similar extent (involving somatic and motor cortical regions) did produce deficits in retention. Now while it is true that Pickett's task was particularly designed to emphasize solely kinaesthetic cues (not only was the maze elevated, but the units of the maze were randomly interchanged, and the maze itself was rotated between trials), that of itself does not explain away the findings: the fact that only one modality is relevant would not seem to alter (and certainly not decrease) the complexity of a maze, in which the principal difficulty would seem to be to organize and retain the appropriate sequence of choices to be made through the maze. In other words, if general intelligence is involved in a multimodal maze, then it should be involved also in a unimodal maze. The difficulty here, of course, is that we have no precise account of which problems involve general intelligence, and which do not, or, to use more neutral terms, which tasks are complex, and which simple. Orbach (1959) has provided further evidence emphasizing this difficulty, by showing that whereas occipital lesions in rhesus monkeys tested in total darkness did not disrupt retention of a stylus maze (in which the subjects had to manipulate a peg along channels in a board), similar lesions did disrupt retention by blinded monkeys of a large locomotor maze of the same formal plan. Here, again, the task difficulty, as distinct from sensorimotor factors, would appear to be the same, and yet the lesions have differential effects.

Until we obtain a better understanding of the ways in which animals master mazes, it will clearly not be possible to provide a full account of the effects of cortical lesions on such problems. At present it seems fair to conclude, from the effects of occipital lesions on blind subjects, that the visual cortex does play a role in certain tasks in which visual processing is not involved, but that it is premature to conclude that visual cortical lesions degrade 'general intelligence' since there are tasks, immune to such lesions, which would seem likely to involve such a general capacity in intact animals.

Consideration of the data from brain damage in humans does not help clarify the above conclusions. Teuber and Weinstein (1956), for example, compared the performance of brain-injured patients with that of control patients on a 'hidden figure' test, in which subjects, shown a simple geometric design, are then required to identify it in more complex designs. Teuber and Weinstein found, first, that all brain-injured groups performed at a lower level than controls, irrespective of the locus of damage, and, second, that brain-injured subjects with aphasic symptoms were worse than non-aphasic subjects. The general decline, irrespective of locus, might be taken as support for the view that the cortex is associated with a general intelligence factor: however, Weinstein and Teuber (1957) found *no* effect of brain damage anywhere except in the left hemisphere (most likely to produce aphasia) on performance in the Army General Classification Test (AGCT), a standardized 'general intelligence' test. Now although the AGCT is, according to Weinstein and Teuber, a predominantly verbal test, it seems odd to suppose, assuming that there is a general intelligence factor, that general intelligence plays no part in verbal intelligence tests.

In summary, there is evidence, from both humans and other mammals, of decrements in performance in certain complex tasks following lesions of diverse cortical areas, and, in particular, evidence that 'sensory' cortical areas do more than process information from their specific modality; on the other hand, there are grave (but possibly not insurmountable) difficulties facing the proposal that general intelligence is a capacity whose physical basis is widely and equally distributed through the neocortex.

Before leaving this topic, two further observations should be made. First, although Lashley's views are currently somewhat out of fashion, it is clear that they might easily be reconciled with ideas on information-processing derived from the development of computers. We might, for example, suppose that there were distributed throughout the neocortex multipurpose logic or memory units which could be called upon in a wide range of different tasks, depending upon their current availability. Gradual loss of such units would lead to a gradual decrease in speed or accuracy (or both), and no particular set of these units would be of any more importance than any other. A given unit in, say, the store of a computer may take part in many different types of computation, and, of course, need not be used identically in the same operation carried out on different occasions; thus the notions of mass action and equipotentiality can be dressed in modern clothes. A second observation is that the implication of Lashley's view would seem to be that intelligence should be correlated with absolute cortical size: that is not the conclusion reached by Lashley himself, who regarded the index of cephalization as 'The only neurological character for which a correlation with behavioural capacity in different animals is supported by significant evidence' (Lashley 1949, p. 33). However, as we have seen, the index of cephalization derives its rationale from the assumption that it reflects the amount of brain tissue not concerned with sensory or motor processing; but Lashley's view is precisely that those areas involved in sensory or motor processes also contribute (and to the same extent as any other area) to general complex processing. Therefore, the more cortical tissue available, the more efficient processing should be. As most workers find it distinctly implausible that intelligence is correlated with absolute brain size, it is perhaps on this ground that we should be most cautious about accepting any simplistic version of Lashley's views.

A further proposal concerning the relationship of neocortex to intelligence, one to which we have alluded in previous chapters, is due to Bitterman and his colleagues. Their claim is that there are qualitative changes in intelligence as we advance from fish to mammals, and that these changes are a consequence of the development of neocortex. We have already had cause to question the very basis of this notion, the assumption that there are qualitative differences between vertebrate classes in intelligence, but here we shall briefly consider evidence relevant to a direct link between the neocortex and behavioural differences.

The most remarkable experimental data are provided by a study of Bresler and Bitterman (1969), in which six African mouthbreeders had extra brain tissue grafted onto the prospective tectal region in an embryonic stage. Subsequent testing of these animals in serial position or colour reversal tasks showed that two subjects showed progressive improvement, two showed very low reversal error scores throughout, and two, no improvement across the series. The final two subjects were found to have had grafts that did not 'take' and so could be regarded as operated controls. Bresler and Bitterman claim that 'the tectum of the fish is regarded as homologous to the cortex of mammals' (p. 591) and interpret their findings as showing that extra 'neocortex' enables fish to show, as normal fish do not, serial reversal improvement. There are, of course, a large number of objections to this interpretation, of which we may highlight two: first, no neuroanatomist supposes that fish tectum is homologous to mammalian neocortex—it is in fact homologous to mammalian superior colliculus, thus this experiment is hardly relevant to any question of the role of the neocortex. Second, we now know (p. 71) that teleosts frequently do show serial reversal improvement, so that we should require very much stronger statistical evidence for a difference between experimental and control subjects than is provided here.

In an experiment having the converse logic to that just described, Gonzalez *et al.* (1964) attempted to see whether removing neocortex from mammals would reduce them to submammalian intellectual status. Rats were extensively decorticated in infancy (more than 50 per cent of all neocortex was removed) and tested when adult on serial reversal or probability learning tasks having either spatial or visual (brightness) cues relevant. The decorticated rats behaved like normals on spatial problems—showing serial reversal improvement and maximizing in the probability tasks—but did not (unlike controls) show improvement across a series of simultaneous brightness discrimination reversals, and matched rather than maximized on the probability tasks. Gonzalez *et al.* regard these findings as evidence for a critical role of neocortex in the qualitative evolutionary advance in intelligence, and note that their across-class behavioural studies also suggest that spatial problems obtain mammal-like performance in animals (reptiles) not capable of similar performance in visual problems (see p. 159). Should we, then, agree that rats deprived of neocortex are reduced to the intellectual level of turtles? A report of Birch, Ferrier, and Cooper (1978) suggests that we should not. Birch *et al.* were concerned that the Gonzalez *et al.* (1964) findings might be a consequence of the degrading of purely visual processes, involved in locating sources of visual stimuli, by lesions that invaded the occipital region of the brain. The authors first showed that lesions confined to visual cortical regions in fact resulted in impaired serial reversal of a simultaneous brightness discrimination, despite having no effect on its original acquisition (that is, they replicated one of the major Gonzalez *et al.* findings, using restricted cortical lesions). In a second experiment, Birch *et al.* minimized the importance of the location of visual cues by testing serial reversal of a successive go no-go brightness discrimination, and found substantial improvement in their lesioned rats in this case. Since the rats now showed improvement in a visual reversal task, Birch *et al.* argued that the impairments seen in their first experiment were attributable not to difficulties in reversal learning *per se*, but to perceptual factors. A final experiment provided a direct test of this proposal. Normal and visually decorticate rats were tested in acquisition and reversals of a simultaneous light–dark discrimination, in which light could be seen through one of two translucent panels on each trial. In one condition, occluders were worn: these are devices which prevent pattern vision, while allowing perception of the average brightness of the visual field. Without occluders, rats may learn to approach (or avoid) the illuminated one of two spatially distinct panels; with occluders, rats may only learn to move in that direction which maximizes) (or minimizes) the brightness level. In the non-occluded condition, normals were superior in both acquisition and across the series of reversals to lesioned rats, who failed to show any reversal improvement. In the occluded condition, there were no significant differences between normals and experimental rats, and both groups showed marked improvement across the series. In fact, the visually decorticate subjects (like the controls) showed more rapid acquisition without occluders but (unlike controls) showed much worse reversal performance; in other words, the presence of additional (spatial) cues caused a selective deterioration in reversal performance in decorticate rats, who could reverse the discrimination perfectly well if only their vision was impaired.

Now although it may not be possible to provide a full explanation of the Birch *et al.* findings, their implications for Bitterman's hypothesis are clear: first, impairment of visual serial reversal in decorticate rats is probably due to invasion of visual cortex; second, the impairment is due, not to specific inability to reverse, but to a disruption of visual perception which in turn manifests itself, for whatever reason, selectively in reversal performance.

A third attempt to relate neocortical tissue in general to a gross behavioural capacity has been the proposal (e.g. Russell 1966) that the neocortex is essential for instrumental conditioning, whereas classical conditioning involves subcortical mechanisms. This view was supported by reports of a number of failures to establish instrumental conditioning in decorticate animals, and of successful acquisition and retention of classical conditioning in decorticates (e.g. Di Cara, Braun, and Pappas 1970; Oakley and Russell 1977). However, there are now in the literature a number of reports of successful training of decorticate mammals in food-rewarded instrumental paradigms, and these have led Oakley (1979) to suggest that at least part of the difficulty experienced by decorticate subjects in the earlier studies lay in hyperemotional responding to the aversive shock stimuli used. A further factor appears to be that decorticates may require extra pretraining to direct 'the animals' attention to the relevant parts of the apparatus' (Oakley 1979, p. 159). In other words, difficulties encountered by decorticate subjects may have been the consequence of procedural rather than intellectual factors. It should be added here that we have argued that, to demonstrate convincingly that an instance of learning is either instrumental or classical, tests involving omission training conditions must be used. At present, no detailed study is available of neodecorticates in an appetitive task with appropriate omission training conditions to allow confident determination of the type (instrumental or classical) of responding observed; what we can say is that neocorticate animals have been shown capable of learning in paradigms (lever pressing; alley-running) that are conventionally assumed to reflect the formation of instrumental (response-reinforcer) connections.

Neither Bitterman nor Russell has, then, provided a convincing role for neocortical tissue, and although Lashley's concepts of mass action and equipotentiality received better support, there were serious difficulties for the hypothesis that there is a relatively simple relationship between neocortex and general intelligence. It should also be borne in mind that that same conclusion holds for any subdivision of neocortex; if the evidence for Lashley's proposals is weak, that for any comparable proposal linked to a specific zone of neocortex is even weaker. It simply does not seem to be the case that there is any specific region (or regions) within the neocortex damage to which impairs performance across a series of tasks having only complexity in common. The findings surveyed do not therefore support the suggestion that mammals may be more intelligent than other vertebrates because they alone possess neocortex.

Association areas and intelligence

It may be that the generally negative conclusions of the preceding subsection were reached because the assumption of the existence of 'general intelligence' is unfounded; perhaps intellectual capacity is a mosaic of relatively independent capacities some of which, at least, might find separate representation in the neocortex. Before considering the evidence on this question, two points may be made. First, the preceding chapters have reviewed behavioural data and found no evidence for any qualitative intellectual differences between mammals and other vertebrates and so, no reason to expect neocortical representation of 'novel' capacities; second, the survey of cortical organization in mammals suggested that it is conservative in mammals, and that there is, for example, little reason to suppose any special development of 'association' cortex in apes and man, as compared to more lowly primates. Given this latter conclusion, the most pertinent question would appear to be – do the functional properties of any particular cortical area vary between groups of mammals, and do these variations imply any differences in intellectual capacity between the groups?

We may approach this question first by briefly considering the effects of damage to the classical sensory association areas. It will be recalled that Diamond (1979) proposes, on the basis of anatomical findings, that these regions should, along with a 'core' area which corresponds to the classical primary sensory area, be regarded as part of a unitary sensory field. We have not, however, considered relevant behavioural studies directed at the possiblility of special involvement of these association or non-core areas in visual learning.

Lashley believed, as a result of his studies using rats, that if the striate cortex (the region to which the relay from the lateral geniculate nucleus of the thalamus projects) was left intact, then the acquisition and retention of both brightness and pattern discriminations was unimpaired. Subsequent work with monkeys, however, has shown that lesions of areas of the visual sensory field excluding the striate cortex (of 'extrastriate' cortex) do cause severe deficits on a number of visual tasks, although they do not cause gross perceptual disturbance – the general visuomotor co-ordination of such animals appears quite normal (for reviews, see Gross 1973; Dean 1976; Wilson 1978*b*). The areas concerned run in an unbroken belt from the striate cortex along the lateral and inferior aspects of the temporal lobe, and form the inferotemporal cortex. All parts of the inferotemporal cortex receive projections from the pulvinar, but from different divisions of that nucleus, some of which receive direct afferents from the superior colliculus, others of which, afferents from the striate cortex; in other words, all parts of the inferotemporal cortex are interconnected with other visual regions.

The question that now arises is, of course, what is the nature of the deficits observed? It will perhaps come as no surprise that there is no simple answer to this question. In the first place, it appears that different regions of the inferotemporal cortex serve different behavioural functions: Cowey and Gross (1970), for example, suggest that the posterior inferotemporal cortex plays a critical role in selective attention within the visual modality, whereas anterior inferotemporal cortex is critical to associative processes in visual learning tasks.

To give a flavour of the type of experimental evidence relevant to this hypothesis, two experiments will be described here. Cowey and Gross (1970) showed that, although monkeys with anterior inferotemporal lesions were superior to animals with posterior lesions in the acquisition of three simultaneous pattern discriminations, presented one after another, subjects with anterior lesions were inferior to those with posterior lesions in a series of concurrent simultaneous discriminations, in which trials of one discrimination, using one pair of discriminanda, were interposed amongst trials of the other discriminations, which involved different discriminanda. Cowey and Gross argue that the concurrent technique maximizes the influence of proactive and retroactive interference, which should put particular stress on the mechanisms responsible for controlling trial-by-trial changes in associative strength. As monkeys with anterior lesions were particularly affected in this procedure, it may be that associative mechanisms have been impaired in them.

Evidence that suggests an attentional source of the impairment in monkeys with posterior inferotemporal lesions is provided by Gross, Cowey, and Manning (1971). In one experiment of their report, three groups of monkeys (unoperated controls, and subjects having either anterior or posterior inferotemporal lesions) were initially trained to the same criterion on a shape discrimination. Once an animal had met criterion, irrelevant information from another dimension (differences in colour or brightness) was added to the display. Addition of the irrelevant dimension adversely affected all three groups to a comparable extent on the first day of testing; however, whereas control monkeys and monkeys with anterior inferotemporal lesions returned to their previously established levels of accuracy on the second day of testing, performance of the posterior inferotemporal group remained severely affected throughout several days of testing. It appears, then, that whereas normals and anterior inferotemporal animals could efficiently 'gate out' the irrelevant information, posterior inferotemporal monkeys remained distracted by it and could not maintain attention to the relevant dimension alone.

There are a number of rival accounts of the nature of extrastriate visual function in monkeys, and it is clear that we should not expect a full understanding of the deficits obtained in the near future. It was argued in Chapter 2 that the value of such investigations lies in the light thrown on the organization of behaviour in normals, and the hypotheses currently advanced do carry implications for that organization. From our comparative point of view, however, what is primarily relevant is that none of these hypotheses is accompanied by any suggestion that the functions specified are peculiar to monkeys. The proposals of Cowey and Gross, for example, imply a distinction between strengths of attention to different dimensions and associative strengths of individual cues: in the preceding chapters, we have seen evidence supporting the existence of that distinction in such distantly related vertebrate groups as teleosts and birds.

Although the extrastriate visual cortex has received more attention in monkeys than in any other group, there are data showing disturbances in visual problems following damage to extrastriate sites in other primates (e.g. in the bushbaby—Atencio, Diamond, and Ward 1975) and in some non-primates (in the tree-shrew—Killackey, Snyder, and Diamond 1971, in the rat—McDaniel, Wildman, and Spears 1979). The deficits found in these studies are, as in the monkey, not easily characterized but do appear to be confined to visual tasks and include disruption of both acquisition and retention of pattern discriminations, and impairment in serial reversal of brightness discriminations. How much such deficits have in common with those seen in rhesus monkeys is not yet clear, but one would not expect identity of extrastriate visual function across the species described, as there are major differences in anatomical organization amongst them; for example, although there are, in both the tree-shrew and the bush baby, projections from all parts of the pulvinar to the cortical visual field, there are in the bushbaby but not in the tree-shrew parts of the pulvinar which do not receive afferents from the superior colliculus. It is important to note, however, that these parts are nevertheless connected with visual brain areas in that they receive projections from the striate cortex (Diamond 1979).

Up to this point we have considered reports which have attempted to analyse extrastriate visual function by exploring visual capacity in the absence of extrastriate visual cortex. An alternative approach is to damage striate cortex, with a view to establishing how much vision is possible when only extrastriate visual cortex is available. As is reasonably well known, this approach does suggest important species distinctions in cortical organization, since whereas monkeys and other higher primates, including man, appear to the casual observer to be quite blind without striate cortex, certain other mammals (including the tree-shrew) appear remarkably unaffected by its absence. This contrast is indeed one of the grounds for supposing a greater encephalization of vision in higher primates. There are now reasons, however, for questioning this interpretation of the differences seen. In the first place, it has recently become clear that a remarkable amount of visual information is available to destriate monkeys and humans, given appropriate training and testing conditions (see, for example, Weiskrantz 1980, who describes the extent of vision in humans who explicitly deny having any visual experience whatever—a phenomenon Weiskrantz labels 'blindsight'). And secondly, the fact that in 'higher' primates much of the extrastriate region depends on striate input (either directly, or via those parts of the pulvinar that do not receive collicular afferents) means that striate removal in higher primates may disrupt normal extrastriate function in a way not found in those mammals in which the entire pulvinar receives collicular projections.

This survey of the functions of the 'association areas' related to vision, the modality for which by far the most extensive data are available, indicates that those areas do indeed play an important role in visual performance. There has, however, not been any evidence for the existence in some species but not others of special capacities related to problem-solving. There is a marked increase in differentiation of extrastriate visual areas in the monkey as compared with the tree-shrew, and there are significant changes in anatomical organization also; but it is reasonable to suppose that these differences may be related to developments in visual capacities, and they do not seem to provide any ground for expecting differences in intelligent behaviour.

Given the generally negative conclusions of the search for species differences in sensory association area function, our attention must now shift to the classical non-sensorimotor association area of the cortex, the frontal cortex. It has already been argued that this area is best defined as the projection field of the dorsomedial nucleus of the thalamus. In monkeys that nucleus has two divisions, a lateral (parvocellular) division and a medial (magnocellular) division. These divisions project to different areas within the frontal cortex, the lateral projecting to the dorsolateral frontal cortex, and the medial, to orbitofrontal cortex, and these two frontal zones in turn have contrasting efferent subcortical projections (Nauta 1964). As there is good evidence for differences in function between the dorsolateral and orbitofrontal areas in monkeys, an important preliminary step here is to ask whether comparable subdivisions have been established in other species.

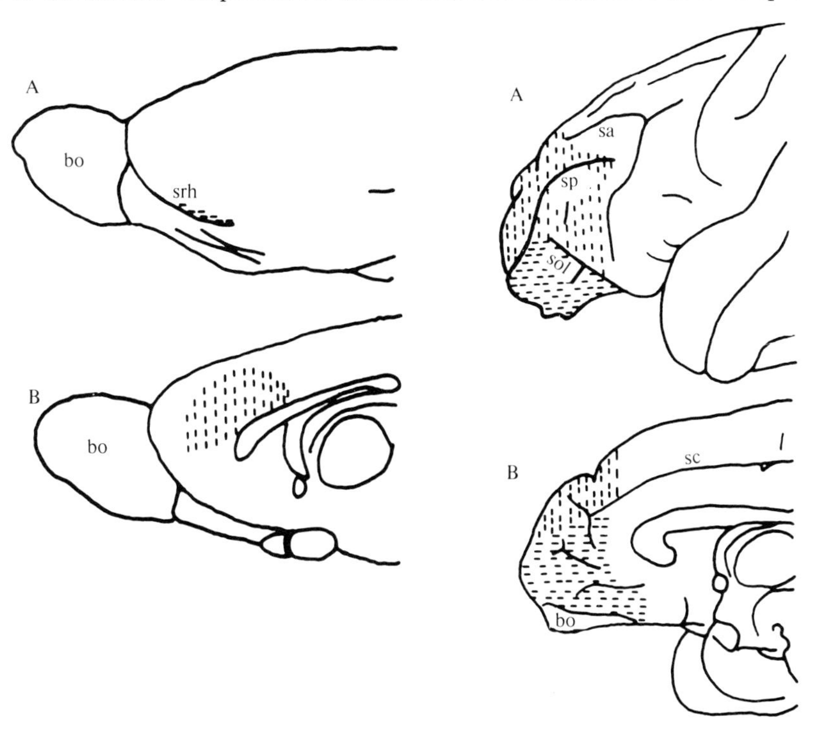

Fig. 7.10. Location of frontal cortical areas in (left) the rat and (right) the monkey. (From Markowitsch and Pritzel 1977.)

Key: A – Lateral view; B – Medial view; bo – olfactory bulb; sa – arcuate sulcus; sc – cingulate sulcus; sol – orbitolateral sulcus; sp – principal sulcus; srh – rhinal sulcus. The orbitofrontal region is dashed horizontally, the dorsolateral vertically.

The answer is encouraging, at least for three species, rats (Leonard 1969), cats (Markowitsch, Pritzel, and Petrović-Minić 1980), and dogs (Narkiewicz and Brutkowski 1967). It is important to note that the location of frontal cortex, defined in terms of connections with the dorsomedial nucleus, is not necessarily in the same topological position on the brain surface in all these species: Fig. 7.10, for example, shows the location of dorsolateral and orbitofrontal cortex in monkeys and rats; this figure emphasizes that in rats the frontal poles do not consist of frontal cortex, as here defined.

There have been a very large number of studies using monkeys with frontal lesions, and the data accumulated have led to a correspondingly large number of hypotheses of their functions. These hypotheses have considerable potential significance for theories of cognitive function in non-human mammals, but need not be explored in detail here as our concerns are equally satisfied by considering the less contentious question whether similar tasks obtain similar deficits across different species.

The best-known paradigm for demonstrating impairments in frontal monkeys is the delayed-response task, in which a monkey is shown food being placed under one of two identical cups behind a restraining screen which is raised, after a delay, to allow the subject to choose one of the cups. Monkeys with extensive frontal cortex damage are severely and permanently impaired in such tasks, even with relatively short delays (5 s or so), although they are capable of unimpaired performance at zero delay and are quite capable of acquiring conventional spatial discriminations (e.g. Butter 1969). There is by now general agreement that the cortical subdivision primarily responsible for this deficit lies in the dorsolateral frontal cortex, and that the deficits are not caused by delay *per se*, since if tasks involving similar delays but having discriminanda that are not spatially defined (e.g. delayed object or colour matching tasks), then monkeys with dorsolateral frontal lesions show relatively minor impairments (e.g. Mishkin and Manning 1978). It seems, then, that monkeys with dorsolateral frontal cortex lesions may be characterized as having a deficit in 'spatial memory'.

Another set of tasks that uncover deficits in frontal monkeys have in common a requirement that the subjects inhibit or withhold responses to a previously rewarded stimulus. Such problems include discrimination reversal (irrespective of the nature of the discriminanda) and extinction. In such problems, monkeys with orbitofrontal lesions are severely impaired (e.g. Butter 1969), whereas dorsolateral frontal subjects show impairments on spatial reversals only; it may be assumed that spatial reversals tap spatial memory since, as in the delayed response problem, the subjects' task is to respond on the basis of which spatial alternative was rewarded most recently, rather than simply on which alternative is rewarded, and which not. Monkeys with orbitofrontal lesions may therefore be characterized as having a 'perseverative' deficit.

Although the great majority of studies of frontal lobe function have used monkeys, there is good evidence that in both other primates and animals from different mammalian orders, similar deficits can be obtained by frontal damage. Deficits in spatial delayed response tasks (implying spatial memory impairment) have been reported following frontal lesions in dogs, cats, and rats (Lawicka, Mishkin, and Konorski 1966; Divac 1968; Kolb, Nonneman, and Singh 1974); deficits in either spatial delayed alternation or spatial reversal tasks (which could indicate either spatial memory or perseverative impairments) have also been found in cats, rats, and tree-shrews (e.g. Divac 1972; Kolb *et al.* 1974; Passingham 1978*a*); finally, deficits on schedules in which animals must respond slowly in order to be rewarded (differential reinforcement of low rates), deficits that may reflect impairment in the ability to withhold responses, have been reported in cats and rats (e.g. Numan and Lubar 1974; Kolb *et al.* 1974).

There is evidence, moreover, that the spatial memory and perseverative deficits are obtained from lesions in different parts of the frontal region in carnivores and rodents. The best data have been provided for rats, following Leonard's (1969) anatomical studies which clearly established a correspondence between, on the one hand, monkey dorsolateral frontal cortex and rat anteromedial cortex, and on the other hand, monkey orbitofrontal cortex and rat suprarhinal cortex. The Kolb *et al.* (1974) report shows that in rats anteromedial lesions obtain spatial memory impairments, and suprarhinal lesions, perseverative deficits; these findings confirm that the anatomically established correspondences find their parallels in behavioural functions. There is general agreement that the area of frontal cortex known as the gyrus proreus in carnivores corresponds to the monkey's dorsolateral region, and evidence that lesions confined to that area in both cats and dogs obtain spatial memory impairment (e.g. Lawicka *et al.* 1966; Rosenkilde 1978). Although there remain doubts concerning which cortical region in carnivores corresponds anatomically with monkey orbitofrontal cortex, there is evidence to show that lesions in the medial prefrontal region of dogs disrupt performance in a conventional successive go no-go auditory discrimination with either long or short ITIs; disruption of that same task is seen following gyrus proreus lesions, but, as it occurs only at short ITIs, it may be concluded that the impairment is less severe (Brutkowski and Dabrowska 1963). It may, then, be that damage to the medial prefrontal region of dogs results in a perseverative deficit.

In summary, the evidence that we have considered suggests a striking qualitative similarity between the functional organization of monkey frontal cortex and that of those few other mammals for which relevant data are available. It may be added that still other forms of disruption are found following lesions in other subdivisions of monkey frontal cortex (e.g. Goldman, Rosvold, Vest, and Galkin 1971; Passingham 1978*b*); whether such deficits can also be obtained in other groups is not yet known, although there are data indicating at least three separate functional regions within dog frontal cortex (Dabrowska 1971).

There are, as may be imagined, numerous difficulties facing attempts to compare the effects of frontal lesions in humans with those in non-humans. There have not, for example, been any reports that suggest any special difficulties for humans with frontal lobe damage in delayed response tasks. On the other hand, it could well be that humans perform such tasks quite differently from monkeys: they could, for example, bridge any delay by verbal mediation—by rehearsing, overtly or otherwise, a phrase indicating the correct choice throughout the delay interval. Again, it might be that frontal humans, prevented from the 'normal' mode of solution resort to some other technique to bridge the delay. Seen against this background, a report (Milner 1971) of an experiment by Corsi is of interest. Corsi showed humans with unilateral left frontal lobe damage a series of cards, each bearing two words; from time to time a card bearing a question mark between the two words was shown and subjects were asked to say which of those words had occurred most recently in the series of exposed cards. On some occasions, only one of the two words had been shown previously, so that the test was in effect a recognition test; on other occasions, both words had been exposed, and the test was a test of relative recency. Frontal humans showed no impairment of recognition, but were severely impaired on relative recency. It will be recalled that it has been argued above (p. 266) that frontal monkeys may show deficits on delayed response, not because they cannot recall which position is rewarded, but because they cannot recall which position has been rewarded most recently. It could, then, be that frontal humans have a comparable deficit. Similarly, deficits of a perseverative nature have been reported in humans: a number of workers have, for example, noted that frontal humans, while as efficient as normals at adopting a particular strategy (sorting cards, say, into suits), find difficulty in changing from that strategy to another at a signal from the experimenter (e.g. Milner 1971). These parallels are suggestive, but it must be emphasized that there remain substantial differences: there is no suggestion that the deficit in recent memory in humans is confined to spatial information, as in monkeys; it appears also that the human perseverative deficits are particularly associated with dorsolateral, as opposed to orbitofrontal, damage (Milner 1971).

The brief digression into human frontal damage symptoms has not, then, forced a change in the overall conclusions reached to date—namely, that qualitatively the similarities between the effects of frontal cortex damage across groups of mammals are more impressive than the differences. One further possibility remains, however, to be considered, and this is that there are quantitative differences between groups in the severity of the effects of frontal lobe damage (e.g. Divac and Warren 1971); a more specific proposal (Numan, Seifert, and Lubar 1975) is that the severity of deficit correlates with the extent of frontal and subcortical connections in a given species, that extent being greater in monkeys than in cats, and in cats, greater than in rats.

The above proposals have been assessed in a survey by Markowitsch and Pritzel (1977). These authors first rated the degree of impairment caused by frontal damage in three species (monkeys, cats, and rats) on four tasks (spatial delayed response, delayed alternation, spatial reversal, and differential reinforcement of low-rates), the range of species and tasks chosen being simply those for which a reasonable number of reports were available. They then evaluated statistically the differences between the species in degree of impairment on the tasks. Pooling together the results of all the first three (spatial) tasks, Markowitsch and Pritzel found that monkeys were significantly more impaired than either rats or cats, which did not differ significantly from each other. The DRL task showed a different pattern, cats now being significantly more impaired than either rats or monkeys, which did not differ from each other. Markowitsch and Pritzel's second analysis concerned the extent of afferent and efferent subcortical connections of frontal areas in the three species. In general, their survey of the anatomical literature did find somewhat more reports of subcortical connections in monkeys than in cats, and in cats than in rats. There have, for example, been reports of connections between frontal cortex and about 66 per cent of component structures of the limbic system in the monkey, 42 per cent in the cat, and only 27 per cent in the rat. On the other hand, it is not possible to equate the number of studies or range of methods used across the species; as there were many fewer reports (seven) on rats than on monkeys (nineteen), it may be that the low level of reported connections in the rat reflects simply the low level of investigation.

The hypotheses that Markowitsch and Pritzel explored received very weak support, and their survey allows very little in the way of conclusions on quantitative interspecies differences in frontal cortex, particularly on anatomical differences in connectivity. It seems that monkeys are more impaired than cats or rats on spatial problems and that cats are more impaired than rats or monkeys on one problem (DRL) that is presumably sensitive to perseverative responding. We have failed to find any evidence for qualitative distinctions, and these quantitative contrasts do not conform to any general hypothesis that would enable us to rank other groups. Our conclusion is, like that reached from consideration of 'sensory association' areas, a negative one: analysis of frontal cortex function does not suggest any specific difference that we might expect to see in intelligent behaviour between different mammalian groups.

We shall end this discussion of frontal cortex function by referring briefly to a report which, while using a different approach, shows the same logic as that underlying the Markowitsch and Pritzel (1977) survey. Masterton and Skeen (1972) suggest that, if frontal cortex is in some way involved in efficiency of performance on tasks involving delays, then animals with more frontal cortex should be more efficient on such tasks. Their report contrasted animals from three groups (two species of hedgehogs, *Hemiechinus auritus* and *Paraechinus hindei*, tree-shrews, and bush babies) in terms of, first, amount of frontal cortex (and associated structures) and, second, performance in a delayed alternation task. Absolute size of frontal cortex rose from the hedgehogs to the tree-shrews, and from the tree-shrews to the bush babies; efficiency of performance, measured by longest delay successfully tolerated, increased similarly, hedgehogs being the poorest, and bush babies the best, performers. However, our previous discussions of the relative size of different brain regions have shown that the absolute size of structures is of little meaning, unless presented relative to the size of the brain itself (just as the size of the brain has to be viewed in the light of body size). There is in fact nothing in the Masterton and Skeen anatomical data to suggest that size differences in the brains of the species used were in any sense peculiar to the frontal cortex. Moreover, we shall see in the section on delayed response tasks (p. 273) that the relative efficiency of species may vary according to the apparatus used; there is, therefore, no necessity to suppose that the ordering obtained in this study reflects some 'absolute' ordering of efficiency of the species concerned on delayed alternation tasks. This report does not, therefore, give convincing support to the notion that performance on delayed-response tasks should vary according to the absolute amount of frontal cortex available.

Comparative studies of learning in mammals

Introduction

A major change between the organization adopted for this section and those adopted in preceding chapters will be the omission of detailed consideration of 'simpler' forms of learning, under the headings 'habituation', 'classical conditioning', and 'instrumental conditioning'. This is not, of course, because such information is lacking, but rather because it holds little of comparative interest: we have already seen that these paradigms do not appear to discriminate between mammals and other vertebrates, and no current theory proposes that important differences within the various groups of mammals may be found in these tasks. References to studies of habituation in mammals may be found in Thompson and Spencer (1966), and the case for supposing that both classical and instrumental conditioning are to be found in mammals is made out by Dickinson and Mackintosh (1978), and Mackintosh and Dickinson (1979). It should be noted that although some authors have argued that only classical conditioning occurs, and others, that only instrumental conditioning occurs, none has argued that mammals are capable of only one type, whereas other vertebrate classes are capable of both, or that, within mammals, certain groups are capable of only one of those types of learning.

Similarly, we shall omit consideration of those issues (probability learning, reward shift effects, and mechanisms of attention) that have been considered by Bitterman to discriminate qualitatively between mammals and non-mammals since those claims have already been considered, and within-mammal differences on those tasks have not been proposed. Reversal learning will, however, be considered, not least because it affords some information on the capacities of the two non-placental orders of mammals, the monotremes and the marsupials.

This section will, then, consider within-mammal contrasts and essentially concerns comparisons of mammals other than man. The various tasks discussed are simply those which have enjoyed most attention and for which sufficient data are available to make discussion worthwhile. In the absence of a generally accepted theory of intelligence, no less arbitrary selection of material seems possible. The next chapter will concern the question whether the capacity for language is peculiar to humans, and will consider in that context the broader issue of contrasts in intelligence between humans and non-humans.

Serial reversal

A number of studies have investigated the possibility that serial reversal performance might discriminate quantitatively between groups of mammals. Gossette, Kraus, and Speiss (1968), for example, compared the spatial reversal performance of two primate species (capuchin monkeys and squirrel monkeys, both of which belong to the same family, Cebidae) and five carnivore species, four of which belong to various genera within the raccoon family (Procyonidae), the fifth being the striped skunk, *Mephitis mephitis*, which is a mustelid (weasel family). Although the capuchin monkeys showed fewer errors than any of the other species across a series of 19 reversals, squirrel monkeys, despite being so closely related, were ranked sixth out of the seven species. In a subsequent experiment, Gossette and Kraus (1968) tested four of the same seven species (using in fact the same animals) on reversals of a brightness discrimination. This experiment now showed squirrel monkeys to be superior to the three carnivore species used. In particular, although squirrel monkeys had shown significantly more errors than cacomistles (*Bassariscus astutis*) in the spatial study, they showed fewer errors than cacomistles on the brightness task. It appears, then, that contextual variables affect the rankings of species in these tasks. A similar conclusion emerges from a report by Doty and Combs (1969) on mink (*Mustela vison*), ferrets (*M. furo*), and skunks (all mustelids), performing reversals of either spatial or object discriminations: whereas the skunks were superior to mink and ferrets on spatial reversals, they were inferior to the other two species on object discrimination reversals. As we have observed previously (p. 223), the fact that alterations in the relevant dimension change the rankings of species argues that serial reversal performance is not in itself a reliable index of intelligence.

The discussion above considered overall performance levels, rather than asymptotic performance; it might be thought, however, that although all mammalian species can show substantial reversal improvement, only the more intelligent species could attain a level at which perfect reversal performance was achieved following a single error—could, in other words, successfully adopt a 'win-stay, lose-shift' strategy.

The first point to be made here is that it is clear that whether one-trial reversals occur in a given experiment depends strongly on contextual variables, although precisely which variables are critical remains unclear: whereas North (1950*a*, *b*), using rats, and Warren (1966), using rhesus monkeys, failed, despite extensive training, to obtain one-trial reversals, other workers have succeeded, using the same species (e.g. Dufort, Guttman, and Kimble 1954 with rats; and Schrier 1966 with rhesus monkeys).

A second observation is that one-trial reversals have been obtained in both marsupials (Buchmann and Grecian 1974, using brown short-nosed bandicoots, *Isoodon obesulus*) and monotremes (Saunders, Chen, and Pridmore 1971; Buchmann and Rhodes 1979; both of these studies used echidnas). Although there have been one or two reports of difficulty in obtaining efficient learning in marsupials (e.g. Livesey and Di Lollo, cited by Buchmann and Grecian 1974, who used quokkas, *Setonix brachyurus*, and Munn 1964, using grey kangaroos, *Macropus giganteus*), the Buchmann and Grecian report, along with another account of efficient learning in the Virginia opossum (Friedman and Marshall 1965) does indicate that marsupials should not in general be considered poor learners. As these are virtually the only behavioural data available relevant to the intelligence of marsupials and monotremes, there are in fact no data to suggest that their intellect is in any way different from that of any other non-human animals.

The proposal (Rajalakshmi and Jeeves 1965) that the Reversal Index, the ratio of the number of errors (or trials) in the first reversal of a discrimination to the number of errors (or trials) that occurred in original acquisition, might prove a useful measure for interspecies comparison, has been discussed previously, in relation to birds (p. 224). It was pointed out there that the ranking of birds (as of mammals) according to the RI, using data collected by Gossette and Gossette (1967) made little sense, by either phylogenetic or brain-size criteria. Also mentioned were Warren's (1967*a*) observations, that RIs within a species (cats) were highly variable, and that RIs obtained for individual cats in one apparatus showed no correlation with those obtained in another. These were problems that had to some extent been anticipated by Rajalakshmi and Jeeves, who argued that the RI for a species should be the mean from a large number of individuals of that species, and indicated also that RIs for a species varied with such factors as apparatus and type of discrimination. It seems, therefore, that even at the outset, this measure could not have provided an assessment of general intelligence, and so would not be relevant to our interests; it need only be added that, in any case, very few RI data have in fact been reported.

If there is by now fairly general agreement that reversal tasks do not discriminate quantitatively between mammalian groups, there have nevertheless been suggestions that the modes of performance adopted may show qualitative differences; in particular, it has been suggested that whereas certain primates solve the repeated reversal task by adopting a generalized win-stay, lose-shift strategy, mammals from other groups do not. The proposal is not that non-primates do not adopt a win-stay lose-shift strategy (the occurrence of one-trial reversals provides strong evidence that they do), but rather that they do not generalize that strategy to stimuli other than those used in the reversal series in question.

The evidence for the adoption of a generalized strategy consists in demonstrations of positive transfer from the repeated reversal task to a standard multiple-discrimination learning set task (p. 210); Schusterman (1962), for example, showed that chimpanzees given extensive serial reversal training showed positive transfer to a learning set task; transfer was both immediate and complete, in the sense that performance of the first 30 problems of that task was at a high level, and that no improvement in that performance was seen over five further blocks of 30 problems. Warren (1966) confirmed that extensive positive transfer was also found in rhesus monkeys, but found no transfer, using the same design, with cats.

Although some authors have interpreted the above findings in terms of species differences, it does not seem that this is a necessary conclusion. The difficulty lies in the fact that transfer between the two tasks can be a graded affair, rather than either complete or entirely absent; there is, moreover, evidence that the amount of transfer obtained within a given species varies with contextual variables. In a second experiment, for example, Schusterman (1964) found that chimpanzees given rather more than twice as much serial reversal training, although showing positive transfer to a learning set task, showed less transfer than was found in the 1962 study, and did improve across the series of 200 discriminations. Schusterman (1964) points to a number of differences between the reversal training procedures used, and suggests in particular that the use of a correction technique (in which chimpanzees were allowed, after a brief delay, to correct wrong choices) in the earlier report may have improved the transfer observed (the 1964 study used a non-correction procedure). Schrier (1966) found positive transfer from serial reversal training to learning sets in two species of macaques (*Macaca fascicularis*, the Phillipine cynomolgus monkey, and *M. arctoides*, the stump-tailed macaque), but the amount of transfer was not large: on Day 1 of learning set training, monkeys given serial reversal pretraining scored 62 per cent correct choices as compared to 52 per cent correct in a control group which had no previous discrimination experience. In a subsequent report, Schrier (1974) suggested that the low level of transfer obtained may have been due to the poor terminal level of serial reversal performance: whereas the macaques in his 1966 study averaged almost 12 errors on their final reversals, the chimpanzees in Schusterman's 1962 and 1964 reports averaged, by the end of reversal training, only just over one error per reversal. Accordingly, Schrier (1974) trained stump-tailed macaques on serial reversals until they achieved 15 consecutive reversals with two or less errors, and then transferred them to a learning-set task. The monkeys now showed substantial amounts of transfer which was, however, not complete, since they did improve over the series of discriminations (of which there were a minimum of 150).

With these results in mind, we may return to Warren's (1966) negative findings using cats (which, it will be recalled, showed no transfer from serial reversal to learning set formation). First, we observe that the animals were trained using a non-correction technique, and we now know that the only report that found complete transfer (Schusterman 1962) used a correction technique. Second, the cats were, at the end of serial reversal training, making an average of approximately twice as many errors as were the rhesus monkeys of the same study: Schrier's (1974) report suggests that the terminal level of serial reversal performance is a critical variable influencing the amount of transfer seen. It therefore appears that the transfer performance of both the cats and the rhesus monkeys might have been improved had a correction procedure been employed and that the performance of the cats relative to that of the monkeys might, again, have been improved had the groups been trained to comparable levels of serial reversal performance.

Now the argument above does not, of course, alter the fact that positive transfer from serial reversal to learning-sets has not been found in cats (or in any non-primate mammal—it has, of course, been found in birds—see p. 210). It does, however, seem reasonable that we should withhold judgement on the capacities of cats until further experiments have been carried out in the attempt to see whether they might not under more favourable circumstances show transfer.

Delayed response

It will be recalled that this task was introduced (p. 266) during the discussion of frontal cortex function, where our interest centred on the sensitivity of delayed-response performance to dorsolateral frontal cortical damage; our interest here, of course, centres on the performance solely of intact animals.

Hunter (1913) presented the first account of results obtained using a delayed response technique and observed first, that some species could master longer delays than others, and, second, that some species maintained orientation towards the appropriate choice site throughout the delay interval, whereas others did not. Hunter argued that where animals (including humans) perform successfully without the use of overt orientation, some central symbolic process must have been employed, and the length of delay survived in that way could give a quantitative estimate of the stability of symbolic processes in various species. These proposals stimulated a series of experiments by early comparative psychologists, and the results of these studies were well summarized by Maier and Schneirla (1935). Their conclusions were that a wide range of species—rats, cats, dogs, raccoons (*Procyon*), and various primates—were capable of successful delayed response performance without overt orientation and that within one species a wide range of maximum delays was obtained, depending on the testing technique used. These conclusions, which have not been seriously challenged, suggest that it will be difficult if not impossible to establish with any confidence a value for the maximum delay which can (under optimal conditions) be bridged by a given species; it would, therefore, be correspondingly difficult to attempt to rank species according to maximum tolerable delays.

An alternative suggestion has been that, where species are tested in the same kind of apparatus and using similar testing procedures, then the relative tolerance of increases in delay with other variables held constant across the species may provide a rank of comparative significance. Figure 7.11 for example, shows the performance of humans, rhesus monkeys, marmosets, and cats, all tested using a similar apparatus and testing procedures (Miles 1971). It can be seen that the ranking agrees with that expected from the supposition that closeness of relationship to man should correlate with intelligence (and with a ranking based on encephalization quotients—see Table 7.1). Now this method of ranking might be worth pursuing if it could be assumed that the same ranking would emerge, whatever apparatus was used. There is, however, good evidence that this is not the case: Divac and Warren (1971), for example, found that although the delayed-response performance of rhesus monkeys was superior to that of cats when tested in the Wisconsin General Test Apparatus, this relationship was reversed when the Nencki testing apparatus was used. We may reasonably assume, therefore, that much of the variation between species illustrated on Fig. 7.11 reflects the differential influence of contextual variables on the species selected. There are many reports of the influence of such variables on delayed-response performance in monkeys (see Fletcher 1965) and Miles and his colleagues, discussing an experiment in which cats and marmosets were compared, wrote: 'Numerous unquantified observations conveyed a strong impression that performances were influenced by characteristic modes of behavior which have little direct relevance to learning ability. During delay intervals the cats showed a tendency to remain near the center of the restraining compartment, whereas the smaller and more active marmosets followed the reinforcement to the baited food well and then concentrated their activity in that area' (Meyers, McQuiston, and Miles 1962, pp. 516–17). It seems, then, that delayed-response tasks will not provide unique rankings of species and cannot be used as a measure of general intelligence.

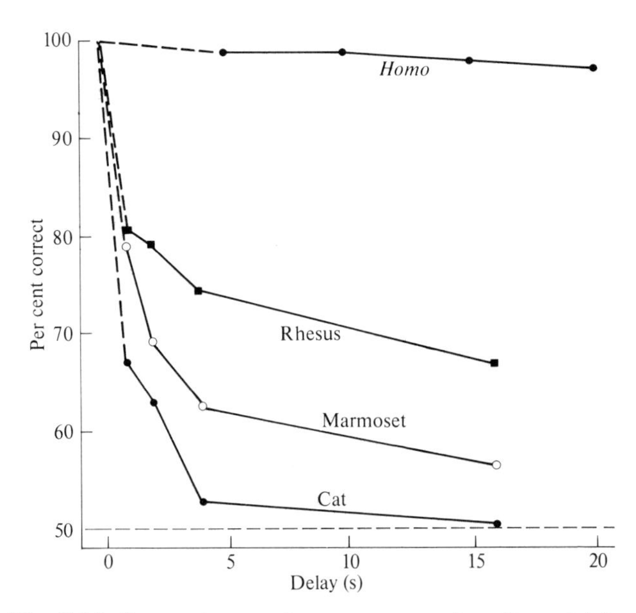

Fig. 7.11. Per cent correct responses as a function of delay interval in four mammalian species. (From Miles 1971.)

Double alternation

An animal performing a double-alternation task is required to choose one of two spatially defined alternative responses twice in succession, and then to choose the other alternative for two consecutive trials. Typically, animals are trained using sequences of four choices having a relatively brief (e.g. 5 s) inter-choice interval, the inter-sequence interval being considerably longer (e.g. 30 s). The correct side of the first choice in a sequence does not alter, for a given individual, from series to series, so that the animal must learn, first, which side to choose at the start of a sequence (after, that is, a long inter-trial interval) and, second, to switch sides after two choices of the side first chosen. Since the correct choice cannot be based on feedback from the immediately preceding trial alone—an animal performing an LLRR sequence must choose left following a correct left choice on one occasion, and right on another—early workers believed that successful performance would require symbolic processes, and so expected that 'higher' animals would be discriminated from 'lower' animals by their performance in this task.

The results of the initial studies of double alternation did appear to concur with expectations, as monkeys were reported to be superior to various carnivores, which were in turn superior to rats (Warren 1965). Warren goes on, however, to point out that most early reports used very few subjects, and could not be supposed to give an estimate of the range of ability for any given species. More recent studies have arrived at very different conclusions, as can be seen from Fig. 7.12 (Livesey 1969), which summarizes the results of a number of different reports each of which used similar apparatus and testing procedures. The mean performance of the rabbits and cats is similar, and superior to that of rats and rhesus monkeys—an outcome which is not predicted by either phylogenetic (closeness of relationship to man) or brain-size considerations. It should, however, be pointed out that there is no reason to suppose that the difference between

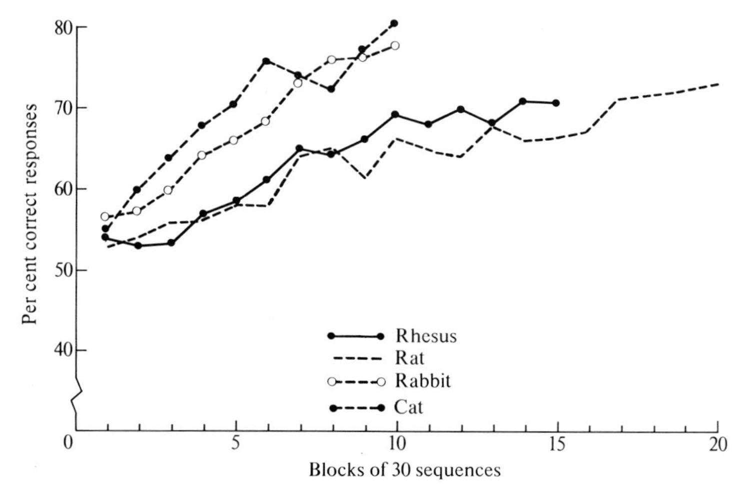

Fig. 7.12. Acquisition of double-alternation in four mammalian species. (From Livesey 1969.)

cats and rabbits, on the one hand, and rats and monkeys on the other, is reliable. Recent workers have strongly emphasized the very considerable amount of variation between individuals of one species in double alternation performance (e.g. Warren 1967*b*; Livesey 1969), and have pointed out that that variation far exceeds the inter-species variation, so that there is no evidence at present that any one species reliably differs from any other on this task.

A further variant of this problem has been to determine whether double-alternation responding, established in four-response sequences, can be extended to longer sequences, of eight, twelve, or more responses. It might be supposed that successful transfer to longer sequences would indicate that subjects had grasped some general principle, and Livesey (1969) has claimed that 'the double-alternation problem is successful in differentiating the performance of primates from that of other mammals in terms of their ability to extend the sequence' (p. 529). The evidence, however, is sparse: in Livesey's study, three out of five monkeys achieved criterion (80 per cent correct sequences) on eight-response sequences, and two of these three achieved criterion on twelve-response sequences.

Stewart and Warren (1957) reported that although as a group their cats (eight in all) did not transfer from a four-response sequence to an eight-response sequence, at least some cats were performing better than chance, and one cat in fact made 30 perfect sequences out of the 90 on which he was tested. Finally, Johnson (1961) using raccoons (*Procyon lotor*) (whose mean performance on four-response sequences was somewhat better than that reported for cats) showed very little transfer to eight-response sequences although, again, some 'scattered' errorless sequences did occur. Now although the general procedures used in these reports were comparable, they did differ in one potentially significant respect—the amount of training on the four-response sequences prior to transfer to the longer sequences: whereas Livesey's (1969) monkeys experienced from 450 to 860 sequences of four-response sequence training, the cats in Stewart and Warren's (1967) study each received 300 sequences, and Johnson's (1961) raccoons, from 90 to 270 sequences. It can be seen that in fact the ranking of the species in amount of transfer agrees with that for amount of four-response sequence training and, given that one report (Leary, Harlow, Settlage, and Greenwood 1952, using rather different procedures), failed to establish mastery of eight-response sequences in monkeys, we should clearly reserve judgement at present on whether the small quantitative differences between monkeys, cats and raccoons on extension of the double alternation principle are of significance.

Learning-set formation

Learning-set formation—the inter-problem improvement in performance seen in subjects given a series of discriminations involving different pairs of stimuli (p. 207)—was originally studied by Harlow (e.g. 1949), who concluded that efficiency of learning-set formation was related to phylogenetic position, those species closest to man being optimally efficient. Figure 7.13 summarizes data from a series of experiments, carried out in various laboratories, that give good support to that claim: the rhesus monkey, an Old World or platyrrhine monkey, is superior to both the squirrel monkey and the marmoset, each of which is a New World, catarrhine, species. All three primate species are superior to the cat, a carnivore, and to the rat and squirrel (*Sciurus*), both rodents. It should, perhaps, be noted that this ordering does not agree with that predicted from relative brain size: Table 7.1 shows that EQ for the squirrel monkey is larger than that for the rhesus, and that the squirrel EQ exceeds that of the cat.

Subsequent findings have, however, muddied this well-ordered picture by showing, first, that closely-related species may show widely-divergent performance and, second, that at least some 'lower' species may equal or excel the performance 'higher' species. Two striking examples from the former category are provided by Manocha (1967) and Doty, Jones, and Doty (1967), each of whom compared species within one family, using the same apparatus and training procedures. Manocha compared the performance of four langurs (*Presbytis entellus*) and six rhesus monkeys (both species belonging to the cercopithecid family of Old World monkeys) in a series of 128 six-trial object discrimination problems; at the end of training, rhesus monkeys were scoring, over trials 2–6 of each problem, an average of 78 per cent correct, the langurs' average being 98 per cent, a significantly higher level of performance. An additional reason for interest in the contrast between their levels of performance is that the langur EQ, of 1.29 (Jerison 1973) is considerably lower than that reported for the rhesus monkey (2.09, see Table 7.1). Doty *et al.* (1967) compared performances of three species from the mustelid family of carnivores (the mink, the ferret, and the striped skunk) including as well the cat, from a different carnivore family. Their results are presented in Fig. 7.14, which shows that mink and ferrets were strikingly superior to skunks (and cats).

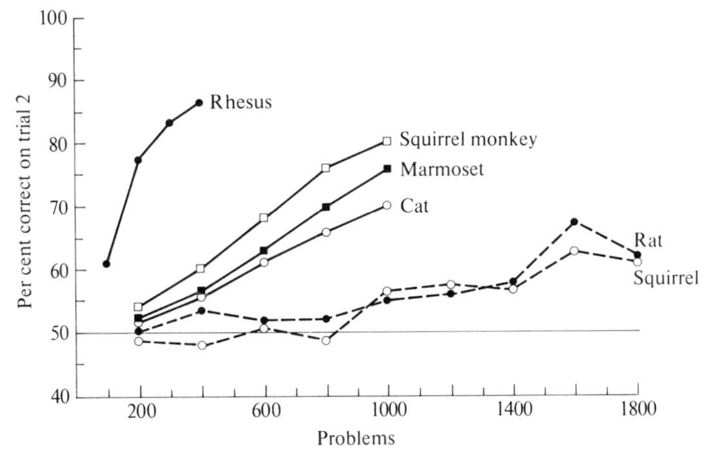

Fig. 7.13. Learning-set formation in six mammalian species. (From Warren 1965.)

As there are clearly wide ranges of learning-set achievement within species of the same family, it is not surprising that there are species whose performance excels that found in species more closely related to man. One good example of this is, of course, provided by the Doty *et al.* data: comparison of Figs. 7.13 and 7.14 shows that the mink and ferrets out-performed two primate species (the marmoset and the squirrel monkey) and fell little short of that shown for rhesus monkeys. Further examples are by now available: Johnson and Michels (1958), for example, found that the performance of raccoons (carnivores of the procyonid family) exceeded that of marmosets; Herman and Arbeit (1973) found that a bottlenosed dolphin (*Tursiops truncatus*) achieved, after extended training, a trial 2 performance of 86.5 per cent correct, using auditory stimuli as discriminanda – a level which, once again, exceeds that observed in squirrel monkeys and marmosets; Cooper (1974) using three Malagasy lemurs (*Lemur macaco*) found trial 2 accuracy after some 270 problems to be 84 per cent correct, so that the prosimian lemur appears to outperform the anthropoid squirrel monkey.

It seems, then, that performance in learning set tasks does not vary reliably with distance of relationship to man (nor, as we have seen, with relative brain size).

Might it not, nevertheless, be a good indicator of intelligence? Are rhesus monkeys not superior to cats by virtue of their superior intelligence, and are mink, perhaps, unexpectedly intelligent? Manocha (1967) implies such a view in his suggestion that, since langurs outperform rhesus monkeys, langurs should be accorded a higher phylogenetic rank.

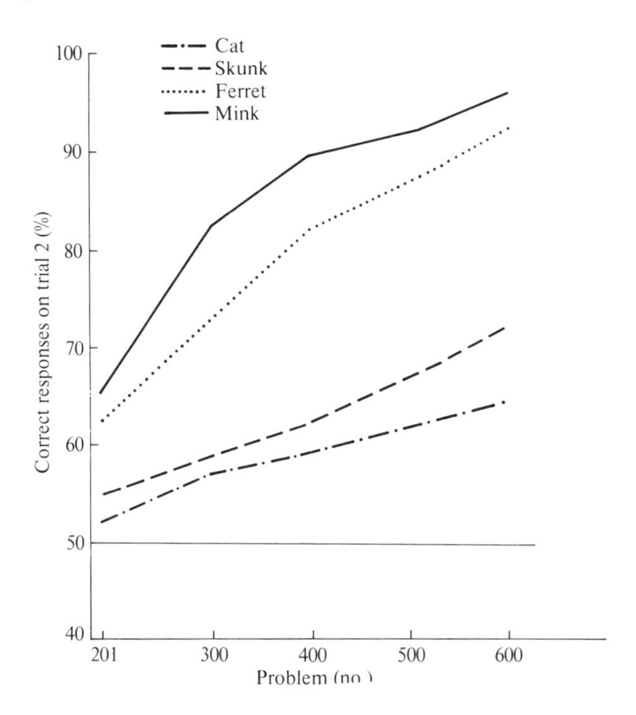

Fig. 7.14. Learning-set formation in four species of carnivores. (From Doty *et al.* 1967).

One initial observation is that, although Manocha himself did show that the interspecies difference in performance in his experiment was statistically significant, such a demonstration has been very much the exception. The curves shown in Fig. 7.13 represent, as Warren (1973) points out, data from very few subjects (from three to six, in each case) and none of the original reports demonstrates significant interspecies differences. In this connection, it is relevant to note, first, that the report of Doty *et al.* (1967)—see Fig. 7.14—does not contain statistical evidence for interspecies differences and, second, that there is wide intraspecies variation in learning-set performance: Warren (1973), for example, reports that of the four cats in the Warren and Baron (1956) study, one achieved a trial 2 accuracy of 84 per cent correct over problems 141 to 340, a level that appears to be superior to the average reported for any of the species shown on Fig. 7.13. As we have no statistical evidence for the reliability of the rankings shown on either Fig. 7.13 or Fig. 7.14, we should in any case, given Warren's evidence of intraspecific variation, withhold judgement on the proper ranking of species in these tasks.

The major objection to the proposal that learning-set performance provides a measure of general intelligence is, however, simply that there is no reason to suppose that there is, for a given species, any single valid measure of 'learning-set ability': both rate of acquisition of learning set and asymptotic performance level may be expected to vary within a species, depending on the specific values of the contextual variables associated with an experiment.

Although there are in the literature only single reports on performance of a number of the species discussed above, there is good evidence for at least some species that estimates of ability do vary with such variables: the performance of rhesus monkeys, for example, varies with reward size (Schrier 1958) and with a number of factors associated with the discriminanda—with, for example, their size, location, modality, and number of cues (Meyer, Treichler, and Meyer 1965; Warren 1973). The sensitivity of performance to the nature of the cues provided suggests that the sensory capacity of the species tested may be of considerable significance. Support for this unsurprising notion comes from a study by Devine (1970) which investigated the performance of a New World monkey (*Cebus albifrons*) whose colour vision differs significantly from that of the Old World rhesus monkey, whose colour vision is essentially identical to man's. Devine found that although the mean rate of acquisition of learning sets in cebus monkeys was lower than that in rhesus monkeys in the standard design, in which each problem was presented for six trials and coloured objects were used, cebus monkeys were superior to rhesus when each problem was taken to a criterion, and the objects were painted a uniform grey. The results suggest that both criterion training and the absence of colour cues improved the performance of the cebus monkeys relative to that of rhesus monkeys, and emphasize the potential importance of procedural and, in particular, of sensory factors. It must be added that, although Devine's results have been presented in terms of average scores, without taking account of variance (in accordance with learning-set traditions), no interspecies differences in this study, which used small numbers of subjects, were in fact statistically significant; what Devine's statistical analysis did demonstrate, nevertheless, was that the two species did not differ significantly in the acquisition of learning sets, and that both colour (or its absence) and the trials procedure were significant determinants of performance.

There are some data from studies of non-primates to show that their level of attainment is influenced by similar factors. Although Herman and Arbeit (1973) found efficient auditory learning-set formation in a bottlenosed dolphin, Herman, Beach, Pepper, and Stalling (1969) found rather poor learning-set formation in another bottlenosed dolphin (*Tursiops gilli*), using as discriminanda plywood shapes, painted white with red edges: interestingly enough, Herman *et al.* observed that the dolphin tended to nose the centre of each stimulus (which was invariably white), and that optimal performance was found for those shapes that had contours whose minimal distance from the centre of the shape was relatively short.

There is evidence, too, that the lowly position of rodents (Fig. 7.13) may be improved: whereas some early reports of poor learning-set formation in rats (e.g. Koronakos and Arnold 1957; Wright, Kay, and Sime 1963) used two-dimensional stimuli, later studies using rats and Mongolian gerbils (*Meriones unguiculates*) used three-dimensional discriminanda, and obtained rather better results. Kay and Oldfield-Box (1965) using rats, for example, found that after 40 problems, each taken to a criterion of 80 per cent correct, performance over trials 2 to 20 of each new problem had risen to well over 80 per cent correct (and was still improving when the experiment ended), and Blass and Rollin (1969), using gerbils and a fixed number of trials (six) per problem, found an asymptotic level of trial 2 accuracy of approximately 70 per cent correct. More dramatic results have been reported by experimenters using olfactory stimuli as discriminanda for rats: Jennings and Keefer (1969) report a trial 2 accuracy of more than 80 per cent correct over problems 22 to 28 of their series, and Slotnick and Katz (1974) report trial 2 accuracy of more than 90 per cent correct after training on 16 problems in which odours were presented successively, in a go no-go discrimination. There are, unfortunately, problems in each of these two reports. The Jennings and Keefer study used only eight odours in all for 28 problems, so that extensive re-use of the same odours in different discriminations was necessary, thus ruling out a direct comparison of these findings with more conventional data. Slotnick and Katz not only used an unconventional (successive) method of discrimination, but reported later (Nigrosh, Slotnick, and Nevin 1975) that where such olfactory discriminations are *not* preceded by a series of brightness discriminations (as they were in the Slotnick and Katz report) then original acquisition of olfactory discriminations proceeds with, in any case, only about 10 per cent erroneous responses; it seems, therefore, that the improvement reported by Slotnick and Katz may represent, not the formation of a learning set, but the gradual elimination of negative transfer from brightness discrimination training.

Finally, there is some indirect evidence that the ordering of the mustelid species shown on Fig. 7.14 may be due to sensory factors: Warren (1973) has pointed out that the skunks of Doty and Combs' (1969) report acquired a position discrimination more rapidly, and an object discrimination, less rapidly, than did mink and ferrets. The implication is, therefore, that skunks are less responsive to visual cues than either mink or ferrets; this lack of visual responsiveness may in turn be the cause of the ordering of these species' performance in learning sets.

In summary, then, reports to date of mean learning-set performance of various mammalian species do not show a ranking that is in good accord with presuppositions derived from either phylogenetic or brain-size data. There are, moreover, generally insufficient data to determine the statistical reliability of differences between the mean levels reported, and considerable evidence that such levels are in any case influenced markedly by procedural details. One final observation is that there is clearly no evidence here for qualitative differences between species: primates, carnivores, and rodents are all capable, under certain conditions, of achieving a trial 2 accuracy of better than 80 per cent correct. This in turn indicates a substantial degree of 'one-trial' learning, and provides good *prima facie* evidence for the adoption of a 'win-stay, lose-shift' strategy in all those groups. The failure of cats to show positive transfer from serial reversal training to learning-set formation (p. 273) need not, therefore, be due to inability to form such strategies. While that failure (if confirmed in studies using appropriate procedures) may indicate that cats do not generalize these strategies from the one pair of objects used in serial reversal to objects in general, it is not clear how significant such an absence of generalisation would be: cats (or, at least, one cat) do, after all, show by their learning set performance that they can adopt a win-stay lose-shift strategy with reference to objects in general.

Sameness–difference concept

There have been relatively few comparative studies of sameness-difference concepts in mammals, and most have concerned oddity-learning tasks; before discussing those results, we should note that highly efficient matching performance (on trial 1 of problems using novel stimuli) has been established in chimpanzees (Nissen, Blum, and Blum 1948), in rhesus monkeys (Mishkin, Prockop, and Rosvold 1962), and, using successive presentation of auditory stimuli, in *Tursiops truncatus* (Herman and Gordon 1974). It will be recalled (p. 211) that oddity problems are soluble in ways that do not require the use of an oddity concept (by, for example, configurational conditioning or conditional discrimination formation), and that the safest guarantee that genuine oddity has been learned is a demonstration of immediate transfer to novel stimulus configurations. Successful performance on trial 1 of novel oddity problems has been reported in chimpanzees, rhesus monkeys, cebus monkeys, and squirrel monkeys (Moon and Harlow 1955; Strong and Hedges 1966; Thomas and Boyd 1973). All of these species are capable of performance at 80 per cent correct or better, and Strong and Hedges (1966) showed that chimpanzees reached a 90 per cent criterion more rapidly than rhesus monkeys, using identical training procedures.

The few reports available using non-primates indicate quantitatively inferior performance: Strong and Hedges (1966) found that neither cats nor raccoons achieved their 90 per cent criterion within the maximum number of sessions allowed (100). Other reports on cats (Boyd and Warren 1957; Warren 1960) agree that most cats cannot master oddity problems. Finally, a report on rats (Wodinsky and Bitterman 1953), while giving incomplete details, does not suggest that they are capable of attaining the levels of accuracy achieved by primates.

The significance of quantitative differences on this task is not yet clear: there are many differences in procedure amongst the studies cited, and there is every reason to suppose that quantitative species differences could be manipulated at least to some extent by variations in testing procedures. In this context, it may be borne in mind that pigeons' performance on oddity tasks was markedly improved by increasing the number of 'matching' stimuli (p. 213); while we do not yet know what influence this procedure might have on transfer to novel stimuli, it does emphasize the potential importance of procedural variations that might not intuitively seem likely to have much effect. In the absence of any substantial body of investigation on effects of contextual variables in mammals, there is little alternative but to reserve judgement on these quantitative differences. A more critical question is whether a qualitative distinction between primates and other mammals may not exist: are non-primates in fact capable of forming an oddity-concept? It will be argued here that although the evidence may not be conclusive, there is good reason to suppose that cats, and perhaps rats, can. Warren (1960) found that one of a total of five cats achieved good performance on a series of oddity problems, when a new set of discriminanda was introduced each time criterion was met. As was pointed out above, the difficulty with a design of this kind is that it does not guarantee that an oddity concept has been formed: it could be that the subject is learning either configurational or conditional discriminations, and that the positive transfer observed is either some general adaptation to the apparatus or formation of the appropriate discrimination learning set. These alternative accounts are, however, unconvincing: Warren's cat made an average of only 4.5 errors a problem over problems 11 to 30, and such a score seems very much too low to allow either configurational or conditional discrimination to have occurred. This is particularly the case since in Warren's study the odd object could appear in any one of three positions, so that either alternative to oddity-concept formation would require a complex series of discriminations to be formed. A similar argument applies, but with less force, to the Wodinsky and Bitterman (1953) report: one rat, for example, showed 13 correct choices out of 18 trials of Day 1 of a novel oddity problem (as all three positions were used, the chance level was six out of 18 correct.) There are too few data given to allow any statistical analysis, but it does seem, in this case also, reasonable to infer that positive transfer was mediated through oddity-concept formation, rather than by the much more complex alternatives.

It is apparent that the data currently available are sufficient for only tentative conclusions: there are reports of quantitative differences both within primates (Strong and Hedges 1966; Davis, Leary, Stevens, and Thompson 1967) and between primates and non-primates (Strong and Hedges 1966), and there is no conclusive proof that oddity-concept formation is possible in non-primates (including, as we saw in Chapter 6, birds). On the other hand, the quantitative differences could presumably be influenced by procedural variables, and there is reason to doubt that non-primates are incapable of genuine oddity learning.

Number concepts

There have been few satisfactory investigations of counting in (non-human) mammals (see Wesley 1961), and the results that have been obtained may be summarized fairly rapidly. Hayes and Nissen (1971) provided evidence that Viki, their home-raised chimpanzee, could discriminate between cards bearing three spots (of various sizes, in various patterns) from cards bearing two spots, and from cards bearing five spots; she could not discriminate between three and four spots, or between four and five, or five and six spots. Hicks (1956) reports some success in training rhesus monkeys to select a card bearing three shapes from cards bearing one, two, four, or five shapes. The location and size of the shapes were varied, although initial training used rectangular shapes only; positive transfer was obtained when new shapes were introduced, although the best monkey was capable of only 70 per cent correct choice on Trial 1 of new problems. As was observed previously, these levels of performance are not impressive, and inferior to those reported in birds (p. 215); it is of interest to add that Viki's performance was comparable to that of human children (3½ years old) who had not yet learned to count verbally.

A recent study (Thomas, Fowlkes, and Vickery 1980) has found higher levels of attainment in squirrel monkeys (which belong to the platyrrhine suborder of primates, generally regarded as less advanced than the catarrhine suborder, to which both rhesus monkeys and chimpanzees belong). Thomas *et al.* used two subjects, and showed that they were capable of discriminating between two simultaneously presented cards on the basis of the number of circles on the cards (using circles of different sizes, in various locations); one monkey succeeded in discriminating between cards bearing 7 and cards bearing 8 circles, and the other successfully discriminated 8 from 9. The performance of these animals in fact exceeds that reported in birds (p. 215) in studies using simultaneous presentation techniques.

One experiment on rats, using a successive training technique, is relevant here. Chen (1967) trained rats to run in a keyhole-shaped runway, consisting of a circular part and a radial stem, which served as start and goal-box. The rats had to learn to run round the circular part a fixed number of times before re-entering the stem, and nine (of an initial group of 12) rats succeeded in mastering this task when three 'laps' were required; one rat eventually learned to run six laps, but could not master the seven-lap problem. The rats showed positive transfer in tests using alterations in the length and shape of the circular setting, and this suggests that kinaesthetic feedback could not (in any simple way) have served as a cue.

It appears, then, that at least some non-human mammals are capable of counting to some degree, and that this level may in fact not differ from that achieved in humans whose linguistic counting skills are undeveloped.

Complex visual concepts

There appear to have been no investigations of visual concepts in non-primates, and even in primates, there are few reports comparable to those available for birds. Hayes and Nissen (1971) found that Viki had little difficulty in selecting between colour pictures according to whether they contained, for example, animals versus inanimate objects, male versus female humans, or human children versus adults. There is also a brief report by Rosenfeld and Van Hoesen (1979) showing that rhesus monkeys can discriminate between photographs of faces of other monkeys (unknown to them), and that such discriminations, once established, show very good transfer across changes in posture, size, colour, and illumination. Thus, although data are sparse, there is no reason to suppose that mammals with adequate vision cannot, like birds, form complex visual concepts.

Cross-modal perception

By cross-modal perception is meant the transfer of information about a source of stimulation between one sense modality and another. A subject may, for example, be trained to discriminate between two rates of light-flashes and subsequently transfer the discrimination to the same rates of, say, sound-clicks; alternatively, a subject might feel an object in the dark, and subsequently recognize the same object seen, but not felt, in the light.

A number of authors have suggested that a capacity for cross-modal perception might be related to language: Ettlinger (1967), for example, suggested that the cross-modal transfer of a discrimination might occur only with the aid of language (although language might, in his view, not be necessary for cross-modal recognition). An alternative proposal (e.g. Geschwind 1964) has been that such a capacity might be a necessary prerequisite for language, a capacity that we should expect to find emerging in those primates most closely related to man.

Two generalizations underlying the notion that cross-modal perception and language were closely linked were, first, that cross-modal transfer did not occur in children unless verbalization occurred (Blank and Bridger 1964), and, second, that cross-modal transfer was not obtained in non-human primates (e.g. Ettlinger 1960). Subsequent research has, however, undermined these generalizations. There are by now reports showing significant amounts of cross-modal transfer in babies less than one month old (Meltzoff and Borton 1979), and in monkeys and apes. Jarvis and Ettlinger (1977), for example, showed transfer from touch to vision and vice versa in rhesus monkeys and chimpanzees, finding little difference between the amount of transfer shown by those species; Elliott (1977) found significant transfer from touch to vision in chimpanzees, rhesus monkeys, and cebus monkeys (*C. apella*). Elliott noted that all those species showed similar amounts of transfer, and that their performance was indeed comparable to that reported in human infants.

Finally, significant auditory-visual and visual-auditory transfer has been obtained in the prosimian bush baby (*Galago senegalensis*) (Ward, Yehle, and Doerflein 1970), the rabbit (Yehle and Ward 1969), and the rat (Over and Mackintosh 1969). Although these data do not, of course, demonstrate equivalence of capacity in all species tested (there have, for example, been so far no reports of transfer to or from the tactile modality in either prosimians or non-primates), they do not encourage the view that cross-modal studies reflect either qualitative differences between mammals, or steps on the road to human language.

Concepts marking stages of cognitive development

This section will consider experiments on non-human animals that have been designed with explicit reference to Piaget's well-known theory of cognitive development. According to this theory, the ontogeny of human intelligence proceeds through a series of qualitatively different stages, each stage being characterized by ability to perform certain tasks and inability to perform others. This view of intelligence prompts the question whether ontogeny recapitulates phylogeny, whether, that is, these 'stages' represent the adult level of attainment of some ancestor of humans; this possibility in turn invites the investigation of living non-human animals (and primates in particular) to see whether their intelligence may be characterized as being at the equivalent of one of the human stages.

A full discussion of the application of Piagetian theory to non-humans would require detailed consideration of the validity of the theory as applied to human development, and that would be out of place here. But as there have been attempts to see whether certain critical concepts (object permanence and conservation, attainment of which are taken to mark important advances in cognitive growth) are formed in non-humans, those experiments may usefully be reviewed in relative isolation.

Object permanence, the notion that objects persist independently of perception by the observer, is said by Piaget to be a concept whose formation is one of the marks that the first stage of intellectual development, the sensorimotor stage, is complete. A number of reports on non-human primates (e.g. squirrel monkeys—Vaughter, Smotherman, and Ordy 1972; rhesus monkeys—Wise, Wise, and Zimmerman 1974; chimpanzees—Wood, Moriarty, Gardner, and Gardner 1980) agree in concluding not only that their subjects achieve object permanence, but also that monkeys and apes in their development pass through the same substages (six in all, according to Piaget—see Piaget and Inhelder 1969) on the way to its achievement, as human children. The reports on monkeys concluded that whereas children take about 18 months to complete the sensorimotor stage, progress in monkeys appears to be much more rapid: Wise *et al.* (1974), for example, found that, depending on the method of testing used, all stages were complete in rhesus monkeys within from 29 to 103 days. Wood *et al.* (1980), on the other hand, in the only study which has used human child controls performing identical tests, found that performances of one age-matched pair of child and chimpanzee were very similar.

An earlier report using kittens and cats (Gruber, Girgus, and Banuazizi 1971) found that they too go through stages similar to those reported in humans, that they appear to progress more rapidly than humans, but that in this case, the final stages are not completed: in their experiment, for example, although a cat (3½ months old) would search under a piece of cloth for an object with which it had been playing if the cloth had been placed there while the cat was actually playing with the object, the same cat would not search for the object if it had been hidden by the cloth while its attention was distracted. Whether this represents a real limitation on cats remains doubtful: Gruber *et al.* themselves observe that their distractor (a piece of string) was very much too effective, and that the cats were extremely quick to react to and strike at it. They suggest that further research on cats should use a different technique to overcome this problem, and until such work is carried out, there is no compelling reason to suppose that cats are inferior to either monkeys or children in these tasks.

The second stage in Piaget's theory of cognitive development is the concrete operational stage, which lasts, on average, from the age of 18 months to 11 years. This stage is divided into the 'preoperational period', which is completed at about the age of seven, and the 'concrete operations' period. One of the marks of the transition from the former to the latter period is the acquisition of conservation, the ability to perceive preservation of quantity in an object through a number of transformations of shape. Two recent reports are relevant to the question whether conservation is obtained in non-humans.

Pasnak (1979) succeeded in training rhesus monkeys to discriminate between, on the one hand, addition to or subtraction from an object and, on the other hand, transformations in shape of the object: the operations concerned were carried out in full view of the monkeys, which could not otherwise distinguish between the results of the procedures. Although the monkeys' achievement may not be directly comparable to the 'spontaneous' conservation of children in the concrete operational stage, it is not unreasonable to argue, as Pasnak does, that his monkeys possess at least some of the prerequisites for conservation—in particular, the ability to distinguish between addition and subtraction by the experimenter and other transformations. An experiment on a chimpanzee, Sarah (Woodruff, Premack, and Kennel 1978), previously trained to make 'same' and 'different' judgements of two simultaneously presented objects showed that Sarah spontaneously gave 'same' judgements to pairs of objects (which might be solid or liquid) one of which had been transformed in her sight, while giving 'different' judgements when a quantity had been added or subtracted (again, in her sight). This report provides a convincing demonstration of conservation of quantity in a chimpanzee, and evidence, therefore, that chimpanzees are capable of attaining the concrete operational stage of intelligence.

Now while these reports may not provide conclusive evidence that non-humans do achieve this or that stage of cognitive development—no one test can provide such evidence on its own—they do not suggest that the intelligence of non-humans is 'arrested' at any particular stage, since the results have been generally positive. This section has confined itself to studies designed within the Piagetian framework. Many other reports are also relevant to the place of non-humans within that framework, although any general attempt at a Piagetian interpretation of non-human data would be both complex and controversial, and no such attempt will be made here. It should, however, be added that some evidence that is clearly relevant to the question whether non-humans can attain Piaget's third and final stage, the stage of formal operations, will be discussed in the section on reasoning (p. 304), in which the possibility of logical inference by non-humans will be considered.

Summary and conclusions

The line of descent leading to mammals diverged from that leading to modern reptiles some 300 Ma ago. Of modern mammals, the monotremes, followed by the marsupials, are most distantly related to man; with the exception of the other members of the primate order to which man belongs, there is relatively little difference between the placental orders in degree of relationship to man. Of primates, the great apes are the most closely related to humans, and it appears that man ceased to share common ancestors with apes somewhere between 4 and 10 Ma ago.

There are wide differences in relative brain-size amongst mammals, and man does indeed possess the largest 'encephalization quotient'; the other 'higher' primates also show a high degree of encephalization, but differences in brain-size amongst other mammalian groups do not appear to form an orderly pattern, and do not correspond to degree of relationship to man.

The general plan of the mammalian telencephalon conforms to that of other tetrapods, although the gross appearance is strikingly altered by the development of the six-layered neocortex. Within mammals, gross differences in forebrain anatomy—presence versus absence of corpus callosum or of cortical folding, for example—do not appear to be of major functional significance. The ratio of neocortex to whole brain volume shows surprisingly little variation in mammals, and in primates in particular; indeed, it is possible to predict with fair accuracy the relative size of most telencephalic structures in primates (including man) from knowledge of brain-size alone. For a primate, man does not possess an unusually large (or small) volume of 'association' cortex.

There does not appear to be any conclusive evidence for a special role of the neocortex (or any subdivision of it) in general intelligence; while there is evidence that some neocortical areas do have relatively specific functions (of what might be regarded as an intellectual kind) there is no reason to suppose that these functions are to be found in some mammals and not in others, or even that they are to be found only in mammals. In summary, it is not clear that the emergence of neocortex carries any implications relevant to the evolution of intelligence.

The surveys of behavioural studies, although finding a tendency for primates to show optimal performance (particularly in oddity-learning), did not find any orderly relationship between behavioural capacity and either relative brain-size or degree of relationship to man. It is not clear that any of the differences in performance observed are due to differences in intellectual capacity, and it is clear that, as is inevitable, procedural variables have a large role to play. No systematic theoretical interpretation is available of differences in performance between various mammalian groups, so that even where quantitative (or qualitative) differences in intellect may appear plausible, it is not clear what the precise nature of those differences might be.

This chapter points to the conclusion that there is no necessity to assume differences in intellect between the various mammalian groups; as the preceding chapters have come to similar conclusions regarding contrasts between non-mammals and mammals, it appears that there is no compulsion to assume variations in intellectual capacity throughout (non-human) vertebrates in general. This far, however, we have excluded man from consideration: Chapter 8 will consider differences between the human and the non-human intellect and, in particular, between human and non-human capacity for language.

8. Language and intelligence

Introduction

Very little has been said this far of differences between non-human and human intellectual capacities. It is quite clear that the tasks described in preceding chapters may all easily be mastered by humans, and that humans can in turn solve a huge range of problems that are quite beyond the capacity of any non-human (so clear that no evidence in support of the assertion need be provided). It may therefore appear that we must conclude that the general intelligence of humans is either quantitatively or qualitatively superior to that of other animals. There is, however, one difficulty in accepting this conclusion at face value, and this is, that human beings may solve these problems with the aid of language, either directly or indirectly; the superiority of humans might simply reflect the possession of language, and the capacity for language in turn might be independent of general intelligence.

Non-humans, on the other hand, while they do communicate with each other in a variety of ways, have not developed anything of comparable versatility to human language, which may be used internally as a mode of thinking about problems, and also as a way of teaching others how to solve problems. The ubiquitous influence of language is such that there may not be any problems the solution of which proceeds in adult humans without any contribution from language. Although humans do solve an extensive range of problems, many of which are not presented verbally, and do not require overtly verbal solutions, it is invariably plausible to suggest that the mode of solution adopted owes much to previous experiences involving language, if not to direct use of language at the time of solution.

It is clear, then, that any consideration of contrasts between human and non-human intelligence must assess the acquisition of language. There seem to be three main possibilities of interest. The first is, that humans possess the same complement of learning mechanisms as do non-humans, but that they are in some sense quantitatively superior, and so allow the acquisition of language: in other words, that humans acquire language and non-humans do not because humans are more intelligent than non-humans. A second possibility is that humans possess the same complement of learning mechanisms, which are not quantitatively different from those of non-humans, but possess in addition a mechanism or mechanisms whose sole function is the acquisition and mastery of language: in other words, humans are more intelligent simply because they possess language. The third possibility is that humans possess the same set of learning mechanisms (not quantitatively different) but in addition a mechanism or mechanisms which allow both the acquisition of language and the solution in novel ways of a number of other problems; this possibility is similar to the second, except that, like the first, it suggests that humans are, quite independent of language, more intelligent than non-humans. There are, of course, many other possibilities, but it is hoped that by singling out these three, the nature of the questions involved may be brought clearly into focus.

What evidence may be brought to bear on these three possibilities? One obvious approach might be to look at human versus non-human comparisons on tasks in which language is not involved: but we have already argued that, for humans, there are no such tasks (or, at least, that it is not possible currently to show that there are). An alternative approach reduces to the following 'thought experiment': how intelligent would a normal adult human be who, through no fault of his own, had not acquired language? What behavioural tests would we expect such a person to master that could not be mastered by a non-human (excepting, of course, language-acquisition tasks)?

Although we have referred to the exploration of the capacities of non-linguistic humans as constituting a 'thought-experiment', it might be believed that such 'experiments' have been carried out—there are, after all, reports of 'feral children' who have been raised apart from humans, and of children who have been raised by deranged humans who did not speak to them (see, for example, Brown 1958). Three points may be made about such reports: first, interest has, very naturally, centred largely on the question whether it is possible to teach the children to talk (in general, this has proved difficult, if not impossible); secondly, the children concerned do usually seem to be of 'low' intelligence, although sophisticated non-verbal test procedures have rarely been used; finally, and this relates back directly to the second point, poor performance by such individuals on tests of general intelligence would not in any case demonstrate that language was necessary for intellectual advance—these children may have been rejected by their parents precisely because they had shown behavioural deficits in early life.

A further source of evidence that might be relevant to the 'thought-experiment' concerns congenitally deaf children. Although much effort now goes into (successfully) teaching language (either spoken or signed) to the congenitally-deaf, this was not always the case and in the last century deaf-mutes (mute simply because they were deaf) were sent to institutions in which they were treated as imbeciles: there were, of course, no proper behavioural studies of the general intellectual capacity of these individuals, but it may be of interest to note that, at least to the untutored eye, humans without language did appear markedly unintelligent.

A third category of children that might be of interest here consists of children suffering from developmental dysphasia—children whose linguistic capabilities are considerably worse than normal. It would appear that, if language does play a major role in overall intellectual capacity in humans, then dysphasic children in whom language development is severely retarded should show severely reduced intelligence; a similar argument applies, of course, to deaf children in whom the growth of language is also retarded. Now there are in fact many reports of both dysphasic and deaf children with low linguistic ability who nevertheless show normal performance on 'non-verbal' intelligence tests (e.g. Cromer 1978). It might appear, therefore, that the intellectual capacity of humans is independent of language and, further, that differences between humans and non-humans on performance in those same non-verbal intelligence tests could be used as an indication of differences in general intelligence (as distinct from linguistic ability). The position is, however, not that straightforward. First, both the dysphasics and the deaf children in the reports referred to above do show levels of linguistic ability which are clearly superior to any shown spontaneously by any non-human group. The very possession of language, albeit at an unsophisticated level, could allow a quantal leap in problem-solving capacity, refinements in the use of language contributing particularly to tasks that involve a salient verbal component (e.g. standard intelligence tests, on which both dysphasic and deaf children perform relatively badly). Second, most 'non-verbal' intelligence tests do in fact have a verbal component—particularly in the instructions—which would rule out their use, without drastic modification, in any non-linguistic organism.

It appears, then, that we cannot empirically divorce human intelligence from language; although we have seen a suggestion that humans without language may seem unintelligent, the reports concerned are in no sense systematic or reliable, and the inclusion of that suggestion shows no more than this author's bias to the belief that such humans would appear much less intelligent than most of us might expect.

An alternative approach to the problem of language and intelligence, and to assessing the possibilities outlined above, is to explore more deeply the question whether non-humans can acquire language. If the differences between humans and other animals are merely quantitative, and humans and non-humans possess the same complement of learning mechanisms (whose quantitative superiority in humans allows us to acquire language), then perhaps it is possible to teach some non-humans at least a rudimentary language; if, on the other hand, humans use qualitatively unique mechanisms for the acquisition of language, then it should prove impossible to teach any language possessing the essential features of human language to a non-human species.

The sections that follow will consider the results of efforts made to teach language to non-humans. The subjects of these studies have invariably been apes (chimpanzees, in most cases), and this reflects the natural belief that, if any non-human species is capable of some capacity for language, it is most likely that a species closely related to man should possess that capacity. The survey cannot be exhaustive, but it is hoped that the aspects selected will be sufficient to bring out the general tenor of the apes' performance, and to allow a decision on the question whether results to date encourage the belief that apes can master at least the essentials of language. A further concern in the sections that follow will be the question whether the achievements reported in apes excel any that might be found in other vertebrates. In other words, whatever we decide concerning apes' capacity for language, is there evidence in the reports for capacities beyond those established in other groups?

This introduction will have served its purpose if it has emphasized the critical importance of the question of language in non-humans to comparisons between human and non-human intelligence. Suppose, for example, it is concluded that non-humans cannot acquire language, and that successful acquisition requires mechanisms which are species-specific, found only in humans. In that case, it would be possible to maintain the view that humans are in fact no more 'intelligent' than non-humans, but possess a mechanism of restricted application (like the navigational ability of the frog—p. 4) which normally inflates their intellectual capacity dramatically; if this mechanism was not engaged (as in the thought-experiment), the true (and unexpectedly low) general intellectual capacity of humans would stand revealed.

Linguistic capacities of non-human mammals

Words

Two early investigations (Hayes 1951; Kellogg and Kellogg 1933) attempted to teach chimpanzees to talk by bringing up very young animals in a human household, treated essentially as human children. Neither study had much success in obtaining vocalization: the Hayes study was the more successful, but even so achieved a vocabulary of only three or four words—sounds, that is, that approximated to the sounds of human words and were uttered on appropriate occasions.

A possible reason for the apes' failure to talk is that their vocal tract is not physically capable of producing the range of sounds required for imitating human speech; Lieberman (1975) reports that, contrary to earlier beliefs, the chimpanzee's vocal cords (the larynx) are not inadequate for speech-production (although the chimpanzee larynx would produce a rather hoarse voice), but that the chimpanzee's supralaryngeal tract (which consists of the pharyngeal and the oral cavities) is inadequate to produce many of the sounds used in human speech.

It seems, therefore, that attempts to obtain speech-production from apes are bound to fail, for a theoretically uninteresting reason. It may, however, be noted that similar arguments do not appear to apply to speech-comprehension: differences between the gross auditory perception of man and ape appear to be minor, and of no evident relevance to the perception of speech sounds (see, for example, Prestrude 1970). Lieberman (1979) has argued that humans have evolved specific neural auditory mechanisms 'that match the quantal sounds of the human supralaryngeal vocal tract' (p. 123). His evidence consists of reports of 'categorical perception' of speech sounds in, for example, human infants. Certain perceptual systems make qualitative distinctions within dimensions the physical stimuli for which vary in a simple quantitative way—as, for example, in colour vision, where a given quantitative change in wavelength will generate either minor (quantitative) or dramatic (qualitative) changes in hue, depending upon the point on the wavelength continuum from which the change is made. Similarly, in speech perception, humans perceive marked differences between stop consonants (plosives) having phonation- (or voice-) onset delays of less than approximately 20 ms (which are perceived as b, d, and g) and those having longer delays (which are perceived as p, t, and k); humans discriminate accurately between consonants having phonation-onset delays of 10 and 30 ms, but poorly between those with delays of 50 and 70 ms and it has been argued that this pattern of results derives from the possession of innate analysing mechanisms that categorize acoustic signals according to phonetic criteria. Three difficulties face this argument: the first is that 'categorical perception' has been observed within auditory dimensions that appear to bear no relation to speech-processing (e.g. Cutting, Rosner, and Foard 1976); the second is that categorical perception of human speech sounds has been reported in both rhesus monkeys (Morse and Snowdon 1975), and rodents (*Chinchilla laniger*–Kuhl and Miller 1975) (but for a failure to obtain categorical perception in a study using three species of Old World monkeys, see Sinnott, Beecher, Moody, and Stebbins 1976); the final difficulty is a general concern that reports of categorical perception may in fact reflect covert acoustic properties of the signals, rather than primarily phonetic boundaries (e.g. Bailey 1979). There is currently, therefore, no compelling reason to suppose that the chimpanzee lacks the auditory mechanisms required for the perception of speech.

If we suppose that chimpanzees can accurately perceive speech, should we expect their inability to produce speech to hamper their comprehension of speech? To answer this question, we may consider a case examined by Lenneberg (1967) of a human boy who was, due to malfunctioning of the vocal tract muscles, incapable of producing intelligible speech sounds. Of this child, first seen by Lenneberg at the age of four, Lenneberg (1967) writes that it was 'obvious that he had a normal and adequate understanding of human language' (p. 307). Now investigations of language in chimpanzees have often reported that speech comprehension is superior to speech-production: Kellogg (1968) for example reports that after nine months home-rearing, his chimpanzee (Gua), by then 16 months old, understood 58 spoken phrases (although she uttered no words at all). However, Gua was by that stage already falling behind her 'human control' child, and Kellogg reports that 'had the comparison continued for a longer period, all indications are that the human subject would have left the animal far behind in the comprehension of words' (Kellogg 1968, p. 425), We shall not consider Gua's comprehension in detail here, as subsequent reports of language-production appear to have produced more impressive results, and are more susceptible to analysis.

The major recent approaches to language in apes have used modes of production other than speech. A number of investigators (e.g. Gardner and Gardner 1969; Terrace, Petitto, Sanders, and Bever 1979, who used chimpanzees, and Patterson 1978, using a gorilla) have attempted to teach American Sign Language (ASL or Ameslan) to their subjects. ASL is used as a first language by many deaf people, and so may be regarded as a 'natural' language; in the version of ASL used in these studies, each sign stands for a word or concept–not, as in some versions of manual languages for the deaf, for individual letters. Two other investigators have used 'artificial' languages: Premack (1971) has used plastic shapes of various shapes and colours which may be magnetically attached to a board as 'words' in a study using chimpanzees, and Rumbaugh (1977) has trained chimpanzees to press lit panels displaying a number of arbitrary stimuli in order to obtain various rewards.

The results of these investigations are rarely described in 'neutral' terms; rather, the authors use linguistic concepts to describe the achievements of their subjects, and this practice has a seductive effect. When a chimpanzee learns to make the ASL sign for, say, 'cat', and produces that sign when a cat is seen, is it not natural to say that the ape has learned the word 'cat'? It may be useful, in place of answering that question directly, to consider an alternative description—that the animal has learned to associate a novel (and no doubt arbitrary) response with reward. In the ASL studies, new signs are learned either by imitation by the chimpanzee of signs performed by humans or by direct 'moulding' of the sign by the human's moving the chimp's limbs appropriately. In either case, it seems clear that signs are 'welcomed' by experimenters in a variety of ways—tickling, giving desirable food objects, or simply 'social encouragement'—which can all be seen as rewarding (and similar encouragements no doubt accompany the early word-acquisition of human children). The question is, therefore (and this is, of course, clearer in the case of the artificial language studies), does this process differ in any important way from the way in which, say, a goldfish learns to press a green key for food—and if not, should we say that the goldfish has learned a word for food, or that neither the goldfish nor the ape has (necessarily) learned a 'word'? In a way, it does not matter how we answer this question: the mere ability to acquire a sign and to use it appropriately does not require any more than the capacity to form novel and arbitrary associations, and that capacity is to be found throughout vertebrates. Whether the same capacity is actually used by humans is early word-acquisition would be an extraordinarily difficult question to answer, and it is more profitable to suggest that an essential property of words is that they form constituents of sentences—in this sense, we may say that any creature that cannot learn words.

The most appropriate data for considering the question whether apes can form sentences are those concerning the production of sequences of signs, and they will be discussed in the following section. One study has, however, attempted to determine whether chimpanzees are capable of one essential prerequisite of sentence-formation, the distinction between words according to their grammatical category. In this report (Gardner and Gardner 1975), chimpanzee Washoe (then about five years old, at the end of four years' training) was asked ten sorts of 'Wh' question, and her replies were scored according to whether they were correct, and, if not, whether they were grammatically appropriate. For example, questions formed by (the sign for) 'Who' followed by a pronoun called for a proper name, the question 'What colour' called for a colour, the question beginning 'Where' called for a locative sign. Gardner and Gardner report that Washoe showed very good agreement between the type of question asked and the grammatical category of the answer given. They compared Washoe's performance with reports of replies of young children to similar questions (during spontaneous conversation) and concluded that 'If Washoe had been a preschool child', then her replies 'would place her at a relatively advanced level of linguistic competence' (Gardner and Gardner 1978, p. 67).

A major problem with the Gardner and Gardner (1975) data concerns the control of the experiments; we are not given full details of the precise way in which each question occurred (this is, perhaps, an inevitable problem in the relatively unstructured approach used in the 'natural' language studies) but it appears that in many, if not all cases, the appropriate answer was indicated by the interrogator's pointing at it. Now it might be that this is of no significance, that one could point out a person, ask 'Who that?', and obtain a name, or ask 'What colour?' and obtain the colour of his shirt. But suppose that Washoe was asked the colour of an object whose name she did not know, or that (habitually) when questions about colour were involved, objects were pointed at in a different way (e.g. by pointing very closely at, say, the shirt, rather than more generally at the person)? Similarly, we do not know the nature of the signs produced (by either party) immediately preceding the question: suppose the sign for 'out' has been used several times (e.g. 'We go out'), and then the experimenter asks 'Where we go?' Now this is not, of course, to say that anything of this kind occurred: what does, however, have to be said is that we do not know whether it did or not. There are, as the Gardners have repeatedly pointed out, very good reasons for the training procedures that they have employed which do, for example, clearly allow comparison with human language-acquisition. On the other hand, it seems that such a technique calls for very full presentation of results, which ideally would be backed up by (unedited) videotapes of sessions in which test questions were put.

Before leaving consideration of 'words', note should be made of the impressive size of vocabulary achieved by Washoe, Terrace's chimpanzee Neam Chimpsky (or Nim), and Patterson's gorilla, Koko. All three animals are reported to have learned 100 signs or more by the end of training, all of these signs evidently being available at the same time. These observations raise the interesting question whether animals of other groups could accurately maintain so many arbitrary associations simultaneously: at present, there are simply no data available which might allow us to answer that question.

Sentences

The ASL investigators have all reported the spontaneous production by their subjects of multisign strings, and the artificial language studies have similarly succeeded in teaching chimpanzees to produce or respond appropriately to strings of symbols in specified orders in order to obtain reward. The interpretation of the results of the two sets of studies raises quite different problems, so that they will be considered separately. The Gardners (e.g. Gardner and Gardner 1971), Patterson (1978), and Terrace *et al.* (1979) all agree that the two-sign combinations of their subjects showed, to some degree, consistent sign-ordering. The critical question is whether this ordering is of semantic significance or reveals merely 'idiosyncratic lexical position habits' (Terrace *et al.* 1979, p. 896), a question which can be answered only by demonstrating a relationship between the presumed meaning of a sign-combination and the order of the signs within the combination. Do chimpanzees, for example, tend to place signs for actions after signs for agents but before signs for objects (as do young children, and human adults, in the normal expanded form of the sentence which the child is presumed to be uttering)? The investigators generally agree that where it is possible to interpret the sign-strings against the context of their occurrence, then the ordering of the constituents of the strings does vary systematically with semantic context; Patterson (1978), for example, reports that for 82 per cent of combinations in which a verb preceded a pronoun, the pronoun indicated the object of some action (e.g. 'swing me'), whereas, for 76 per cent of pronoun–verb combinations, the pronoun indicated the agent (e.g. 'you tickle'). The most extensive analysis of such data has been provided by Terrace *et al.*, who interpreted 1262 two-sign combinations produced by their chimpanzee Nim, finding 'several instances of significant preferences for placing signs expressing a particular semantic role in either the first or the second positions. Agent, attribute, and recurrence (more) were expressed most frequently in the first position, while place and beneficiary roles were expressed most frequently by second-position signs' (p. 896); these orderings, of course, parallel those found in English speech. It should be noted that the relationship between 'meaning' and sign-order was not invariably obtained: Nim (unlike Koko—Patterson 1978) when apparently referring to an action and an object, produced the signs for each as often in one order as in the other.

Terrace *et al.* (1979) note two difficulties facing the proposal that Nim is showing, by his word-order, the beginnings (at least) of adherence to grammatical rules. The first difficulty is that some 'classes' of semantic rule are represented by very few, or even only one, sign: for example, the only sign used for recurrence was the sign 'more'—in the absence of any information concerning the position that would be occupied by some other sign for recurrence, it can hardly be concluded with any confidence that the initial position occupied by 'more' is of semantic significance (this same general criticism appears to apply to the few data provided by Patterson and the Gardners).

The second difficulty raised by Terrace *et al.* arises from their analysis of video-tape recordings of Nim's training sessions, from which it emerged that a large number of Nim's signings were imitations of signs produced recently (or simultaneously) by his teachers. Following this discovery, Terrace *et al.* studied films and video-tapes available of Washoe and Koko, and claim to have identified the same marked tendency of the apes to produce signs and, more importantly, combinations of signs, recently produced by the experimenters. Now these films and tapes show, of course, only a minute fraction of the total performance of Washoe and Koko, so that it may be that Terrace's conclusions will not, in the long term, prove justified, What is clear, however, is that the resolution of the questions considered here requires very much more detailed information than has previously been available from the Gardners and Patterson: 'In the absence of a permanent record of an ape's signing, and the context in which that signing occurred, even an objectively assembled corpus of an ape's utterances does not provide a sufficient basis for drawing conclusions about the grammatical regularities of those utterances' (Terrace *et al.* 1979, p. 897).

The Terrace *et al.* (1979) investigations of sign-order do not, then, encourage the conclusion that apes are grammatically competent. A similar caution has been expressed by the Gardners, who suggested (Gardner and Gardner 1971) that the limited degree of consistent sign-order seen in Washoe might originate simply through imitation of her human models. The Gardners have, however, gone on to argue that in fact inconsistency in word-order is much more common in young children than was at one time believed, concluding that: 'In fact, the chimpanzee data are like the child data in showing a certain degree of consistent word order, a certain degree of varied word order correlated with varied contexts, and a residue of free variation easily attributable to the immaturity of the subjects and the ambiguity of some of the context notes' (Gardner and Gardner 1978, p. 56). The problem is that we know that children go on to be consistent in word-order in the production of wholly novel sentences, so that we suspect that whatever degree of consistent word order is seen in them at an early stage is not an artefact of, say, imitation or simple habit. If we are wrong in that assumption, then children too are not yet grammatically competent at the early stages of production of word-strings, and the demonstration of comparable performance in child and ape does not affect the case for arguing that linguistically competent humans differ qualitatively from apes. The general point is that similarity between apes' language and young children's language does not demonstrate that differences between apes and adult humans are merely quantitative: if, for example, chimpanzees' 'manual babbling' had resembled the (vocal) babbling of human infants (which, apparently, it does not, since it occurs much less frequently—Gardner and Gardner 1971), we need not have concluded, simply because babbling is seen as a critical stage in human speech-acquisition, that subsequent differences between the species are only quantitative.

We shall conclude this discussion of sentence-production in ASL by considering multisign utterances, where differences between the apes' performance and that of young children have been reported. Terrace *et al.* (1979) have analysed Nim's three- and four-sign combinations, and have shown that these contrast strikingly with the early three- and four-word utterances of children, primarily in the predominance of repetition of signs within utterances. Nim shows a strong tendency to use in a three-sign combination one of his most common two-sign combinations, with one sign repeated, as if for emphasis: he does not (as young children do) give any indication of using longer strings to convey more complex information (Terrace *et al.* 1979). A further contrast between apes and children lies in the manner in which the average number of signs (or words) in a single combination (the mean length of utterance, or MLU) grows. The MLU of an individual child increases fairly regularly from the age at which the first few words are uttered (and that age, of course, varies considerably from child to child). The Gardners (1971) and Patterson (1978) both found a gradual increase in MLU in their subjects, not quantified in the Gardners' report, but, in Koko's case, rising over an 11-month period from 1.37 to 1.82; Terrace *et al.* (1979), on the other hand, found no increase in Nim's MLU across a 19-month period. Even if we accept Patterson's figures as giving the more representative view of apes' ability, increase in MLU is proceeding very slowly in comparison to that typically seen in children (Brown 1970).

Brown (1970) has argued, however, that progress by apes from single utterances to multiple utterances is in itself highly suggestive: 'If they were only signs strung out in time and not interacting semantically and grammatically then one would think they might be of any length at all, and that there would be no reason for them to start short and become long' (Brown 1970, p. 225). One finding that goes against this very reasonable suggestion is that both Nim and Koko appear to produce very long sign-strings (up to 11 and 16 signs, respectively), given their MLUs (well below 2.0 in each case); children, by contrast, with an MLU of 2.0 will show a maximal utterance-length of 5 (±2) (Brown, quoted by Terrace *et al.* 1979). Now this observation, coupled with the Terrace *et al.* report of the use of sign-strings for repetition and emphasis rather than elaboration, does suggest that the strings observed are not unitary constructions but are indeed 'only signs strung out in time'. Our conclusion from this discussion of multisign strings is, like that from the discussion of two-sign strings, that these strings are in all probability not genuine combinations of signs and do not, therefore, resemble sentences in any critical sense.

The studies which have used artificial languages leave no doubt that their subjects do discriminate, in both production and comprehension, between sign strings in different orders. Premack's (1971) chimpanzee Sarah, for example, learned to produce the symbols for 'Mary', 'give', 'apple', and 'Sarah' in the appropriate order (from top to bottom of the vertical board to which the tokens were attached). Initially, Sarah learned to place the 'apple' token on the board in order to obtain an (out of reach) apple; then, she was trained to place both the 'give' and the 'apple' tokens in order to be given the apple, with the constraint that the 'give' token should appear above the 'apple' token. Subsequently, she was taught to expand the 'sentence' to include the name of the experimenter at the beginning, and her own name at the end. Does Sarah's achievement suggest the capacity for at least the beginnings of syntax? Sarah does, after all, know that 'give apple' has a very different significance from 'apple give'. This is, again, a very difficult question to answer—it must, after all, be true that the human capacity for language does depend on the ability to produce words in particular orders, and how are we to decide whether the ability to generate ordered sequences of words in humans is or is not related to Sarah's ability to produce the sequence 'give apple'? Two observations may be relevant to answering the question, at least in part. In the first place, the ability to order responses sequentially is by no means limited (outside humans) to chimpanzees: Straub, Seidenberg, Bever, and Terrace (1979) have shown that pigeons can learn (for food reward) to peck four coloured keys in a particular sequence (green, white, red, blue) independent of the keys on which the colours appear. Sarah's achievement need not, therefore, be seen as indicating a step on the road to language in higher primates. Second, the distinction between 'give apple' and 'apple give' is hardly of semantic significance: Sarah is not discriminating between certain sequences of signs which mean one thing, other sequences which mean another, and yet other sequences which are meaningless. What she is doing is discriminating between sequences which obtain reward, and sequences which do not. There is, then, as we saw with the ASL studies, no reason here to ascribe semantic significance to sign order.

Premack (1971) himself agrees that 'properly ordered strings of words' (p. 814) need not form sentences, arguing that sentences must possess internal organization . To show that chimpanzees can appreciate internal organization, Premack carried out a series of tests using the sign for 'on'. Sarah was trained, first, to place a red card on top of a green card when shown the tokens for 'red', 'on' and 'green' (in that order), and to place a green card on top of a red card when shown the string 'green', 'on', 'red'. She was then tested for the degree of transfer to two new coloured cards (blue and yellow), the symbols for these colours having been learned previously: although we are given no details, it appears that she showed good, perhaps perfect, transfer to the new cards. Finally, she experienced a series of trials on which she was given three symbols ('on', plus two colours), and her ability to produce appropriate sentences in response to the experimenter's arranging two cards was tested. Her responses were correct on eight of the first ten of these trials. The difficulty with accepting this experiment as demonstrating the possession of the rudiments of syntax required for sentence construction is that the rules that Sarah needed to detect were so simple: she need learn only that the symbol for 'on' must occur in the middle of the three symbols she is shown or given, and that the cards (or symbols) must be arranged so that the colour of the top card agrees with the colour associated with the first symbol. Once again, it seems unlikely that only chimpanzees could solve such problems (although no direct analogues using other species have been conducted), and there seems no reason to suppose semantic significance in the word-order: Sarah has, in a sense, learned the winning combinations (as of a fruit-machine) and that achievement has no necessary relationship to the understanding of a sentence.

Similar considerations apply to the sequential responding established by the chimpanzee Lana in the Rumbaugh (1977) study, and it would be repetitous to consider that aspect of Rumbaugh's work in detail. It is of interest in this context to note that Thompson and Church (1980) have described a computer model of Lana's sentence-production which, relying on two basic processes, association formation and conditional discrimination learning, provides a reasonably good simulation of Lana's performance. The success of this model encourages the view that relatively simple and widespread processes of vertebrate learning are indeed sufficient to generate much of the behaviour seen in the artificial language studies using chimps. Since no model incorporating these same simple principles has been even remotely successful in simulating human language-production, we are led again to the conclusion that what Lana is doing differs from what humans do. We shall consider another aspect of Rumbaugh's results in the following section, which will consider use by apes of language as a medium of communication.

This section has considered the question of whether apes are capable of deriving syntactic rules akin to those of the grammars of human languages, and has concluded that they are not. It should, on the other hand, be admitted that this review has (necessarily) been highly selective, and that the question remains a topic of active investigation (and of controversy, tending towards acrimony). The following sections will adopt a neutral attitude towards this central problem and ask, instead, whether apes use language (as humans do) to convey information to one another, and whether they can use it constructively in reasoning.

Communication

One of the central uses of language by humans is the conveying of information from one person to another. Without this means of transfer of information, our general education would be very much impoverished, and our problem-solving ability, which no doubt relies in part on heuristics passed on to us by others, might be much reduced. It is therefore germane to the question of intelligence in non-humans to ask whether non-human subjects can communicate information to one another using arbitrary symbols; there is, of course, no doubt that non-humans can communicate information to one another in a variety of ways, where the interpretation of the signal is stereotyped and species specific.

Although there are anecdotal references to chimpanzees signing to each other in ASL (Gardner and Gardner 1978), there have not yet been any substantial reports from such studies, and the most explicit demonstration of symbolic communication between chimpanzees is that described by Savage-Rumbaugh, Rumbaugh, and Boysen (1978). In this study, which used the same lit-panel technique as did the original Lana project, two chimpanzees (Sherman and Austin) were trained, first, to press keys showing different symbols when shown different types of food (being rewarded either by social encouragement or by a subsequent opportunity to obtain some other food by panel-pressing). They were then trained to respond to the symbol for a type of food (hidden in a container) by 'requesting' that type of food by appropriate panel-responses; in this case, correct responses obtained the food in the container. Finally, one chimpanzee was shown one of 11 kinds of food or drink being hidden in a container in an adjoining room, and was asked to indicate (via the symbol-key panel) what type of food had been hidden; the second chimpanzee could observe the symbol produced and, if he went on to request that type of food or drink, both chimpanzees were rewarded with the contents of the container. In one variation of the task, the second chimpanzee was asked, not simply to produce the appropriate symbol (which required him in fact to match the symbol produced by the first chimpanzee) but to point instead to the appropriate photograph (from three shown on each trial) of a food or drink.

Now although the Savage-Rumbaugh *et al.* report can be described as showing inter-animal communication by symbols, it is clear that this communication is of a distinctly impoverished nature: the sender has been trained to respond in such a way as to produce a symbol which in turn acts as a stimulus to which the receiver has learned an appropriate response. There is an absence of spontaneity or genuine flexibility in the communication, and 'information' is transferred in a very restricted sense: the receiver does not acquire anything comparable to new knowledge from the transaction. It is, therefore, not surprising that an analogous form of symbolic communication has been described in pigeons (Epstein, Lanza, and Skinner 1980); in this case, the sender pigeon was shown one of three colours (hidden from the receiver), whereupon it pecked one of three keys (not coloured), and this enabled the receiver to select that one of the three keys available to it on which the appropriate colour was displayed. Epstein *et al.* conclude (p. 545): 'We have thus demonstrated that pigeons can learn to engage in a sustained and natural conversation without human intervention, and that one pigeon can transmit information to another entirely through the use of symbols.' They do, however, go on to agree that: 'It has not escaped our notice that an alternative account of this exchange may be given in terms of the prevailing contingencies of reinforcement'—but then, of course, Skinner and his followers believe that 'A similar account may be given of . . . comparable human language' (*ibid*., p. 454). Now the question whether general principles of learning are adequate to account for the acquisition of human language (of grammar in particular) is, of course, the central issue of these sections (and if they are, it is not easy to see why chimpanzees do not generate sentences). What can be agreed here is that the forms of symbolic communication described thus far do not require the supposition of any more than learning mechanisms that we already know to be widely distributed among vertebrates.

A further demonstration that pigeons are capable of a form of communication that might at first sight seem highly sophisticated has been provided by Boakes and Gaertner (1977). These authors were following up a study of communication between two bottlenosed dolphins (*Tursiops*). In this experiment (carried out by Bastian, and described in Wood 1973), the first dolphin was shown one of two stimuli, which were not visible to the second dolphin. The second dolphin, which could hear but not see the first dolphin, had to choose to respond between one of two paddles, and the correct paddle varied according to which stimulus had been shown to the first dolphin. If the second dolphin responded correctly, both animals were rewarded. Neither dolphin had received specific training before beginning the task, but the second dolphin was able to choose appropriately, so that it must have used auditory information provided by the first dolphin. Boakes and Gaertner conducted an essentially identical experiment using pigeons, the one significant change being that the birds could see each other (although, again, the receiver could not see the stimuli—red or green light—shown to the sender).

The pigeons, like the dolphins, solved the problem, but Boakes and Gaertner show that the mode of solution is less complicated than might have been expected. In essence, what happens is this: the receiver pigeon (which has to choose between two white keys) initially adopts a position habit, responding consistently to, say, the left key. Suppose that the red light (shown to the sender) is the stimulus that signifies that the receiver's left key is correct; then, if the receiver adopts a left position habit, the red stimuli (for the sender) will be followed by reward, green stimuli will not. As a consequence of autoshaping (see p. 193), the sender now begins to peck the red (but not the green) light. The receiver then detects that, when the sender does not peck, his own pecks are not rewarded, so that while the sender's pecking predicts rewards for left-key pecks, the absence of pecks predicts non-reward; this has the result that when the sender does not peck, the receiver avoids the left key, and now (correctly) pecks the right key. This elegant study emphasizes that, in whatever terms experiments are described, we should adopt the simplest plausible explanation—we should, that is, abide by Lloyd Morgan's canon (p. 1).

A striking contrast between the studies reviewed above and conversation between humans is that there is no reason to suppose that the sender 'knows' that he is communicating with another being, that he has any 'understanding' of the means by which his behaviour leads to reward. It is clearly extremely problematical whether in general such terms have application to non-humans, and if so, how their application might be demonstrated; but we may briefly consider here one attempt, in the general area of communication, to show that chimpanzees have intentions. This report (Woodruff and Premack 1979) argues that the use of communication to mislead as well as to inform would provide good evidence for intentionality: this is a matter which, fortunately, we need not debate unless we are given good evidence for deception in the first place. In the Woodruff and Premack report, chimpanzees were shown food being hidden under one of two containers which they could not reach. Human trainers (who did not know the location of the food) were required to infer from the chimpanzees' behaviour which container hid the food; one trainer (the 'co-operative' trainer) gave the hidden food to the chimpanzee if he (the trainer) chose the baited container; if he chose the wrong container, the chimpanzee was 'consoled', but not given the food. The other ('competitive') trainer kept the food himself if he chose the correct container, but if he chose the wrong container, the chimpanzee then did receive the food. The two trainers were made highly discriminable from one another by a number of differences in both their appearance and their behaviour.

Initially, all four chimpanzees used in the Woodruff and Premack study clearly oriented towards the baited container, and both trainers used that cue to enable them to choose the correct container. As training progressed, however, the animals stopped orienting towards the baited container in the presence of the 'competitive' trainer (while continuing to do so for the 'co-operative' trainer); this had the effect that the competitive trainer could now only guess which container to choose. Finally, two of the chimpanzees learned to orient towards the unbaited container, so that the competitive trainer (who was instructed not to change his mode of interpretation, once one had been successfully adopted) obtained very few rewards.

Now although Woodruff and Premack prefer to interpret the behaviour of the two most successful chimpanzees as showing intentional deception, this is hardly a compelling interpretation: orienting towards the baited container was rewarded in the presence of the co-operative trainer, but not the competitive trainer (therefore orienting should drop out in the latter condition); orienting towards the non-baited container (which the animals did not do spontaneously) was consistently rewarded in the presence of the competitive trainer—therefore, if any subject happened to orient towards that container (perhaps by mistake) on one or two trials, that rule could be acquired. In a subsequent phase of the experiment, the roles were reversed, so that the chimpanzees had to choose containers on the basis of the orientations of the trainers (and in this case, the competitive trainer oriented towards the unbaited container). One of the two chimpanzees that exhibited 'deceit' in the first phase of the study never learned to avoid the container indicated by the competitive trainer (and none of the animals initially avoided the containers so indicated). It seems very odd to suppose that an animal that can 'intentionally deceive' cannot learn that he is being 'intentionally deceived' (and vice versa). Our conclusion is, then, that these results have been subjected to an over-rich interpretation: non-humans can learn to produce signs and symbols where these are consistently rewarded, and they can respond appropriately to signs and symbols, again where such responding is rewarded. We do not, however, need to assume (and so, should not assume) that they are in any wider sense communicating, or that any of the host of concepts associated with communication in humans are properly applicable to communication in non-humans.

Reasoning

We shall consider both inductive and deductive reasoning under this heading, the topic of reasoning being introduced at this point because, as will be seen, there are grounds for supposing a close relationship between reasoning and language.

As is well known, the inductive mode of reasoning is that in which generalizations are formed through the perception of regularities in past events, these generalizations being used to predict similar regularities in the future. Such predictions will not necessarily hold good; in general, their utility depends upon there being a genuine causal relationship underlying the regularities observed, and our confidence in generalizations as predictors varies according to our perceiving some plausible causal link.

Now since the associations formed by non-humans depend upon there being a contingency (and not simple contiguity) between the events associated (Rescorla 1967), it appears that they do in that sense form generalizations and apply them to future occurrences of the predictive event (the CS) in the association. This is likely to prove a satisfactory procedure, since in fact where a contingency is detected, it will normally reflect a genuine causal relationship. There is, moreover, evidence that certain types of event enter preferentially into associations in such a way as to enhance the likelihood of detecting causally related events. Testa (1974), for example, has drawn attention to evidence which indicates that where two stimuli both predict a UCS equally well, then similarity (in, say, location or temporal pattern) between one of those stimuli and the UCS will enhance the likelihood that it, rather than the other stimulus, will come to function as a CS. Similarly, there is a well-known body of findings (e.g. Garcia, Hankins, and Rusiniak 1974) which shows that rats tend to associate internal (e.g. taste) stimuli with internal UCSs (such as sickness), and external (e.g. visual) stimuli preferentially with external UCSs (such as foot-shock). In these cases, it seems that the animal selects as the functional CS that stimulus which is in fact most likely to be causally related to the UCS.

It may, of course, be objected that the ability to detect regularities involving causally related events is not the same as possession of a concept of causality: one may construct a clock without knowing that the earth rotates round its own axis. Premack (1976) has, however, argued that chimpanzees do indeed grasp that certain events are not merely associated, but causally linked. In support of this claim, Premack cites an experiment in which chimpanzees were shown on each trial two items, one a familiar object, and the other, that same object following some alteration (e.g. a dry sponge and a wet sponge; a ball and the two halves of a cut ball). The animals were then asked to replace an interrogative marker placed between the two versions of the object with one of three further objects, which could function as instruments of change (these objects being, for example, a container of water, a knife, and a pencil). Three of Premack's four subjects consistently selected the object which could in fact have caused the alteration between the two versions of the original object. Premack argues that this performance could not have arisen simply through associations between the objects and the instruments—the ball, for example, had been seen by the animals very much more often in water than in association with a knife, but, shown a whole ball and the two halves of a ball, the knife and not water was selected.

The difficulty with Premack's argument here is that he is using a somewhat impoverished concept of association: the chimpanzees, in order to infer a causal relationship, must have detected a contingency between (e.g.) knives and cut objects, and that is precisely the conditon for the formation of an association. If the two halves of the ball are seen, not as a ball, but as a cut object, then we should expect, on the basis of associations alone, the knife rather than the water to be selected. In order to demonstrate a concept of causality, it would be necessary to show that the animal discriminates between cause and correlation, that, for example, there is for it a distinction between a pencil causing a mark and a buzzer which signals (but does not cause) a shock. It can be seen that this is a well-nigh impossible requirement: a chimpanzee might well have a concept of causality and still believe (mistakenly) that buzzers cause shocks. Unless, however, it can be shown that the animals do treat some correlations as not being causal, then the notion of cause seems redundant to an account of their behaviour—they simply detect contingencies. Similar arguments apply, of course, to humans; but in our case, the possession of a concept of causality is apparent from the way in which we talk about relationships between events, and from our (verbal) explanations of happenings. It is because chimpanzees do not provide evidence of that kind that we need not assume a concept of causality in them.

We shall preface the discussion of deductive reasoning with a brief look back at two early experimental paradigms. Maier (e.g. 1932) introduced a series of tasks which he described as requiring reasoning for their solution. By reasoning Maier meant the capacity to integrate information derived from two separate experiences to derive new information, not available from either experience alone. One of Maier's reasoning tasks was the three-table problem, in which three elevated runways radiate from a common central point to three tables (A, B, and C) each of which may be entered through a small hole in a screen which otherwise conceals the table. Hungry rats are initially allowed simply to explore the apparatus, running freely from table to table without explicit rewards (and this is Experience 1); they are then placed on one of the tables (Table A, say), and given food there for a limited period (Experience 2). Finally, they are placed on, say, Table B, and their behaviour is observed: if the majority of rats run to Table A, then they have integrated the two experiences (reasoned), but if no preference between Tables A and C is shown, no reasoning has occurred. Since rats do solve such problems, Maier concludes that they do reason, demonstrating a capacity, in his view, beyond that required for association-formation.

Problems such as those devised by Maier did pose difficulties for the Hullian school of behaviourism which was gaining strength at that time, since that version of associationism held that stimuli were associated with responses as a consequence of reinforcement; as no response in the reasoning task had been differentially reinforced, it was difficult to see how learning could have occurred. Modern theories of association formation are, however, very much more liberal in allowing a range of types of event (stimuli, responses, and reinforcers) to serve as terms in associations, and Maier's reasoning can be accounted for by appealing to well-established phenomena of classical conditioning: Experience 1, for example, could allow the association of the stimuli (CS_1) associated with the runway leading to Table A with the stimuli (CS_2) of Table A itself; Experience 2 would allow the conditioning of Table A stimuli (CS_2) to the food (UCS) stimuli. The test asks, in effect, whether CS_1 is now associated with the UCS, as such an association would elicit approach (a component of the UCR to food). It can be seen that the whole task parallels the standard sensory preconditioning design (see p. 63): Stage 1 pairs two neutral stimuli, CS_1 and CS_2, Stage 2 pairs CS_2 and UCS; the final stage tests for successful preconditioning by assessing the CR to CS_1. This is not the only possible account of Maier's findings, but does serve to illustrate the remarkable ease with which associationist theory can accommodate a superficially complex performance. Tests of reasoning will have to be of sophisticated design if they are to rule out alternative, associationist, accounts.

A second well-known early attempt to demonstrate processes in non-humans beyond those posited by associationists was the work of Köhler (1925) on insightful problem-solving by chimpanzees. Köhler described the use by chimpanzees of tools, such as crates and sticks, to reach rewards otherwise unobtainable, and believed that the abrupt nature of the solutions observed (the sudden insertion of one stick into another to make a longer stick, for example) argued against a trial-and-error account, and for some insightful process. Subsequent work has cast serious doubt on Köhler's interpretation, primarily by showing, first, that previous experience with sticks plays a crucial role in solving problems in which rewards must be drawn in with sticks (Birch 1945) and, second, that the probability of insightful solution depends to a considerable extent on the degree to which spontaneous stick manipulations occur (Schiller 1952). These later studies indicate that close control of potentially relevant previous experience is an important prerequisite to the analysis of problem-solving and that its absence in Köhler's work seriously weakens the force of his argument.

Very little was written on the subject of reasoning in non-humans during the long predominance of the views of Hull and his followers, this partly due to the widespread confidence that once the basic laws of stimulus–response association had been clearly established (and this was the primary task), more complicated types of intelligent performance would be shown to be reducible to simple principles. The current decline in Hull's influence (marked, for example, by the wide interest in classical conditioning) may account for a resurgence of interest in reasoning; it remains, however, the case that no coherent view of reasoning in non-humans has emerged, so that in what follows we shall discuss three disparate approaches.

One remarkable experiment (McGonigle and Chalmers 1977) has explored the possibility that squirrel monkeys are capable of transitive inference—of, that is, deductions such as the following: if A is greater than B, and B is greater than C, then A is greater than C. McGonicle and Chalmers used a variation of a non-verbal task (involving sticks of different lengths) originally designed for children (Bryant and Trabasso 1971); in the McGonigle and Chalmers study the stimuli were five metal cylinders, each a different colour, which may be labelled Y, B, G, R, W. The monkeys first learned a series of four simultaneous discriminations, Y v B, B v G, G v R, R v W, with the first (or for half the subjects, the second) member of each pair being consistently rewarded. The correct cylinder was always heavy (light, for half the animals) although this could not be detected until the choice was made. So Cylinder Y would always be rewarded and heavy, and Cylinder W would always be unrewarded and light (and vice versa for half the animals). But the other stimuli would be rewarded on 50 per cent of trials, and heavy on 50 per cent of trials (depending on the other stimulus present on a given trial). When the monkeys had thoroughly mastered all four discriminations, they were tested (all choices now being rewarded) for choice between all possible pairs of stimuli, the critical test pair being B v R. This pairing is critical because each stimulus has previously served as the rewarded and non-rewarded stimulus equally often, and because it is a novel choice (no other pairing meets each of these requirements).

The monkeys showed a marked preference for cylinder B, which they selected on 90 per cent of trials. The monkeys' performance on test trials was closely comparable to that reported by Bryant and Trebasso in four-year-old children, a performance which, according to Bryant 'demonstrates conclusively that young children are capable of making genuine transitive inferences' (Bryant 1974, p. 47). Should we, then, conclude that squirrel monkeys too are capable of transitive inference, that they reason that since B is heavier (or, perhaps, better) than G, and G is heavier (or better) than R, B is heavier (better) than R? If we do, then we shall also have to accept (see p. 287) that squirrel monkeys achieve the third stage of Piaget's scheme of cognitive development, the formal operations stage (given of course, that Piaget's account is acceptable in general).

There are both empirical and theoretical grounds for withholding a decision at present. The empirical ground is to be found in the outcome of a subsequent test of McGonigle and Chalmers, in which all possible triplets from the five stimuli were tested (e.g. B v G v R, Y v B v W), the subject being allowed to select only one cylinder on each trial: over all these tests, the 'correct' stimulus was selected on only 67 per cent of trials and, in particular, when B and R were presented with either Y or W, the monkeys showed no preference for B over R (selecting each on 17 per cent of trials, where Y was the third alternative, and on 50 per cent of trials where W was the alternative). The authors' own explanation of the choice-pattern on these tests, although ingenious, is too complex and insufficiently plausible to warrant detailed exposition here; what can be agreed is that the pattern is not what should have been expected, had transitive inferences been occurring.

The first theoretical ground for not necessarily ascribing transitive inference to monkeys is that there are alternative accounts available which, while not especially plausible, might seem more plausible than the existence of transitive inference. It could, for example, be argued that training on the original discriminations of Y versus B and of R versus W resulted in some association between the members of each pair so that B came to be associated with Y (which was always rewarded) and R with W (which was never rewarded); such associations could result in a preference for B over R. The suggestion is, of course, implausible: we are being asked to believe that a stimulus signifying non-reward gains approach strength through its association in a simultaneous discrimination with the positive stimulus (whereas there is good evidence that such training actually reduces approach strength—e.g. Macphail 1972). Why, then, should we explore such apparently implausible accounts? The answer to this question is to be found in the second theoretical ground for reserving judgement, and that is, that the notion that transitive inference might occur without the mediation of language is itself thoroughly implausible. Unless, that is, we believe that the monkey is saying to itself, B is better than G, G is better than R, and so on, (and no alternative account has been proposed), it is difficult to see what is meant by transitive inference. The derivation of the inference relies on there being formal (or at least formalisable) rules for the use of the words or symbols involved, for the ordering of the symbols within each string, and so on. To suppose that monkeys possessed such a system would, then, be to suppose that they indeed possessed a rather sophisticated language, one which they do not apparently use in communication with one another.

For a second modern investigation of logical operations in non-humans, we turn again to Premack's work, and to his report of training a chimpanzee (Sarah) to comprehend the logical operator 'if-then' (Premack 1976). At an early stage of training to use this symbol, Sarah was shown, on each trial, two 'sentences', either: 'Sarah take apple if-then Mary give Sarah chocolate' and 'Sarah take banana if-then Mary no give Sarah chocolate', or: 'Sarah take apple if-then Mary no give Sarah chocolate' and 'Sarah take banana if-then Mary give Sarah chocolate'. Sarah, required to choose between an apple and a banana, learned (with some difficulty) to take the apple in response to the first pair of sentences, and the banana, in response to the second pair (Sarah much preferred chocolate to either apples or bananas). Premack then tested Sarah to see whether she had grasped the rule that the truth of the first statement of the string obtained fulfilment of the second (in Premack's studies, if the first statement was false, the second also was not fulfilled, so that the 'truth-table' for his connective differed from that of the logical operator usually symbolized by '$\supset$'; this is, however, a minor matter—it would be enough to know whether Sarah can grasp any set of rules analogous to those associated with logical operators). Premack's first test consisted of (single) sentences such as: 'Sarah eat apple if-then Mary give candy Sarah'; 'Sarah no eat apple if-then Mary give candy Sarah', Sarah again being given a choice between apple and banana on each trial. Sarah made no errors on five such sentences; but as each had the same consequence, all Sarah had to do was to 'obey' the first part of the sentence in the way that she had learned previously (avoid fruit symbolized in strings containing the symbol for 'no')—a rule which applies equally well to her if-then training sentences. A precisely similar account may be given of a further set of trials, in which Sarah had, for example, to give either a red or a green card to Mary: as the consequence was the same (and rewarding to Sarah), simply responding to the antecedent, as she had been trained to do, obtained correct responses. Further tests showed that Sarah performed very well on a series of sentences such as: 'Debby give apple Mary if-then Sarah insert cracker dish' (where Sarah was given a choice between a cracker and chocolate for insertion to the dish). Here again, however, as the antecedent was inevitably fulfilled, Sarah merely had to obey the consequent string. Premack appreciated the weakness in the foregoing tests, and designed more stringent versions which did require, for appropriate responding, attention to both antecedent and consequence; unfortunately, one set of these tests was so forbidding that Sarah refused to attempt them and another, less demanding, set was presented for eight trials only (in which Sarah made two errors)—too few results for any confident conclusions to be drawn.

Premack (1976) also reports the introduction of other symbols associated with logical inference such as those for 'or', 'and', and 'not'. There is not space to discuss these symbols here, and much of what might be said about them has been raised in earlier discussions. Sarah's goal is to solve problems involving discriminations between complex sets of symbols: labelling each symbol with a word does not alter the fact that when alternative modes of processing those symbols are available, we should ascribe the simplest of those modes to Sarah.

It is probably true that for most of us the essence of logical deduction is to be found in the syllogism (e.g. all As are B, all Bs are C, therefore all As are C); the ability to perform valid syllogisms depends on the appropriate use of quantifiers –of the concepts 'all', 'some', and 'none'. It is therefore of interest to enquire whether non-humans can acquire such concepts.

Brown, Lenneberg, and Ettlinger (1978) have reported a study on the acquisition of quantifiers in chimpanzees, rhesus monkeys, and children. The subjects were shown a number of objects (from three to nine) on each trial, all the objects on a given trial being the same colour. When the objects were all red, then the subject was rewarded if he touched each object before signalling (by touching a central lever) that his selection was complete. If the objects were all blue, then reward was obtained if no object was touched; if all the objects were green, reward followed a response to one object only, and if they were yellow, at least two, but not all the objects had to be touched. Red therefore symbolized all, blue symbolized none, green symbolized one, and yellow, some (more than one, but not all). It may be noted that none of these concepts is identical to the 'some' of the syllogism, which means 'at least one' and is not contradicted by 'all'–that is, if 'all As are B' is true, then 'some As are B' is also true.

In the Brown *et al.* (1978) report, subjects were trained with one set of objects until relatively stable, and then transferred to new sets, and it is those transfer data that are presented in most detail. The results showed that the colours for 'all' and 'none' transferred well in all three species, but that the non-humans did not show good transfer for the colour indicating 'some', and showed a lesser impairment for 'one'. It also appears that the absolute levels of performance achieved for the 'some' and 'one' symbols were considerably lower in non-humans than those for 'all' and 'none'. The Brown *et al.* data show, then, that the grasp of a concept akin to 'some' was, in chimpanzees and monkeys, very much less stable than it was in children and, also, that chimpanzees were no different from monkeys in this respect. The children (aged four to nine years) could, of course, all talk and, as the use of the symbol for 'some' approximates to the word 'some' in ordinary speech (a slightly narrower definition than that appropriate for syllogistic reasoning), it seems likely that some of the children's superiority lay in prior relevant verbal experience.

Premack (1976) reported rather more successful training on the same set of concepts with Sarah, who learned, first, to select the appropriate symbol according to whether five crackers shown to her were all a given shape, none that shape, just one that shape, or from two to four that shape. She transferred this ability to new sets of five objects where the number that were a particular colour or size was asked; finally, she responded fairly well when asked to insert (from a large number available) a certain quantity of items into a dish (although it is not clear here how she indicated that she had finished inserting items, so that there is the danger that she may have reacted to unintentional signals from her trainer).

Although there is a discrepancy between the results achieved by Brown *et al.* (1978) and Premack (1976), it may not be unreasonable to assume that chimpanzees could, with appropriate training, learn to discriminate successfully between 'all', 'none', 'one' and 'more than one but not all', and that it should be possible to train them to respond to 'at least one'. The question that arises, then, is whether this (putative) capacity would show the beginning of logical thought, a step on the road to a syllogism? One part of an answer to this question is to point out that we do not know whether this capacity is peculiar to monkeys and apes, and so have no reason to see it as a step in the sense that it is a primate specialization on which man has capitalized. A further response is that, while it may be impossible to decide whether such concepts are acquired by non-humans in the same way as by humans, their appropriate use in reasoning depends on their deployment in arguments; it will be recalled that a similar attitude was taken to the question whether non-humans acquire words—the critical question, it was argued, is: do they use words to form sentences? There is currently no evidence that chimpanzees (or any other non-humans) could use quantifiers appropriately in arguments; no-one has yet asked (experimentally) whether chimpanzees can discriminate valid from invalid syllogisms, or such inferences as 'if all spheres are blue then some spheres are blue' from, for example, 'if all spheres are blue then no spheres are blue'. We should therefore conclude that the acquisition of concepts akin to quantifiers does not in itself demonstrate a capacity for reasoning, and may not discriminate primates from any other vertebrate group.

Conclusions

This survey has found no compelling evidence to support either the proposition that non-human primates can construct sentences (or arguments) or the proposition that non-human primates possess capacities of linguistic relevance that are less well-established in other vertebrate groups (mammalian or non-mammalian). It will be observed, however, that many of the arguments deployed above have been founded on premises that are themselves highly contentious and which have exerted strong influence on the overall conclusions reached; in some cases, these premises may seem indeed to dispense in advance with any need to look at empirical data. It has been argued, for example, that the ability to acquire words (or quantifiers) does not show even the beginnings of the capacity for language (or reasoning) unless they can be shown to be used effectively in sentences (or arguments). Similarly, it has been argued that to demonstrate (and by implication, to possess) a concept of causality, an organism must possess language, and that deductive reasoning (as exemplified, for example, by transitive inference) cannot proceed except through the use of language (or some other system whose formal properties would closely resemble those of language).

No attempt will be made here to justify the assumptions listed above, as such a course would lead us deep into issues, largely philosophical in nature, concerned with the essential nature of language and of reasoning and even further from the body of experimental evidence to which we have attempted to adhere in previous chapters. It is hoped, however, that the selection and description of experiments has not been severely distorted by these assumptions and that, perhaps, their bald assertion may provoke careful evaluation of them, as well as of potential alternatives.

While, therefore, it is by no means claimed that there is proof that non-humans cannot acquire a language, we shall proceed on the assumption that our survey of the evidence has strengthened the case for supposing that humans acquire language (and non-humans do not) not because humans are (quantitatively) more intelligent, but because humans possess some species-specific mechanism (or mechanisms) which is a prerequisite of language-acquisition. The best-known recent proponent of this view is Chomsky (e.g. 1972) who has reached the same conclusion for very different reasons. Although Chomsky's position cannot be considered in detail here, it may be appropriate to draw attention to his central arguments, and, by doing so, to emphasize the fact that viewing language as a specifically human attribute is by no means a consequence simply of the outcome of language acquisition experiments using non-humans.

Analysis of a wide range of natural human languages shows a considerable degree of similarity in the grammatical principles used: 'In attempting to construct generative grammars for languages of widely varied kinds, investigators have repeatedly been led to rather similar assumptions as to form and organization of such generative systems' (Chomsky 1972, pp. 112–13). This observation points to the conclusion that, while an individual language has a particular grammar, each of those particular grammars is constrained by a 'universal grammar' whose rules must be observed by all natural languages.

Since speakers of a natural language can discriminate between grammatical and ungrammatical sentences of that language, those speakers must, at least in some sense, 'know' the rules of the grammar of their language. How could these rules be acquired? Clearly, part of the answer must lie in the speech heard by the speaker at an early age: a child learns, after all, the language spoken by those amongst whom he is reared. On the other hand, according to Chomsky, 'The native speaker has acquired a grammar on the basis of very restricted and degenerate evidence; the grammar has empirical consequences that extend far beyond the evidence' (*ibid.*, p. 27). In other words, the nature of the 'input' to the language-learner is in itself insufficient to define the particular grammar of the language unambiguously; a rule-generating device of general application (such as that which may, for example, be involved in learning-set formation) would not be able to detect reliably those rules in fact applicable to the language heard. The language-acquisition device must possess 'a very rich system of constraints on the form of a possible grammar; otherwise, it is impossible to explain how children come to construct grammars of the kind that seem empirically adequate under the given conditions of time and access to data' (*ibid.*, p. 113). The device must, that is, be structured, and structured in such a way as to reflect the universal grammar—since a child, whatever the language of its parents, can acquire any human language.

Chomsky's views have proved both stimulating—they have led directly to an entirely new field of research, now known as psycholinguistics—and controversial. We shall not pursue those views further here, although it may be fair to record that, just as traditional theories of learning of have so far failed (pace Skinner 1957) to produce any plausible theory of language acquisition, so have Chomsky and his followers failed to specify the rules of the universal grammar.

If we assume that the structural properties of the language acquisition device are 'unlearned', and that that device is species-specific, then we must suppose that the development of the device is programmed by information carried in human genes, and that both its normal development and disturbances in that development might occur relatively independent of specific environmental experiences. Such a 'biological' approach to language was adopted by Lenneberg (e.g. 1967, 1969), and two of this principal conclusions will be briefly summarized here. First, language develops in an orderly and predictable way in all human cultures, and proceeds at a comparable rate in widely differing linguistic environments. Second, there is a critical age for language acquisition, so that, for example, damage to the left hemisphere of right-handed children does not prevent acquisition (or re-acquisition) of language if the damage occurs early in life (before the teens) but later does have a severe and lasting effect; similarly, Lenneberg argues that while second-language acquisition proceeds with relative ease in young children, it becomes very much more difficult after the early teens, when much interference from the mother tongue is evident.

The views advanced by Chomsky and Lenneberg do, of course, conform to the general theme of this chapter, and have been presented in a positive light. It is, however, important to note that their account is controversial on not only theoretical, but also empirical grounds. There is evidence, first, that the speech input of children is not as 'restricted' and 'degenerate' as Chomsky has supposed (Vorster 1975); second, that there may be no 'critical' age for language acquisition, whether based on hemispheric damage data (Dennis and Whitaker 1976) or on second language studies (McLaughlin 1977). The issues discussed in this chapter remain the subjects of active investigation, and it should not be imagined that any unambiguous account is yet available.

The introduction of the relationship between brain and language leads on to what will be the final considerations of this chapter: given the possibility that language-acquisition is a specifically human capacity which qualitatively differentiates us from other vertebrates, it now makes sense to ask whether that capacity is marked by qualitative changes in functional organization of the brain in humans as opposed to non-humans.

The following two sections will consider two aspects of brain organization which may show contrasts between humans and non-humans; the first to be considered, cerebral lateralization, has a clear link with language. The second, hippocampal function, is connected with language indirectly, through the potential role in language of a discrete STM. This notion in turn leads on to the final section, which considers whether behavioural data provide support for an STM/LTM dichotomy in non-human mammals.

Cerebral lateralization

Certain basic generalizations concerning the representation of speech in the human hemispheres are well-established. First, the great majority (95 per cent or more) of right-handed people (dextrals) show left cerebral dominance for speech in the sense that they show speech disorders following interference with or damage to the left, but not the right hemisphere; left-handers (sinistrals), and people of mixed hand and foot preferences (as well as some dextrals with sinistral relatives), are more likely to show bilateral speech representation (seen in about 10 per cent of sinistrals) or right cerebral dominance (seen in about 20 per cent of sinistrals) (e.g. Hicks and Kinsbourne 1978). Second, there is within the dominant hemisphere a broad band of neocortex, extending from the frontal lobe back via the parietal lobe to the temporal lobe, which is concerned with speech function (e.g. Geschwind 1970). The anterior part of this band is known as Broca's area, and is concerned primarily with the production of speech; the posterior part, Wernicke's area, is concerned rather with perception and comprehension of speech. The location of these areas is shown on Fig. 8.1.

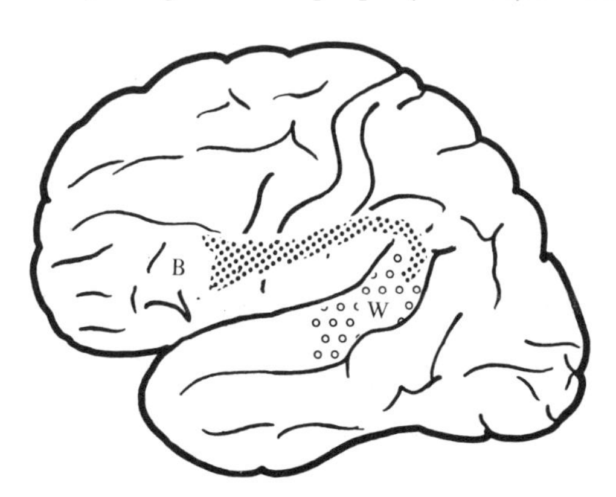

Fig. 8.1. Lateral view of the human left hemisphere, showing Broca's area (B), which lies anterior to the lower end of the motor cortex, and Wernicke's area (W). (From Geschwind 1970.)

It is pertinent to our interest in the question whether language is a specifically human innovation to ask if any aspects of the neural representation of speech are foreshadowed in non-humans. As non-humans do not talk (and none of the subjects of the various language-acquisition programmes has been subjected to cerebral damage), the only directly relevant evidence concerns lateralization in non-humans. Before discussing the evidence, one point should be made about the localization of speech processes: the survey in Chapter 7 of the effects of cortical lesions on general intelligence suggested that, if there is indeed such a thing as general intelligence, then it is not localized in any particular cortical region in any mammal; if, therefore, language-acquisition is a consequence solely of the possession of high general intelligence, it may seem odd that relatively small left-hemisphere lesions in dextral adults lead to profound and lasting aphasias.

Consistent hand or paw preferences have been reported in a number of non-human mammalian species, but there appear to be a series of major differences between these preferences and handedness in humans. First, the reports agree in finding that left-and right-hand (or paw) preferences are equally common within a species (whereas only about 10 per cent of humans are left-handed) (Annett 1972); second, there is no evidence that any animal is more skilled with one hand (or paw) than with the other; third, non-human preferences may readily be modified by experience (e.g. Warren 1977, using monkeys; Collins 1975, using mice, *Mus musculus*) and, finally, while such preferences may be consistent for a given non-human individual across a series of tests using the same task, non-humans (or monkeys, at least) do not show a consistent hand-preference across a series of different tasks (Warren 1977).

Doubts about the comparability of handedness between humans and non-humans suggest that we might not expect to find any correlation in non-humans between handedness and the effects of cerebral lesions. In monkeys, at least, this negative expectation is, to date, realized: although unilateral frontal cortex lesions do disrupt delayed response tasks and unilateral inferotemporal lesions, visual discrimination tasks, the magnitude of the disruption does not vary according to whether the lesion is of the hemisphere ipsilateral or contralateral to the preferred hand (Warren and Nonnemann 1976). There is, however, a report (Webster 1977) that retention of preoperatively acquired visual discriminations in split-brain cats (cats whose corpus callosum and optic chiasma had been vertically severed) was superior when tested with the hemisphere ipsilateral to the preferred paw; in this study, each hemisphere was tested independently by temporarily occluding one eye, and paw preference was established preoperatively. This non-dominant hemisphere superiority might be thought to parallel the human non-dominant hemisphere's superiority in certain aspects of non-verbal visuospatial tasks (e.g. Nebes 1974). Although Webster's results are of considerable potential significance, they should perhaps be treated with some caution at this stage: the study used only eight cats in all, and a non-dominant hemisphere superiority was evident on only five of the eight problems presented. Moreover, a report by Hamilton (1977) using a similar design with rhesus monkeys failed to find any indication of dominance by either hemisphere (ipsilateral or contralateral to the preferred hand) in retention of easy or difficult visual discriminations, or in the acquisition of new perceptual discriminations.

There are a few other reports suggesting asymmetry in non-human hemispheric function, without touching on the question of hand or paw preference: we shall consider, first, three recent reports on rodents. Glick, Weaver, and Meibach (1980) report that rats with electrodes implanted in each lateral hypothalamus (left and right) tended to circle during self-stimulation in a preferential direction, independent of the side stimulated; thresholds for self-stimulation were lower on the side contralateral to the direction of rotation, and Glick *et al*. suggest that the two sides of the brain may be specialized for high and low mood. There was no evidence in this report of any species (as opposed to individual) preference for a particular side of stimulation. The two other reports do suggest a species asymmetry in rats. Robinson (1979) found that ligation of the right middle cerebral artery resulted in a restricted lesion of right cerebral cortex, and in hyperactivity and a reduction (bilaterally) in cortical noradrenalin; ligation of the left middle cerebral artery did not cause either hyperactivity or noradrenalin-depletion, although it did cause a lesion (of comparable size) in the left cortex. This report implies, then, that the rat's right cerebral cortex is (unlike the left cortex) involved in the inhibition of activity and in the maintenance of cortical noradrenalin levels. Finally, Sherman, Garbanati, Rosen, Yutzey, and Denenberg (1980) provide support for a somewhat complex series of propositions: first, that the rat right cerebral cortex is dominant for the determination of spatial preferences; second, that this dominance is normally counteracted by an inhibitory action of the left cortex; third, that handling rats in infancy removes the inhibitory influence of the left cortex. Sherman *et al*. also cite evidence for a similar lateralization (also normally obscured by the inhibitory influence on the left cortex) of fear and aggressive motivation in the right cortex. Although all three reports agree in finding cerebral lateralization in rats, it does not yet seem possible to integrate the findings so as to provide a unified account of dominance in rodents.

Two studies which suggest parallels in lateralization between monkeys and man have been reported by Dewson (1977) and Petersen, Beecher, Zoloth, Moody, and Stebbins (1978). Dewson trained rhesus monkeys preoperatively on a delayed conditional discrimination, in which the subject first heard either a tone or a burst of noise and then, after a delay, had to choose between a red and a green light, the correct choice depending on which auditory stimulus had been presented. The subjects (five in all) underwent surgery in which either the left or the right superior temporal gyrus (thought to be the homologue of part of Wernicke's area in humans) was damaged. Subsequent tests showed that the monkeys with left cerebral damage, while discriminating normally at zero delay, showed marked impairment, relative to those with right hemisphere damage, as the delay increased: in other words, there is evidence (necessarily preliminary, in view of the very small numbers) that the monkey left hemisphere may be dominant for some aspect of auditory recent memory.

Petersen *et al*. (1978) presented species-specific vocalizations of Japanese macaques (*Macaca fuscata*) monaurally to either the left or the right ear of five *M. fuscata* and of five other monkeys (one vervet, *Cercopithicus aethiops*, two pigtail macaques, and two bonnet macaques, *M. radiata*). Two types of call were used, both 'smooth coos'; the communicative significance of these for *M. fuscata* varies according to whether the peak frequency of the coo is early (the smooth early—SE—call) or late (SL). The initial pitch of the coos, which is not of communicative significance, varied in an unpredictable way, and the monkeys' task was to respond differentially, for food reward, according to whether an SE or an SL coo was heard. All five Japanese macaques showed a significant right-ear advantage (REA) on this task—were, that is, more accurate when the coos were presented to the right than to the left ear, and this is taken to indicate more efficient left cerebral processing of the stimuli. Four of the five other monkeys showed no systematic ear advantage, although the vervet also showed an REA. In an attempt to find out whether this result was due to the communicative significance of the calls used, Petersen *et al.* went on to run a further discrimination using the same calls, but in which the latency of the peak was now irrelevant, and the initial pitch of the coos was related to the correct response. The two Japanese macaques tested on this task each showed a significant shift in ear advantage, one of them now exhibiting a significant LEA; the two other monkeys tested (one *M. nemestrina*, one *M. radiata*) showed no change from the peak latency discrimination task, both continuing to show no significant ear advantage.

The Petersen *et al.* (1978) results come closer than any others that we have considered to suggesting a parallel in non-humans to the cerebral lateralization of language in humans. The numbers are, of course, very small: we do not yet know whether vervets in general would show an REA for SE versus SL calls, or whether the REA found in the solitary vervet studied would have been eliminated in the pitch-discrimination version of the task. Moreover, we cannot be sure that Japanese macaques in general will show a significantly larger REA for SE versus SL discriminations than for pitch discriminations (given data for only two subjects so far). If subsequent work confirms the pattern of these early findings, they are, of course, bound to have a considerable influence on theories of the evolution of cerebral dominance.

A final set of observations which may be relevant to the phylogeny of the human 'speech areas' concerns gross anatomical asymmetries in the brain. One such asymmetry in the human brain is found between the left and right planum temporale (a cortical region within the superior temporal area which forms part of Wernicke's area), the left planum being larger in 65 per cent of human brains (and smaller in only 11 per cent) (Galaburda, LeMay, Kemper, and Geschwind 1978); it is natural to associate this asymmetry with left cerebral dominance for language although, as Galaburda *et al.* point out, there is the difficulty that whereas approximately 4 per cent of aphasics have right hemisphere damage, 11 per cent of humans have a larger right planum temporale. A related asymmetry has been reported by Yeni-Komshian and Benson (1976), who measured the length of the Sylvian fissure (which varies with the size of the planum temporale, whose precise extent is not unambiguously measurable in non-humans), finding a longer left Sylvian fissure in both humans (a 15 per cent difference) and chimpanzees (a 5 per cent difference), but not in rhesus monkeys (a non-significant difference of 1 per cent).

Although it appears from this survey that cerebral asymmetries of function and anatomy have been found in non-humans, it is clear that no very general conclusions have yet emerged. There is some evidence for an association between handedness and cerebral dominance in cats (Webster 1977), but not in monkeys (Hamilton 1977; Warren 1977); there is evidence for left hemispheric dominance for complex auditory tasks in monkeys (Dewson 1977; Petersen *et al.* 1978); on the other hand, monkeys, unlike chimpanzees and humans, do not show anatomical asymmetry in the temporal region. Finally, there is evidence for cerebral asymmetry of function in rodents.

It is not only unclear to what extent cerebral specialization occurs in non-humans, but also unclear what implications such asymmetry (or lack of it) would have for language in non-humans. It may be tempting to suppose that language imposes such demands on information-processing that it is necessary to increase efficiency by processing language in one hemisphere, while processing other types of information in the other. But it should be remembered that humans who do not exhibit cerebral dominance nevertheless acquire language efficiently, and are not impaired in general intelligence (Newcombe and Ratcliff 1973). Cerebral dominance may or may not be a peculiarly human characteristic, and it may or may not be a prerequisite for the evolution of language. The data available at present do not, however, make a decisive contribution to the question whether steps on the road to language are to be seen in other vertebrates, and, in particular, in primates closely related to man.

The hippocamal formation and memory

Both this and the following section will be concerned with memory processes, and in particular with the concept of an STM/LTM dichotomy and with active rehearsal processes. It was argued in Chapter 6 that investigation of recent memory storage is relevant to analysis of intellectual processes, for two main reasons: first, any limitations to the amount of information that could be manipulated simultaneously might impose limitations on general problem-solving processes; second, possession of a mechanism enabling the active maintenance of the strength of recent traces could contribute to the solution of problems for which retention of recent trial outcomes would be beneficial. A further reason for our renewed investigation of memory processes here is that STM may be particularly relevant to language: it is clear that the comprehension of sentences requires the retention and simultaneous processing of a series of words, since, for example, the meaning of a given word may not become unambiguous until the occurrence of some other word at a much later stage in a sentence. It could be that one role of a discrete STM might be to hold word-strings (of, say, phrase- or sentence-length) until the semantic interpretation of the string is complete, and then to transfer that interpretation (rather than the words themselves) to LTM. Such a view could be supported by the observation that the retention of words is subject to acoustic confusions (errors involving words that sound similar) in immediate recall, and to semantic confusions (involving words having similar meanings) in delayed recall (e.g. Baddeley 1966*a,b*).

This section discusses the nature of hippocampal function in human and non-human mammals, since there appears, superficially at least, to be a contrast between the effects of hippocampal damage in humans and non-humans, and there are reasons for supposing that the effects of human hippocampal damage may provide evidence for a distinction between STM and LTM.

It will be recalled that a brief description was given in Chapter 2 (p. 33) of the nature of the anterograde amnesia found in the patient H.M. following bilateral hippocampal resection. To recapitulate, it appeared that H.M. could recall events that occurred prior to his operation but could not recall for more than a few seconds events that occurred subsequently (unless given the opportunity to maintain those memories by constant rehearsal). Although subsequent research has made it clear that H.M.'s deficit is, as might have been anticipated, very much more complicated than at first appeared (see Milner, Corkin, and Teuber 1968, for a review), it is nevertheless clearly the case that his long-term retention of everyday events is severely impaired, and no comparable deficit has been reported in non-humans having hippocampal damage. The appropriate characterization of deficits that are found in non-human hippocampal mammals is a matter of controversy, but two types of task that reveal deficits in hippocampal animals are, first, tasks that appear to involve the inhibition of prepotent responses—as in extinction, reversal learning, and punishment, and , second, maze-learning problems; whether these deficits, perseverative and spatial, are independent or both aspects of a common underlying deficit is a matter of dispute, but what is not in dispute is that long-term retention of discriminations is not disrupted in non-human hippocampal animals. It is of interest to know whether the differences in effects of hippocampal damage can be reconciled since, if not, the possibility will emerge that the differences reflect contrasts in the organization of memory in humans and non-humans, and this, given the potential role of STM in comprehension of speech, is relevant to our current interest in language.

There have been attempts to reconcile the human and non-human data on anatomical grounds, by arguing that the critical area for the human amnesic syndrome is not simply the hippocampus (Mishkin 1978; Horel 1978). Mishkin, for example, provides evidence that although damage confined to the hippocampus in rhesus monkeys does not disrupt retention (in a delayed recognition task), combined lesions of the hippocampus and amygdala do, and suggests that the human data be re-examined to see to what extent the amygdala has been involved. Horel argues that the 'temporal stem' (or 'albal stalk'), a bundle of fibres in the medial temporal lobe that contains fibres to and from the amygdala and temporal neocortex (but not the hippocampus), is the critical site for the amnesic syndrome in humans, and that memory deficits do occur in monkeys with comparable damage.

The anatomical arguments are complex, and hindered by the fact that for many cases (including H.M.) the necessary pathological information is not available. One complication is that, once the consequences of bilateral temporal lesions in humans were discovered, the operation was no longer carried out and in fact the majority of humans in which the amnesic syndrome has been investigated are victims of Korsakoff's psychosis, generally the consequence of prolonged alcoholism. Patients with Korsakoff's psychosis usually show damage to the mammillary bodies (in the posterior hypothalamus) and to dorsomedial thalamic regions, without displaying overt hippocampal damage. As, however, both the mammillary bodies and the dorsomedial thalamus are targets of hippocampal efferents (via the post-commissural fornix, which originates in the subiculum, a region of allo-cortex adjoining the hippocampus from which it receives a rich innervation), and as the amnesic syndrome seen in humans with hippocampal damage appears closely similar to that found in Korsakoff's psychosis, most workers have taken the view that all of these amnesic humans should be seen as exhibiting the consequences of disruption of normal hippocampal function. It is clear that the behavioural syndrome may emerge without overt damage to the hippocampus, the amygdala, or the temporal stem; it is not, however, yet clear whether the syndrome is found in cases where pathology indicates damage to a region having connections with only one (or even two) of these regions, and so, not possible to resolve the anatomical dispute at this stage. It also seems likely in fact that what would most convincingly resolve the dispute would be the demonstration of a syndrome in experimental non-human subjects that would be generally accepted as closely comparable to the human amnesic syndrome; although Mishkin (1978) and Horel (1978) do report some suggestive data, they are as yet far from conclusive. We shall therefore adopt what is the majority current view, and go on to examine the human amnesic syndrome in more detail, to see whether its symptoms can in fact be reconciled with those of hippocampal damage in non-humans.

The description provided this far of the amnesic syndrome would allow the interpretation that hippocampal humans can enter information into a short-term store, with a life of a few seconds, can retrieve information from both STM and LTM, but cannot transfer information from STM to LTM (e.g. Baddeley 1972). Such an account might be reconciled with the work on hippocampal lesions in non-humans on the assumption that there is, in non-humans, no comparable distinction between STM and LTM, and this is a possibility that we shall consider in the following section. There are, however, serious problems in that interpretation of the human data, since recent research has shown that in fact amnesic humans can, under certain circumstances, store information over long intervals.

One powerful technique for the demonstration of long-term retention in amnesics is the cueing technique, developed by Warrington and Weiskrantz (e.g. 1968). This method requires subjects to recall stored information when given cues or partial information, about the nature of the item to be recalled. The cues may take the form, for example, of fragments of a to-be-recognized picture, or of the first two or three letters of a word. Weiskrantz (1978), in a review of this work, points out that the beneficial effects of cueing are significantly more marked in amnesics than in control subjects. It is clear that cueing could not be effective at all unless the relevant information had been stored originally; why, however, should that original storage be inadequate for normal recall in amnesics? Weiskrantz (1978) argues that cues should be of use where more than one alternative response is available and a person cannot discriminate between the alternatives—where, in other words, distractor items are competing or interfering with the correct item. One possibility is, then, that amnesics have not so much a storage as a retrieval problem, occasioned by excessive interference from irrelevant information. Weiskrantz cites other results favouring this interpretation, such as, for example, the finding that amnesics in a free-recall task (on which they performed very badly) made several intrusion errors, in which the incorrect response was a word that had occurred in a list some days previously.

Weiskrantz goes on to argue that many of the phenomena associated with hippocampal damage in non-humans can be interpreted as being the consequence of excessive interference between prior and current experiences; in general, for example, the tasks which demonstrate 'perseverative' deficits are tasks in which initial experiences of rewards (or safety) are now contradicted by experiences of non-reward (or punishment). It may be that the hippocampal animal is not incapable of inhibiting his response so much as incapable of overriding the earlier experience, which carries conflicting implications for response.

The 'interference' account of hippocampal function has led to a series of important observations in human amnesics, but it may still be doubted whether it is sufficient to close the gap between humans and non-humans. One problem faced by the account is that most (but not all) humans with an amnesic syndrome show a fair degree of retention of events that occurred prior to the onset of the syndrome, at least in comparison to their retention of events subsequent to its onset. It is difficult to see why early experiences should, at retrieval, be released from the effects of interference which so drastically affects recall of later events. A second problem is that at least some of the features of the syndrome seen in non-humans do not seem to be explicable in any simple way as being a consequence of interference. Mahut and Zola (1973), for example, have reported that although spatial reversals are disrupted by lesions within the hippocampal system, object discrimination reversals are not; O'Keefe and Nadel (1978), in a review of the reversal literature point out, moreover, that where non-spatial reversal deficits are obtained they do not appear to be caused by persistent choice of the formerly rewarded stimulus (perseveration), so much as by the adoption of an (inappropriate) position habit. There is no evidence in the human literature that interference effects are restricted to spatial information (and very good evidence that they are not). Neither of these difficulties for the interference hypothesis rules out its possible application to both humans and non-humans with hippocampal damage; they may, however, help to indicate why the hypothesis has yet to gain universal acceptance. At present, it still seems possible to argue that consolidation (that is, transfer from STM to LTM), while severely impaired in amnesic humans, is not entirely prevented, and that cueing simply helps to retrieve the rather weak traces so formed. It may be objected (Weiskrantz 1978) that, if this were so, then amnesics would not be differentially affected by cueing procedures, aided, that is, relatively more than controls: but a similar argument applies to the interference hypothesis, unless it is supposed that proactive interference is not a major factor in normal humans' forgetting also.

A further account of the amnesic syndrome proposes that it is a consequence of inadequate encoding of new information by amnesics, that encoding being superficial, relying on surface features, rather than deep, or semantic, in nature (e.g. Butters and Cermak 1975). According to this view, both short-term and long-term retention are defective in amnesics, and although one report (Baddeley and Warrington 1970) found a normal rate of decay of information in amnesics over delays up to 18 seconds (with rehearsal prevented), Butters and his colleagues have on a number of occasions found significantly more rapid decay in amnesics (Butters and Cermak 1975). It is not clear why the discrepancy between these studies has arisen, but the reason may lie in differences in the aetiologies of the patients used, or in techniques employed for the prevention of rehearsal. While a deficit in short-term retention would not weaken an interference hypothesis, it would, of course, contradict the notion that the amnesics' principal difficulty lay in transferring information from STM to LTM.

Butters and Cermak cite a number of studies which suggest that semantic encoding is deficient in amnesics, and, to give a flavour of the nature of that deficit, one of these reports will be briefly described here. When given a series of short-term recall trials using a given category of material, subjects (normals) show a progressively poorer level of recall; Wickens (1970) has shown that performance may be restored to its initial (high) level by abruptly changing the category from which the material is selected, a phenomenon known as release from proactive inhibition. The extent to which release is obtained may, therefore, give a measure of the degree to which the stimulus material is differentially categorized. Cermak, Butters, and Moreines (1974) showed that Korsakoff patients, like controls (alcoholics without an amnesic syndrome), showed release from proactive inhibition when the to-be-recalled stimuli were changed from consonant trigrams (e.g. RQF) to three-digit numbers (e.g. 813); Korsakoff patients did not, however (unlike controls), show any release when the material was altered from one semantic category (e.g. animals) to another (e.g. vegetables). It therefore appears that, while the Korsakoff patients did categorize the stimulus material at a gross level, allowing discrimination between consonants and digits, they did not encode the material at the deeper level required to discriminate between words from different semantic classes. This encoding deficit, according to Butters and Cermak, is in itself sufficient to explain poor retention and high susceptibility to interference since they take the view that the retention of information varies according to the depth to which it is processed (e.g. Craik and Lockhart 1972)—a view which rejects the notion that there are distinct short-term and long-term memory stores.

There appear to be two ways in which the encoding hypothesis might regard the non-human data. The first is simply to argue that since non-humans do not possess language, there is no sense in which they might be supposed to process information at a deep or semantic level and so, no reason why hippocampal damage should affect either their encoding or their memory. In other words, the contrast between the effects of hippocampal damage would reflect a real difference in psychological processing. A difficulty with this interpretation is that there is good evidence that hippocampal damage in man disrupts retention of non-verbal material and, indeed, patients with unilateral damage to the right hippocampus, while showing normal retention of verbal material, are impaired on retention of non-verbal material (e.g. Milner 1971). A report by Dricker, Butters, Berman, Samuels, and Carey (1978) of impaired face recognition by Korsakoff patients and by patients having right hemispheric damage, suggests that deficits in non-verbal retention may again reflect differences in encoding: it appeared that these patients 'matched faces on the basis of superficial features such as paraphernalia and expression rather than the deeper configurational characteristics of faces' (Dricker *et al*. 1978, p. 683). The absence of language *per se* does not, then, rule out a distinction between superficial and deep processing of information.

An alternative approach has been proposed by O'Keefe and Nadel (1978), whose principal thesis is that the hippocampus functions as a spatial mapping system, by means of which an animal may instantly apprehend the spatial relationships holding between a given place (recognized from a variety of cues) and other places known to the animal. Support for this view is derived both from the evidence of spatial disruption following hippocampal lesions, and from electrophysiological data, showing that there are in the hippocampus individual neurones which fire at certain spatial locations, relatively independent of the stimuli present at those locations. O'Keefe and Nadel draw an analogy between spatial mapping and the nature of the process, of central interest to psycholinguistics, involved in mapping superficial structures of sentences onto their underlying deep structures; they suggest that this 'cognitive' mapping also takes place in the hippocampus, so that humans with hippocampal damage would not only be spatially disoriented (unable, like amnesic patients, to learn their way about hospitals) but also deficient in encoding information at a deep level—and so, according to the Craik and Lockhart view, incapable of long-term retention of novel information.

The interpretations advanced by O'Keefe and Nadel for the non-human data have been immensely stimulating, but remain controversial (see, for example, Moore 1979; Solomon 1979) and currently rank as perhaps the most intriguing of a number of alternative hypotheses concerning hippocampal function in non-humans. No attempt will be made here to evaluate those theories—what concerns us is whether the argument by analogy from spatial to cognitive maps is a convincing one. To this writer, it is not, and this principally because the human amnesics' problems appear to be entirely with memory, and not with understanding. Although some amnesics show an impairment in IQ scores, such deficits are presumed to be a consequence of widespread organic damage, and to be independent of the amnesia. It is quite common to find amnesics whose IQ is unimpaired, and these patients show perfectly good comprehension of speech; it is hard to see how such comprehension would be possible in the absence of a capacity to derive deep structure from surface structure. A similar argument applies to the spatial disorientation of amnesics: H.M., although he could not learn the location of his new home, frequently did find his way back to his old home (the place in which he lived prior to his operation). H.M.'s difficulty appears to lie, not in the loss of a map, but in the inability to make new entries onto the map.

Neither the interference hypothesis nor the encoding hypothesis at present provides a wholly convincing account of the disparity between the effects of hippocampal lesions in man and non-humans; it therefore is of interest to inquire whether that disparity may be due to there being a radical difference between humans and non-humans in the organization of recent memory storage. We have seen that there are serious difficulties with the view that the human amnesic syndrome simply reflects a disruption of the transfer of information from STM to LTM, but we shall pursue the point since there is an interest quite beyond that of the interpretation of hippocampal lesions in the question whether non-human mammals show evidence of possessing discrete short-term and long-term memory stores.

Before leaving this section, mention must be made of a feature of amnesics' performance which Weiskrantz (1978) has emphasized, namely, the fact that amnesics frequently exhibit memories of which they themselves are quite unaware. In Weiskrantz's words: 'By now it is known that amnesic subjects can learn and remember a variety of tasks when they are tested in particular ways. The only person who remains unconvinced about this is the amnesic subject himself, who persists in failing to acknowledge that his performance is based on specific past experience, or may occasionally confabulate about such a basis' (Weiskrantz 1978, p. 382). This dissociation between the amnesic's performance and his conscious knowledge suggests that his failure may result from a lack of communication between different 'levels' of processing, rather than from a failure within any particular level. Weiskrantz concludes his discussion, as we shall ours, with this challenging question: 'How would we ourselves recognize such a dissociation in an animal when it occurred?' (*ibid*., p. 385).

Memory in non-human mammals

This section will discuss behavioural studies of recent memory in non-human mammals, with particular interest in whether these animals show evidence for possession of a discrete STM, and whether they are capable of active maintenance of recent traces. It will be recalled that the corresponding survey in Chapter 6 found rather little support for either a discrete STM or rehearsal in birds, and the findings in birds will be contrasted here with those in mammals; it will also be borne in mind that the preceding section has raised the possibility that differential effects of hippocampal damage in humans and non-humans may reflect differential modes of memory organization.

Pigeons, unlike humans, did not appear to show a reasonably well-defined limit to the number of items that they were able to remember accurately over a short interval, and this was taken as evidence counter to the notion that there is a pigeon STM with limited capacity. Non-human mammals given lists of items to retain give results similar to those found in pigeons. Thompson and Herman (1977) explored short-term retention of lists (from two to six items) of novel sounds in bottlenosed dolphins (*Tursiops truncatus*) finding that performance decreased in a relatively orderly way across the lists, there being no evidence of any sudden drop in performance as if some critical number of items had been exceeded. Olton (e.g. 1978) has described a radial maze in which each arm has a baited goal box, and rats are trained to visit each goal-box once only on a given trial (since once the bait is removed, it is not replaced until the following trial). Rats are remarkably efficient in this task which requires, for its performance, short-term retention of all the arms that have previously been visited on a given trial. In general, performance declines gradually as the number of arms is increased but no precipitous drop is observed at any point, and, indeed, performance remains above chance levels even when 17 arms are used (Olton, Collison, and Werz 1977). Non-human mammals have not, then, exhibited anything akin to the memory span of humans.

A second apparent distinction between pigeons and humans lay in the absence in pigeons of a 'primacy' effect in recall (or recognition) of lists of items. Here, too, dolphins and rats appear to resemble pigeons rather than humans, as neither Thompson and Herman (1977) nor Olton (1978) has found any sign of primacy in their studies, each of which showed strong recency effects: that is, the more recently a sound had been heard (or an arm entered) the better it was remembered, and the first item in a list enjoyed no advantage over its successor items.

A primacy effect has, however, been obtained in a study of recognition memory in a rhesus monkey (Sands and Wright 1980). This animal was shown, in succession, 10 photographs of objects or scenes (from an ensemble of 211 photographs) on each trial, and then, after a brief delay, shown one test photograph; the monkey's task was to press a lever in one direction if the photograph was one which had been shown in the list, and in another direction if it was not. The animal showed a high overall level of accuracy (86 per cent correct) and showed (on trials on which the test photograph was drawn from the exposed list), a serial position curve having marked primacy and recency effects. If the occurrence of primacy is taken to reflect rehearsal processes (and, as Sands and Wright point out, other interpretations are available) then these findings support the notion that monkeys rehearse. It is premature to conclude that the contrasts between these results and those found in dolphins and rats are due to species (rather than procedural) differences; humans, for whom evidence of rehearsal is stronger than in any non-human, do not, after all, usually show primacy in retention of lists of non-verbalized material (see p. 220) and there are reports (e.g. Gaffan and Weiskrantz 1980) of list-learning in rhesus monkeys in which primacy has not been observed. Nevertheless, it is the case that a primacy effect in retention of a recent list of items has so far been demonstrated only in a primate (and in man). We shall now, however, consider two types of evidence (the effects of spacing, and of procedures designed to disrupt rehearsal) that favour the existence of an active or rehearsal-like process in non-primate mammals.

Roberts (1974) trained rats on a version of delayed alternation, in which the animals were given one, two, or four forced trials on which they were compelled (by a barrier) to go on one side of a T-maze, followed, after a delay, by a free choice between the two arms of the maze, the correct arm being that which they had not been forced to enter. Roberts found that increasing the number of forced trials improved performance, and that optimal performance was obtained when those trials were spaced (with a 60 s ITI) rather than massed (immediate replacement). It therefore appears that the spaced repetition of information may benefit retention in rats, as in humans (Bjork 1970). A facilitatory effect of spaced repetition (of a sample presentation) in pigtail macaques performing DMTS has also been reported (Medin 1974), although Medin found an advantage in spaced repetition only where the stimuli used were (relatively) novel on each trial. As one interpretation of the effects of spacing in humans is that it increases the opportunity for rehearsal (e.g. Rundus 1971), these results can be taken to support the view that rehearsal occurs in non-human mammals also.

The most direct evidence for a rehearsal-like process in non-humans comes from a series of experiments conducted by Wagner, Rudy, and Whitlow (1973), on eyelid conditioning in rabbits. These experiments explored the effects of various post-trial events (PTEs) on the course of acquisition of a CR to a novel CS. The PTEs concerned were presentations of CSs which, as a result of previous training, were either conditioned excitors (signalling shock) or conditioned inhibitors (signalling absence of the shock). Each presentation could be followed by shock or absence of shock. Where the excitatory CS was followed by shock, or the inhibitory CS was followed by no shock, the PTE was described as congruent; where the opposite relation held (the inhibitor, for example, being followed by shock) the PTE was incongruent. The PTEs could occur at various times (from 3 to 300 s) after each pairing of the novel CS and shock, so that their retroactive effect on the learning engendered by those pairings could be assessed. The results showed that incongruent PTEs disrupted the course of acquisition of the CR to the novel CS, and that the degree of disruption varied according to the interval between the pairings and the PTE, being greater, the shorter that interval; congruent PTEs, which contained the same elements but in their customary relation to one another, did not effect learning. In other words, incongruent (surprising or unexpected) events which occur shortly after some novel episode disrupt learning about that episode.

One interpretation of the Wagner *et al.* data is that novel episodes initiative active (rehearsal-like) processes which, if allowed to proceed without interruption, lead to the storage of information about those episodes. Congruent PTEs, containing no novel information, do not themselves require active processing, and so do not compete with processing already under way; incongruent PTEs, themselves novel, displace items currently being processed, and are themselves rehearsed instead. Such an explanation implies both limited capacity of the rehearsal mechanism and (since no effects are found where long intervals precede PTEs) decay of information within that mechanism—the two basic properties conventionally ascribed to human STM.

These results and their interpretation form but a part of a very much larger (and still developing) body of theorizing originated by Wagner and his colleagues (see, for example, Wagner 1976, 1981), which has broad application to many areas of learning and memory, and not simply to the distinction between STM and LTM. There is space here only to recommend examination of the original papers, and to note that there are, of course, contrary views and troublesome findings. Two such findings, directly relevant to the work described above, will be noted here. First, where extraneous intra-delay stimulation disrupts DMTS performance in monkeys, it appears immaterial whether the interfering stimulation occurs early or late in the delay (e.g. Moise 1970; Etkin 1972). Now if such interference was due to the disruption of rehearsal and the displacement of relevant material from STM, then its effects should be greater at the beginning rather than at the end of the delay interval; this is because successful rehearsal at the beginning of the interval should allow material to be retained in LTM, immune to any subsequent interference with rehearsal processes. Second, on at least some occasions, 'surprising' events facilitate processing of a preceding episode (e.g. Dickinson, Hall, and Mackintosh 1976), and this has led Kremer (1979) to the following conclusion: 'As an empirical generalization, it appears that a PTE is likely to produce interference with conditioning when substantial associative learning would otherwise occur but that it may produce facilitation of conditioning when less associative learning would otherwise occur' (Kremer 1979, p. 140). Now it is, of course, very difficult to see how any event, competing for access to a rehearsal mechanism, could facilitate the use by any other event of that same mechanism.

Before leaving the general topic of PTEs, it should be noted that one class of PTEs having retroactive amnestic effects, physiological manipulations such as electroconvulsive shock and administration of certain drugs, has been excluded from consideration here (and a similar exclusion occurred in the chapters on fish and birds, as experiments using such manipulations have also been conducted using animals of those classes). In brief, these physiological techniques will not be discussed largely because it does not appear that the effects obtained are relevant to the identification of any store which might be comparable to human STM. Two main reasons for this conclusion are, first, that the temporal gradients of the effects concerned seem much longer than those appropriate to STM (e.g. Schneider, Kapp, Aron, and Jarvik 1969; Squire and Cohen 1979) and, second, that the amnesias seen may reflect failure of retrieval rather than of rehearsal or consolidation (e.g. Miller and Springer 1973).

This section can pretend to do no more than point to the STM–LTM distinction as an area of interest and of active current investigation. There is as yet no evidence for a fixed memory span in non-human mammals (or birds); there are, however, a number of findings that suggest the existence of active rehearsal in non-human mammals and these provide much better evidence for rehearsal than was found in birds.

The study of recent memory is, as has been argued, of considerable potential importance to general theories of intelligence (and language), and it is encouraging to see that there has been a considerable increase in research in this area over recent years. At present, however, it is clear that no conclusions on potential species differences in recent memory processes are yet well-established; finally, of course, it is still very possible that there is in any case no real dichotomy in humans between STM and LTM.

Summary and conclusions

It has been argued in this chapter that language plays a large role in human problem-solving, a role so important and pervasive that it is not possible to estimate human general intellectual capacity independent of linguistic capacity, and so, not possible to compare directly human intelligence with that of non-linguistic animals. A critical question is whether language is a consequence of superior general intelligence, or whether it is independent of that capacity, combining with it to produce a qualitatively superior problem-solving ability. An indirect approach to this question is the investigation of language-acquisition in non-humans; if language is a product of general mechanisms of intelligence, then it might be expected that non-humans should show some ability for the mastery of language. The survey of language-learning by apes came, however, to generally negative conclusions, particularly concerning the acquisition of grammar; the evidence currently available did not appear to support the view that apes are capable of language acquisition in any important sense.

Similar conclusions were reached when two related issues, intentional communication and reasoning, were considered: although it was argued that neither of these phenomena has yet been demonstrated in non-humans, it should be pointed out that the conclusions were in the main arrived at, not because of failure by the animals to perform appropriately, but because alternative accounts of their performance, appealing to simpler processes, were available. What is clear is that it is in fact extremely difficult to conceive experiments that could conclusively show communication or reasoning but would not overtly involve language.

The apparent failure of apes to master language lent support to the view that humans possess a species-specific language-acquisition device, and consideration of this possibility led on to the question whether the human brain showed specializations that were related to language. Two topics were discussed, cerebral lateralization and hippocampal function. Evidence for some degree of cortical lateralization of function was found in a number of species including not only primates, but also cats and rats, although the most striking parallels with human lateralization were found in macaques. The question whether hippocampal lesions obtain similar deficits in man and non-humans depends heavily on the interpretation of the deficits seen in man, and that remains a controversial issue: if it is assumed that the principal difficulty of humans with hippocampal damage in fact lies in an inability to transfer information from an STM to an LTM, then it is difficult to reconcile the human with the non-human syndrome.

The final section surveyed behavioural studies of recent memory in non-human mammals, finding stronger evidence for a process of rehearsal than had been the case for birds; however the data are at present inadequate to allow a definite conclusion on the existence or otherwise of an STM/LTM dichotomy in non-humans.

The general tenor of this chapter has been in accordance with the view that language is a peculiarly human phenomenon, and it may be appropriate to introduce here a brief reference to an issue relevant to the plausibility of that view. If we conclude that grammar is wholly absent in apes, then there are no behavioural data to show how language evolved: modern languages are all highly sophisticated (no known language appears to be more primitive than any other—see Lyons 1970), and we do not know at what stage humans (or australopithecines) began to talk, or, of course, anything of the nature of that language. We do, however, have to assume that the evolution of language must have proceeded very rapidly, particularly if the line leading to modern man diverged from that leading to modern apes as recently as, say, 4 M a ago (p. 241). The following question is, then, posed by the conclusions reached here: is it plausible that the 'universal grammar' could have developed so rapidly, and that it is now fully developed in man and wholly absent in chimpanzees, despite the fact that 'The fine structure and genetic organization of the chromosomes of man and chimpanzee are so similar that it is difficult to account for their phenotypic differences'? (Yunis, Sawyer, and Dunham 1980, p. 1145).

Reverting now to the three possibilities introduced at the beginning of the chapter (p. 290), we see that the first of those possibilities—that humans acquire language because they are more intelligent than non-humans—has been rejected. However, there has been no discussion of the issue that discriminates between the second two possibilities, namely whether the device that allows the acquisition of language is specific to language, or whether it allows a number of novel processes of which language is only one. In general, we have favoured the view that the device is indeed language-specific, partly because this view takes clear form in the light of Chomsky's arguments, and partly because it is currently the most parsimonious account: we must posit a device that allows language acquisition in humans, but whether there are further human intellectual attainments that are not to be explained as a secondary consequence of the possession of language is a question which, while of central importance, finds no convincing answer at present.

9. Intelligence in vertebrates: two hypotheses

Introduction

This final chapter will not attempt a detailed summary of the data and arguments presented in earlier chapters. Instead, we shall assume familiarity with those surveys and crystallize their conclusions by stating explicitly two general hypotheses that have emerged concerning intelligence in vertebrates. These hypotheses, which are based primarily on behavioural data, will then be assessed in the light of other relevant information—phylogenetic relationships and variations in size or organization of the brain, for example—and their broader implications will be briefly considered.

The first chapter of this book began by raising the question whether there existed contrasts in the behavioural capacities of various vertebrate groups which might reflect qualitative or quantitative differences in intellectual mechanisms underlying those capacities. The preceding six chapters have surveyed evidence relevant to that question, and the conclusions of these surveys may be baldly stated as follows. First, when we examined contrasts in intelligent performance between groups of vertebrates excluding man, we found no compelling grounds for supposing any qualitative differences between those groups and no necessity either to assume quantitative differences (as there has consistently been evidence to suggest that the performance differences seen might be reduced by variations in contextual variables). Second, when we considered the contrast between man and other vertebrates, we found that there were good reasons for supposing a qualitative distinction, and that the most economical account of that distinction was that it consisted in man possessing some device or devices whose principal function is concerned with language acquisition. Those findings may now be stated more positively to provide us with our two hypotheses: first, that there are no differences, either quantitative or qualitative, among the mechanisms of intelligence of non-human vertebrates; second, that man's intellect is distinguished from that of non-human vertebrates by his possession of the capacity for language. It should be added, as a rider to this second hypothesis, that we have in fact seen no reason to suppose that there are quantitative differences in the mechanisms of intelligence that are presumed to be common to both man and other vertebrates. The view taken here has been that man's possession of language leads to qualitative differences in his approach to problems of all kinds, and that this makes it virtually impossible to assess the comparative effectiveness of contributions from non-linguistic devices.

Before going on to assess these hypotheses, some remarks on their status are appropriate. One preliminary qualification is, of course, that they can be regarded as no more than working hypotheses, hypotheses that currently await disproof.

It need hardly be emphasized, for example, that absence of proof of difference in intelligence between species is a very different thing from proof of identical mechanisms of intelligence in those species, and the attention given here to the first hypothesis is motivated partly by the fact that it is an interpretation which at first sight may seem so unexpected and implausible that without some (perhaps exaggerated) emphasis it may not receive the attention it surely deserves.

A similar caution applies to the second hypothesis. Its best current support lies in the failure (itself contested) of apes to show a capacity for the acquisition of grammar—but, of course, the basic thrust of much of the argument supporting our first hypothesis has been that failures by a species to master some tasks may well be due to inappropriate values of contextual variables. Although this survey has taken the view that it is more plausible to assume a (specified) difference in mechanisms of language than a (so far unspecified) difference in sensitivity to contextual variables, it is by no means claimed that it is impossible that subsequent explorations could find grammar in non-humans.

The sections that follow will concentrate primarily on assessment of our first hypothesis, that the intellects of non-human vertebrates do not differ. This is largely because the evidence relevant to that proposition is to be found distributed throughout all the preceding six chapters, whereas that for the second hypothesis, the peculiarity of language to man, was concentrated in Chapter 8, in which it received an overall assessment. Special attention to the first hypothesis is also, however, due to the fact that although the proposal that man's intellect is qualitatively different from that of other vertebrates may be in accordance with intuitive views of man that are fairly widespread, it seems much less likely that the suggestion of identity in intellectual mechanisms throughout all non-human vertebrates would easily obtain general acceptance. It is with the anticipated resistance to this hypothesis in mind that its assessment will proceed by considering a number of grounds on which it might be held to be implausible.

Phylogenetic considerations

One reason that might be advanced for scepticism regarding our hypothesis concerning non-human vertebrates is that some vertebrates (fish, for example) are of lower phylogenetic status than others (like mammals). It has, however, been argued in earlier chapters that there is no sense to the notion that there are 'lower' and 'higher' vertebrates, at least where intelligence is concerned. Modern fish (teleosts in particular) are radically different from those fish that were the common ancestors of both living fish and mammals, and living fish have enjoyed as much time in which to evolve as have mammals. One cannot conclude that fish are likely to be of low intellectual capacity because they are distantly related to mammals without assuming that mammals are more intelligent in the first place—but that is precisely what is at issue. It is therefore not implausible on phylogenetic grounds that there are no differences in intelligence amongst the vertebrate classes (or, of course, amongst the various subdivisions of those classes).

Before leaving this topic, we should refer briefly to our second hypothesis. It will be recalled (from Chapter 7) that the line of descent leading to modern apes may have diverged from that leading to man as recently as 4 M a ago. This is a more recent divergence than might be predicted on the assumption that language has developed *de novo* in the specifically human line. The second hypothesis does not, therefore, gain positive support from phylogenetic considerations although, given our ignorance of the basic requirements for a device capable of grammar, such considerations do not require us to reject the proposal.

General adaptiveness

A more compelling reason for doubting that all non-human vertebrates are of equal intelligence is the fact that there are clear differences in what might be termed the 'general adaptiveness' of various vertebrate groups; the behavioural repertoire and apparent inventiveness of monkeys, for example, appear to be much wider than those of goldfish, and this might well be seen as a reflection of differences in intelligence. Now this general observation does, of course, find its counterpart in the results of controlled laboratory investigations, in which we find many instances of tasks which have been successfully performed by monkeys and either not even attempted with goldfish, or attempted and performed at a quantitatively inferior level.

How should we respond to problems which have been solved by one group but not attempted by another? In effect, we are faced with a choice between two null hypotheses: the first is that, in the absence of proof to the contrary, we should assume that all vertebrates possess identical mechanisms of intelligence; the second is that, in the absence of proof to the contrary, we should assume that a species that has not mastered a particular problem (or class of problems) is incapable of mastering it. We have adopted the former hypothesis here, rejecting the latter; choice between these null hypotheses must inevitably be to a large degree arbitrary, and perhaps the best defence for the choice advocated here is that one trend in comparative psychology to date has in fact been the eventual demonstration of phenomena (e.g. serial reversal improvement by fish; cross-modal transfer by non-human mammals) originally supposed to be beyond the capacities of the group in question.

The behavioural data on tasks that have been attempted by members of different vertebrate groups have, of course, invariably revealed quantitative differences between groups of animals (both from within the same class and from different classes), but these differences have been interpreted as reflecting, not quantitative differences in intelligence, but differential effects of contextual variables. It has, in almost all cases, been possible to show that the performance of one or other of the groups concerned can indeed be influenced by variations of some kind in the experimental procedure, although it has not generally been possible to show that the gap between groups can actually be closed by such manipulations (and, in particular, that the low-performance group could be raised to the level of the high-performance group). It might seem that the strategy adopted should be characterized as follows: where a performance difference on a given task emerges between two groups of animals, see whether there is any variation in testing procedure which affects performance of either group; if there is any such variation, conclude that the difference is due to contextual variables, and, implicitly, that it could be eliminated by (unspecified) manipulations of those variables. Since it is inconceivable that there is any behavioural performance that is not influenced to some degree by procedural factors, it may seem that the adoption of that strategy must inevitably lead to the conclusion that there are no quantitative differences in intellectual capacity.

If the above account of the strategy followed in this survey was valid, then clearly we should place little value on our conclusions. It will, however, be argued here both that the account distorts the true nature of the strategy and that there is no justifiable alternative to the strategy that has in fact been adopted.

The distortion concerned lies in the suggestion that the interest in studying variations lies solely in establishing whether such variations induce changes in performance levels, whereas the real interest goes deeper: it was argued in Chapter 1 that, given that one cannot equate (for example) sensory and motor demands across species, the method of systematic variation should be employed to see whether performance levels could be equated using any values of those variables that might plausibly be supposed to affect performance. Now the fact is that the technique of systematic variation has simply not been applied in an exhaustive way to the tasks that we have considered. It would, then, be preferable to have available a much larger body of parametric explorations of performance in a given task in various species. As we do not have such studies available, we must decide upon the best interpretation of the data that have been reported.

Upon what grounds, then, is it urged that the observed quantitative differences in performance should be ascribed to effects of contextual variables, rather than to differences in intellectual capacity? Both accounts presumably, could accommodate the data, so why should the former account be preferred?

The basic reason for rejecting quantitative differences in intelligence is that it is most parsimonious to do so. We already know that there are procedural variations which do affect performance levels, and there is no difficulty in conceiving ways in which such factors as, for example, reward size and discriminability or location of stimuli might affect performance in learning tasks (although it may not be easy to specify those ways in detail). In other words, it will not be denied that contextual variables do play at least a part in generating quantitative differences. On the other hand, we do not know that there are quantitative differences in intellectual capacity, nor is it clear how such differences are to be understood. Only one theoretical hypothesis (Mackintosh 1969*a*) of the nature of quantitative intellectual differences between species has been advanced in a systematic way, and that proposal—that there are quantitative differences in the stability of attention—has not received convincing experimental support (see the relevant surveys in Chapters 3 and 6). We must decide, therefore, whether to assume that the performance differences observed are to be understood wholly in terms of the effects of contextual variables, or in terms of the effects of contextual variables plus some other, unspecified, quantitative difference in intelligence. As there is no evidence to show that contextual variables are inadequate to account for the whole of any performance difference observed to date, it is clearly most parsimonious to reject the hypothesis that there are quantitative differences in intelligence between the various species. It may be objected that since there is effectively no limit to the number of contextual variations that might be tested, there is no possibility of a conclusive demonstration that a given performance difference could not be eliminated by some such variation; this is, of course, a valid objection—but the requirement is only that sufficient attempt at systematic variation is made that it is at least implausible to suggest such variations might be effective.

It may well be felt that while it is parsimonious to deny that there are differences in intelligence, it is nevertheless implausible—that even the tentative advocacy of our first hypothesis is in effect *a reductio ad absurdum* which merely indicates that comparative psychology has followed a systematically incorrect route. The conviction may persist that, while it may not be possible to prove, for example, that goldfish are less intelligent than, say, rats, this reflects, not the success of the fish, but the failure of the comparative psychologist. A specific form of this objection might be that major differences in 'general adaptiveness' have been overlooked by comparative psychologists because the tests used to date have been inappropriate, and have tapped only 'low-level' processes (e.g. of association) that are indeed common to all vertebrates. This is a proposal that is very hard to refute; it does seem perfectly plausible that the intellectual capacities of non-human vertebrates are not exhaustively explored through the use of procedures such as serial reversal training, or discrimination learning-set formation. The difficulty is, of course, to know how to devise tests which would test potential 'higher' capacities. Without specific theoretical accounts of such capacities, procedures that tap them are hardly likely to emerge. Unfortunately, such accounts, specified sufficiently well so that behavioural predictions are generated, are not currently available and until they are put forward, appropriate tests cannot be designed. It may be added, moreover, that until specific theoretical proposals are available, the existence of untapped 'higher' capacities is no more than an article of faith: what aspect of (say) a chimpanzee's behaviour requires us to suppose that intellectual mechanisms are engaged which are not engaged by any of the laboratory techniques currently employed? It is not impossible that there are such capacities—but as there is currently no compelling evidence for their existence it is most economical to assume that there are not.

This section concludes, then, that although there are indeed major differences in the general adaptiveness of vertebrate groups, it is not necessary to assume that these differences reflect differences in mechanisms of intelligence, rather than differences in perceptual, motor and motivational mechanisms. It should be noted that this generalization applies also to our surveys of the organization of memory, and that this poses some problems for the second hypothesis, the peculiarity of language to man. The difficulties arise because it has been argued (p. 318) that the organization of human memory may have developed to cope with the comprehension of speech. If this is so, then we might expect to see a different organization of memory in non-humans. However, the relevant surveys of Chapters 6 and 8, although they did throw up findings which might suggest differences in organization—particularly between birds and man—fell very short of any proof of such differences. Comparative studies of memory remain a promising area of research, but the emphasis on parsimony that has been adopted elsewhere clearly requires at present that no differences in memory organization among vertebrates should be assumed.

Differences in brain size

The fact that the teleost brain is in general about ten times smaller than that of a bird or a mammal of comparable weight may appear to be a good reason for supposing a greater measure of intelligence in birds and mammals. A preliminary observation relevant to this belief is that we have found no convincing behavioural evidence, from scrutiny of both between- and within-class comparisons, that brain-size measures do predict intellectual capacity. The (relative) brain-size rankings that we have seen do, however, show some measure of agreement with popular conceptions of intelligence-rankings: man has the largest brain, primate brains are larger than those of rodents, mammalian brains are larger than those of reptiles, amphibians, and bony fish—is this measure of agreement simply coincidence, or does it reflect a real trend, not adequately reflected in the tests used by comparative psychologists? Now the rankings cited are, of course, relative brain-size rankings since in general, the larger the animal, the larger its brain, and we do not suppose that larger animals are consistently more intelligent than smaller animals. We must therefore consider again the derivation of relative brain-size. In Chapter 2 we introduced the equation of simple allometry, $E = kP^{a}$ (where E is brain weight and P is body weight), and advanced Jerison's (1973) argument that a should be taken to be $\frac{2}{3}$, since that both gave a good fit to a wide range of vertebrate data, and allowed the rationale that a somatic factor in brain size might well vary with body surface; the constant k was labelled the 'index of cephalization', the notion being that differences in this index would reflect differences in some non-somatic factor, which could be, for example, general intelligence.

One objection to the rigid use of the formula of simple allometry (with a taken to be $\frac{2}{3}$) is that it does not satisfactorily implement its own rationale—that the somatic brain factor should vary with body surface: the agreement could at best only be approximate, and will vary considerably with body shape, so that, for example, fish, with streamlined bodies, would be expected (other things being equal) to have very much less body surface for a given body weight than would, say, monkeys. Similarly, it is not even to be expected that body weight will give a good measure of body volume across different groups—some groups, like birds, being specialized for lightness (with light bones) and others, aquatic vertebrates in particular, being animals for which weight is a relatively minor consideration. It could be, then, that the degree of cephalization in fish is underestimated by ignoring their low surface to volume ratio, and (certainly in comparison to birds) their lower volume to weight ratio; both factors would overestimate the amount of body surface requiring brain volume in fish (relative to birds) and so underestimate the 'residual' brain volume available for intellectual processing.

A further difficulty with the notion that the somatic demand for brain volume should be estimated by using a measure of body surface across a wide range of vertebrates is that it makes no allowance for differences in the degree of refinement of sensory processing, or of the development of motor skills. It may not be possible to quantify the differences between species in these matters, but it is surely odd to suppose that creatures with very different perceptual worlds and repertoires of responses should devote exactly the same amount of brain tissue (per unit of body surface) to such somatic factors.

An alternative account of the brain size data which supposes that vertebrates do not differ in intellectual capacity is, then, that variations in brain size may be accounted for entirely by variations in requirements for somatic processing. Such requirements do vary roughly in accordance with body surface in groups of animals having bodies of comparable shape and constitution, and similar perceptual and motor capacities. If two distantly related species of similar body weight are compared, however, it is to be expected that the species having the higher surface to volume ratio and (or) the greater elaboration of sensory and motor information should have the larger brain. It should be noted that this proposal, (according to which k might be interpreted as an 'index of somatic elaboration'), may obviate a difficulty facing the interpretation of k as an index of cephalization. The difficulty lies in the fact that k interacts with P, the somatic factor, in a multiplicative and not in an additive way. If the equation which best fitted the data had been $E = k+P^{\frac{2}{3}}$, then animals with the same value of k would have had the same (absolute) amount of brain tissue available for intelligent processing. But as the formula is in fact $E = kP^{\frac{2}{3}}$, then animals with the same value of k would possess different absolute amounts of 'extra' brain volume according to their body weight, the difference increasing with the $\frac{2}{3}$ power of body weight. The equation of simple allometry, that is, postulates that the size of the brain is directly proportional to the $\frac{2}{3}$ power of body weight, so that if the amount of brain tissue devoted to somatic processes increases according to the $\frac{2}{3}$ power of body weight, so must the residual brain volume. It is, of course, very hard to see why, on the assumption that intellectual and somatic processing are independent, the amount of brain volume required for purely intelligent processing should increase with body surface.

The nature of the formula of simple allometry does, however, pose some problems for the hypothesis of equal intelligence amongst vertebrates: given that some brain tissue is given over to intellectual activities, then that amount should appear as an additive factor, so distorting the fit of the equation. There are two ways in which this problem might be resolved. First, it could be that the amount of tissue devoted to intelligence is so small that it is simply lost in the general variance of the data; this is an option not open to the alternative 'cephalization' account, since variations between fish and mammals in intellect alone are supposedly sufficient to cause tenfold changes in brain size—the amount of mammalian brain devoted to intelligence, according to this view, must be substantial (at least 90 per cent).

A second account is to suppose that each unit of brain volume given over to somatic (perceptual or motor) analysis has associated with it a unit of tissue given over to some aspect of the intellectual processing of information passing through the somatic unit. The 'intellectual' unit might be concerned with, for example, mnemonic storage, or with associations specific to that unit. Such units would not be generally available to other sources of perceptual information, and so increases in the numbers of such units would not increase the general intellectual capacity of a species although they would, of course, imply that species with larger brains would be able to make a larger absolute number of associations involving perceptual units. Now this in fact would not be surprising: animals with larger brains, according to this account, have either larger body surfaces or more elaborate perceptual processes. In either case, the number of discriminations possible for them should be larger than that for smaller-brained species—and this is indeed one way in which contextual variables might have their effects. This second account must, of course, incorporate the assumption of the first account, that the amount of brain tissue devoted to mechanisms of intelligence having general application across all the senses remains small. The argument might be that larger brains possess larger memory stores, and that a small central processor is common to all brains; the increased memory stores would reflect only the fact that all species, if capable of perceptual differentiation of events, are equally capable of forming associations involving representations of those events. Animals with larger brains would be capable of more discriminations (between sensory inputs or between various motor responses) and so, would indeed be capable of forming more associations than animals with smaller brains. However, since all groups could form associations involving all the sensory and motor events between which they are capable of discriminating, the long-term memory of all groups would in effect be unlimited—or limited only by the sensory and motor capacities of the species.

We may conclude this consideration of brain size by pointing to what has been implicit this far: the agreement between our 'intuitive' rankings of intelligence and relative brain size is not simply a coincidence. Those rankings in each case reflect sizeable differences in behavioural repertoire brought about largely by changes in perceptual and motor skills; those skills allow monkeys (for example) to behave adaptively in a far wider range of situations than is possible for a fish, and it may be that it is for such reasons that we are inclined to suppose that monkeys are so much more intelligent. This conclusion is, in fact, not far from that reached by Jerison, whose argument, supported by an extensive range of brain-size data (derived from endocasts of extinct species as well as from living groups) is that brain size in vertebrates increased as new ecological niches were invaded, niches which required the development of new perceptual and motor skills. Jerison in fact writes (1973, p. 17) that: 'biological intelligence may be nothing more (or less) than the capacity to construct a perceptual world', so that, if there is disagreement, it may rest on the question whether that capacity should be seen as distinct from and relatively independent of some other set of somatic demands on brain capacity. The view taken here is that the most economical account is to suppose that biological intelligence, as defined above by Jerison, represents the major factor in determining brain size, but that it is independent of general intelligence, as measured by general problem-solving capacity. This proposal incorporates the assumption that there is indeed something that may properly be labelled 'general intelligence', and this is a subject to which we shall return after consideration of one final ground for objection to our first general hypothesis, which refers to variations in vertebrate brain organization.

Vertebrate brain organization

The preceding sections have been largely concerned with the proposal that there are no quantitative differences in intelligence amongst non-human vertebrates, and in this section we turn to the other proposal contained in our first hypothesis, that there are no qualitative differences either. The surveys of Chapters 3 to 7 explored a number of hypotheses concerning potential qualitative species-differences amongst non-human vertebrates, but did not find conclusive evidence in favour of any of them. Can the null hypothesis, that there are no qualitative differences in intelligence amongst non-human vertebrates, be reconciled with the considerable differences in brain anatomy seen in these animals?

There are good grounds for supposing that such a reconciliation may be achieved. Despite the major differences in anatomical organization, striking parallels in functional organization across vertebrate classes have been observed, particularly in systems of transmission of sensory information to the telencephalon. We have seen no compelling evidence that the telencephalon of any one group of vertebrates is 'dominated' by any single sense (specifically by olfaction), and much evidence that non-olfactory afferents do reach the telencephalon in all groups for which modern data are available. There is, therefore, no reason to suppose that any vertebrate group lacks the facility to co-ordinate information drawn from a variety of sense modalities. We have, moreover, seen that the early notion of progressive encephalization of sensory processing, supposedly reflected in the expansion of visual neocortex and in the relative diminution of midbrain visual areas, lacks support. It appears rather that the collicular and geniculate routes for visual information carry similar types of analyses of the retinal input in different vertebrate groups, although the relative development of the two routes, both of which do project to the telencephalon, does clearly differ across those groups. These differences in relative reliance on the two routes do no doubt signify important differences in modes of perceptual processing, but they do not suggest that there is, in those groups with well-developed 'collicular' routes, a dependence on midbrain processes, the outputs of which are unavailable to mechanisms in the telencephalon.

Just as there is no reason to suppose domination by any one sense, so there is no reason to suppose that the entire telencephalon of any species is given over to sensorimotor processing: indeed, quite the reverse—research to date on the telencephalon of fish and amphibians leaves many regions that appear to be quite 'silent' when probed for sensory involvement. If, therefore, it is supposed that intellectual activities are carried out in regions distinct from sensorimotor regions (a supposition for which very little support can in fact be found), then the brain organization of 'lower' vertebrates allows more than adequate 'space' for such intellectual activities. A similar argument applies to the generally low level of cerebral representation of motor activity found in all non-mammalian vertebrates: the implication for intelligence would seem to be only that all the more space is available for intellectual processes.

Studies of the functional organization of vertebrate brains do not, then, lead one to expect systematic differences in intelligence between various groups. Moreover, those studies, along with embryological studies, have led us to believe that the differences in anatomical organization may be much less than was previously believed. It can be argued that some structure homologous with mammalian neocortex is to be found in all vertebrates, although precise homologues are obscured by very different modes of development of the telencephalon in different groups. This is a case that has, for example, been clearly made out for the telencephalons of reptiles and birds by Nauta and Karten (1970), who argue that homologous neurons are to be found in the neocortex of mammals and in 'external striatum' of non-mammals (primarily the DVR in reptiles, the neostriatal and hyperstriatal regions in birds). We have not been concerned to discuss homologies in detail, but it is of course relevant to our interests to find that the apparently gross differences in anatomical organisation may conceal real parallels in the embryonic derivation of structures.

This summary, while it has emphasized similarities in brain organization across vertebrate groups, should not, of course, be taken to imply that there are no differences in functional organization. There clearly are: there is evidence, for example, that some mammalian groups which have well-developed visual capacities possess visual cortical areas (subdivisions of the total cortical mass primarily devoted to vision) which may have no counterpart in mammals with less well-developed vision (Kaas 1980). Similarly, there is evidence that the number of fibres in the pyramidal tract of a wide range of mammals (69 species in all) correlates reasonably well with digital dexterity (given that the variance in fibre number due to body-weight variations is taken into account) (Heffner and Masterton 1975); the pyramidal tract contains fibres that project from motor cortex to spinal regions, and variations in its size should be expected to correlate with variations in the extent of primary motor cortex. Examples such as these, however, serve to support our general conclusions: there are wide variations in perceptual capacities and motor skills across the various groups of vertebrates, and these are indeed reflected in the functional organization of their brains. However to the extent that variations in brain organization can be attributed to differences in sensorimotor capacity, there is no need to suppose that differences in intellectual capacity need be invoked.

Turning now to consideration of the relevance of brain organization to our second hypothesis, it will be recalled (from Chapter 8) that there was no very good evidence to show that the development of language in man had been accompanied by any gross re-organization of brain functions. The apparent contrasts between the effects of hippocampal lesions in man and those in other vertebrates can be reconciled without necessarily supposing a radical difference in hippocampal function; similarly, there are indications of some degree of cerebral lateralization of function in a number of non-human mammals. The proposal that humans possess a species-specific language-acquisition device stands, then, on its own as it gains no positive support from phylogenetic considerations, from behavioural studies of memory organization, or from brain studies.

General intelligence

We introduced earlier the suggestion that comparative psychologists may have made some systematic error, and explored the possibility that they may have been using inappropriate behavioural tests. We turn now to another potential source of error – the possibility that there is in fact no such thing as 'general intelligence', so that comparisons between species are meaningless.

This proposition was originally considered in Chapter 1, but may now be briefly assessed again in the light of the behavioural data that have been surveyed. These data have shown that a large variety of vertebrate species behave in similar ways when confronted with similar problems, albeit in entirely artificial physical surroundings. It seems reasonable to suppose, therefore, that the animals use similar mechanisms to solve those problems. Now if the mechanisms engaged were in fact mechanisms that had evolved for use in much more restricted contexts, then it would seem likely that different groups, each having specialized mechanisms peculiar to themselves, would engage different mechanisms under at least some of these arbitrary circumstances. We might, then, have expected (as argued in Chapter 1) that some qualitative differences in performance would have emerged. We have, however, not observed conclusive evidence for qualitative differences and to that extent the proposal that laboratory tasks are solved by context-specific devices is not supported.

It would, of course, be preferable to obtain somewhat more positive evidence for the existence of mechanisms of general intelligence, and it is appropriate, therefore, to consider briefly evidence that has been brought forward to support such a notion in humans. The first major proponent of the view that humans do possess general intelligence was Spearman, who wrote in 1904 that 'all branches of intellectual activity have in common one fundamental function (or group of functions), whereas the remaining or specific elements seem in every case to be wholly different from that in all the others' (Spearman 1904, p. 284). The evidence used to support this view is statistical in nature, and consists essentially in analysis of the degree of correlation between performances of individuals in different sets of test questions. The major statistical technique that has come to be employed is known as factor analysis. This technique, whose details will not be examined here, attempts to account for correlations between scores on various different questions in terms of a relationship between the questions and a hypothetical underlying factor or set of factors. According to Spearman, the postulation of a single general factor was sufficient to account for the data: each item of an intelligence test provides a more or less accurate measure of general intelligence, and the additional variation in scores on that item reflects only variance specific to that question.

Although Spearman's account provides an intuitively satisfying interpretation of the widespread observation of correlations between a wide range of measures of intellectual performance in humans, his is not the only account. Other workers have supposed that the variance across test items may best be explained by positing one general factor, relevant to all test items, and a number of more specific factors (verbal or spatial ability, for example), relevant to particular subsets of items. Finally, some interpretations dispense altogether with the notion of any general factor and suppose instead a set of specific independent factors, each relevant to some subset of items. This divergence of views reflects the unfortunate fact that factor analysis does not provide a unique account of the optimal interpretation of the variance. In general, the more factors posited, the more variance can be accounted for; according to Butcher (1968), British workers have tended to prefer a small set of factors, including a general factor, American workers tending to opt for larger sets of factors. It appears, then, that no unequivocal decision can yet be taken on whether the human material encourages the view that 'general intelligence' should be ascribed to humans and so, no clear implication that we should expect to find (or not find) such a capacity in other vertebrates.

Although the human data are inconclusive, or so, at least, it would seem from a cursory survey, it might nevertheless be the case that the methodology might be applicable to non-humans. There are obvious problems: it is not easy to test large numbers of non-human animals in dozens of different tasks, and it appears likely that it would be much more difficult to control for transfer between different types of task in them than in humans. It may be for such reasons that there have in fact been very few attempts to analyse the performance of non-humans in a series of tests. What attempts there have been have produced little in the way of general conclusions: Commins, McNemar and Stone (1932) and Campbell (1935) agree that rats tested in a series of maze problems show a good degree of correlation in terms of error scores across those mazes: that is, rats that are good learners in one maze tend to be good learners in another. However, neither report found any correlation between maze ability and performance in other tasks (a 'triple-platform problem box' and a visual discrimination in the Commins *et al.* report, and Maier's three-table 'reasoning' apparatus, described on p. 306, in the Campbell report). On the other hand, Rajalakshmi and Jeeves (1968) did find some degree of correlation between maze-learning ability and visual discrimination performance in rats. Very few other studies of this nature have been attempted so that there is currently no good evidence that a non-human of good ability in one task will tend to show good ability in other tasks and so no support from that source for the notion of general intelligence in non-humans.

It should, perhaps, be pointed out that even if subsequent investigations using non-humans fail to produce evidence for a factor of general intelligence in them, such an outcome need not be damaging to the notion that comparable mechanisms of general intelligence are available to all non-human vertebrates. It could, for example, be the case that variations in general intelligence are so small that their effects are lost in the much larger variance brought about by task-specific variables—such as, for example, the tendency to attend to visual as opposed to kinaesthetic cues. A second possibility is that although all non-human vertebrates possess identical mechanisms of general intelligence, not all of these mechanisms are engaged in all problems (or not equally engaged); there may be, in other words, specific factors but no general factor in non-humans (and in this case, we would expect to find correlations between some sets of tasks—as was indeed the case for the reports of maze-learning ability described previously). Finally, we should reiterate our earlier concern that performance on all human intelligence tests may reflect linguistic competence to a substantial degree, so that it may be the case that the human 'general intelligence' factor (if accepted) might be a consequence of the possession of language; in this connection, it is of interest to note that there are in fact rather low correlations between estimates of general intelligence in humans and performance in conditioning or simple associative learning tasks (Estes 1970; Jensen 1980).

Implications

In this final section, we shall discuss briefly the implications of the two working hypotheses for the evolution of intelligence, and for the direction that future research might take.

It might be thought that if there are no differences in intelligence amongst non-human vertebrates, then intelligence did not evolve during vertebrate history until the abrupt emergence of language. However, it should be pointed out that to suppose that all living non-human vertebrates possess comparable mechanisms of intelligence is not necessarily to assert that intelligence has not developed during the evolution of vertebrates. As we cannot assess the intelligence of extinct vertebrates, we cannot decide whether they were of lower or different intellectual status from living vertebrates. If mechanisms of general intelligence have evolved primarily to detect regularities between causally related events, then the same selection pressures have been operative throughout the evolution of the animal kingdom, and it may be that those mechanisms developed to a high level of efficiency at an early stage in evolution. The further exploration of this possibility would, of course, lead to a consideration of invertebrate learning, which is outside the scope of this book. Evidence drawn from living vertebrates is, however, necessarily of indirect relevance to capacities of extinct vertebrates; what can be agreed is that it appears to have contributed relatively little to our understanding of the evolution of mechanisms of intelligence.

It is clear that subsequent investigations in comparative psychology should probe further those many cases that we have seen of quantitative differences in performance to see whether the present hypothesis—that those differences may be attributed to contextual variables alone—may be conclusively ruled out; such a demonstration would clearly have an immediate and considerable impact on theories of intelligence. At the same time, it may be argued that serious and imaginative consideration should be given to the broader implications of the notion that no intellectual differences exist: would this indicate that some animals (fish, for example) are much more intelligent than has generally been believed in the past—or would it indicate, on the contrary, that only simple mechanisms are available to all species, but that the power of these devices, when co-ordinated with sophisticated perceptual and motor mechanisms, is very much greater than might be imagined?

Just as we can say little, from the study of non-human vertebrates, of the evolution of intelligence, so there is little that can be said with any degree of assurance on the evolution of language. As was pointed out in Chapter 8, we do not know at what stage language first emerged, at what pace it developed, or through what stages it passed. It was argued there that its impact on intellectual activity is of incalculable importance, that not only is reasoning impossible without language, but that many other activities—such as communication (in the sense in which humans communicate)—might also be dependent on language. It is, perhaps, this avenue of speculation that may best repay imaginative effort. If rational man is unique because he possesses language, then it should be the case that the successful implementation in a computer of a system of language comparable to that of humans (incorporating the 'universal grammar') should yield a rational computer. Further questions would arise: would the computer be conscious, as a result of the ability to talk? It might seem that—to the extent that the machine used words like 'I', 'you', and so on appropriately—then at least it would be self-conscious. But—conscious? Capable, for example, of experiencing pleasure and pain? And if we were to suppose that it was conscious—as a consequence of language—then should we suppose that non-human animals—incapable of language—are not conscious? Or, on the other hand, if we assume that language does not confer consciousness, then what would be the appropriate procedure for building a machine that would be conscious?

We have pointed to a series of questions to which there may in fact never be answers, but which clarify the context against which investigations of the capacity for language in non-humans should be seen. It is hardly surprising that the interpretation of language experiments in non-humans has become so controversial, or that so many non-psychologists take a part in discussing their results: these studies force us to clarify what it is that we mean by language and to consider anew the nature of the human mind.

References

Abbie, A. A. and Adey, W. R. (1950). Motor mechanisms of the anuran brain. *J. comp. Neurol.* **92**, 241–91.

Adler, N. and Hogan J. A. (1963). Classical conditioning and punishment of an instinctive response in *Betta splendens. Anim. Behav.* **11**, 351–4.

Amiro, T. W. and Bitterman M. E. (1980). Second-order appetitive conditioning in goldfish. *J. exp. Psychol: Anim. Behav. Processes* **6**, 41–8.

Amsel, A. (1962). Frustrative nonreward in partial reinforcement and discrimination learning. *Psychol. Rev.* **69**, 306–28.

Andrew R. J. (1967), Intracranial self-stimulation in the chick. *Nature, Lond.* **213**, 847–8.

— (1969). Intracranial self-stimulation in the chick and the causation of emotional behavior. *Ann. N.Y. Acad. Sci.* **159**, 625–39.

Andrews O. (1915). The ability of turtles to discriminate between sounds, *Bull. Wisconsin nat. hist. Soc.* **13**, 189–95.

Annett, M. (1972). The distribution of manual asymmetry. *Br. J. Psychol.* **63**, 343–58.

Aronson, L. R. (1970). Functional evolution of the forebrain in lower vertebrates. In *Development and evolution of behavior: essays in memory of T. C. Schneirla* (ed. L. R. Aronson, E. Tobach, D. S. Lehrman, and J. S. Rosenblatt) pp. 75–107. W. H. Freeman, San Francisco.

— and Kaplan, H. (1968). Function of the teleostean forebrain. In *The central nervous system and fish behavior* (ed. D. Ingle) pp. 107–25. University of Chicago Press.

Antencio, F. W., Diamond, I. T., and Ward, J. P. (1975). Behavioral study of the visual cortex of *Galago senegalensis. J. comp. physiol. Psychol.* **89**, 1109–35.

Bacon, H. R., Warren, J. M., and Schein, M. W. (1962). Non-spatial reversal learning in chickens. *Anim. Behav.* **10**, 239–43.

Baddeley, A. D. (1966*a*). The influence of acoustic and semantic similarity on long-term memory for word sequences. *Q.J. exp. Psychol.* **18**, 302–9.

— (1966*b*). Short-term memory for word sequences as a function of acoustic, semantic and formal similarity. *Q.J. exp. Psychol.* **18**, 362–5.

— (1972). Human memory. In *New horizons in psychology*, Vol. 2 (ed. P. C. Dodwell) pp. 36–61. Penguin, Harmondsworth.

— and Warrington, E. K. (1970). Amnesia and the distinction between long- and short-term memory. *J. verbal Learn. verbal Behav.* **9**, 176–89.

Baenninger, R. (1966). Waning of aggressive motivation in *Betta splendens. Psychon. Sci.* **4**, 241–2.

Baerends, G. P. (1957). The ethological analysis of fish behavior. In *The physiology of fishes* (ed. M. E. Brown) Vol 2, pp. 229–70. Academic Press, New York.

Bailey, P. J. (1979). Aspects of categorical processing of speech sounds in man: on the determinants of phonetic category boundaries. In *Hearing mechanisms and speech* (ed. O. Creutzfeld, H. Scheich, and C. Schreiner) pp. 301–17. *Experimental Brain Research*. Suppl. II.

Baker-Cohen, K. F. (1968). Comparative enzyme histochemical observations on submammalian brains. *Ergebn. Anat. EntwGesch.* **40**, 1–70.

— (1969). Comparative enzyme histochemical observations on submammalian brains. III. Hippocampal formation in reptiles. *Brain Res.* **16**, 215–25.

Bakker, R. T. (1975). Dinosaur renaissance. *Scient. Am.* **232**, 58–78.

Barlow, H. B., Blakemore, C., and Pettigrew, J. D. (1967). The neural mechanism of binocular depth discrimination. *J. Physiol., Lond.* **193**, 327–42.

Bass, A. H. (1977). Effects of lesions of the optic tectum on the ability of turtles to locate food stimuli. *Brain Behav. Evol.* **14**, 251–60.

— Pritz, M. B., and Northcutt, R. G. (1973). Effects of telencephalic and tectal ablations on visual behavior in the side-necked turtle, *Podocnemis unifilis. Brain Res.* **55**, 455–60.

Bateson, P. P. G. (1966). The characteristics and context of imprinting. *Biol. Rev.* **41**, 177–220.

— Horn, G., and McCabe, B. J. (1978). Imprinting: the effect of partial ablation of the medial hyperstriatum ventrale of the chick. *J. Physiol., Lond.* **285**, 23P.

Bauer, J. H. and Cooper, R. M. (1964). Effects of posterior cortical lesions on performance of a brightness-discrimination task. *J. comp. physiol. Psychol.* **58**, 84–92.

Behrend, E. R. and Bitterman, M. E. (1961). Probability-matching in the fish. *Am. J. Psychol.* **74**, 542–51.

— — (1962). Avoidance conditioning in the goldfish; exploratory studies of the CS–US interval. *Am. J. Psychol.* **75**, 18–34.

— — (1963). Sidman avoidance in the fish. *J. exp. Anal. Behav.* **6**, 47–52.

— — (1966). Probability-matching in the goldfish. *Psychon. Sci.* **6**, 327–8.

— Domesick, V. B., and Bitterman. M. E. (1965). Habit reversal in the fish. *J. comp. physiol. Psychol.* **60**, 407–11.

— Powers, A. S., and Bitterman, M. E. (1970). Interference and forgetting in bird and fish. *Science, N.Y.* **167**, 389–90.

Belekhova, M. G. and Kosareva, A. A. (1971). Organisation of the turtle thalamus: visual, somatic and tectal zones. *Brain Behav. Evol.* **4**, 337–75.

Benowitz, L. (1972). Effects of forebrain ablations on avoidance learning in chicks. *Physiol. Behav.* **9**, 601–8.

— and Karten J. J. (1976). The tractus infundibuli and other afferents to the parahippocampal region of the pigeon. *Brain Res.* **102**, 174–80.

Bernstein, J. J. (1961). Brightness discrimination following forebrain ablation in fish. *Expl Neurol.* **3**, 297–306.

— (1962). Role of the telencephalon in color vision of fish. *Expl Neurol.* **6**, 173–85.

Berryman, R., Cumming, W. W., Cohen, L. R., and Johnson, D. F. (1965). Acquisition and transfer of simultaneous oddity. Psychol. Rep. **17**, 767–75.

Bessemer, D. W. and Stollnitz, F. (1971). Retention of discriminations and an analysis of learning set. In *Behavior of nonhuman primates: modern research trends*, Vol. 4 (ed. A. M. Scrier and F. Stollnitz) pp. 1–58. Academic Press, New York.

Bicknell, A. T. and Richardson, A. M. (1973). Comparison of avoidance learning in two species of lizards, *Crotaphytus collaris* and *Dipsosaurus dorsalis. Psychol. Rep.* **32**, 1055–65.

Bintz, J. (1971). Between- and within-subject effect of shock intensity on avoidance in goldfish (*Carassius auratus*). *J. comp. physiol. Psychol.* **75**, 92–7.

Birch, H. G. (1945). The relation of previous experience to insightful problem solving. *J. comp. Psychol.* **38**, 367–83.

Birch, M. P., Ferrier, R. J., and Cooper, R. M. (1978). Reversal set formation in the visually decorticate rat. *J. comp. physiol. Psychol.* **92**, 1050–61.
Birnberger, K. L. and Rovainen, C. M. (1971). Behavioral and intracellular studies of a habituating fin reflex in the sea lamprey. *J. Neurophysiol.* **34**, 983–9.
Bitterman, M. E. (1964*a*). Classical conditioning in the goldfish as a function of the CS–US interval. *J. comp. physiol. Psychol.* **58**, 359–66.
— (1964*b*). An instrumental technique for the turtle. *J. exp. Anal. Behav.* **7**, 189–90.
— (1965*a*). The evolution of intelligence. *Scient. Am.* **212**, 92–100.
— (1965*b*). Phyletic differences in learning. *Am. Psychol.* **20**, 396–410.
— (1968). Reversal learning and forgetting. *Science, N.Y.* **160**, 99–100.
— (1969). Habit reversal and probability learning: rats, birds and fish. In *Animal discrimination learning* (ed. R. Gilbert and N. S. Sutherland) pp. 163–75. Academic Press, London.
— (1971). Visual probability learning in the rat. *Psychon. Sci.* **22**, 191–2.
— (1975). The comparative analysis of learning. *Science, N.Y.* **188**, 699–709.
— Wodinsky, J., and Candland, D. K. (1958). Some comparative psychology. *Am. J. Psychol.* **71**, 94–110.
Bjork, R. A. (1970). Repetition and rehearsal mechanisms in models for short term memory. In *Models of human memory* (ed. D. A. Norman) pp. 307–30. Academic Press, New York.
Blank, M. and Bridger, W. H. (1964). Cross-modal transfer in nursery school children. *J. comp. physiol. Psychol.* **58**, 277–82.
Blankenagel, F. (1931). Untersuchungen über die Grosshirnfunktionen von *Rana temporaria. Zool. Jber Abt. Allgem. Zool. Physiol. Tiere* **49**, 271–322.
Blass, E. M. and Rollin, A. R. (1969). Formation of object discrimination learning sets by Mongolian gerbils (*Meriones unguiculates*). *J. comp. physiol. Psychol.* **69**, 519–21.
Boakes, R. A. and Gaertner, I. (1977). The development of a simple form of communication. *Q. J. exp. Psychol.* **29**, 561–75.
Boice, R. (1970). Avoidance learning in active and passive frogs and toads. *J. comp. physiol. Psychol.* **70**, 154–6.
Boitano, J. J. and Foskett, M. D. (1968). Effects of partial reinforcement on speed of approach responses in goldfish (*Carassius auratus*). *Psychol. Rep.* **22**, 741–4.
Bok, S. T. (1959). *Histonomy of the cerebral cortex.* Van Nostrand-Reinhold, Princeton, N.J.
Bolles, R. C. (1970). Species-specific defense reactions and avoidance learning. *Psych. Rev.* **77**, 32–48.
Bols, R. J. (1977). Display reinforcement in the Siamese fighting fish, *Betta splendens*: aggressive motivation or curiosity? *J. comp. physiol. Psychol.* **91**, 233–44.
Boord, R. L. (1969). The anatomy of the avian auditory system. *Ann. N.Y. Acad. Sci.* **167**, 186–98.
Bottjer, S. W., Scobie, S. R., and Wallace, J. (1977). Positive behavioral contrast, autoshaping, and omission responding in the goldfish (*Carassius auratus*). *Anim. Learn. Behav.* **5**, 336–42.
Boycott, B. B. and Guillery, R. W. (1962). Olfactory and visual learning in the red-eared terrapin, *Pseudemys scripta elegans* (Wied). *J. exp. Biol.* **39**, 569–77.
— and Young, J. Z. (1950). The comparative study of learning. In *Physiological mechanisms in animal behaviour,* Symposia of the Society of Experimental Biology, Vol. 4, pp. 432–53. London.

Boyd, B. O. and Warren, J. M. (1957). Solution of oddity problems by cats. *J. comp. physiol. Psychol.* **50**, 258–60.
Bradley, P. and Horn, G. (1978). Afferent connexions of the hyperstriatum ventrale in the chick brain. *J. Physiol., Lond.* **278**, 46P.
Brandon, S. E. and Bitterman, M. E. (1979). Analysis of autoshaping in goldfish. *Anim. Learn. Behav.* **7**, 57–62.
Braun, J. J., Meyer, P. M., and Meyer, D. R. (1966). Sparing of a brightness habit in rats following visual decortication. *J. comp. physiol. Psychol.* **61**, 79–82.
Brauth, S. E., Ferguson, J. L., and Kitt, C. A. (1978). Prosencenphalic pathways related to the paleostriatum of the pigeon (*Columba livia*). *Brain Res.* **147**, 205–21.
— and Kitt, C. A. (1980). The paleostriatal system of *Caiman crocodilus*. *J. comp. Neurol.* **189**, 437–65.
Bremer, F., Dow, S., and Morruzzi, G. (1939). Physiological analysis of the general cortex in reptiles and birds. *J. Neurophysiol.* **2**, 473–87.
Bresler, D. E. and Bitterman, M. E. (1969). Learning in fish with transplanted brain tissue. *Science, N.Y.* **163**, 590–2.
Breuning, S. E. and Wolach, A. H. (1977). Successive negative contrast effects with goldfish (*Carassius auratus*). *Psychol. Rec.* **27**, 565–75.
— — (1979). Successive negative contrast effects for activity-conditioned goldfish (*Carassius auratus*) as a function of housing conditions. *Psychol. Rec.* **29**, 245–54.
Broadbent, D. E. (1958). *Perception and communication*. Pergamon, London.
Brodkorb, P. (1971). Origin and evolution of birds. In *Avian biology*, Vol. 1 (ed. D. S. Farner and J. R. King) pp. 19–55. Academic Press, New York.
Brookshire, K. H., Warren, J. M., and Ball, G. G. (1961). Reversal and transfer learning following overtraining in rat and chicken. *J. comp. physiol. Psychol.* **54**, 98–102.
Brower, L. P., Brower, J. Z., and Westcott, P. W. (1960). Experimental studies of mimicry. 5. The reactions of toads (*Bufo terrestris*) to bumblebees (*Bombus americanorum*) and their robberfly mimics (*Mallophora bomboides*) with a discussion of aggressive mimicry. *Am. Nat.* **94**, 343–55.
Brown, D. P. F., Lenneberg, E. H., and Ettlinger, G. (1978). Ability of chimpanzees to respond to symbols of quantity in comparison with that of children and of monkeys. *J. comp. physiol. Psychol.* **92**, 815–20.
Brown, M. W. and Horn, G. (1978). Effects of visual experience on unit responses in hyperstriatum of chick brain. *J. Physiol., Lond.* **278**, 48P.
Brown, P. L. and Jenkins, H. M. (1968). Auto-shaping of the pigeon's key-peck. *J. exp. Anal. Behav.* **11**, 1–8.
Brown, R. (1958). *Words and things*. Free Press, New York.
— (1970). *Psycholinguistics*. Free Press, New York.
Brownlee, A. and Bitterman, M. E. (1968). Differential reward conditioning in the pigeon. *Psychon. Sci.* **12**, 345–6.
Bruckmoser, P. and Dieringer, N. (1973). Evoked potentials in the primary and secondary olfactory projection areas of the forebrain in elasmobranchia. *J. comp. Physiol.* **A87**, 65–74.
Brutkowski, S. and Dabrowska, J. (1963). Disinhibition after prefrontal lesions as a function of duration of intertrial intervals. *Science, N.Y.* **139**, 505–6.
Bryant, P. E. (1974). *Perception and understanding in young children: an experimental approach*. Methuen, London.
— and Trabasso, T. (1971). Transitive inferences and memory in young children. *Nature, Lond.* **232**, 456–8.

Buchmann, O. L. K. and Grecian, E. A. (1974). Discrimination-reversal learning in the marsupial *Isoodon obesulus* (Marsupialia, Peramelidae). *Anim. Behav.* 22, 975–81.

— and Rhodes, J. (1979). Instrumental discrimination: reversal learning in the monotreme *Tachyglossus aculeatus setosus. Anim. Behav.* **27**, 1048–53.

Bullock, D. H. and Bitterman, M. E. (1962*a*). Habit reversal in the pigeon. *J. comp. physiol. Psychol.* **55**, 958–62.

— — (1962*b*). Probability-matching in the pigeon. *Am. J. Psychol.* **75**, 634–9.

Bullock, T. H. and Corwin, J. T. (1979). Acoustic evoked activity in the brain in sharks. *J. comp. Physiol.* **A129**, 223–34.

Burghardt, G. M. (1969). Comparative prey-attack studies in newborn snakes of the genus *Thamnophis. Behaviour* **33**, 77–114.

— (1977). Learning processes in reptiles. In *Biology of the Reptilia*, Vol. 7 (ed. C. Gans and D. Tinkle) pp. 555–681. Academic Press, London.

— and Hess, E. H. (1966). Food imprinting in the snapping turtle, *Chelydra serpentina. Science, N.Y.* **151**, 108–9.

Burnett, T. C. (1912). Some observations on decerebrate frogs with special reference to the formation of associations. *Am. J. Physiol.* **30**, 80–7.

Burns, A. H. and Goodman, D. C. (1967). Retinofugal projections of *Caiman sclerops. Expl Neurol.* **18**, 105–15.

Burns, R. A., Woodard, W. T., Henderson, T. B., and Bitterman, M. E. (1974). Simultaneous contrast in the goldfish. *Anim. Learn. Behav.* 2, 97–100.

Burt, D. E. and Wike, E. L. (1963). Effects of alternating partial reinforcement and alternating delay of reinforcement on a runway response. *Psychol. Rep.* **13**, 439–42.

Butcher, H. J. (1968) *Human intelligence: its nature and assessment.* Methuen, London.

Butler, A. B. and Northcutt, R. G. (1971*a*). Retinal projections in *Iguana iguana* and *Anolis carolinensis. Brian Res.* **26**, 1–13.

— — (1971*b*). Ascending tectal efferent projections in the lizard *Iguana iguana. Brain Res.* **35**, 597–601.

— — (1978). New thalamic visual nuclei in lizards. *Brain Res.* **149**, 469–76.

Butter, C. M. (1969). Perseveration in extinction and in discrimination reversal tasks following selective frontal ablations in *Macaca mulatta. Physiol. Behav.* **4**, 163–71.

Butters, N. and Cermak, L. S. (1975). Some analyses of amnesic syndromes in brain damaged patients. In *The hippocampus*, Vol. 2 *Neurophysiology and behavior* (ed. R. L. Isaacson and K. H. Pribram) pp. 377–409. Plenum Press, New York.

Campbell, A. A. (1935). Community of function in the performance of rats on alley mazes and the Maier reasoning apparatus. *J. comp. Psychol.* **19**, 69–76.

Campbell, C. B. G. and Boord, R. L. (1974). Central auditory pathways of non-mammalian vertebrates. In *Handbook of sensory physiology*, Vol. 5 *Part 1: Auditory system* (ed. W. D. Keidel and W. D. Neff) pp. 337–62. Springer, Berlin.

— and Ebbesson, S. O. E. (1969). The optic system of a teleost: *Holocentrus* re-examined. *Brain Behav. Evol.* **2**, 415–30.

Cannon, R. E. and Salzen, E. A. (1971). Brain stimulation in newly-hatched chicks. *Anim. Behav.* **19**, 375–85.

Capaldi, E. J. (1967). A sequential hypothesis of instrumental learning. In *The psychology of learning and motivation*, Vol. 1 (ed. K. W. Spence and J. T. Spence) pp. 67–156. Academic Press, New York.

Carter, D. E. and Eckerman, D. A. (1975). Symbolic matching by pigeons: Rate of learning complex discriminations predicted from simple discriminations. *Science, N.Y.* **187**, 662–4.

— and Werner, T. J. (1978). Complex learning and information processing by pigeons: a critical analysis. *J. exp. Anal. Behav.* **29**, 565–601.

Cermak, L. S., Butters, N., and Moreines, J. (1974). Some analyses of the verbal encoding deficit of alcoholic Korsakoff patients. *Brain Lang.* **1**, 141–50.

Charig, A. J. (1976). 'Dinosaur monophyly and a new class of vertebrates': a critical review. in *Morphology and biology of reptiles* (ed. A. D'A. Bellairs and C. B. Cox) pp. 65–104. Academic Press, London.

Chen, C.-S. (1967). Can rats count? *Nature, Lond.* **214**, 15–17.

Cherkin, A. (1969). Kinetics of memory consolidation: role of amnesic treatment parameters. *Proc. nat. Acad. Sci. U.S.A.* **63**, 1094–101.

Chiszar, D., Carter, T., Knight, L., Simonsen, L., and Taylor, S. (1976). Investigatory behavior in the plains garter snake (*Thamnophis radix*) and several additional species. *Anim. Learn. Behav.* **4**, 273–8.

Chomsky, N. (1972). *Language and mind*, enlarged edn. Harcourt Brace Jovanovich, New York.

Clark, E. (1959). Instrumental conditioning of lemon sharks. *Science, N.Y.* **130**, 217–18.

Clayton, F. L. and Hinde, R. A. (1968). The habituation and recovery of aggressive display in *Betta splendens*. *Behaviour* **30**, 96–106.

Cochrane, T. L., Scobie, S. R., and Fallon, D. (1973). Negative contrast in goldfish (*Carassius auratus*). *Bull. Psychon. Soc.* **1**, 411–13.

Cogan, D. and Capaldi, E. J. (1961). Relative effects of delayed reinforcement and partial reinforcement on acquisition and extinction. *Psychol. Rep.* **9**, 7–13.

Cohen, D. H. (1975). Involvement of the avian amygdalar homoloque (archistriatum posterior and mediale) in defensively conditioned heart rate change. *J. comp. Neurol.* **160**, 13–36.

— Duff, T. A., and Ebbesson, S. O. E. (1973). Electrophysiological identification of a visual area in shark telecephalon. *Science, N.Y.* **182**, 492–4.

— and Durkovic, R. G. (1966). Cardiac and respiratory conditioning, differentiation, and extinction in the pigeon. *J. exp. Anal. Behav.* **9**, 681–8.

— and Goff, D. M. (1978). Effect of avian basal forebrain lesions, including septum, on heart rate conditioning. *Brain Res. Bull.* **3**, 311–18.

— and Karten, H. J. (1974. The structural organization of avian brain: an overview. In *Birds: brain and behavior* (ed. I. J. Goodman and M. W. Schein) pp.29–73. Academic Press, New York.

— and MacDonald, R. L. (1971). Some variables affecting orienting and conditioned heart-rate responses in the pigeon. *J. comp. physiol. Psychol.* **74**, 123–33.

Colbert, E. H. (1955). *Evolution of the vertebrates: a history of the backboned animals through time*. Wiley, New York.

Coleman, S. R. (1975). Consequences of response-contingent change in unconditioned stimulus intensity upon the rabbit (*Oryctolagus cuniculus*) nictitating membrane response. *J. comp. physiol. Psychol.* **88**, 591–5.

Collias, N. E. (1980). Basal telecephalon suffices for early socialization in chicks. *Physiol. Behav.* **24**, 93–7.

Collins, R. L. (1975). When left-handed mice live in right-handed worlds. *Science, N.Y.* **187**, 181–4.

Commins, W. D., McNemar, Q., and Stone, C. P. (1932). Intercorrelations of measures of ability in the rat. *J. comp. Psychol.* **14**, 225–35.

Cooper, H. (1974). Learning set in *Lemur macaco*. In *Prosimian biology* (ed. R. D. Martin, G. A. Doyle, and A. C. Walker) pp. 293–300. Duckworth, London.
Couvillon, P. A., Tennant, W. A., and Bitterman, M. E. (1976). Intradimensional vs. extradimensional transfer in the discriminative learning of goldfish and pigeons. *Anim. Learn. Behav.* **4**, 197–203.
Cowey, A. and Cross. C. G. (1970). Effects of foveal prestriate and inferotemporal lesions on visual discrimination by rhesus monkeys. *Expl Brain Res.* **11**, 128–44.
Craik, F. I. M. and Lockhart, R. S. (1972). Levels of processing: a framework for memory research. *J. verbal Learn. verbal Behav.* **11**, 671–84.
Crawford, F. T. and Holmes, C. E. (1966). Escape conditioning in snakes employing vibratory stimuli. *Psychon. Sci.* **4**, 125–6.
Crespi, L. P. (1942). Quantitative variation of incentive and performance in the white rat. *Am. J. Psychol.* **55**, 467–517.
Crile, G. and Quiring, D. P. (1940). A record of the body weight and certain organ and gland weights of 3690 animals. *Ohio J. Sci.* **40**, 219–59.
Cromer, R. F. (1978). The basis of childhood dysphasia: a linguistic approach. In *Developmental dysphasia* (ed. M. A. Wyke) pp. 85–134. Academic Press London.
Crowder, R. G. (1976). *Principles of learning and memory*. Erlbaum, Hillsdale, NJ.
Cruce, W. L. R. (1975). Termination of supraspinal descending pathways in the spinal cord of the Tegu lizard (*Tupinambis nigropunctatus*). *Brain Behav. Evol.* **12**, 247–69.
Cumming, W. W. and Berryman, R. (1961). Some data on matching behavior in the pigeon. *J. exp. Anal. Behav.* **4**, 281–4.
Cutting, J. E., Rosner, B. S., and Foard, C. F. (1976). Perceptual categories for musiclike sounds: implications for theories of speech perception. *Q.J. exp. Psychol.* **28**, 361–78.
Czaplicki, J. A. (1975). Habituation of the chemically elicited prey-attack response in the diamond-backed water snake, *Natrix rhombifera rhombifera*. *Herpetologica* **31**, 403–9.
— Porter, R. H., and Wilcoxon, H. C. (1975. Olfactory mimicry involving garter snakes and artificial models and mimics. *Behaviour* **54**, 60–71.
Dabrowska, J. (1971). Dissociation of impairment after lateral and medial prefrontal lesions in dogs. *Science, N.Y.* **171**, 1037–8.
Dafters, R. (1975). Active avoidance behavior following archistriatal lesions in pigeons. *J. comp. physiol. Psychol.* **89**, 1169–79.
— (1976). Effect of medial archistriatal lesions on the conditioned emotional response and on auditory discrimination performance of the pigeon. *Physiol. Behav.* **17**, 659–65.
D'Amato, M. R. and Jagoda, H. (1960). Effects of extinction trials on discrimination reversal. *J. exp. Psychol.* **59**, 254–60.
Darwin, C. (1866). *On the origin of species by means of natural selection or the preservation of favoured races in the struggle for life*. 4th edn. John Murray, London.
Davidson, R. E. and Richardson, A. M. (1970). Classical conditioning of skeletal and autonomic responses in the lizard (*Crotaphytus collaris*). *Physiol. Behav.* **5**, 589–94.
Davidson, R. S. (1966). Operant stimulus control applied to maze behavior: heat escape conditioning and discrimination reversal in *Alligator mississippiensis*. *J. exp. Anal. Behav.* **9**, 671–6.

Davis, J. L. and Coates, S. R. (1978). Classical conditioning of the nictitating membrane response in the domestic chick. *Physiol. Psychol.* **6**, 7–10.

Davis, R. T., Leary, R. W., Stevens, D. A., and Thompson, R. (1967). Learning and perception of oddity problems by lemurs and 7 species of monkeys. *Primates* **8**, 311–22.

De Fina, A. V. and Webster, D. B. (1974). Projections of the intraotic ganglion to the medullary nuclei in the Tegu lizard *Tupinambis nigropunctatus. Brain Behav. Evol.* **10**, 197–211.

Dean, P. (1976). Effects of inferotemporal lesions on the behavior of monkeys. *Psychol. Bull.* **83**, 41–71.

Deets, A. C., Harlow, H. F., and Blomquist, A. J. (1970). Effects of intertrial interval and Trial 1 reward during acquisition of an object-discrimination learning set in monkeys. *J. comp. physiol. Psychol.* **73**, 501–5.

Delius, J. D. and Bennetto, K. (1972). Cutaneous sensory projections to the avian forebrain. *Brain Res.* **37**, 205–21.

— Runge, T. E., and Oeckinghaus, H. (1979). Short-latency auditory projection to the frontal telencephalon of the pigeon. *Expl Neurol.* **63**, 594–609

Demski, L. S. and Knigge, K. M. (1971). The telencephalon and hypothalamus of the bluegill (*Lepomis macrochirus*): evoked feeding, aggressive and reproductive behavior with representative frontal sections. *J. comp. Neurol.* **143**, 1–16.

Dennis, M. and Whitaker, H. A. (1976). Language acquisition following hemidecortication: superiority of the left over the right hemisphere. *Brain Lang.* **3**, 404–33.

Devine, J. V. (1970). Stimulus attributes and training procedures in learning-set formation of rhesus and cebus monkeys. *J. comp. physiol. Psychol.* **73**, 62–7.

Dewsbury, D. A. and Bernstein, J. J. (1969). Role of the telencephalon in performance of conditioned avoidance responses by goldfish. *Expl Neurol.* **23**, 445–56.

Dewson, J. H. (1977). Preliminary evidence of hemispheric asymmetry of auditory function in monkeys. In *Lateralization in the nervous system* (ed. S. Harnad, R. W. Doty, L. Goldstein, J. Jaynes, and G. Krauthamer) pp. 63–71. Academic Press, New York.

Di Cara, L. V., Braun, J. J., and Pappas, B. A. (1970). Classical conditioning and instrumental learning of cardiac and gastrointestinal responses following removal of neocortex in the rat. *J. comp. physiol. Psychol.* **73**, 208–16.

Di Lollo, V. and Beez, V. (1966). Negative contrast effects as a function of magnitude of reward decrements. *Psychon. Sci.* **5**, 99–100.

Diamond, I. T. (1979). The subdivisions of neocortex: a proposal to revise the traditional view of sensory, motor, and association areas. In *Progress in psychobiology and physiological psychology*, Vol. 8 (ed. J. M. Sprague and A. N. Epstein) pp. 1–43. Academic Press, New York.

Dickinson, A., Hall. G., and Mackintosh, N. J. (1976). Surprise and the attenuation of blocking. *J. exp. Psychol: Anim. Behav. Processes* **2**, 313–22.

— and Mackintosh, N. J. (1978). Classical conditioning in animals. *A. Rev. Psychol.* **29**, 587–612.

Dicks, D., Myers, R. E., and Kling, A. (1969). Uncus and amygdala lesions: Effects on social behavior in the free-ranging rhesus monkey. *Science, N.Y.* **165**, 69–71.

Diebschlag, E. (1934). Zur Kenntnis der Grosshirnfunktionen einiger Urodelen und Anuren. Z. vergleich. Physiol., **21**, 343–94.

— (1938). Beobachtungen und Versuche an intakten und grosshirnlosen Eideschsen und Ringelnattern. *Zool. Anz.* **124**, 30–40.
Dijkgraaf, S. (1949). Lokalisationsversuche am Fischgehirn. *Experientia* **5**, 44–5.
Distel, H. (1978). Behavior and electrical brain stimulation in the green iguana, *Iguana iguana L.* II. Stimulation effects. *Expl Brain Res.* **31**, 353–67.
Divac, I. (1968). Effects of prefrontal and caudate lesions on delayed response in cats. *Acta biol. exp.* **28**, 149–67.
— (1972). Delayed alternation in cats with lesions of the prefrontal cortex and the caudate nucleus. *Physiol. Behav.* **8**, 519–22.
— and Warren, J. M. (1971). Delayed response by frontal monkeys in the Nencki testing situation. *Neuropsychologia* **9**, 209–17.
Donovan, W. J. (1978). Structure and function of the pigeon visual system. *Physiol. Psychol.* **6**, 403–37.
Doty, B. A. and Combs, W. C. (1969). Reversal learning of object and positional discriminations by mink, ferrets and skunks. *Q.J. exp. Psychol.* **21**, 58–62.
— Jones, C. N., and Doty, L. A. (1967). Learning-set formation by mink, ferrets, skunks and cats. *Science, N.Y.* **155**, 1579–80.
Dricker, J., Butters, N., Berman, G., Samuels, I., and Carey, S. (1978). The recognition and encoding of faces by alcoholic Korsakoff and right hemisphere patients. *Neuropsychologia* **16**, 683–95.
Dubois, E. (1897). Sur le rapport du poids de l'encephale avec la grandeur du corps chez mammiferes. *Bull. Soc. anthropol., Paris (4)* **8**, 337–76.
Dufort, R. H., Guttman, N., and Kimble, G. A. (1954). One-trial discrimination reversal in the white rat. *J. comp. physiol. Psychol.* **47**, 248–9.
Ebbesson, S. O. E. (1968). Retinal projections in two teleost fishes (*Opsanus tau* and *Gymnothorax funebris*). An experimental study with silver impregnation methods. *Brain Behav. Evol.* **1**, 134–54.
— (1972). New insights into the organization of the shark brain. *Comp. Biochem. Physiol.* **42A**, 121–9.
— and Heimer, L. (1970). Projections of the olfactory tract fibres in the nurse shark. (*Ginglymostoma cirratum*). *Brain Res.* **17**, 47–55.
— and Northcutt, R. G. (1976). Neurology of anamniotic vertebrates. In *Evolution of brain and behavior in vertebrates* (ed. R. B. Masterton, M. E. Bitterman, C. B. G. Campbell, and N. Hotton) pp. 115–46. Erlbaum, Hillsdale, NJ.
— and Vanegas, H. (1976). Projection of the optic tectum in two teleost species. *J. comp. Neurol.* **165**, 161–80.
— and Voneida, T. J. (1969). The cytoarchitecture of the pallium in the Tegu lizard (*Tupinambis nigropunctatus*). *Brain Behav. Evol.* **2**, 431–66.
Ebner, F. F. (1969). A comparison of primitive forebrain organization in metatherian and eutherian mammals. *Ann. N.Y. Acad. Sci.* **167**, 241–57.
— (1976). The forebrain of reptiles and mammals. In *Evolution of brain and behavior in vertebrates* (ed. R. B. Masterton, C. B. G. Campbell, M. E. Bitterman, and N. Hotton) pp. 147–67. Erlbaum, Hillsdale, NJ.
Eikmanns, K. (1955). Verhaltensphysiologische Untersuchungen über den Beutefang und das Bewegungssehen der Erdkrote (*Bufo bufo* L.). *Z. Tierpsychol.* **12**, 229–53.
Elias, H. and Schwartz, D. (1969). Surface areas of the cerebral cortex of mammals determined by stereological methods. *Science, N.Y.* **166**, 111–13.
Elliott, R. C. (1977). Cross-modal recognition in three primates. *Neuropsychologia* **15**, 183–6.
Engelhardt, F., Woodard, W. T., and Bitterman, M. E. (1973). Discrimination

reversal in the goldfish as a function of training conditions. *J. comp. physiol. Psychol.* **85**, 144–50.
Epstein, R., Lanza, R. P., and Skinner, B. F. (1980). Symbolic communication between two pigeons (*Columba livia domestica*). *Science, N.Y.* **207**, 543–5.
Eskin, R. M. and Bitterman, M. E. (1961). Partial reinforcement in the turtle. *Q.J. exp. Psychol.* **13**, 112–16.
Estes, W. K. (1970). *Learning theory and mental development*. Academic Press, New York.
Etkin, M. W. (1972). Light produced interference in a delayed matching task with capuchin monkeys. *Learn. Motiv.* **3**, 313–24.
Ettlinger, G. (1960). Cross-modal transfer of training in monkeys. *Behaviour* **16**, 56–65.
— (1967). Analysis of cross-modal effects and their relationship to language. In *Brain mechanisms underlying speech and language* (ed. C. H. Millikan and F. L. Darley) pp. 53–60. Grune and Stratton, New York.
Evans, E. F. and Whitfield, I. C. (1964). Classification of unit responses in the auditory cortex of the unanaesthetized and unrestrained cat. *J. Physiol., Lond.* **171**, 476–93.
Everett, N. B. (1965). *Functional neuroanatomy*, 5th edn. Lea and Febiger, Philadelphia.
Ewert, J. P. (1967). Untersuchungen uber die Anteile zentralnervoser Aktionen an der taxisspezifischen Ermudung beim Beutefang der Erdkrote (*Bufo bufo* L.). *Z. vergl. Physiol.* **57**, 263–98.
— (1970). Neural mechanisms of prey-catching and avoidance behavior in the toad (*Bufo bufo* L.). *Brain Behav. Evol.* **3**, 36–56.
— (1971). Single unit responses of the toad's (*Bufo americanus*) caudal thalamus to visual objects. *Z. vergl. Physiol.* **74**, 81–102.
— (1976). The visual system of the toad: behavioral and physiological studies on a pattern recognition system. In *The amphibian visual system* (ed. K. V. Fite) pp. 141–202. Academic Press, New York.
— and Ingle, D. (1971). Excitatory effects following habituation of prey-catching activity in frogs and toads. *J. comp. physiol. Psychol.* **77**, 369–74.
Fankhauser, G., Vernon, J. A., Frank, W. H., and Slack, W. V. (1955). Effect of size and number of brain cells on learning in larvae of the salamander, *Triturus viridescens*. *Science, N. Y.* **122**, 692–3.
Fantino, E., Weigele, S., and Lancy, D. (1972). Aggressive display in the Siamese fighting fish (*Betta splendens*). *Learn. Motiv.* **3**, 457–68.
Farel, P. B. and Buerger, A. A. (1972). Instrumental conditioning of leg position in chronic spinal frog: before and after sciatic section. *Brain Res.* **47**, 345–51.
Farris, H. E. and Breuning, S. E. (1977). Post-conditioning habituation and classically conditioned head withdrawal in the red-eared turtle (*Chrysemys scripta elegans*). *Psychol. Rec.* **27**, 307–13.
Fearing, F. S. (1926). Post-rotational head nystagmus in adult pigeons. *J. comp. Psychol.* **6**, 115–31.
Finger, T. E. (1980). Nonolfactory sensory pathway to the telencephalon in a teleost fish. *Science, N.Y.* **210**, 671–3.
and Karten, J. J. (1978). The accessory optic system in teleosts. *Brain Res.* **153**, 144–9.
Flaherty, C. F., Riley, E. P., and Spear, N. E. (1973). Effects of sucrose concentration and goal units on runway behavior in the rat. *Learn. Motiv.* **4**, 163–75.

Fletcher, H. J. (1965). The delayed-response problem. In *Behavior of nonhuman primates: modern research trends*, Vol. 1 (ed. A. M. Schrier, H. F. Harlow, and F. Stollnitz) pp. 129–65. Academic Press, New York.

Flood, N. B. (1975). Effect of forebrain ablation on long-term retention of an appetitive discrimination by goldfish. *Psychol. Rep.* **36**, 783–6.

— and Overmier, J. B. (1971). Effect of telencephalic and olfactory lesions on appetitive learning in goldfish. *Physiol. Behav.* **6**, 35–40.

— — and Savage, G. E. (1976). Teleost telencephalon and learning: an interpretative review of data and hypotheses. *Physiol. Behav.* **16**, 783–98.

Foree, D. D. and LoLordo, V. M. (1970). Signalled and unsignalled free-operant avoidance in the pigeon. *J. exp. Anal. Behav.* **13**, 283–90.

Foster, R. E. and Hall, W. C. (1978). The organization of central auditory pathways in a reptile, *Iguana iguana*. *J. comp. Neurol.* **178**, 783–832.

Fox. R., Lehmkuhle, S. W., and Bush, R. C. (1977). Stereopsis in the falcon. *Science, N.Y.* **197**, 79–81.

Frank, A. H., Flood, N. B., and Overmier, J. B. (1972). Reversal learning in forebrain-ablated and olfactory tract sectioned teleost, *Carassius auratus*. *Psychon. Sci.* **26**, 149–51.

Franzisket, L. (1963). Characteristics of instinctive behaviour and learning in reflex activity of the frog. *Anim. Behav.* **11**, 318–24.

Friedman, H. and Marshall, D. A. (1965). Position reversal training in the Virginia opposum: evidence for the acquisition of a learning set. *Q. J. exp. Psychol.* **17**, 250–4.

Fuenzalida, C. E. Ulrich, G., and Ichikawa, B. T. (1975). Response decrement to repeated shadow stimuli in the garter snake, *Thamnophis radix*. *Bull. psychon. Soc.* **5**, 221–2.

Fujita, O. and Oi, S. (1969). Effect of forebrain ablation and inter-trial intervals upon conditioned avoidance responses in the goldfish *Ann. Anim. Psychol.* **19**, 39–47.

Gaffan, D. and Weiskrantz, L. (1980). Recency effects and lesion effects in delayed non-matching to randomly baited samples by monkeys. *Brain Res.* **196**, 373–86.

Galaburda, A. M., LeMay, M., Kemper, T. L., and Geschwind, N. (1978). Right-left asymmetries in the brain. *Science, N.Y.* **199**, 852–6.

Gallon, R. L. (1972). Effect of pretraining with fear and escape conditioning on shuttlebox avoidance acquisition by goldfish. *Psychol. Rep.* **31**, 919–24.

Gallup, G. G. (1974). Animal hypnosis: factual status of a fictional concept. *Psychol. Bull.* **81**, 836–53.

Gamble, H. J. (1952). An experimental study of the secondary olfactory connexions in *Lacerta viridis*. *J. Anat., Lond.* **86**, 180–96.

— (1956). An experimental study of the secondary olfactory connexions in *Testudo graeca*. *J. Anat., Lond.* **90**, 15–29.

Gamzu, E. and Schwartz, B. (1973). The maintenance of key-pecking by stimulus-contingent and response-independent food presentation. *J. exp. Anal. Behav.* **19**, 65–72.

Garcia, J., Hankins, W. G., and Rusiniak, K. W. (1974). Behavioral regulation of the milieu interne in man and rat. *Science, N.Y.* **185**, 824–31.

Gardner, B. T. and Gardner, R. A. (1971). Two-way communication with an infant chimpanzee. In *Behavior of nonhuman primates: Modern Research Trends*, Vol. 4 (ed. A. M. Schrier and F. Stollnitz) pp. 117–84. Academic Press, New York.

— — (1975). Evidence for sentence constituents in the early utterances of child and chimpanzee. *J. exp. Psychol: Gen.* **104**, 244–67.
Gardner, R. A. and Gardner, B. T. (1969). Teaching sign language to a chimpanzee. *Science, N.Y.* **165**, 664–72.
— — (1978). Comparative psychology and language acquisition. *Ann. N.Y. Acad. Sci.* **309**, 37–76.
Geller, I. (1963). Conditioned 'anxiety' and punishment effects on operant behavior of goldfish (*Carassius auratus*). *Science, N.Y.* **141**, 351–3.
Geschwind, N. (1964). The development of the brain and the evolution of language. In *Monograph series on language and linguistics*, Vol. 17 (ed. C. I. J. M. Stuart) pp. 155–69. Georgetown University Press, Washington.
— (1970). The organization of language and the brain. *Science, N.Y.* **170**, 940–4.
Glanzer, M. and Cunitz, A. R. (1966). Two storage mechanisms in free recall. *J. verbal Learn. verbal. Behav.* **5**, 351–6.
Glanzman, D. L. and Schmidt, E. C. (1979). Habituation of the nictitating membrane reflex response in the intact frog. *Physiol. Behav.* **22**, 1141–8.
Glick, S. D., Weaver, L. M., and Meibach, R. C. (1980). Lateralization of reward in rats: differences in reinforcing thresholds. *Science, N.Y.* **207**, 1093–5.
Goddard, G. V. (1964). Functions of the amygdala. *Psychol. Bull.* **62**, 89–109.
Goldberg, M. E. and Robinson, D. L. (1978). Visual system: superior colliculus. In *Handbook of behavioral neurobiology*, Vol. 1 (ed. R. B. Masterton) pp. 119–64. Plenum Press, New York.
Goldby, F. (1937). An experimental investigation of the cerebral hemispheres of *Lacerta viridis*. *J. Anat., Lond.* **71**, 332–55.
— and Gamble, H. J. (1957). The reptilian cerebral hemispheres. *Biol. Rev.* **32**, 383–420.
Goldman, P. S., Rosvold, H. E., Vest, B., and Galkin, T. W. (1971). Analysis of the delayed-alternation deficit produced by dorsolateral frontal lesions in the rhesus monkey. *J. comp. physiol. Psychol.* **77**, 212–20.
Goldstein, A. C., Spies, G., and Sepinwall, J. (1964). Conditioning of the nictitating membrane in the frog, *Rana p. pipiens*. *J. comp. physiol. Psychol.* **57**, 456–8.
Gonzalez, R. C. (1972). Patterning in goldfish as a function of magnitude of reinforcement. *Psychon. Sci.* **28**, 53–5.
— Behrend, E. R., and Bitterman, M. E. (1965). Partial reinforcement in the fish: experiments with spaced trials and partial delay. *Am. J. Psychol.* **78**, 198–207.
— — — (1967). Reversal learning and forgetting in bird and fish. *Science, N.Y.* **158**, 519–21.
— Berger, B. D., and Bitterman, M. E. (1966*a*). A further comparison of key-pecking with an ingestive technique for the study of discriminative learning in pigeons. *Am. J. Psychol.* **79**, 217–25.
— — — (1966*b*). Improvement in habit-reversal as a function of amount of training per reversal and other variables. *Am. J. Psychol.* **79**, 517–30.
— and Bitterman, M. E. (1962). A further study of partial reinforcement in the turtle. *Q. J. exp. Psychol.* **14**, 109–12.
— — (1967). Partial reinforcement effect in the goldfish as a function of amount of reward. *J. comp. physiol. Psychol.* **64**, 163–7.
— — (1969). Spaced-trials partial reinforcement effect as a function of contrast. *J. comp. physiol. Psychol.* **67**, 94–103.
— and Champlin, G. (1974). Positive behavioral contrast, simultaneous contrast

and their relation to frustration in pigeons. *J. comp. physiol. Psychol.* **87**, 173–87.
— Eskin, R. M., and Bitterman, M. E. (1961). Alternating and random partial reinforcement in the fish, with some observations on asymptotic resistance to extinction. *Am. J. Psychol.* **74**, 561–8.
— — — (1962), Extinction in the fish after partial and consistent reinforcement with number of reinforcements equated. *J. comp. physiol. Psychol.* **55**, 381–6.
— — — (1963). Further experiments on partial reinforcement in the fish. *Am. J. Psychol.* **76**, 366–75.
— Ferry, M., and Powers, A. S. (1974). The adjustment of goldfish to reduction in magnitude of reward in massed trials. *Anim. Learn. Behav.* 2, 23–6.
— Holmes, N. K., and Bitterman, M. E. (1967*a*). Asymptotic resistance to extinction in fish and rat as a function of interpolated training. *J. comp. physiol. Psychol.* **63**, 342–4.
— — — (1967*b*). Resistance to extinction in the goldfish as a function of frequency and amount of reward. *Am. J. Psychol.* **80**, 269–75.
— Longo, N., and Bitterman, M. E. (1961). Classical conditioning in the fish: exploratory studies of partial reinforcement. *J. comp. physiol. Psychol.* **54**, 452–6.
— Milstein, S., and Bitterman, M. E. (1962). Classical conditioning in the fish: further studies of partial reinforcement. *Am. J. Psychol.* **75**, 421–8.
— Potts, A., Pitcoff, K., and Bitterman, M. E. (1972). Runway performance of goldfish as a function of complete and incomplete reduction in amount of reward. *Psychon. Sci.* **27**, 305–7.
— and Powers, A. S. (1973). Simultaneous contrast in goldfish. *Anim. Learn. Behav.* **1**, 96–8.
— Roberts, W. A., and Bitterman, M. E. (1964). Learning in adult rats with extensive cortical lesions made in infancy. *Am. J. Psychol.* **77**, 547–62.
Goodman, D. A. and Weinberger, N. M. (1973). Habituation in 'lower' tetrapod vertebrates: amphibia as vertebrate model systems. In *Habituation*, Vol. 2 *Physiological substrates* (ed. H. V. S. Peeke and M. J. Herz) pp. 85–140. Academic Press, New York.
Goodman, I. J. (1970). Approach and avoidance effects of central stimulation: an exploration of the pigeon fore- and midbrain. *Psychon. Sci.* **19**, 39–40.
Gordon, D. (1979). Effects of forebrain ablation on taste aversion in goldfish (*Carassius auratus*). *Expl Neurol.* **63**, 356–66.
Gormezano, I. (1965). Yoked comparisons of classical and instrumental conditioning of the eyelid response; and an addendum on 'voluntary responders'. In *Classical conditioning: a symposium* (ed. W. F. Prokasy) pp. 48–70. Appleton-Century-Crofts, New York.
— (1972). Investigations of defense and reward conditioning in the rabbit. In *Classical conditioning II: current research and theory* (ed. A. H. Black and W. F. Prokasy) pp. 151–81. Appleton-Century-Crofts, New York.
— Schneiderman, N., Deaux, E., and Fuentes, I. (1962). Nictitating membrane: classical conditioning and extinction in the albino rabbit. *Science, N.Y.* **138**, 33–4.
Gossette, R. L. (1967). Successive discrimination reversal (SDR) performance of four avian species on a brightness discrimination task. *Psychon. Sci.* 8, 17–18.
— and Cohen, H. (1966). Error reduction by pigeons on a spatial successive reversal task under conditions of non-correction. *Psychol. Rep.* **18**, 367–70.

— and Gossette, M. F. (1967). Examination of the reversal index (RI) across fifteen different mammalian and avian species. *Percept. mot. Skills* **24**, 987–90.

— — and Inman, N. (1966). Successive discrimination reversal performance by the greater hill myna. *Anim. Behav.* **14**, 50–3.

— — and Riddell, W. (1966). Comparison of successive discrimination reversal performances among closely and remotely related avian species. *Anim. Behav.* **14**, 560–4.

— and Hombach, A. (1969). Successive discrimination reversal (SDR) performances of American alligators and American crocodiles on a spatial task. *Percept. mot. Skills* **28**, 63–7.

— and Hood, P. (1967). The reversal index (RI) as a joint function of drive and incentive level. *Psychon. Sci.* **8**, 217–18.

— — (1968). Successive discrimination reversal measures as a function of variations of motivational and incentive levels. *Percept. mot. Skills* **26**, 47–52.

— and Kraus, G. (1968). Successive discrimination reversal performance of mammalian species on a brightness task. *Percept. mot. Skills* **27**, 675–8.

— — and Speiss, J. (1968). Comparison of successive discrimination reversal (SDR) performances of seven mammalian species on a spatial task. *Psychon. Sci.* **12**, 193–4.

Gould, S. J. (1966). Allometry and size in ontogency and phylogeny. *Biol. Rev.* **41**, 587–640.

Graeber, R. C. (1978), Behavioral studies correlated with central nervous integration of vision in sharks. In *Sensory biology of sharks, skates, and rays* (ed. E. S. Hodgson and R. F. Mathewson) pp. 195–225. Office of Naval Research, Arlington, Va.

— and Ebbesson, S. O. E. (1972). Retinal projections in the lemon shark (*Negaprion brevirostris*). *Brain Behav. Evol.* **5**, 461–77.

— — and Jane, J. A. (1973). Visual discrimination in sharks without optic tectum. *Science, N.Y.* **180**, 413–15.

Graf, C. L. (1972). Spaced-trial partial reward in the lizard. *Psychon. Sci.* **27**, 153–4.

Graf, V. and Bitterman, M. E. (1963). General activity as instrumental: application to avoidance training. *J. exp. Anal. Behav.* **6**, 301–5.

— Bullock, D. H., and Bitterman, M. E. (1964). Further experiments on probability-matching in the pigeon. *J. exp. Anal. Behav.* **7**, 151–7.

— and Tighe, T. J. (1971). Subproblem analysis of discrimination shift learning in the turtle (*Chrysemys picta picta*). *Psychon. Sci.* **25**, 257–9.

Granda, A. M., Matsumiya, Y., and Stirling, C. E. (1965). A method for producing avoidance behavior in the turtle. *Psychon. Sci.* **2**, 187–8.

Gray, J. A. (1971). *The psychology of fear and stress*. McGraw-Hill, New York.

— (1975). *Elements of a two-process theory of learning;* Academic Press, London.

Gregory, R. L. (1961). The brain as an engineering problem. In *Current problems in animal behaviour* (ed. W. H. Thorpe and O. L. Zangwill) pp. 307–30. Cambridge University Press.

Gross, C. G. (1973). Inferotemporal cortex and vision. In *Progress in physiological psychology*, Vol. 5 (ed. E. Stellar and J. M. Sprague) pp. 77–123. Academic Press, New York.

— Cowey, A., and Manning, F. J. (1971). Further analysis of visual discrimination deficits following foveal prestriate and inferotemporal lesions in monkeys. *J. comp. physiol. Psychol.* **76**, 1–7.

Gruber, H. E., Girgus, J. S., and Banuazizi, A. (1971). The development of object permanence in the cat. *Devl Psychol.* **4**, 9–15.

Gruber, S. H. and Schneiderman, N. (1975). Classical conditioning of the nictitating membrane response of the lemon shark (*Negaprion brevirostris*). *Behav. Res. Meth. Instrum.* 7, 430–4.

Gruberg, E. R. and Ambros, V. R. (1974). A forebrain visual projection in the frog (*Rana pipiens*). *Exp Neurol.* **44**, 187–97.

Grüsser, O.-J. and Grüsser-Cornehls, U. (1976). Neurophysiology of the anuran visual system. In *Frog neurobiology: a handbook* (ed. R. Llinás and W. Precht) pp. 297–385. Springer, Berlin.

Gubernick, D. J. and Wright, J. W. (1979). Habituation of two response systems in the lizard, *Anolis carolinensis. Anim. Learn. Behav.* **7**, 125–32.

Gusel'nikov, V. I., Morenkov. E. D., and Pivovarov, A. S. (1971). Unit responses of the turtle forebrain to visual stimuli. *Neurosci. Behav. Physiol.* **5**, 235–42.

Hainsworth, F. R., Overmier, J. B., and Snowdon, C. T. (1967). Specific and permanent deficits in instrumental avoidance responding following forebrain ablation in the goldfish. *J. comp. physiol. Psychol.* **63**, 111–16.

Hale, E. B. (1956). Social facilitation and forebrain function in maze performance of green sunfish, *Lepomis cyanellus. Physiol. Zool.* **29**, 93–107.

Hall, W. C. (1972). Visual pathways to the telencephalon in reptiles and mammals. *Brain Behav. Evol.* **5**, 95–113.

— and Ebner, F. F. (1970*a*). Thalamotelencephalic projections in the turtle (*Pseudemys scripta*). *J. comp. Neurol.* **140**, 101–22.

— — (1970*b*). Parallels in the visual afferent projections of the thalamus in the hedgehog (*Paraechinus hypomelas*) and the turtle (*Pseudemys scripta*). *Brain Behav. Evol.* **3**, 135–54.

Halpern, M. (1972). Some connections of the telencephalon of the frog, *Rana pipiens. Brain Behav. Evol.* **6**, 42–68.

— (1976). The efferent connections of the olfactory bulb and accessory olfactory bulb in the snakes, *Thamnophis sirtalis* and *Thamnophis radix. J. Morph.* **150**, 553–78.

— and Frumin, N. (1973). Retinal projections in a snake, *Thamnophis sirtalis. J. Morph.* **141**, 359–82.

Hamilton, C. R. (1977). Investigations of perceptual and mnemonic lateralization in monkeys. In *Lateralization in the nervous system* (ed. S. Harnad, R. W. Doty, L. Goldstein, J. Jaynes, and G. Krauthamer) pp. 45–62. Academic Press, New York.

Harlow, H. F. (1939). Forward conditioning, backward conditioning and pseudoconditioning in the goldfish. *J. genet. Psychol.* **55**, 49–58.

— (1949). The formation of learning sets. *Psychol. Rev.* **56**, 51–65.

Harman, P. J. (1957). Paleoneurologic, neoneurologic, and ontogenetic aspects of brain phylogeny. James Arthur Lecture on the Evolution of the Human Brain. American Museum of Natural History, New York.

Hayes, C. (1951). *The ape in our house*. Harper, New York.

Hayes, K. J. and Nissen, C. H. (1971). Higher mental functions of a home-raised chimpanzee. In *Behavior of nonhuman primates: modern research trends*, Vol. 4 (ed. A. M. Schrier and F. Stollnitz) pp. 59–115. Academic Press, New York.

Hayes, W. N. and Hertzler, D. R. (1967). Role of the optic tectum and general cortex in reptilian vision. *Psychon. Sci.* **9**, 521–2.

— — and Hogberg, D. K. (1968). Visual responsiveness and habituation in the turtle. *J. comp. physiol. Psychol.* **65**, 331–5.

— and Ireland, L. C. (1972). A study of visual orientation mechanisms in turtles and guinea pigs. *Brain Behav. Evol.* **5**, 226–39.

— and Saiff, E. I. (1967). Visual alarm reactions in turtles. *Anim. Behav.* **15**, 102–6.

Hearst, E. and Franklin, S. R. (1977). Positive and negative relations between a signal and food: approach-withdrawal behavior. *J. exp. Psychol: Anim. Behav. Processes* **3**, 37–52.

Heffner, R. and Masterton, B. (1975). Variation in the form of the pyramidal tract and its relationship to digital dexterity. *Brain Behav. Evol.* **12**, 161–200.

Heier, P. (1948). Fundamental principles in the structure of the brain: a study of the brain of *Petromyzon fluviatilis. Acta anat.* **5**, Suppl. 8, 1–213.

Herman, L. M. and Arbeit, W. R. (1973). Stimulus control and auditory discrimination learning sets in the bottlenose dolphin. *J. exp. Anal. Behav.* **19**, 379–94.

— Beach, F. A. Pepper, R. L., and Stalling, R. B. (1969). Learning-set formation in the bottlenose dolphin. *Psychon. Sci.* **14**, 98–9.

— and Gordon, J. A. (1974). Auditory delayed matching in the bottlenose dolphin. *J. exp. Anal. Behav.* **21**, 19–26.

Herrick, C. J. (1924). *Neurological foundations of animal behavior.* Henry Holt, New York.

— (1948). *The brain of the tiger salamander.* University of Chicago Press.

Herrnstein, R. J. and Loveland, D. H. (1964). Complex visual concept in the pigeon. *Science, N.Y.* **146**, 549–51.

— — and Cable, C. (1976). Natural concepts in pigeons. *J. exp. Psychol: Anim. Behav. Processes* **2**, 285–302.

Hershkowitz, M. and Samuel, D. (1973). The retention of learning during metamorphosis of the crested newt (*Triturus cristatus*). *Anim. Behav.* **21**, 83–5.

Hertzler, D. R. (1972). Tectal integration of visual input in the turtle. *Brain Behav. Evol.* **5**, 240–55.

— and Hayes, W. N. (1967). Cortical and tectal function in visually guided behavior of turtles. *J. comp. physiol. Psychol.* **63**, 444–7.

Hicks. L. H. (1956). An anlaysis of number-concept formation in the rhesus monkey. *J. comp. physiol. Psychol.* **49**, 212–18.

Hicks, R. E. and Kinsbourne, M. (1978). Human handedness. In *Asymmetrical function of the brain* (ed. M. Kinsbourne) pp. 523–49. Cambridge University Press.

Hinde, R. A. (1970). Behavioural habituation. In *Short-term changes in neural activity and behaviour* (ed. G. Horn and R. A Hinde) pp. 3–40. Cambridge University Press.

Hineline, P. N. and Rachlin, H. (1969). Escape and avoidance of shock by pigeons pecking a key. *J. exp. Anal. Behav.* **12**, 533–8.

Hines, D. (1975). Immediate and delayed recognition of sequentially presented random shapes. *J. exp. Psychol: Hum. Learn. Mem.* **1**, 634–9.

Hodos, W. (1976). Vision and the visual system: a bird's eye view. In *Progress in psychobiology and physiological psychology*, Vol. 6 (ed. J. M. Sprague and A. N. Epstein) pp. 29–62. Academic Press, New York.

— and Bonbright, J. C. (1974). Intensity difference thresholds in pigeons after lesions of the tectofugal and thalamofugal visual pathways. *J. comp. physiol. Psychol.* **87**, 1013–31.

— and Karten, H. J. (1966). Brightness and pattern discrimination deficits in the pigeon after lesions of the nucleus rotundus. *Expl Brain Res.* **2**, 151–67.

— — (1970). Visual intensity and pattern discrimination deficits after lesions of ectostriatum in pigeons. *J. comp. Neurol.* **140**, 53–68.

— — (1974). Visual intensity and pattern discrimination deficits after lesions of the optic lobe in pigeons. *Brain Behav. Evol.* **9**, 165–94.

— — and Bonbright, J. C. (1973). Visual intensity and pattern discrimination after lesions of the thalamofugal visual pathway in pigeons. *J. comp. Neurol.* **148**, 447–68.

Hoffman, H. S. (1965). The stimulus generalization of conditioned supression. In *Stimulus generalization* (ed. D. I. Mostofsky) pp. 356–72. Stanford University Press.

— and Fleshler, M. (1959). Aversive control with the pigeon. *J. exp. Anal. Behav.* **2**, 213-18.

— and Ratner, A. M. (1973). A reinforcement model of imprinting: implications for socialization in monkeys and man. *Psychol. Rev.* **80**, 527–44.

Hoffmann, A. and Lico, M. C. (1972*a*). Autonomic and motor responses to electrical stimulation of the toad's septum. *Physiol. Behav.* **8**, 1035–7.

— — (1972*b*). Autonomic and motor responses to electrical stimulation of the toad's brain. *Physiol. Behav.* **8**, 1039–43.

Hollard, V. and Davison, M. C. (1978). Histological data: Hollard and Davison (1971). *J. exp. Anal. Behav.* **29**, 149.

Holmes, P. A. and Bitterman, M. E. (1966). Spatial and visual habit reversal in the turtle. *J. comp. physiol. Psychol.* **62**, 328–31.

Honigmann, H. (1942). The number conception in animal psychology. *Biol. Rev.* **17**, 315–37.

Hoogland, P. V. (1977). Efferent projections of the striatum in *Tupinambis nigropunctatus. J. Morph.* **152**, 229–46.

Horel, J. A. (1978). The neuroanatomy of amnesia: a critique of the hippocampal memory hypothesis. *Brain* **101**, 403–45.

Horn, A. L. D. and Horn, G. (1969). Modification of leg flexion in response to repeated stimulation in a spinal amphibian (*Xenopus mullerei*). *Anim. Behav.* **17**, 618–23.

Horn, G., McCabe, B. J. and Bateson, P. P. G. (1979). An autoradiographic study of the chick brain after imprinting. *Brain Res.* **168**, 361–73.

Horridge, G. A. (1962). Learning of leg position by the ventral nerve cord in headless insects. *Proc. R. Soc.* **B157**, 33–52.

Hubel, D. H. and Wiesel, T. N. (1962). Receptive fields, binocular interaction and functional architecture in the cat's visual cortex. *J. Physiol., Lond.* **160**, 106–54.

Hull, C. L. (1945). The place of innate individual and species differences in a natural-science theory of behavior. *Psychol. Rev.* **52**, 55–60.

— (1951). *Essentials of behavior*. Yale University Press, New Haven.

Humphrey, G. (1933). *The nature of learning*. Harcourt, Brace, New York.

Hunter, M. W. and Kamil, A. C. (1971). Object-discrimination learning set and hypothesis behavior in the northern blue-jay (*Cyanocitta cristata*). *Psychon. Sci.* **22**, 271-3.

— — (1975). Marginal learning-set formation by the crow (*Corvus brachyrhynchos*). *Bull. psychon. Soc.* **5**, 373–5.

Hunter, W. S. (1913). The delayed reaction in animals and children. *Behav. Monogr.* 2, No. 1 (Serial No. 6).

Ince, L. P., Brucker, B. S., and Alba, A. (1978). Reflex conditioning in a spinal man. *J. comp. physiol. Psychol.* **92**, 796–802.

Ingle, D. (1973). Two visual systems in the frog. *Science, N.Y.* **181**, 1053–5.

— (1977). Detection of stationary objects by frogs (*Rana pipiens*) after ablation of optic tectum. *J. comp. physiol. Psychol.* **91**, 1359–64.
Ireland, L. C., Hayes, W. N., and Laddin, L. H. (1969). Relation between frequency and amplitude of visual alarm reactions in *Pseudemys scripta. Anim. Behav.* **17**, 386–8.
Ito, H., Morita, Y., Sakamoto, N., and Ueda, S. (1980). Possibility of telencephalic visual projection in teleosts, Holocentridae. *Brain Res.* **197**, 219–22.
Iwai, E., Saito, S., and Tsukahara, S. (1970). Analysis of central mechanism in visual discrimination learning of goldfish. *Tohoku J. exp. Med.* **102**, 135–42.
Jacobs, D. W. and Popper, A. N. (1962). Stimulus effectiveness in avoidance behavior in fish. *Psychon. Sci.* **12**, 109–10.
Jarvis, C. D. (1974). Visual discrimination and spatial localization deficits after lesions of the tectofugal pathway in pigeons. *Brain Behav. Evol.* **9**, 195–228.
Jarvis, M. J. and Ettlinger, G. (1977). Cross-modal recognition in chimpanzees and monkeys. *Neuropsychologia* **15**, 499–506.
Jenkins, H. M. and Moore, B. R. (1973). The form of the autoshaped response with food or water reinforcers. *J. exp. Anal. Behav.* **20**, 163–81.
Jennings, J. W. and Keefer, L. H. (1969). Olfactory learning set in two varieties of domestic rat. *Psychol. Rep.* **24**, 3–15.
Jensen, A. R. (1980). *Bias in mental testing*. Methuen, London.
Jerison, H. J. (1969). Brain evolution and dinosaur brains. *Am. Nat.* **103**, 575–88.
— (1973). *Evolution of the brain and intelligence*. Academic Press, New York.
— (1976). Paleoneurology and the evolution of mind. *Scient. Am.* **234**, 90–101.
Johanson, D. C. and White, T. D. (1979). A systematic assessment of early African hominids. *Science, N.Y.* **203**, 321–30.
Johnson, J. I. (1961). Double alternation by raccoons. *J. comp. physiol. Psychol.* **54**, 248–51.
— and Michels, K. M. (1958). Learning sets and object-size effects in visual discrimination learning by raccoons. *J. comp. physiol. Psychol.* **51**, 376–9.
Jones, A. W. and Levi-Montalcini, R. (1958). Patterns of differentiation of the nerve centers and fiber tracts in the avian cerebral hemispheres. *Archs ital. Biol.* **96**, 231–84.
Juorio, A. V. (1969). The distribution of dopamine in the brain of a tortoise, *Geochelone chilensis* (Gray). *J. Physiol., Lond.* **204**, 503–9.
Kaas, J. H. (1980). A comparative survey of visual cortex organization in mammals. In *Comparative neurology of the telencephalon* (ed. S. O. E. Ebbesson) pp. 483–502. Plenum Press, New York.
Källén, B. (1951*a*). Embryological studies on the nuclei and their homologization in the vertebrate forebrain. *K. fysiogr. Sällsk. Lund handling.* **62**, No. 5.
— (1951*b*). On the ontogeny of the reptilian forebrain. Nuclear structures and ventricular structures. *J. comp. Neurol.* **95**, 307–47.
— (1962). Embryogenesis of brain nuclei in the chick telencephalon. *Ergebn. Anat. EntwGesch.* **36**, 62–82.
Kamil, A. C. and Hunter, M. W. (1970). Performance on object-discrimination learning set by the greater hill myna (*Gracula religiosa*). *J. comp. physiol. Psychol.* **73**, 68–73.
— Jones, T. B., Pietrewicz, A., and Mauldin, J. E. (1977). Positive transfer from successive reversal training to learning set in blue jays (*Cyanocitta cristata*). *J. comp. physiol. Psychol.* **91**, 79–86.
— Lougee, M., and Shulman, R. I. (1973). Learning-set behavior in the learning-set experienced blue-jay (*Cyanocitta cristata*). *J. comp. physiol. Psychol.* **82**, 394–405.

— and Mauldin, J. E. (1975). Intraproblem retention during learning-set acquisition in blue jays (*Cyanocitta cristata*). *Anim. Learn. Behav.* 3, 125–30.

Kaplan, H. and Aronson, L. R. (1967). Effect of forebrain ablation on the performance of a conditioned avoidance response in the teleost fish, *Tilapia h. macrocephala*. *Anim. Behav.* **15**, 438–48.

Kappers, C. U. A., Huber, G. C., and Crosby, E. C. (1960). *The comparative anatomy of the nervous system of vertebrates, including man*, 3 Vols. (Reprint of 1936 edn.) Hafner, New York.

Karamian, A. I. (1962). *Evolution of the function of the cerebellum and cerebral hemispheres*. Israel Program for Scientific Translations, Jerusalem.

— Vesselkin, N. P., Belekhova, M. G., and Zagorulko, T. M. (1966). Electrophysiological characteristics of tectal and thalamo-cortical divisions of the visual system in lower vertebrates. *J. comp. Neurol.* **127**, 559–76.

Karten, H. J. (1968). The ascending auditory pathway in the pigeon (*Columba livia*) II. Telencephalic projections of the nucleus ovoidalis thalami. *Brain Res.* **11**, 134–53.

— and Dubbeldam, J. L. (1973). The organization and projections of the paleostratial complex in the pigeon (*Columba livia*). *J. comp. Neurol.* **148**, 61–90.

— and Hodos, W. (1967). *A stereotaxic atlas of the brain of the pigeon* (Columba livia). Johns Hopkins University Press, Baltimore, Md.

— — (1970). Telecephalic projections of the nucleus rotundus in the pigeon (*Columba livia*). *J. comp. Neurol.* **140**, 35–52.

— — Nauta, W. J. H., and Revzin, A. M. (1973). Neural connections of the 'visual Wulst' of the avian telencephalon. Experimental studies in the pigeon (*Columba livia*) and owl (*Speotyto cunicularia*). *J. comp. Neurol.* **150**, 253–78.

— and Revzin, A. M. (1966). The afferent connections of the nucleus rotundus in the pigeon. *Brain Res.* 2, 368–77.

Kaufman, E. L., Lord, M. W., Reese, T. W., and Volkmann, J. (1949). The discrimination of visual number. *Am. J. Psychol.* **62**, 498–525.

Kay, H. and Oldfield-Box, H. (1965). A study of learning-sets in rats with an apparatus using 3-dimensional shapes. *Anim. Behav.* **13**, 19–24.

Keating, E. G., Kormann, L. A., and Horel, J. A. (1970). The behavioral effects of stimulating and ablating the reptilian amygdala (*Caiman sklerops*). *Physiol. Behav.* **5**, 55–9.

Kelley, D. B. and Nottebohm, F. (1979). Projections of a telencephalic auditory nucleus—Field L—in the canary. *J. comp. Neurol.* **183**, 455–70.

Kellogg, W. N. (1968). Communication and language in the home-raised chimpanzee. *Science, N.Y.* **162**, 423–6.

— and Kellogg, L. A. (1933). *The ape and the child. A study of environmental influence upon early behavior*. Whittlesey House, New York.

Kemp, F. D. (1969) Thermal reinforcement and thermoregulatory behaviour in the lizard *Dipsosaurus dorsalis*: an operant technique. *Anim. Behav.* **17**, 446–51.

Kicliter, E. (1973). Flux, wavelength and movement discrimination in frogs: forebrain and midbrain contributions. *Brain Behav. Evol.* 8, 340–65.

— (1979). Some telencephalic connections in the frog, *Rana pipiens*. *J. comp. Neurol.* **185**, 75–86.

— and Ebbesson, S. O. E. (1976). Organization of 'nonolfactory' telencephalon. In *Frog neurobiology: a handbook* (ed. R. Llinás and W. Precht) pp. 946–72. Springer, Berlin.

Killackey, H., Snyder, M., and Diamond, I. T. (1971). Function of striate and temporal cortex in the tree shrew. *J. comp. physiol. Psychol.* **74**, Monograph No. 1, Part 2.
Kimberly, R. P., Holden, A. L., and Bamborough, P. (1971). Response characteristics of pigeon forebrain cells to visual stimulation. *Vision Res.* **11**, 475–8.
Kimble, D. P. and Ray, R. S. (1965). Reflex habituation and potentiation in *Rana pipiens. Anim. Behav.* **13**, 530–3.
Kirk, K. L. and Bitterman, M. E. (1965). Probability-learning by the turtle. *Science, N.Y.* **148**, 1484–5.
Kirkish, P. M., Fobes, J. L., and Richardson, A. M. (1979). Spatial reversal learning in the lizard *Coleonyx variegatus. Bull. Psychon. Soc.* **13**, 265–7.
Kleinginna, P. R. (1970). Operant conditioning in the indigo snake. *Psychon. Sci.* **18**, 53–5.
Klinman, C. S. and Bitterman, M. E. (1963). Classical conditioning in the fish: The CS–US interval. *J. comp. physiol. Psychol.* **56**, 578–83.
Knapp, H. and Kang, D. S. (1968*a*). The visual pathways of the snapping turtle (*Chelydra serpentina*). *Brain Behav. Evol.* **1**, 19–42.
—— —— (1968*b*). The retinal projections of the side-necked turtle (*Podocnemis unifilis*) with some notes on the possible origin of the pars dorsalis of the lateral geniculate body. *Brain Behav. Evol.* **1**, 369–404.
Knudsen, E. I. (1977). Distinct auditory and lateral line nuclei in the midbrain of catfishes. *J. comp. Neurol.* **173**, 417–32.
—— and Konishi, M. (1978). A neural map of auditory space in the owl. *Science, N.Y.* **200**, 795–7.
Koehler, O. (1950). The ability of birds to 'count'. *Bull. Anim. Behav.* **9**, 41–5.
Köhler, W. (1925). *The mentality of apes*. Routledge and Kegan Paul, London.
Kokoros, J. J. and Northcutt, R. G. (1977). Telencephalic efferents of the tiger salamander *Ambystoma tigrinum tigrinum* (Green). *J. comp. Neurol.* **173**, 613–28.
Kolb, B., Nonneman, A. J., and Singh, R. K. (1974). Double dissociation of spatial impairments and perseveration following selective prefrontal lesions in rats. *J. comp. physiol. Psychol.* **87**, 772–80.
Koronakos, C. and Arnold, W. J. (1957). The formation of learning sets in rats. *J. comp. physiol. Psychol.* **50**, 11–14.
Krayniak, P. F. and Siegel, A. (1978*a*). Efferent connections of the hippocampus and adjacent regions in the pigeon. *Brain Behav. Evol.* **15**, 372–88.
—— —— (1978*b*). Efferent connections of the septal area in the pigeon. *Brain Behav. Evol.* **15**, 389–404.
Krekorian, C. O., Vance, V. J., and Richardson, A. M. (1968). Temperature-dependent maze learning in the desert iguana, *Dipsosaurus dorsalis. Anim. Behav.* **16**, 429–36.
Kremer, E. F. (1979). Effect of post-trial episodes on conditioning in compound conditioned stimuli. *J. exp. Psychol: Anim. Behav. Processes* **5**, 130–41.
Krompecher, S. and Lipak, J. A. (1966). A simple method for determining cerebralization: brain weight and intelligence. *J. comp. Neurol.* **127**, 113–20.
Kruger, L. (1969). Experimental analyses of the reptilian nervous system. *Ann. N. Y. Acad. Sci.* **167**. 102–17.
—— and Berkowitz, E. C. (1960). The main afferent connections of the reptilian telencephalon as determined by degeneration and electrophysiological methods. *J. comp. Neurol.* **115**, 125–42.
Kuczka, H. (1956). Verhaltenphysiologische Untersuchungen über die Wischhandlung der Erdkröte (*Bufo bufo* L.) *Z. Tierpsychol.* **13**, 185–207.

Kuhl, P. K. and Miller, J. D. (1975). Speech perception by the chinchilla: voiced-voiceless distinction in alveolar plosive consonants. *Science, N.Y.* **190**, 69–72.

Lacey, D. J. (1978). The organization of the hippocampus of the fence lizard: a light microscopic study. *J. comp. Neurol.* **182**, 247–64.

Lashley, K. S. (1929). *Brain mechanisms and intelligence: a quantitative study of injuries to the brain*. University of Chicago Press.

— (1935). The mechanism of vision. XII. Nervous structures concerned in the acquisition and retention of habits based on reactions to light. *Comp. Psychol. Mongor.* **11**(2), 43–79.

— (1943). Studies of cerebral function in learning. XII. Loss of the maze habit after occipital lesions in blind rats. *J. comp. Neurol.* **79**, 431–62.

— (1949). Persistent problems in the evolution of mind. *Q. Rev. Biol.* **24**, 28–42.

— (1950). In search of the engram. *Soc. exp. Biol. Symp.* **4**, 454–82.

Lawicka, W., Mishkin, W., and Konorski, J. (1966). Delayed response deficit in dogs after selective ablation of the proreal gyrus. *Acta biol. exp., Vars* **26**, 309–22.

Le Gros Clark, W. E. (1945). Deformation patterns in the cerebral cortex. In *Essays on growth and form presented to D'Arcy Wentworth Thompson* (ed. W. E. Le Gros Clark and P. B. Medawar) pp. 1–22. Clarendon Press, Oxford.

Lea, S. E. G. and Harrison, S. N. (1978). Discrimination of polymorphous stimulus sets by pigeons. *Q. J. exp. Psychol.* **30**, 521–37.

Leakey, R. E. F. and Walker, A. (1980). On the status of *Australopithecus afarensis. Science, N.Y.* **207**, 1103.

Leary, R. W., Harlow, H. F., Settlage, P. H., and Greenwood, D. D. (1952). Performance on double-alternation problems by normal and brain-injured monkeys. *J. comp. physiol. Psychol.* **45**, 576–84.

Lende, R. A. (1969). A comparative approach to the neocortex: localization in monotremes, marsupials and insectivores. *Ann. N.Y. Acad. Sci.* **167**, 262-76.

Lenneberg, E. H. (1967). *Biological foundations of language*. Wiley, New York.

— (1969). On explaining language. *Science, N.Y.* **164**, 635–43.

Leonard, C. M. (1969). The prefrontal cortex of the rat. I. Cortical projections of the mediodorsal nucleus. II. Efferent connections. *Brain Res.* **12**, 321–43.

Leppelsack, H.-J. (1978). Unit responses to species–specific sounds in the auditory forebrain center of birds. *Fedn Proc. Fedn Am. Socs exp. Biol.* **37**, 2336–41.

— and Vogt, M. (1974). Responses of auditory neurons in the forebrain of a song bird during stimulation with species-specific sounds. *J. comp. Physiol.* **A107**, 263–74.

Levine, B. A. (1974). Effects of drive and incentive magnitude on serial discrimination reversal learning in pigeons and chickens. *J. comp. physiol. Psychol.* **86**, 730–5.

Levine, M. (1959). A model of hypothesis behavior in discrimination learning set. *Psychol. Rev.* **66**, 353–66.

— (1965). Hypothesis behavior. In *Behavior of nonhuman primates: modern research trends*, Vol. 1 (ed. A. M. Schrier, H. F. Harlow, and F. Stollnitz) pp. 97–127. Academic Press, New York.

Lieberman, P. (1975). *On the origins of language: an introduction to the evolution of human speech*. Macmillan, New York.

— (1979). Hominid evolution, supralaryngeal vocal tract physiology, and the fossil evidence for reconstructions. *Brain Lang.* **7**, 101–26.

Liege, B. and Galand, G. (1972). Single-unit visual responses in the frog's brain. *Vision Res.* **12**, 609–22.

Likely, D., Little, L., and Mackintosh, N. J. (1971). Extinction as a function of magnitude and percentage of food or sucrose reward. *Can. J. Psychol.* **25**, 130–7.

Livesey, P. J. (1969). Double- and single-alternation learning by rhesus monkeys. *J. comp. physiol. Psychol.* **67**, 526–30.

Lockard, R. B. (1971). Reflections on the fall of comparative psychology: is there a message for us all? *Am. Psychol.* **26**, 168–79.

Lockhart, J. M., Parks, T. E., and Davenport, J. W. (1963). Information acquired in one trial by learning-set experienced monkeys. *J. comp. physiol. Psychol.* **56**, 1035–7.

Lohman, A. H. M. and Van Woerden-Verkley, I. (1976). Further studies on the cortical connections of the Tegu lizard. *Brain Res.* **103**, 9–28.

— — (1978). Ascending connections to the forebrain in the Tegu lizard. *J. comp. Neurol.* **182**, 555–94.

Longo, N. and Bitterman, M. E. (1960). The effects of partial reinforcement with spaced practice on resistance to extinction in the fish. *J. comp. physiol. Psychol.* **53**, 169–72.

— Klempay, S., and Bitterman, M. E. (1964). Classical appetitive conditioning in the pigeon. *Psychon. Sci.* **1**, 19–20.

Loop, M. S. (1976). Auto-shaping—a simple technique for teaching a lizard to perform a visual discrimination task. *Copeia* 574–6.

Losey, G. S. and Margules, L. (1974). Cleaning symbiosis provides a positive reinforcer for fish. *Science, N.Y.* **184**, 179–80.

Lowes, G. and Bitterman, M. E. (1967). Reward and learning in goldfish. *Science, N.Y.* **157**, 455–7.

Lubow, R. E. (1973). Latent inhibition. *Psychol. Bull.* **79**, 398–407.

— (1974). High-order concept formation in the pigeon. *J. exp. Anal. Behav.* **21**, 475–83.

Ludvigson, H. W. and Gay, R. A. (1967). An investigation of conditions determining contrast effects in differential reward conditioning. *J. exp. Psychol.* **75**, 37–42.

Lyons, J. (1970). *Chomsky*. Collins, London.

Macadar, A. W., Rausch, L. J., Wenzel, B. M., and Hutchison, L. V. (1980). Electrophysiology of the olfactory pathway in the pigeon. *J. comp. Physiol.* **A137**, 39–46.

McCabe, B. J. Horn, G., and Bateson, P. P. G. (1979). Effects of rhythmic hyperstriatal stimulation on chicks' preferences for visual flicker. *Physiol. Behav.* **23**, 137–40.

McDaniel, W. F., Wildman, L. D., and Spears, R. H. (1979). Posterior association cortex and visual pattern discrimination in the rat. *Physiol. Psychol.* **7**, 241–4.

McGill, T. E. (1960). Response of the leopard frog to electric shock in an escape-learning situation. *J. comp. physiol. Psychol.* **53**, 443–5.

McGonigle, B. O. and Chalmers, M. (1977). Are monkeys logical? *Nature, Lond.* **267**, 694–6.

McIlwain, J. T. and Fields, H. L. (1971). Interactions of cortical and retinal projections on single neurons of the cat's superior colliculus. *J. Neurophysiol.* **34**, 763–72.

Mackintosh, N. J. (1965). Overtraining, reversal, and extinction in rats and chicks. *J. comp. physiol. Psychol.* **59**, 31–6.
— (1969*a*). Comparative studies of reversal and probability learning: rats, birds, and fish. In *Animal discrimination learning* (ed. R. Gilbert and N. S. Sutherland) pp. 137–62. Academic Press, London.
— (1969*b*). Habit-reversal and probability learning: rats, birds, and fish. In *Animal discrimination learning* (ed. R. Gilbert and N. S. Sutherland) pp. 175–85. Academic Press, London.
— (1969*c*). Further analysis of the overtraining reversal effect. *J. comp. physiol. Psychol.* **67**, Monograph No. 2, Part 2.
— (1970). Attention and probability learning. In *Attention: contemporary theory and analysis* (ed. D. I. Mostofsky) pp. 173–91. Appleton-Century-Crofts, New York.
— (1971). Reward and aftereffects of reward in the learning of goldfish. *J. comp. physiol. Psychol.* **76**, 225–32.
— (1973). Stimulus selection: learning to ignore stimuli that predict no change in reinforcement. In *Constraints on learning* (ed. R. A. Hinde and J. Stevenson-Hinde) pp. 75–96. Academic Press, London.
— (1974). *The psychology of animal learning.* Academic Press, London.
— and Cauty, A. (1971). Spatial reversal learning in rats, pigeons, and goldfish. *Psychon. Sci.* **22**, 281–2.
— and Dickinson, A. (1979). Instrumental (Type II) conditioning. In *Mechanisms of learning and motivation* (ed. A. Dickinson and R. A. Boakes) pp. 143–69. Erlbaum, Hillsdale, NJ.
— and Holgate, V. (1967). Effects of several pre-training procedures on brightness probability learning. *Percept. mot. Skills* **25**, 629–37.
— and Honig, W. K. (1970). Blocking and attentional enhancement in pigeons. *J. comp. physiol. Psychol.* **73**, 78–85.
— and Little, L. (1969*a*). Selective attention and response strategies as factors in serial reversal learning. *Can. J. Psychol.* **23**, 335–46.
— — (1969*b*). Intradimensional and extradimensional shift learning by pigeons. *Psychon. Sci.* **14**, 5–6.
— — (1970). An analysis of transfer along a continuum. *Can. J. Psychol.* **24**, 362–9.
— and Lord, J. (1973). Simultaneous and successive contrast with delay of reward. *Anim. Learn. Behav.* **1**, 283–6.
— — and Little, L. (1971). Visual and spatial probability learning in pigeons and goldfish. *Psychon. Sci.* **24**, 221–3.
— Mackintosh, J., Safriel-Jorne, O., and Sutherland, N. S. (1966). Overtraining, reversal and extinction in the goldfish. *Anim. Behav.* **14**, 314–18.
McLaughlin, B. (1977). Second-language learning in children. *Psychol. Bull.* **84**, 438–59.
Macphail, E. M. (1967). Positive and negative reinforcement from intracranial stimulation in pigeons. *Nature, Lond.* **213**, 947–8.
— (1968). Avoidance responding in pigeons. *J. exp. Anal. Behav.* **11**, 629–32.
— (1969). Avian hyperstriatal complex and response facilitation. *Commun. behav. Biol.* **A4**, 129–37.
— (1970). Serial reversal performance in pigeons: role of inhibition. *Learn. Motiv.* **1**, 401–10.
— (1971). Hyperstriatal lesions in pigeons: effects on response inhibition, behavioral contrast, and reversal learning. *J. comp. physiol. Psychol.* **75**, 500–7.

— (1972). Inhibition in the acquisition and reversal of simultaneous discriminations. In *Inhibition and learning* (ed. R. A. Boakes and M. S. Halliday) pp. 121–51. Academic Press, London.
— (1975*a*). The role of the avian hyperstriatal complex in learning. In *Neural and endocrine aspects of behaviour in birds* (ed. P. Wright, P. G. Cargyl, and D. M. Vowles) pp. 139–62. Elsevier, Amsterdam.
— (1975*b*). Hyperstriatal function in the pigeon: response inhibition or response shift? *J. comp. physiol. Psychol.* **89**, 607–18.
— (1976*a*). Effects of hyperstriatal lesions on within-day serial reversal performance in pigeons. *Physiol. Behav.* **16**, 529–36.
— (1976*b*). Evidence against the response-shift account of hyperstriatal function in the pigeon (*Columba livia*) *J. comp. physiol. Psychol.* **90**, 547–59.
— (1980). Short-term visual recognition memory in pigeons. *Q. J. exp. Psychol.* **32**, 521–38.
Mahut, H. and Zola, S. M. (1973). A non-modality specific impairment in spatial learning after fornix lesions in monkeys. *Neuropsychologia* **11**, 244–69.
Maier, N. R. F. (1932). The effect of cerebral destruction on reasoning and learning in rats. *J. comp. Neurol.* **54**, 45–75.
— and Schneirla, T. C. (1935). *Principles of animal psychology*. McGraw-Hill, New York.
Mandriota, F. J., Thompson, R. L., and Bennett, M. V. L. (1968). Avoidance conditioning of the rate of electric organ discharge in mormyrid fish. *Anim. Behav.* **16**, 448–55.
Manley, J. A. (1971). Single unit studies in the midbrain auditory area of *Caiman*. *Z. vergl. Physiol.* **71**, 255–61.
Manocha, S. N. (1967). Discrimination learning in langurs and rhesus monkeys. *Percept. mot. Skills* **24**, 805–6.
Markowitsch, H.J. and Pritzel, M. (1977). Comparative analysis of prefrontal learning functions in rats, cats, and monkeys. *Psychol. Bull.* **84**, 817–37.
— — and Petrović-Minić, B. (1980). On the extent of the lateral prefrontal cortex of the cat. *Neuroscience* **5**, 1143–50.
Marx, M. H. (1967). Interaction of drive and reward as a determiner of resistance to extinction. *J. comp. physiol. Psychol.* **64**, 488–9.
Maser, J. D., Klara, J. W., and Gallup, G. G. (1973). Archistriatal lesions enhance tonic immobility in the chicken (*Gallus gallus*). *Physiol. Behav.* **11**, 729–33.
Masterton, B. and Skeen, L. C. (1972). Origins of anthropoid intelligence: prefrontal system and delayed alternation in hedgehog, tree shrew, and bush baby. *J. comp. physiol. Psychol.* **81**, 423–33.
Matthews, T. J., McHugh, T. G., and Carr, L. D. (1974). Pavlovian and instrumental determinants of response suppression in the pigeon. *J. comp. physiol. Psychol.* **87**, 500–6.
Matyniak, K. A. and Stettner, L. J. (1970). Reversal learning in birds as a function of amount of overtraining. *Psychon. Sci.* **21**, 308–9.
Mayr, R. (1969). *Principles of systematic zoology*. McGraw-Hill, New York.
Medin, D. L. (1974). The comparative study of memory. *J. hum. Evol.* **3**, 455–63.
Meier, R. E., Mihailović, J., and Cuénod, M. (1974). Thalamic organisation of the retino-thalamo-hyperstriatal pathway in the pigeon (*Columba livia*). *Expl Brain Res.* **19**, 351–64.
Melton, A. W. (1970). The situation with respect to the spacing of repetitions and memory. *J. verbal Learn. verbal Behav.* **9**, 596–606.
Meltzoff, A. N. and Borton, R. W. (1979). Intermodal matching by human neonates. *Nature, Lond.* **282**, 403–4.

Melzack, R. (1961). On the survival of mallard ducks after 'habituation' to the hawk-shaped figure. *Behaviour* **17**, 9–16.
Meyer, D. R. Treichler, F. R., and Meyer, P. M. (1965). Discrete-trial training techniques and stimulus variables. In *Behavior of nonhuman primates: modern research trends*, Vol. 1 (ed. A. M. Schrier, H. F. Harlow, and F. Stollnitz) pp. 1–49. Academic Press, New York.
Meyers, W. J., McQuiston, M. D., and Miles, R. C. (1962). Delayed-response and learning-set performance of cats. *J. comp. physiol. Psychol.* **55**, 515–17.
Miles, C. G. and Jenkins, H. M. (1973). Overshadowing in operant conditioning as a function of discriminability. *Learn. Motiv.* **4**, 11–27.
Miles, R. C. (1957). Learning-set formation in the squirrel monkey. *J. comp. physiol. Psychol.* **50**, 356–7.
— (1971). Species differences in 'transmitting' spatial location information. In *Cognitive processes of nonhuman primates* (ed. L. E. Jarrard) pp. 165–79. Academic Press, New York.
— and Meyer, D. R. (1956). Learning sets in marmosets. *J. comp. physiol. Psychol.* **49**, 219–22.
Miller, J. T., Hansen, G., and Thomas, D. R. (1972). Effects of stimulus similarity and response criterion on successive discrimination reversal learning. *J. comp. physiol. Psychol.* **81**, 434–40.
Miller, R. R. and Berk, A. M. (1977). Retention over metamorphosis in the African claw-toed frog. *J. exp. Psychol: Anim. Behav. Processes* **3**, 343–56.
— and Springer, A. D. (1973). Amnesia, consolidation, and retrieval. *Psychol. Rev.* **80**, 69–79.
Milner, B. (1971). Interhemispheric differences in the localization of psychological processes in man. *Br. med. Bull.* **27**, 272–7.
— Corkin, S., and Teuber, H. L. (1968). Further analysis of the hippocampal amnesic syndrome: 14-year follow-up study of H.M. *Neuropsychologia* **6**, 215–34.
Mishkin, M. (1978). Memory in monkeys severely impaired by combined but not by separate removal of amygdala and hippocampus. *Nature, Lond.* **273**, 297–8.
— and Manning, F. J. (1978). Non-spatial memory after selective prefrontal lesions in monkeys. *Brain Res.* **143**, 313–23.
— Prockop, E. S., and Rosvold, H. E. (1962). One-trial object-discrimination learning in monkeys with frontal lesions. *J. comp. physiol. Psychol.* **55**, 178–81.
Moise, S. L. (1970). Short-term retention in *Macaca speciosa* following interpolated activity during delayed matching from sample. *J. comp. physiol. Psychol.* **73**, 506–14.
Moon, L. E. and Harlow, H. F. (1955). Analysis of oddity learning by rhesus monkeys. *J. comp. physiol. Psychol.* **48**, 188–94.
Moore, A. R. and Welch, J. C. (1940). Associative hysteresis in larval amblystoma. *J. comp. Psychol.* **29**, 283–92.
Moore, B. R. (1973). The role of directed Pavlovian reactions in simple instrumental learning in the pigeon. In *Constraints on learning* (ed. R. A. Hinde and J. Stevenson-Hinde) pp. 159–86. Academic Press, London.
Moore, G. P. and Tschirgi, R. D. (1962). Nonspecific responses of reptilian cortex to sensory stimuli. *Expl Neurol.* **5**, 196–209.
Moore, J. W. (1979). Information processing in space–time by the hippocampus. *Physiol. Psychol.* **7**, 224–32.
Morgan, C. L. (1894). *An introduction to comparative psychology*. Walter Scott, London.

Morgan, M. J. Fitch, M. D., Holman, J. G., and Lea, S. E. G. (1976). Pigeons learn the concept of an 'A'. *Perception* **5**, 57–66.

Morlock, H. C. (1972). Behavior following ablation of the dorsal cortex of turtles. *Brain Behav. Evol.* **5**, 256–63.

Morse, P. A. and Snowdon, C. T. (1975). An investigation of categorical speech discrimination by rhesus monkeys. *Percept. Psychophys.* **17**, 9–16.

Mrosovsky, N. (1964). Modification of the diving-in response of the red-eared terrapin, *Pseudemys ornata callirostris. Q. J. exp. Psychol.* **16**, 166–71.

Mudry, K. M. and Capranica, R. R. (1980). Evoked auditory activity within the telencephalon of the bullfrog (*Rana catesbeiana*). *Brain Res.* **182**, 303–11.

— Constantine-Paton, M., and Capranica, R. R. (1977). Auditory sensitivity of the diencephalon of the leopard frog. *R. p. pipiens. J. comp. Physiol.* **A114**, 1–13.

Munn, N. L. (1940). Learning experiments with larval frogs: a preliminary report. *J. comp. Psychol.* **29**, 97–108.

— (1964). Discrimination-reversal learning in kangaroos. *Aust. J. Psychol.* **16**, 1–8.

Muntz, W. R. A. (1962*a*). Microelectrode recordings from the diencephalon of the frog (*Rana pipiens*), and a blue sensitive system. *J. Neurophysiol.* **25**, 699–711.

— (1962*b*). Effectiveness of different colors of light in releasing the positive phototactic behavior of frogs, and a possible function of the retinal projection to the diencephalon. *J. Neurophysiol.* **25**, 712–20.

Murillo, N. R., Diercks, J. K., and Capaldi, E. J. (1961). Performance of the turtle, *Pseudemys scripta troostii*, in a partial-reinforcement situation. *J. comp. physiol. Psychol.* **54**, 204–6.

Narkiewicz, O. and Brutkowski, S. (1967). The organization of projections from the thalamic mediodorsal nucleus to the prefrontal cortex of the dog. *J. comp. Neurol.* **129**, 361–74.

Nash, R. F. and Gallup, G. G. (1976). Habituation and tonic immobility in domestic chickens. *J. comp. physiol. Psychol.* **90**, 870–6.

Nauta, W. J. H. (1964). Some efferent connections of the prefrontal cortex in the monkey. In *The frontal granular cortex and behavior* (ed. J. M. Warren and K. Akert) pp. 397–409. McGraw-Hill, New York.

— and Gygax, P. A. (1954). Silver impregnation of degenerating axons in the central nervous system: a modified technic. *Stain Technol.* **29**, 91–3.

— and Karten, H. J. (1970). A general profile of the vertebrate brain, with sidelights on the ancestry of cerebral cortex. In *The neurosciences: second study program* (ed. F. O. Schmitt) pp. 7–26. Rockefeller University Press, New York.

Nebes, R. D. (1974). Hemispheric specialization in commissurotomized man. *Psychol. Bull.* **81**, 1–14.

Nelson, J. S. (1976). *Fishes of the world*. Wiley, New York.

Newcombe, F. and Ratcliff, G. (1973). Handedness, speech lateralization and ability. *Neuropsychologia* **11**, 399–407.

Nieuwenhuys, R. (1967). Comparative anatomy of olfactory centres and tracts. In *Sensory mechanisms* (*Progress in Brain Research* Vol. 23) (ed. Y. Zotterman) pp. 1–64. Elsevier, Amsterdam.

Nigrosh, B. J., Slotnick, B. M., and Nevin, J. A. (1975). Olfactory discrimination, reversal learning, and stimulus control in rats. *J. comp. physiol. Psychol.* **89**, 285–94.

Nissen, H. W., Blum, J. S., and Blum, R. A. (1948). Analysis of matching behavior in chimpanzee. *J. comp. physiol. Psychol.* **41**, 62–74.

— and McCulloch, T. L. (1937). Equated and nonequated stimulus situations in discrimination learning by chimpanzees: III. Pre-potency of response to oddity through training. *J. comp. Psychol.* **23**, 377–81.

Noble, M., Gruender, A., and Meyer, D. R. (1959). Conditioning in fish (*Mollienisia* sp.) as a function of the interval between CS and US. *J. comp. physiol. Psychol.* **52**, 236–9.

Norgren, R. (1974). Gustatory afferents to ventral forebrain. *Brain Res.* **81**, 285–95.

North, A. J. (1950*a*). Improvement in successive discrimination reversals. *J. comp. physiol. Psychol.* **43**, 442–60.

— (1950*b*). Performance during an extended series of discrimination reversals. *J. comp. physiol. Psychol.* **43**, 461–70.

— (1959). Discrimination reversal with spaced trials and distinctive cues. *J. comp. physiol. Psychol.* **52**, 426–9.

Northcutt, R. G. (1974). Some histochemical observations on the telencephalon of the bullfrog, *Rana catesbeiana* Shaw. *J. comp. Neurol.* **157**, 379–90.

— (1977). Elasmobranch central nervous system organization and its possible evolutionary significance. *Am. Zool.* **17**, 411–29.

— (1978). Forebrain and midbrain organization in lizards and its phylogenetic significance. In *Behavior and neurology of lizards* (ed. N. Greenberg and P. D. MacLean) pp. 11–64. NIMH, Rockville, Md.

— Braford, M. R., and Landreth, G. E. (1975). Retinal projections in the tuatara *Sphenodon punctatus*: an autoradiographic study. *Anat. Rec.* **178**, 428.

— and Butler, A. B. (1974). Retinal projections in the northern water snake *Natrix sipedon sipedon* (L.) *J. Morph.* **142**, 117–36.

— and Heath, J. E. (1971). Performance of caimans in a T-maze. *Copeia* 557–60.

— — (1973). T-maze behavior of the tuatara (*Sphenodon punctatus*). *Copeia* 617–20.

— and Przybylski, R. J. (1973). Retinal projections in the lamprey *Petromyson marinus* L. *Anat. Rec.* **175**, 400.

— and Royce, G. J. (1975). Olfactory bulb projections in the bullfrog *Rana catesbeiana* Shaw. *J. Morph.* **145**, 251–68.

Nott, K. H. (1980). Some effects of hippocampal lesions on the behaviour of pigeons. Unpublished Ph. D. thesis, University of Durham.

Nottebohm, F., Stokes, T. M., and Leonard, C. M. (1976). Central control of song in the canary, *Serinus canarius*. *J. comp. Neurol.* **165**, 457–86.

Numan, R. and Lubar, J. F. (1974). Role of the proreal gyrus and septal area in response modulation in the cat. *Neuropsychologia* **12**, 219–34.

— Seifert, A. R., and Lubar, J. F. (1975). Effects of medio-cortical frontal lesions on DRL performance in the rat. *Physiol. Psychol.* **3**, 390–4.

Oakley, D. A. (1979). Neocortex and learning. *Trends Neurosci.* June, 149–52.

— and Russell, I. S. (1977). Subcortical storage of Pavlovian conditioning in the rabbit. *Physiol. Behav.* **18**, 931–7.

O'Keefe, J. and Nadel, L. (1978). *The hippocampus as a cognitive map.* Clarendon Press, Oxford.

Olton, D. S. (1978). Characteristics of spatial memory. In *Cognitive processes in animal behavior* (ed. S. H. Hulse, H. Fowler, and W. K. Honig) pp. 341–73. Erlbaum, Hillsdale, NJ.

— Collison, C., and Werz, M. A. (1977). Spatial memory and radial arm maze performance of rats. *Learn. Motiv.* **8**, 289–314.

Orbach, J. (1959). 'Functions' of striate cortex and the problem of mass action. *Psychol. Bull.* **56**, 271–92.

Orrego, F. (1961). The reptilian forebrain. I The olfactory pathways and cortical areas in the turtle. *Archs ital. Biol.* **99**, 425–45.

Otis, L. S. and Cerf, J. A. (1963). Conditioned avoidance learning in two fish species. *Psychol. Rep.* **12**, 679–82.

— — and Thomas, G. J. (1957). Conditioned inhibition of respiration and heart rate in the goldfish. *Science, N.Y.* **126**, 263–4.

Over, R. and Mackintosh, N. J. (1969). Cross-modal transfer of intensity discrimination by rats. *Nature, Lond.* **224**, 918–19.

Overmier, J. B. and Curnow, P. F. (1969). Classical conditioning, pseudoconditioning and sensitization in 'normal' and forebrainless fish. *J. comp physiol. Psychol*, **68**, 193–8.

— and Flood, N. B. (1969). Passive avoidance in forebrain ablated teleost fish, *Carassius auratus. Physiol. Behav.* **4**, 791–4.

— and Savage, G. E. (1974). Effects of telencephalic ablation on trace classical conditioning of heart rate in goldfish. *Expl Neurol.* **42**, 339–46.

— and Starkman, N. (1974). Transfer of control of avoidance behavior in normal and telencephalon ablated goldfish (*Carassius auratus*). *Physiol. Behav.* **12**, 605–8.

Parent, A. (1971). Comparative histochemical study of the amygdaloid complex. *J. Hirnforsch.* **13**, 89–96.

— and Olivier, A. (1970). Comparative histochemical study of the corpus striatum. *J. Hirnforsch.* **12**, 73–81.

— and Poitras, D. (1974). The origin and distribution of catecholamingergic axon terminals in the cerebral cortex of the turtle (*Chrysemys picta*). *Brain Res.* **78**, 345–58.

Pasnak. R. (1979). Acquisition of prerequisites to conservation by macaques. *J. exp. Psychol: Anim. Behav. Processes* **5**, 194–210.

Passingham, R. E. (1973). Anatomical differences between the neocortex of man and other primates. *Brain Behav. Evol.* **7**, 337–59.

— (1975). Changes in the size and organisation of the brain in man and his ancestors. *Brain Behav. Evol.* **11**, 73–90.

— (1978*a*). The functions of prefrontal cortex in the tree shrew (*Tupaia belangeri*). *Brain Res.* **145**, 147–52.

— (1978*b*). Information about movements in monkeys (*Macaca mulatta*) with lesions of dorsal prefrontal cortex. *Brain Res.* **152**, 313–28.

— and Ettlinger, G. (1974). A comparison of cortical functions in man and other primates. *Int. Rev. Neurobiol.* **16**, 233–99.

Pasternak, T. and Hodos, W. (1977). Intensity difference thresholds after lesions of the visual wulst in pigeons. *J. comp. physiol. Psychol.* **91**, 485–97.

Pastore, N. (1954). Discrimination learning in the canary. *J. comp. physiol. Psychol.* **47**, 389–90.

Patterson, F. G. (1978). The gestures of a gorilla: language acquisition in another pongid. *Brain Lang.* **5**, 72–97.

Pearson, R. (1972). *The avian brain*. Academic Press, London.

Peden, B. F., Browne, M. P., and Hearst, E. (1977). Persistent approaches to a signal for food despite food omission for approaching. *J. exp. Psychol: Anim. Behav. Processes* **3**, 377–99.

Peeke, H. V. S. and Peeke, S. C. (1972). Habituation, reinforcement and recovery of predatory responses in two species of fish (*Carassius auratus* and *Macropodus opercularis*). *Anim. Behav.* **20**, 268–73.

—— (1973). Habituation in fish with special reference to intraspecific aggressive behavior. In *Habituation*, Vol. 1. *Behavioral studies* (ed. H. V. S. Peeke and M. J. Herz) pp. 59–83. Academic Press, New York.

—— and Williston, J. S. (1972). Long-term memory deficits for habituation of predatory behavior in the forebrain ablated goldfish (*Carassius auratus*). *Expl Neurol.* **36**, 288–94.

Pert, A. and Bitterman, M. E. (1970). Reward and learning in the turtle. *Learn. Motiv.* **1**, 121–8.

— and Gonzalez, R. C. (1974). Behavior of the turtle (*Chrysemys picta picta*) in simultaneous, successive, and behavioral contrast situations. *J. comp. physiol. Psychol.* **87**, 526–38.

Peters, R. I. and Wirth, M. C. (1976). Shock avoidance conditioning and H-leucine uptake in a simple vertebrate system. *Physiol. Behav.* **16**, 365–9.

Petersen, M. R. Beecher, M. D., Zoloth, S. R., Moody, D. B., and Stebbins, W. C. (1978). Neural lateralization of species-specific vocalizations by Japanese macaques (*Macaca fuscata*). *Science, N.Y.* **202**, 324–6.

Petrinovich, L. and Peeke, H. V. S. (1973). Habituation to territorial song in the white-crowned sparrow (*Zonotrichia leucophyrs*). *Behav. Biol.* **8**, 743–8.

Pettrigrew, J. D. and Konishi, M. (1976). Neurons selective for orientation and binocular desparity in the visual Wulst of the barn owl (*Tyto alba*). *Science, N.Y.* **193**, 675–7.

Peyrichoux, J., Repérant, J., and Weidner, C. (1978). Les centres visuels primaires chez la Grenouille (*Rana esculenta* L.) et le problème des projections ipsilatérales. Étude radioautographique. *C. r. hebd. Séance. Acad. Sci., Paris* **D287**, 37–40.

Phillips, R. E. (1964). 'Wildness' in the mallard duck: effects of brain lesions and stimulation on 'escape behavior' and reproduction. *J. comp. Neurol.* **122**, 139–55.

— (1968). Approach-withdrawal behavior of peach-faced lovebirds, *Agapornis roseicolis*, and its modification by brain lesions. *Behaviour* **31**, 163–84.

— and Youngren, O. M. (1971). Brain stimulation and species typical behavior: activities evoked by electrical stimulation of the brains of chickens (*Gallus gallus*). *Anim. Behav.* **19**, 757–79.

Phillips, W. A. and Christie, D. F. M. (1977). Components of visual memory. *Q. J. exp. Psychol.* **29**, 117–33.

Piaget, J. and Inhelder, B. (1969). *The psychology of the child.* Routledge and Kegan Paul, London.

Pickett, J. M. (1952). Non-equipotential cortical function in maze learning. *Am. J. Psychol.* **65**, 177–95.

Pilbeam, D. and Gould, S. J. (1974). Size and scaling in human evolution. *Science, N.Y.* **186**, 892–901.

Pinckney, G. A. (1968). Response consequences and Sidman avoidance behavior in the goldfish. *Psychon. Sci.* **12**, 13–14.

Platel, R. (1979). Brain weight–body weight relationships. In *Biology of the Reptilia*, Vol. 9 *Neurology A* (ed. C. Gans, R. G. Northcutt, and P. S. Ulinski) pp. 147–71. Academic Press, London.

Platt, C. J., Bullock, T. H., Czéh, G., Kovačević, N., Konjević, D., and Gojković, M. (1974). Comparison of electroreceptor, mechanoreceptor, and optic evoked potentials in the brain of some rays and sharks. *J. comp. Physiol.* **A95**, 323–55.

Plotnik, R. J. and Tallarico, R. B. (1966). Object-quality learning-set formation in the young chicken. *Psychon. Sci.* **5**, 195–6.

Portmann, A. (1946). Études sur la cérébralisation chez les oiseaux I. *Alauda* **14**, 2–20.

— (1947). Études sur la cérébralisation chez les oiseaux II. Les indices intra-cérébraux. *Alauda* **15**, 1–15.

Potter, M. C. and Levy, E. I. (1969). Recognition memory for a rapid sequence of pictures. *J. exp. Psychol.* **81**, 10–15.

Powell, R. W. and Mantor, H. (1969). Failure to obtain one-way shuttle avoidance in the lizard *Anolis sagrei*. *Psychol. Rec.* **19**, 623–7.

Powell, T. P. S. and Cowan, W. M. (1961). The thalamic projection upon the telencephalon in the pigeon (*Columba livia*). *J. Anat., Lond.* **95**, 78–109.

Premack, D. (1971). Language in chimpanzees? *Science, N.Y.* **172**, 808–22.

— (1976). *Intelligence in ape and man*. Erlbaum, Hillsdale, NJ.

Prestrude, A. M. (1970). Sensory capacities of the chimpanzee: a review. *Psychol. Bull.* **74**, 47–67.

Pribram, K. H. and Kruger, L. (1954). Functions of the 'olfactory brain'. *Ann. N.Y. Acad. Sci.* **58**, 109–38.

Pritz, M. B. (1974), Ascending connections of a thalamic auditory area in a crododile, *Caiman crocodilus*. *J. comp. Neurol.* **153**, 199–214.

— (1975). Anatomical identification of a telencephalic visual area in crocodiles: Ascending connections of nucleus rotundus in *Caiman crocodilus*. *J. comp. Neurol.* **164**, 323–38.

— Mead, W. R., and Northcutt, R. G. (1970). The effects of Wulst ablation on color, brightness, and pattern discrimination. *J. comp. Neurol.* **140**, 81–100.

— and Northcutt, R. G. (1980). Anatomical evidence for an ascending somatosensory pathway to the telencephalon in crocodiles, *Caiman crocodilus*. *Expl Brain Res.* **40**, 342–5.

Pubols, B. H. (1957). Successive discrimination reversal learning in the white rat: a comparison of two procedures. *J. comp. physiol. Psychol.* **50**, 319–22.

Putkonen, P. T. S. (1967). Electrical stimulation of the avian brain. *Ann. Acad. sci. fenn.* **130**, 1–95.

Rabinovitch, M. S. and Rosvold, H. E. (1951). A closed-field intelligence test for rats. *Can. J. Psychol.* **5**, 122–8.

Rachlin, H. and Hineline, P. N. (1967). Training and maintenance of keypecking in the pigeon by negative reinforcement. *Science, N.Y.* **157**, 954–5.

Rajalakshmi, R. and Jeeves, M. A. (1965). The relative difficulty of reversal learning (reversal index) as a basis of behavioural comparisons. *Anim. Behav.* **13**, 203–11.

— — (1968). Performance on the Hebb–Williams maze as related to discrimination and reversal learning in rats. *Anim. Behav.* **16**, 114–16.

Rakover, S. S. (1979). Fish (*Tilapia aurea*), as rats, learn shuttle better than lever-bumping (press) avoidance tasks: a suggestion for functionally similar universal reactions to a conditioned fear-arousing stimulus. *Am. J. Psychol.* **92**, 489–95.

Raymond, B., Aderman, M., and Wolach, A. H. (1972). Incentive shifts in the goldfish. *J. comp. physiol. Psychol.* **78**, 10–13.

Razran, G. (1971). *Mind in evolution: an East–West synthesis of learned behavior and cognition*. Houghton Mifflin, Boston.

Reese, H. W. (1964). Discrimination learning set in rhesus monkeys. *Psychol. Bull.* **61**, 321–40.

Reid, R. L. (1958). Discrimination-reversal learning in pigeons. *J. comp. physiol. Psychol.* **51**, 716–20.

Reiner, A. J. and Karten, H. J. (1978). A bisynaptic retinocerebellar pathway in the turtle. *Brain Res.* **150**, 163–9.

— and Powers, A. S. (1978). Intensity and pattern discriminations in turtles after lesions of nucleus rotundus. *J. comp. physiol. Psychol.* **92**, 1156–68.

— — (1980). The effects of extensive forebrain lesions on visual discrimination performance in turtles (*Chrysemys picta picta*). *Brain Res.* **192**, 327–37.

Rensch, B. (1956). Increase of learning capability with increase of brain size. *Am. Nat.* **90**, 81–95.

Repérant, J. (1973). Nouvelles données sur les projections visuelles chez le pigeon (*Columba livia*). *J. Hirnforsch.* **14**, 151–88.

— and Lemire, M. (1976). Retinal projections in cyprinid fishes: a degeneration and radioautographic study. *Brain Behav. Evol.* **13**, 34–57.

— and Rio, J.-P. (1976). Retinal projections in *Vipera aspis*. A reinvestigation using light radioautographic and electron microscopic degeneration techniques. *Brain Res.* **107**, 603–9.

Rescorla, R. A. (1967). Pavlovian conditioning and its proper control procedures. *Psychol. Rev.* **74**, 71–80.

— (1973). Effect of US habituation following conditioning. *J. comp. physiol. Psychol.* **82**, 137–43.

— (1977). Pavlovian second-order conditioning: some implications for instrumental behavior. In *Operant–Pavlovian interactions* (ed. H. Davis and H. M. B. Hurwitz) pp. 133–64. Erlbaum, Hillsdale, NJ.

— and Solomon, R. L. (1967). Two-process learning theory: relationships between Pavlovian conditioning and instrumental learning. *Psychol. Rev.* **74**, 151–82.

— and Wagner, A. R. (1972). A theory of Pavlovian conditioning: variations in the effectiveness of reinforcement and nonreinforcement. In *Classical conditioning II: current research and theory* (ed. A. H. Black and W. F. Prokasy) pp. 64–99. Appleton-Century-Crofts, New York.

Restle, F. (1958). Toward a quantitative description of learning set data. *Psychol. Rev.* **65**, 77–91.

Revzin, A. M. (1969). A specific visual projection area in the hyperstriatum of the pigeon (*Columba livia*). *Brain Res.* **15**, 246–9.

— (1970). Some characteristics of wide-field units in the brain of the pigeon. *Brain Behav. Evol.* **3**, 195–204.

Reynolds, G. S. and Limpo, A. J. (1965). Selective resistance of performance on a schedule of reinforcement to disruption by forebrain lesions. *Psychon. Sci.* **3**, 35–6.

Richardson, A. M. and Julian, S. M. (1974). Avoidance learning in the lizard *Dipsosaurus dorsalis*. *Psychol. Rep.* **35**, 35–40.

Riege, W. H. and Cherkin, A. (1971). One-trial learning and biphasic time course of performance in the goldfish. *Science, N.Y.* **172**, 966–8.

Rieke, G. K. (1980). Kainic acid lesions of pigeon paleostriatum: a model for the study of movement disorders. *Physiol. Behav.* **24**, 683–7.

— and Wenzel, B. M. (1978). Forebrain projections of the pigeon olfactory bulb. *J. Morph.* **158**, 41–56.

Riopelle, A. J., Francisco, E. W., and Ades, H. W. (1954). Differential first-trial procedures and discrimination learning performance. *J. comp. physiol. Psychol.* **47**, 293–7.

Robbins, D. O. (1972). Coding of intensity and wavelength in optic tectal cells of the turtle. *Brain Behav. Evol.* **5**, 124–42.

Roberts, D., Heckel, R. V., and Wiggins, S. L. (1962). Effect of a preferred color in modifying maze running in the *Bufo terrestris*. *Percept. mot. skills* **15**, 736–8.

Roberts, W. A. (1972). Short-term memory in the pigeon: effects of repetition and spacing. *J. exp. Psychol.* **94**, 74–83.

— (1974). Spaced repetition facilitates short-term retention in the rat. *J. comp. physiol. Psychol.* **86**, 164–71.

— Bullock, D. H., and Bitterman, M. E. (1963). Resistance to extinction in the pigeon after partially reinforced instrumental training under discrete trial conditions. *Am. J. Psychol.* **76**, 353–65.

— and Grant, D. S. (1976). Studies of short-term memory in the pigeon using the delayed matching to sample procedure. In *Processes of animal memory* (ed. D. L. Medin, W. A. Roberts, and R. T. Davis) pp. 79–112. Erlbaum, Hillsdale, NJ.

Robinson, R. G. (1979). Differential behavioral and biochemical effects of right and left hemispheric cerebral infarction in the rat. *Science, N.Y.* **205**, 707–10.

Rodgers, W. L., Melzack, R., and Segal, J. R. (1963). 'Tail-flip' response in goldfish. *J. comp. physiol. Psychol.* **56**, 917–23.

Romanes, G. J. (1882). *Animal intelligence*, 2nd edn. Kegan Paul, Trench, London.

Romer, A. S. (1966). *Vertebrate paleontology*. University of Chicago Press.

— and Parsons, T. S. (1977). *The vertebrate body*, 5th edn. W. B. Saunders, Philadelphia.

Rosenfeld, S. A. and Van Hoesen, G. W. (1979). Face recognition in the rhesus monkey. *Neuropsychologia* **17**, 503–9.

Rosenkilde, C. E. (1978). Delayed alternation behavior following ablations of the medial or dorsal prefrontal cortex in dogs. *Physiol. Behav.* **20**, 397–402.

Rouse, J. E. (1905). Respiration and emotion in pigeons. *J. comp. Neurol.* **15**, 494–513.

Rovainen, C. M. (1979). Neurobiology of lampreys. *Physiol. Rev.* **59**, 1007–77.

Royce, G. J. and Northcutt, R. G. (1969). Olfactory bulb projections in the tiger salamander (*Ambystoma tigrinum*) and the bullfrog (*Rana catesbeiana*). *Anat. Rec.* **163**, 254.

Rozin, P. (1976). The evolution of intelligence and access to the cognitive unconscious. In *Progress in psychobiology and physiological psychology*, Vol. 6 (ed. J. M. sprague and A. N. Epstein) pp. 245–80. Academic Press, New York.

— and Kalat, J. W. (1971). Specific hungers and poison avoidance as adapative specializations of learning. *Psychol. Rev.* **78**, 459–86.

— and Mayer, J. (1961). Thermal reinforcement and thermoregulatory behavior in the goldfish, *Carassius auratus*. *Science, N.Y.* **134**, 942–3.

Rudolph, R. L. and Van Houten, R. (1977). Auditory stimulus control in pigeons: Jenkins and Harrison (1960) revisited. *J. exp. Anal. Behav.* **27**, 327–30.

Rumbaugh, D. M. (ed.) (1977). *Language learning by a chimpanzee: the Lana project*. Academic Press, New York.

Rundus, D. (1971). Analysis of rehearsal processes in free recall. *J. exp. Psychol.* **89**, 63–77.

Russell, I. S. (1966). Animal learning and memory. In *Aspects of learning and memory* (ed. D. Richter) pp. 121–71. Heinemann, London.

— (1979). Brain size and intelligence: a comparative perspective. In *Brain, behaviour, and evolution* (ed. D. A. Oakley and H. C. Plotkin) pp. 126–53. Methuen, London.

Sacher, G. A. (1970). Allometric and factorial analysis of brain structure in insectivores and primates. In *The primate brain* (*Advances in primatology*, Vol. 1) (ed. C. R. Noback and W. Montagna) pp. 245–87. Appleton-Century-Crofts, New York.

Salzen, E. A., Parker, D. M., and Williamson, A. J. (1975). A forebrain lesion preventing imprinting in domestic chicks. *Expl Brain Res.* **24**, 145–57.

Sands, S. F. and Wright, A. A. (1980). Primate memory: Retention of serial list items by a rhesus monkey. *Science, N.Y.* **209**, 938–40.

Saunders, J. C., Chen, C.-S., and Pridmore, P. A. (1971). Successive habit-reversal learning in monotreme *Tachyglossus aculeatus* (Echidna). *Anim. Behav.* **19**, 552–5.

Savage, G. E. (1968*a*). Temporal factors in avoidance learning in normal and forebrainless goldfish (*Carassius auratus*). *Nature, Lond.* **218**, 1168–9.

— (1968*b*). Function of the forebrain in the memory system of the fish. In *The central nervous system and fish behavior* (ed. D. Ingle) pp. 127–38. University of Chicago Press.

— (1969*a*). Some preliminary observations on the role of the telencephalon in food-reinforced behaviour in the goldfish. *Carassius auratus. Anim. Behav.* **17**, 760–2.

— (1969*b*). Telencephalic lesions and avoidance behaviour in the goldfish (*Carassius auratus*). *Anim. Behav.* **17**, 362–73.

— (1971). Behavioural effects of electrical stimulation of the telencephalon of the goldfish (*Carassius auratus*). *Anim. Behav.* **19**, 661–8.

— and Swingland, I. R. (1969). Positively reinforced behaviour and the forebrain in goldfish. *Nature, Lond.* **221**, 878–9.

Savage-Rumbaugh, E. S., Rumbaugh, D. M., and Boysen, S. (1978). Symbolic communication between two chimpanzees (*Pan troglodytes*). *Science, N.Y.* **201**, 641–4.

Scalia, F. (1972). The projection of the accessory olfactory bulb in the frog. *Brain Res.* **36**, 409–11.

— and Ebbesson, S. O. E. (1971). The central projections of the olfactory bulb in a teleost (*Gymnothorax funebris*). *Brain Behav. Evol.* **4**, 376–99.

— Halpern, M., Knapp, H., and Riss, W. (1968). The efferent connexions of the olfactory bulb in the frog: a study of degenerating unmyelinated fibres. *J. Anat., Lond.* **103**, 245–62.

— — and Riss, W. (1969). Olfactory bulb projections in the South American caiman. *Brain Behav. Evol.* **2**, 238–62.

Scarborough, B. B. and Addison, R. G. (1962). Conditioning in fish: effects of X-irradiation. *Science, N.Y.* **136**, 712–14.

Schade, A. F. and Bitterman, M. E. (1965). The relative difficulty of reversal and dimensional shifting as a function of over-learning. *Psychon. Sci.* **3**, 283–4.

— — (1966). Improvement in habit reversal as related to dimensional set. *J. comp. physiol. Psychol.* **62**, 43–8.

Schapiro, H. (1968). Retching responses in *Caiman sklerops* elicited by central nervous stimulation. *Proc. Soc. exp. Biol. Med.* **129**, 917–20.

Schiller, P. H. (1952). Innate constituents of complex responses in primates. *Psychol. Rev.* **59**, 177–91.

Schmajuk, N. A., Segura, E. T., and Reboreda, J. C. (1980). Appetitive conditioning and discriminatory learning in toads. *Behav. neural Biol.* **28**, 392–7.

Schneider, A. M., Kapp, B., Aron, C., and Jarvik, M. E. (1969). Retroactive effects of transcorneal and transpinnate ECS on step-through latencies of mice and rats. *J. comp. physiol. Psychol.* **69**, 505–9.

Schneider, G. E. (1969). Two visual systems: brain mechanisms for localization and discrimination are dissociated by tectal and cortical lesions. *Science, N.Y.* **163**, 895–902.

Schrier, A. M. (1958). Comparison of two methods of investigating the effect of amount of reward on performance. *J. comp. physiol. Psychol.* **51**, 725–31.

— (1966). Transfer by macaque monkeys between learning-set and repeated-reversal. *Percept. mot. Skills* **23**, 787–92.

— (1974). Transfer between the repeated reversal and learning set tasks: a re-examination. *J. comp. physiol. Psychol.* **87**, 1004–10.

Schroeder, D. M. and Ebbesson, S. O. E. (1974). Non-olfactory telencephalic afferents in the nurse shark (*Ginglymostoma cirratum*). *Brain Behav. Evol.* **9**, 121–55.

Schusterman, R. J. (1962). Transfer effects of successive discrimination-reversal training in chimpanzees. *Science, N.Y.* **137**, 422–3.

— (1964). Successive discrimination-reversal training and multiple discrimination training in one-trial learning by chimpanzees. *J. comp. physiol. Psychol.* **58**, 153–6.

Schutz, S. L. and Bitterman, M. E. (1969). Spaced-trials partial reinforcement and resistance to extinction in the goldfish. *J. comp. physiol. Psychol.* **68**, 126–8.

Schwartz, B. and Williams, D. R. (1972). The role of the response-reinforcer contingency in negative automaintenance. *J. exp. Anal. Behav.* **17**, 351–7.

Scobie, S. R. and Fallon, D. (1974). Operant and Pavlovian control of a defensive shuttle response in goldfish (*Carassius auratus*). *J. comp. physiol. Psychol.* **86**, 858–66.

Scull, J. W. and Macphail, E. M. (1976). Behavioural contrast does not occur with continuous reinforcement in the positive component. *Q. J. exp. Psychol.* **28**, 77–82.

Segaar, J. (1961). Telencephalon and behaviour in *Gasterosteus aculeatus. Behaviour* **18**, 256–87.

— (1965). Behavioural aspects of degeneration and regeneration in fish brain: a comparison with higher vertebrates. In *Progress in brain research*, Vol. 14 (ed. M. Singer and J. P. Schadé) pp. 143–231. Elsevier, New York.

Seller, T. J. (1979). Unilateral nervous control of the syrinx in Java sparrows (*Padda oryzivora*). *J. comp. Physiol.* **A129**, 281–8.

Shaffer, W. O. and Shiffrin, R. M. (1972). Rehearsal and storage of visual information. *J. exp. Psychol.* **92**, 292–6.

Shapiro, S., Schuckman, H., Sussman, D., and Tucker, A. M. (1974). Effect of telencephalic lesions on the gill cover response of Siamese fighting fish. *Physiol. Behav.* **13**, 749–55.

Sharma, S. C. (1972). The retinal projection in the goldfish: an experimental study. *Brain Res.* **39**, 213–33.

Sheffield, V. F. (1950). Resistance to extinction as a function of the distribution of extinction trials. *J. exp. Psychol.* **40**, 305–13.

Shepp, B. E. and Eimas, P. D. (1964). Intradimensional and extradimensional shifts in the rat. *J. comp. physiol. Psychol.* **57**, 357–61.

— and Schrier, A. M. (1969). Consecutive intradimensional and extradimensional shifts in monkeys. *J. comp. physiol. Psychol.* **67**, 199–203.

Sherman, G. F., Garbanati, J. A., Rosen, G. D., Yutzey, D. A., and Denenberg, V. H. (1980). Brain and behavioral asymmetries for spatial preference in rats. *Brain Res.* **192**, 61–7.

Shimp, C. P. (1966). Probabilistically reinforced choice behavior in pigeons. *J. exp. Anal. Behav.* **9**, 443–55.

— (1976). Short-term memory in the pigeon: relative recency. *J. exp. Anal. Behav.* **25**, 55–61.
— and Moffitt, M. (1974). Short-term memory in the pigeon: stimulus–response associations. *J. exp. Anal. Behav.* **22**, 507–12.
Simons, E. L. (1977). Ramapithecus. *Scient. Am.* **236**, 28–35.
Singer, W. (1977). Control of thalamic transmission by corticofugal and ascending reticular pathways in the visual system. *Physiol. Rev.* **57**, 386–420.
Sinnott, J. M., Beecher, M. D., Moody, D. B., and Stebbins, W. C. (1976). Speech sound discrimination by monkeys and humans. *J. acoust. Soc. Am.* **60**, 687–95.
Skinner, B. F. (1938). *The behavior of organisms.* Appleton-Century-Crofts, New York.
— (1957). *Verbal behavior.* Appleton-Century-Crofts, New York.
Slotnick, B. M. and Katz, H. M. (1974). Olfactory learning-set formation in rats. *Science, N.Y.* **185**, 796–8.
Smeets, W. J. A. J. (1981). Retinofugal pathways in two chondrichthyans, the shark *Scyliorhinus canicula* and the ray *Raja clavata. J. comp Neurol.* **195**, 1–11.
Smith, M. (1964). *The British amphibians and reptiles*, 3rd edn. Collins, London.
Smith, R. F. and Keller, F. R. (1970). Free-operant avoidance in the pigeon using a treadle response. *J. exp. Anal. Behav.* **13**, 211–14.
Solomon, R. P. (1979). Temporal versus spatial information processing theories of hippocampal function. *Psychol. Bull.* **86**, 1272–9.
Spearman, C. E. (1904). 'General intelligence' objectively determined and measured. *Am. J. Psychol.* **15**, 201–93.
Squier, L. H. (1969). Autoshaping key responses with fish. *Psychon. Sci.* **17**, 177–8.
Squire, L. R. and Cohen, N. (1979). Memory and amnesia: resistance to disruption develops for years after learning. *Behav. neural Biol.* **25**, 115–25.
Stahl, B. J. (1974). *Vertebrate history: problems in evolution.* McGraw-Hill, New York.
Stearns, E. M. and Bitterman, M. E. (1965). A comparison of key-specking with an ingestive technique for the study of discriminative learning in pigeons. *Am. J. Psychol.* **78**, 48–56.
Stephan, H. and Andy, O. J. (1969). Quantitative comparative neuroanatomy of primates: an attempt at a phylogenetic interpretation. *Ann. N.Y. Acad. Sci.* **167**, 370–87.
— — (1970). The allocortex in primates. In *The primate brain* (*Advances in Primatology,* Vol. 1) (ed. C. R. Noback and W. Montagna) pp. 109–35. Appleton-Century-Crofts, New York.
— Bauchot, R., and Andy, O. J. (1970). Data on size of the brain and of various brain parts in insectivores and primates. In *The primate brain* (ed. C. R. Noback and W. Montagna) pp. 289–97. Appleton-Century-Crofts, New York.
Stettner, L. J. (1974). The neural basis of avian discrimination and reversal learning. In *Birds: brain and behavior* (ed. I. J. Goodman and M. W. Schein) pp. 165–201. Academic Press, New York.
— and Schultz, W. J. (1967). Brain lesions in birds: effects on discrimination acquisition and reversal. *Science, N.Y.* **155**, 1689–92.
Stewart, C. N. and Warren, J. M. (1957). The behavior of cats on the double-alternation problem. *J. comp. physiol. Psychol.* **50**, 26–8.
Storer, R. W. (1971). Classification of birds. In *Avian biology*, Vol. 1 (ed. D. S. Farner and J. R. King) pp. 1–18. Academic Press, New York.

Straub, R. O., Seidenberg, M. S., Bever, T. G., and Terrace, H. S. (1979). Serial learning in the pigeon. *J. exp. Anal. Behav.* **32**, 137–48.

Strong, P. N. and Hedges, M. (1966). Comparative studies in simple oddity learning: 1. Cats, raccoons, monkeys and chimpanzees. *Psychon. Sci.* **5**, 13–14.

Sugerman, R. A. and Demski, L. S. (1978). Agonistic behavior elicited by electrical stimulation of the brain in western collared lizards, *Crotaphytus collaris. Brain Behav. Evol.* **15**, 446–69.

Sutherland, N. S. and Mackintosh, N. J. (1971). *Mechanisms of animal discrimination learning.* Academic Press, New York.

Tarr, R. S. (1977). Role of the amygdala in the intraspecies aggressive behavior of the iguanid lizard, *Sceloporus occidentalis. Physiol. Behav.* **18**, 1153–8.

Tauxe, L. (1979). A new date for *Ramapithecus. Nature, Lond.* **282**, 399–401.

Ten Donkelaar, H. J. and De Boer-Van Huizen, R. (1978). Cells of origin of propriospinal and ascending supraspinal fibres in a lizard (*Lacerta galloti*). *Neurosci. Lett.* **9**, 285–90.

Tennant, W. A. and Bitterman, M. E. (1973). Some comparisons of intra- and extradimensional transfer in discriminative learning of goldfish. *J. comp. physiol. Psychol.* **83**, 134–9.

—— —— (1975). Blocking and overshadowing in two species of fish. *J. exp. Psychol: Anim. Behav. Processes* **1**, 22–9.

Terrace, H. S., Petitto, L. A., Sanders, R. J., and Bever, T. G. (1979). Can an ape create a sentence? *Science, N.Y.* **206**, 891–902.

Testa, T. J. (1974). Causal relationships and the acquisition of avoidance responses. *Psychol. Rev.* **81**, 491–505.

Teuber, H. L. and Weinstein, S. (1956). Ability to discover hidden figures after cerebral lesions. *Archs Neurol. Psychiat., Chicago* **76**, 369–79.

Theios, J. (1965). The mathematical structure of reversal learning in a shock-escape T-maze: overtraining and successive reversals. *J. math. Psychol.* **2**, 26–52.

Thomas, G. J., Hostetter, G., and Barker, D. J. (1968). Behavioral functions of the limbic system. In *Progress in physiological psychology*, Vol. 2 (ed. E. Stellar and J. M. Sprague) pp. 229–311. Academic Press, New York.

Thomas, R. K. and Boyd, M. G. (1973). A comparison of *Cebus albifrons* and *Saimiri sciureus* on oddity performance. *Anim. Learn. Behav.* **1**, 151–3.

—— Fowlkes, D., and Vickery, J. D. (1980). Conceptual numerousness judgments by squirrel monkeys. *Am. J. Psychol.* **93**, 247–57.

Thompson, C. R. and Church, R. M. (1980). An explanation of the language of a chimpanzee. *Science, N.Y.* **208**, 313–14.

Thompson, R. F., Groves, P. M., Teyler, T. J., and Roemer, R. A. (1973). A dual-process theory of habituation: theory and behavior. In *Habituation*, Vol. I *Behavioral studies* (ed. H. V. S. Peeke and M. J. Herz) pp. 239–71. Academic Press, New York.

—— and Spencer, W. A. (1966). Habituation: a model phenomenon for the study of neuronal substrates of behavior. *Psychol. Rev.* **73**, 16–43.

Thompson, R. K. R. and Herman, L. M. (1977). Memory for lists of sounds by the bottle-nosed dolphin: convergence of memory processes with humans? *Science, N.Y.* **195**, 501–3.

Thompson, T. (1963). Visual reinforcement in Siamese fighting fish. *Science, N.Y.* **141**, 55–7.

—— and Sturm, T. (1965). Classical conditioning of aggressive display in Siamese fighting fish. *J. exp. Anal. Behav.* **8**, 397–403.

Thorndike, E. L. (1898). Animal intelligence: an experimental study of the associative processes in animals. *Psychol. Rev.* Monograph Suppl. 2, No. 4 (Whole No. 8).

— (1911). *Animal intelligence: experimental studies.* Macmillan, New York.

Thorpe, W. H. (1963). *Learning and instinct in animals,* 2nd edn. Methuen, London.

Tighe, T. J. and Frey, K. (1972). Subproblem analysis of discrimination shift learning in the rat. *Psychon. Sci.* **28**, 129–33.

— and Tighe, L. S. (1967). Discrimination shift performance of children as a function of age and shift procedure. *J. exp. Psychol.* **74**, 466–70.

Tinklepaugh, O. L. (1932). Maze learning of a turtle. *J. comp. Psychol.* **13**, 201–6.

Tranberg, D. K. and Rilling, M. (1978). Latent inhibition in the autoshaping paradigm. *Bull. psychon. Soc.* **11**, 273–6.

Trapold, M. A. and Overmier, J. B. (1972). The second learning process in instrumental learning. In *Classical conditioning II: Current research and theory* (ed. A. H. Black and W. F. Prokasy) pp. 427–52. Appleton-Century-Crofts, New York.

Tsang, Y. C. (1934). The functions of the visual areas of the cerebral cortex of the rat in the learning and retention of the maze. I. *Comp. Psychol. Monogr.* **10**(4), 1–56.

— (1936). The functions of the visual areas of the cerebral cortex of the rat in the learning and retention of the maze. II. *Comp. Psychol. Monogr.* **12**(2), 1–41.

Tucker, D. (1971). Nonolfactory responses from the nasal cavity: Jacobson's organ and the trigeminal system. In *Handbook of sensory physiology*, Vol. 4 *Chemical senses, Part I Olfaction* (ed. L. M. Beidler) pp. 151–81. Springer, Berlin.

Turner, B. H. and Mishkin, M. (1978). A reassessment of the direct projections of the olfactory bulb. *Brain Res.* **151**, 375–80.

Tversky, B. and Sherman, T. (1975). Picture memory improves with longer on and off time. *J. exp. Psychol: Hum. Learn. Mem.* **1**, 114–18.

Ulinski, P. S. (1978). Organization of anterior dorsal ventricular ridge in snakes. *J. comp. Neurol.* **178**, 411–50.

Van Bergeijk, W. A. (1967). Anticipatory feeding behaviour in the bullfrog (*Rana catesbeiana*). *Anim. Behav.* **15**, 231–8.

Van Sommers, P. (1962). Oxygen-motivated behavior in the goldfish. *Carassius auratus. Science, N.Y.* **137**, 678–9.

— (1963). Air-motivated behavior in the turtle. *J. comp. physiol. Psychol.* **56**, 590–6.

Vance, V. J., Richardson, A. M., and Goodrich, R. B. (1965). Brightness discrimination in the collared lizard. *Science, N.Y.* **147**, 758–9.

Vanegas, H. and Ebbesson, S. O. E. (1976). Telencephalic projections in two teleost species. *J. comp. Neurol.* **165**, 181–96.

Vaughter, R. M., Smotherman, W., and Ordy, J. M. (1972). Development of object permanence in the infant squirrel monkey. *Devl Psychol.* **7**, 34–8.

Vesselkin, N. P., Agayan, A. L., and Nomokonova, L. M. (1971). A study of thalamo-telencephalic afferent systems in frogs. *Brain Behav. Evol.* **4**, 295–306.

— Ermakova, T. V., Repérant, J., Kosareva, A. A., and Kenigfest, N. B. (1980) The retinofugal and retinopetal systems in *Lampetra fluviatilis*. An experimental study using radioautographic and HRP methods. *Brain Res.* **195**, 453–60.

Vinogradova, O. S. (1978). Discussion. In *Functions of the septohippocampal system*. Ciba Foundation Symposium 58, pp. 343–9.

Von Bonin, G. (1937). Brain-weight and body-weight of mammals. *J. gen. Psychol.* **16**, 379–89.

— (1941). Side lights on cerebral evolution: brain size of lower vertebrates and degree of cortical folding. *J. gen. Psychol.* **25**, 273–82.

Voneida, T. J. and Ebbesson, S. O. E. (1969). On the origin and distribution of axons in the pallial commissures in the Tegu Lizard (*Tupinambis nigropunctatus*). *Brain Behav. Evol.* 2, 467–81.

— and Sligar, C. M. (1976). A comparative neuroanatomic study of retinal projections in two fishes: *Astyanax hubbsi* (the blind cave fish), and *Astyanax mexicanus*. *J. comp. Neurol.* **165**, 89–106.

— — (1979). Efferent projections of the dorsal ventricular ridge and the striatum in the Tegu lizard, *Tupinambis nigropunctatus*. *J. comp. Neurol.* **186**, 43–64.

Vorster, J. (1975). Mommy linguist: the case for motherese. *Lingua* **37**, 281–312.

Wagner, A. R. (1961). Effects of amount and percentage of reinforcement and number of acquisition trials on conditioning and extinction. *J. exp. Psychol.* **62**, 234–42.

— (1976). Priming in STM: an information-processing mechanism for self-generated or retrieval-generated depression in performance. In *Habituation: perspectives from child development, animal behavior and neurophysiology* (ed. T. J. Tighe and R. N. Leaton) pp. 95–128. Erlbaum, Hillsdale, NJ.

— (1981). SOP: a model of automatic memory processing in animal behavior. In *Information processing in animals: memory mechanisms* (in press) (ed. N. E. Spear and R. R. Miller). Erlbaum, Hillsdale, NJ.

— Rudy, J. W., and Whitlow, J. W. (1973). Rehearsal in animal conditioning. *J. exp. Psychol.* **97**, 407–26.

Walk, R. D. and Gibson, E. J. (1961). A comparative and analytical study of visual depth perception. *Psychol. Monogr.* **75**, No. 15, whole No. 519.

Walker, A. and Leakey, R. E. F. (1978). The hominids of East Turkana. *Scient. Am.* **239**, 44–56.

Ward, J. P., Yehle, A. L., and Doerflein, R. S. (1970). Cross-modal transfer of a specific discrimination in the bushbaby (*Galago senegalensis*). *J. comp. physiol. Psychol.* **73**, 74–7.

Warren, J. L., Bryant, R. C., Petty, F., and Byrne, W. L. (1975). Group training in goldfish (*Carassius auratus*): effects on acquisition and retention. *J. comp. physiol. Psychol.* **89**, 933–8.

Warren, J. M. (1960*a*). Reversal learning by paradise fish (*Macropodus opercularis*). *J. comp. physiol. Psychol.* **53**, 376–8.

— (1960*b*). Oddity learning set in a cat. *J. comp. physiol. Psychol.* **53**, 433–4.

— (1961). The effect of telencephalic injuries on learning by paradise fish, *Macropodus opercularis*. *J. comp. physiol. Psychol.* **54**, 130–2.

— (1965). Primate learning in comparative perspective. In *Behavior of nonhuman primates: modern research trends*, Vol. 1 (ed. A. M. Schrier, H. F. Harlow, and F. Stollnitz) pp. 249–81. Academic Press, New York.

— (1966). Reversal learning and the formation of learning sets by cats and rhesus monkeys. *J. comp. physiol. Psychol.* **61**, 421–8.

— (1967*a*). An assessment of the reversal index. *Anim. Behav.* **15**, 493–8.

— (1967*b*). Double alternation learning by experimentally naive and sophisticated cats. *J. comp. physiol. Psychol.* **64**, 161–3.

— (1973). Learning in vertebrates. In *Comparative psychology: a modern survey* (ed. D. A. Dewsbury and D. A. Rethlingshafer) pp. 471–509. McGraw-Hill, New York.

— (1977). Handedness and cerebral dominance in monkeys. In *Lateralization in the nervous system* (ed. S. Harnad, R. W. Doty, L. Goldstein, J. Jaynes, and G. Krauthamer) pp. 151–72. Academic Press, New York.
— and Baron, A. (1956). The formation of learning sets by cats. *J. comp. physiol. Psychol.* **49**, 227–31.
— and Nonneman, A. J. (1976). The search for cerebral dominance in monkeys. *Ann. N.Y. Acad. Sci.* **280**, 732–44.
Warrington, E. K. and Weiskrantz, L. (1978). A new method of testing long-term retention with special reference to amnesic patients. *Nature, Lond.* **217**, 972–4.
Washburn, S. L. (1978). The evolution of man. *Scient. Am.* **239**, 146–54.
Wasserman, E. A. (1973). Pavlovian conditioning with heat reinforcement produces stimulus-directed pecking in chicks. *Science, N.Y.* **181**, 875–7.
— Hunter, N. B., Gutowski, K. A., and Bader, S. A. (1975). Autoshaping chicks with heat reinforcement: the role of stimulus–reinforcer and response–reinforcer relations. *J. exp. Psychol: Anim. Behav. Processes* **1**, 158–69.
— and Molina, E. J. (1975). Explicitly unpaired key light and food presentations: interference with subsequent auto-shaped keypecking in pigeons. *J. exp. Psychol: Anim. Behav. Processes* **1**, 30–8.
Weaver, G. E. and Stanning, C. J. (1978). Short-term retention of pictorial stimuli as assessed by a probe recognition technique. *J. exp. Psychol: Hum. Learn. Mem.* **4**, 55–65.
Webster, W. G. (1977). Hemispheric asymmetry in cats. In *Lateralization in the nervous system* (ed. S. Harnad, R. W. Doty, L. Goldstein, J. Jaynes, and G. Krauthamer) pp. 471–80. Academic Press, New York.
Weinstein, S. and Teuber, H. L. (1957). Effects of penetrating brain injury on intelligence test scores. *Science, N.Y.* **125**, 1036–7.
Weisbach, W. and Schwartzkopff, J. (1967). Nervöse Antworten auf Schallreit im Grosshirn von Krokodilen. *Naturwissenschaften* **54**, 650.
Weiskrantz, L. (1978). A comparison of hippocampal pathology in man and other animals. In *Functions of the septo-hippocampal system*. Ciba Foundation Symposium 58, pp. 373–87. Elsevier, Amsterdam.
— (1980). Varieties of residual experience. *Q. J. exp. Psychol.* **32**, 365–86.
Weitzman, R. A. (1967). Positional matching in rats and fish. *J. comp. physiol. Psychol.* **63**, 54–9.
Welker, W. I. and Campos, G. B. (1963). Physiological significance of sulci in somatic sensory cerebral cortex in mammals of the family Procyonidae. *J. comp. Neurol.* **120**, 19–36.
— and Lende, R. A. (1980). Thalamocortical relationships in Echidna (*Tachyglossus aculeatus*). In *Comparative neurology of the telencephalon* (ed. S. O. E. Ebbesson) pp. 449–81. Plenum Press, New York.
— and Welker, J. (1958). Reaction of fish (*Eucinostomus gula*) to environmental changes. *Ecology* **39**, 283–8.
Werboff, J. and Lloyd, T. (1963). Avoidance conditioning in the guppy (*Lebistes reticulatus*). *Psychol. Rep.* **12**, 615–18.
Wesley, F. (1961). The number concept: a phylogenetic review. *Psychol. Bull.* **58**, 420–8.
Wesp, R. and Goodman, I. J. (1978). Fixed interval responding by pigeons following damage to corpus striatal and limbic brain structures (paleostriatal complex and parolfactory lobe). *Physiol. Behav.* **20**, 571–7.
Wessels, M. G. (1974). The effects of reinforcement upon the prepecking behaviors of pigeons in the autoshaping experiment. *J. exp. Anal. Behav.* **21**, 125–44.

White, T. D. (1980). Evolutionary implications of Pliocene hominid footprints. *Science, N.Y.* **208**, 175–6.
Wickelgren, W. O. (1977). Post-tetanic potentiation, habituation and facilitation of synaptic potentials in reticulospinal neurones of lamprey. *J. Neurophysiol.* **270**, 115–31.
Wickens, D. D. (1970). Encoding categories of words: an empirical approach to meaning. *Psychol. Rev.* **77**, 1–15.
Wilcoxon, H. C., Dragoin, W. B., and Kral, P. A. (1971). Illness-induced aversions in rat and quail: relative salience of visual and gustatory cues. *Science, N.Y.* **171**, 826–8.
Williams, B. A. (1971). The effects of intertrial interval on discrimination reversal learning in the pigeon. *Psychon. Sci.* **23**, 241–3.
— (1976). Short-term retention of response outcome as a determinant of serial reversal learning. *Learn. Motiv.* **7**, 418–30.
Williams, D. I. (1967). The overtraining reversal effect in the pigeon. *Psychon. Sci.* **7**, 261–2.
— (1968). Transfer along a continuum in the pigeon. *J. comp. physiol. Psychol.* **65**, 369–71.
Williams, D. R. and Williams, H. (1969). Automaintenance in the pigeon: sustained pecking despite contingent non-reinforcement. *J. exp. Anal. Behav.* **12**, 511–20.
Williams, J. T. (1967). Efficient motivation for studying crocodilian learning. *Psychon. Sci.* **8**, 279.
— (1968). Reversal-learning in the spectacled caiman. *Am. J. Psychol.* **81**, 258–61.
— and Albiniak, B. A. (1972). Probability learning in a crocodilian. *Psychon. Sci.* **27**, 165–6.
Wilson, B. J. (1978*a*). Complex learning in birds. Unpublished Ph. D. thesis, University of Sussex.
Wilson, J. J. (1964). Level of training and goal-box movements as parameters of the partial reinforcement effect. *J. comp. physiol. Psychol.* **57**, 211–13.
Wilson, M. (1978*b*). Visual system: pulvinar-extrastriate cortex. In *Handbook of behavioral neurobiology*, Vol. 1 *Sensory integration* (ed. R. B. Masterton) pp. 209–47. Plenum, New York.
Wilson, W. A., Oscar, M., and Bitterman, M. E. (1964*a*). Probability-learning in the monkey. *Q. J. exp. Psychol.* **16**, 163–5.
— — — (1964*b*). Visual probability-learning in the monkey. *Psychon. Sci.* **1**, 71–2.
Wise, K. L., Wise, L. A., and Zimmerman, R. R. (1974). Piagetian object permanence in the infant rhesus monkey. *Devl Psychol.* **10**, 429–37.
Wise, L. M. and Gallagher, D. P. (1964). Partial reinforcement of a discriminative response in the turtle. *J. comp. physiol. Psychol.* **57**, 311–13.
Wodinsky, J., Behrend, E. R., and Bitterman, M. E. (1962). Avoidance-conditioning in two species of fish. *Anim. Behav.* **10**, 76–8.
— and Bitterman, M. E. (1953). The solution of oddity-problems by the rat. *Am. J. Psychol.* **66**, 137–40.
— — (1960). Resistance to extinction in the fish after extensive training with partial reinforcement. *Am. J. Psychol.* **73**, 429–34.
Wolach, A. H., Breuning, S. E., Roccaforte, P., and Solhkhan, N. (1977). Overshadowing and blocking in a goldfish (*Carassius auratus*) respiratory conditioning situation. *Psychol. Rec.* **27**, 693–702.
— Raymond, B., and Hurst, J. W. (1973). Reward magnitude shifts with goldfish. *Psychol. Rec.* **23**, 371–6.

Wolff, J. L. (1967). Concept-shift and discrimination-reversal learning in humans. *Psychol. Bull.* **68**, 369–408.

Wollberg, Z. and Newman, J. D. (1972). Auditory cortex of squirrel monkey: response patterns of single cells to species-specific vocalizations. *Science, N.Y.* **175**, 212–14.

Wood, F. G. (1973). *Marine mammals and man: the Navy's porpoises and sea lions.* R. B. Luce, Washington.

Wood, S., Moriarty, K. M., Gardner, B. T., and Gardner, R. A. (1980). Object permanence in child and chimpanzee. *Anim. Learn, Behav.* **8**, 3–9.

Woodard. W. T., Ballinger, J. C., and Bitterman, M. E. (1974). Autoshaping: further study of 'negative automaintenance'. *J. exp. Anal. Behav.* **22**, 47–51.

— and Bitterman, M. E. (1971*a*). Punishment in the goldfish as a function of electrode orientation. *Behav. Res. Meth. Instrum.* **3**, 72–3.

— — (1971*b*). Classical conditioning of goldfish in the shuttlebox. *Behav. Res. Meth. Instrum.* **3**, 193–4.

— — (1973*a*). Pavlovian analysis of avoidance conditioning in the goldfish (*Carassius auratus*). *J. comp. physiol. Psychol.* **82**, 123–9.

— — (1973*b*). Further experiments on probability learning in goldfish. *Anim. Learn. Behav.* **1**, 25–8.

— — (1974). Autoshaping in the goldfish. *Behav. Res. Meth. Instrum.* **6**, 409–10.

— Schoel, W. M., and Bitterman, M. E. (1971). Reversal learning with singly presented stimuli in pigeons and goldfish. *J. comp. physiol. Psychol.* **76**, 460–7.

Woodruff, G. and Premack, D. (1979). Intentional communication in the chimpansee: the development of deception. *Cognition* **7**, 333–62.

— — and Kennel, K. (1978). Conservation of liquid and solid quantity by the chimpanzee. *Science, N.Y.* **202**, 991–4.

— and Williams, D. R. (1976). The associative relation underlying autoshaping in the pigeon. *J. exp. Anal. Behav.* **26**, 1–13.

Wright, P. L., Kay, H., and Sime, M. E. (1963). The establishment of learning sets in rats. *J. comp. physiol. Psychol.* **56**, 200–3.

Yaremko, R. M., Boice, R., and Thompson, R. W. (1969). Classical and avoidance conditioning of the nictitating membrane in frogs (*Rana pipiens*) and toads (*Bufo americanus*). *Psychon. Sci.* **16**, 162–4.

Yehle, A. L. and Ward, J. P. (1969). Cross-modal transfer of a specific discrimination in the rabbit. *Psychon. Sci.* **16**, 269–70.

Yeni-Komshian, G. H. and Benson, D. A. (1976). Anatomical study of cerebral asymmetry in the temporal lobe of humans, chimpanzees, and rhesus monkeys. *Science, N.Y.* **192**, 387–9.

Yerkes, R. M. (1901). The formation of habits in the turtle. *Pop. Sci. Mon.* **58**, 519–25.

— (1903). The instincts, habits and reactions of the frog. *Psychol. Monogr.* **4**, 579–638.

Yori, J. G. (1978). Active one-way avoidance to a heat aversive stimulus in Tegu lizards (*Tupinambus teguixen*). *Behav. Biol.* **23**, 100–6.

Young, J. Z. (1962). *The life of vertebrates*, 2nd edn. Clarendon Press, Oxford.

Yunis, J. J., Sawyer, J. R., and Dunham, K. (1980). The striking resemblance of high-resolution G-banded chromosomes of man and chimpanzee. *Science, N.Y.* **208**, 1145–8.

Zangwill, O. L. (1961). Lashley's concept of cerebral mass action. In *Current problems in animal behaviour* (ed. W. H. Thorpe and O. L. Zangwill) pp. 59–86. Cambridge University Press.

Zaretsky, M. D. (1978). A new auditory area of the songbird forebrain: a connection between auditory and song control centers. *Expl Brain Res.* **32**, 267–73.

— and Konishi, M. (1976). Tonotopic organization in the avian telencephalon. *Brain Res.* **111**, 167–71.

Zecha, A. (1962). The 'pyramidal tract' and other telencephalic efferents in birds. *Acta morph. neerl.-scand.* **5**, 194–5.

Zeier, H. (1968). Changes in operant behavior of pigeons following bilateral forebrain lesions. *J. comp. physiol. Psychol.* **66**, 198–203.

— (1969). DRL-performance and timing behavior of pigeons with archistriatal lesions. *Physiol. Behav.* **4**, 189–93.

— (1971). Archistriatal lesions and response inhibition in the pigeon. *Brain Res.* **31**, 327–39.

— and Karten, H. J. (1971). The archistriatum of the pigeon: organization of afferent and efferent connections. *Brain Res.* **31**, 313–26.

Zeigler, H. P. (1961). Learning-set formation in pigeons. *J. comp. physiol. Psychol.* **54**, 252–4.

— (1963). Effects of endbrain lesions upon visual discrimination learning in pigeons. *J. comp. Neurol.* **120**, 161–81.

— (1974). Feeding behavior in the pigeon: a neurobehavioral analysis. In *Birds: brain and behavior* (ed. I. J. Goodman and M. W. Schein) pp. 101–32. Academic Press, New York.

— Green, H. L., and Karten, H. J. (1969). Neural control of feeding behavior in the pigeon. *Psychon. Sci.* **15**, 156–7.

— Hollard, V., Wild, J. M., and Webster, D. M. (1978). Intracranial self stimulation from endbrain nuclei in the pigeon (*Columba livia*). *Physiol. Behav.* **21**, 387–94.

Zeiler, M. D. (1971). Eliminating behavior with reinforcement. *J. exp. Anal. Behav.* **16**, 401–5.

Zentall, T. R. and Hogan, D. E. (1974). Abstract concept learning in the pigeon. *J. exp. Psychol.* **102**, 393–8.

— — (1975). Concept learning in the pigeon: transfer to new matching and nonmatching stimuli. *Am. J. Psychol.* **88**, 233–44.

— — (1976). Pigeons can learn identity or difference, or both. *Science, N.Y.* **191**, 408–9.

— — (1978). Same/different concept learning in the pigeon: the effect of negative instances and prior adaptation to transfer stimuli. *J. exp. Anal. Behav.* **30**, 177–86.

— — Edwards, C. A., and Hearst, E. (1980). Oddity learning in the pigeon as a function of the number of incorrect alternatives. *J. exp. Psychol: Anim. Behav. Processes* **6**, 278–99.

Zihlman, A. L., Cronin, J. E., Cramer, D. L., and Sarich, V. M. (1978). Pygmy chimpanzee as a possible prototype for the common ancestor of humans, chimpanzees and gorillas. *Nature, Lond.* **275**, 744–6.

Zych, K. A. and Wolach, A. H. (1973). Resistance to extinction in the goldfish (*Carassius auratus*). *J. comp. physiol. Psychol.* **82**, 115–22.

Author Index

Abbie, A. A. 124, 344
Addison, R. G. 60, 376
Aderman, M. 89, 373
Ades, H. W. 208, 374
Adey, W. R. 124, 344
Adler, N. 68, 344
Agayan, A. L. 123, 380
Alba, A. 133, 360
Albiniak, B. A. 160, 383
Ambros, V. R. 122, 358
Amiro, T. W. 64, 344
Amsel, A. 86, 164, 344
Andrew, R. J. 234, 344
Andrews, O. 157, 344
Andy, O. J. 244, 247, 249, 250, 378
Annett, M. 315, 344
Arbeit, W. R. 278, 280, 359
Arnold, W. J. 280, 363
Aron, C. 327, 376
Aronson, L. R. 53, 104, 105, 106, 108, 344, 362
Atencio, F. W. 263, 344

Bacon, H. R. 224, 344
Baddeley, A. D. 33, 318, 320, 322, 344
Bader, S. A. 192, 194, 382
Baenninger, R. 57, 344
Baerends, G. P. 57, 344
Bailey, P. J. 294, 344
Baker-Cohen, K. F. 140, 142, 174, 345
Bakker, R. T. 168, 345
Ball, G. G. 205, 347
Ballinger, J. C. 63, 384
Bamborough, P. 181, 363
Banuazizi, A. 286, 358
Barker, D. J. 32, 379
Barlow, H. B. 181, 345
Baron, A. 279, 382
Bass, A. H. 146, 345
Bastian, J. 302
Bateson, P. P. G. 183, 192, 232, 345, 360, 365
Bauchot, R. 247, 378
Bauer, J. H. 256, 345
Beach, F. A. 280, 359
Beecher, M. D. 294, 316, 317, 372, 378,
Beez, V. 90, 351
Behrend, E. R. 66, 68, 70, 72, 73, 76, 80, 85, 161, 197, 217, 345, 355, 383
Belekhova, M. G. 122, 146, 148, 149, 345, 362
Bennett, M. V. L. 67, 68, 367
Bennetto, K. 187, 351
Benowitz, L. 176, 231, 345
Benson, D. A. 317, 384
Berger, B. D. 76, 77, 198, 223, 355
Berk, A. M. 130, 131, 368
Berkowitz, E. C. 148, 363
Berman, G. 323, 352
Bernstein, J. J. 102, 104, 105, 345, 351
Berryman, R. 212, 345, 350
Bessemer, D. W. 210, 345
Bever, T. G. 294, 296, 297, 298, 299, 300, 379
Bicknell, A. T. 156, 157, 345
Bintz, J. 66, 345
Birch, H. G. 307, 345
Birch, M. P. 260, 346
Birnberger, K. L. 57, 346
Bitterman, M. E. 7, 13, 31, 60, 61, 62, 63, 64, 65, 66, 67, 68, 69, 70, 71, 72, 73, 74, 75, 76, 77, 78, 79, 80, 81, 82, 83, 84, 85, 86, 87, 88, 89, 90, 91, 92, 93, 94, 95, 96, 97, 98, 99, 105, 155, 158, 159, 160, 161, 162, 163, 164, 192, 195, 197, 198, 200, 201, 203, 204, 205, 216, 217, 223, 259, 260, 261, 270, 282, 283, 344, 345, 346, 347, 348, 350, 352, 353, 355, 356, 357, 360, 363, 365, 372, 375, 376, 377, 378, 379, 383, 384
Bjork, R. A. 326, 346
Blakemore, C. 181, 345
Blank, M. 285, 346
Blankenagel, F. 132, 346
Blass, E. M. 281, 346
Blomquist, A. J. 210, 351
Blum, J. S. 282, 369
Blum, R. A. 282, 369
Boakes, R. A. 302, 303, 346
Boice, R. 127, 128, 130, 131, 157, 346, 384
Boitano, J. J. 85, 346
Bok, S. T. 25, 26, 346
Bolles, R. C. 66, 346
Bols, R. J. 64, 346
Bonbright, J. C. 183, 359, 360
Boord, R. L. 123, 184, 187, 346, 348
Borton, R. W. 285, 367
Bottjer, S. W. 60, 346
Boycott, B. B. 42, 150, 157, 346
Boyd, B. O. 282, 347
Boyd, M. G. 282, 379

Boysen, S. 301, 302, 376
Bradley, P. 183, 347
Braford, M. R. 144, 370
Brandon, S. E. 62, 63, 65, 347
Braun, J. J. 256, 261, 347, 351
Brauth, S. E. 142, 177, 178, 347
Bremer, F. 188, 347
Bresler, D. E. 259, 347
Breuning, S. E. 61, 90, 98, 99, 150, 153, 154, 347, 353, 383
Bridger, W. H. 285, 346
Broadbent, D. E. 217, 347
Brodkorb, P. 169, 347
Brodmann, K. 253
Brookshire, K. H. 205, 347
Brower, J. Z. 131, 347
Brower, L. P. 131, 347
Brown, D. P. F. 310, 311, 347
Brown, M. W. 183, 347
Brown, P. L. 60, 193, 347
Brown, R. 291, 299, 347
Browne, M. P. 193, 371
Brownlee, A. 203, 347
Brucker, B. S. 133, 360
Bruckmoser, P. 54, 347
Brutkowski, S. 265, 267, 347, 369
Bryant, P. E. 217, 307, 308, 347
Bryant, R. C. 66, 381
Buchmann, O. L. K. 271, 348
Buerger, A. A. 132, 133, 353
Bullock, D. H. 197, 200, 201, 203, 348, 357, 375
Bullock, T. H. 54, 348, 372
Burghardt, G. M. 150, 152, 153, 155, 158, 348
Burnett, T. C. 132, 348
Burns, A. H. 144, 348
Burns, R. A. 90, 348
Burt, D. E. 93, 348
Bush, R. C. 182, 354
Butcher, H. J. 341, 348
Butler, A. B. 144, 145, 348, 370
Butter, C. M. 266, 348
Butters, N. 321, 322, 323, 348, 349, 352
Byrne, W. L. 66, 381

Cable, C. 216, 359
Campbell, A. A. 341, 348
Campbell, C. B. G. 55, 123, 184, 348
Campos, G. B. 246, 382
Candland, D. K. 70, 79, 346
Cannon, R. E. 234, 348
Capaldi, E. J. 88, 93, 162, 348, 349, 369
Capranica, R. R. 123, 369
Carey, S. 323, 352
Carr, L. D. 195, 367
Carter, D. E. 212, 213, 349
Carter, T. 150, 151, 349
Cauty, A. 70, 71, 72, 73, 74, 75, 78, 199, 366
Cerf, J. A. 60, 66, 371
Cermak, L. S. 321, 322, 348, 349
Chalmers, M. 307, 308, 365
Champlin, G. 203, 355
Charig, A. J. 168, 349
Chen, C.-S. 271, 284, 349, 376
Cherkin, A. 68, 195, 349, 374
Chiszar, D. 150, 151, 349
Chomsky, N. 312, 313, 329, 349
Christie, D. F. M. 220, 372
Church, R. M. 300, 379
Clark, E. 64, 349
Clayton, F. L. 57, 58, 349
Coates, S. R. 192, 351
Cochrane, T. L. 89, 90, 349
Cogan, D. 93, 349
Cohen, D. H. 54, 176, 184, 187, 188, 191, 192, 205, 227, 233, 349
Cohen, H. 223, 356
Cohen, L. R. 212, 345
Cohen, N. 327, 378
Colbert, E. H. 238, 349
Coleman, S. R. 196, 349
Collias, N. E. 233, 349
Collins, R. L. 315, 349
Collison, C. 325, 370
Combs, W. C. 271, 281, 352
Commins, W. D. 341, 349
Constantine-Paton, M. 123, 369
Cooper, H. 278, 350
Cooper, R. M. 256, 260, 345, 346
Corkin, S. 319, 368
Corsi, P. M. 267
Corwin, J. T. 54, 348
Couvillon, P. A. 96, 97, 203, 204, 205, 350
Cowan, W. M. 187, 373
Cowey, A. 262, 263, 350, 357
Craik, F. I. M. 322, 323, 350
Cramer, D. L. 241, 385
Crawford, F. T. 156, 350
Crespi, L. P. 89, 350
Crile, G. 26, 27, 350
Cromer, R. F. 292, 350
Cronin, J. E. 241, 385
Crosby, E. C. 138, 142, 172, 246, 253, 362
Crowder, R. G. 217, 350
Cruce, W. L. R. 149, 350
Cuénod, M. 179, 367
Cumming, W. W. 212, 345, 350
Cunitz, A. R. 218, 355
Curnow, P. F. 60, 102, 106, 108, 109, 371
Cutting, J. E. 294, 350
Czaplicki, J. A. 150, 151, 350
Czéh, G. 54, 372

Dabrowska, J. 267, 347, 350

Dafters, R. 227, 228, 350
D'Amato, M. R. 96, 350
Darwin, C. 1, 3, 350
Davenport, J. W. 210, 365
Davidson, R. E. 153, 350
Davidson, R. S. 155, 156, 350
Davis, J. L. 192, 351
Davis, R. T. 283, 351
Davison, M. C. 234, 360
De Boer-Van Huizen, R. 149, 379
De Fina, A. V. 147, 351
Dean, P. 262, 351
Deaux, E. 59, 356
Deets, A. C. 210, 351
Delius, J. D. 184, 185, 187, 351
Demski, L. S. 106, 149, 351, 379
Denenberg, V. H. 316, 377
Dennis, M. 313, 351
Descartes, R. 1
Devine, J. V. 211, 280, 351
Dewsbury, D. A. 104, 105, 351
Dewson, J. H. 316, 317, 351
Di Cara, L. V. 261, 351
Di Lollo, V. 90, 271, 351
Diamond, I. T. 252, 253, 262, 263, 264, 344, 351, 363
Dickinson, A. 11, 12, 13, 196, 270, 327, 351, 366
Dicks, D. 143, 351
Diebschlag, E. 132, 165, 351, 352
Diercks, J. K. 162, 369
Dieringer, N. 54, 347
Dijkgraaf, S. 56, 352
Distel, H. 149, 352
Divac, I. 266, 268, 274, 352
Doerflein, R. S. 285, 381
Domesick, V. B. 70, 76, 345
Donovan, W. J. 178, 181, 352
Doty, B. A. 271, 277, 278, 279, 281, 352
Doty, L. A. 277, 278, 279, 352
Dow, S. 188, 347
Dragoin, W. B. 192, 383
Dricker, J. 323, 352
Dubbeldam, J. L. 177, 178, 362
Dubois, E. 30, 242, 352
Duff, T. A. 54, 349
Dufort, R. H. 271, 352
Dunham, K. 329, 384
Durkovic, R. G. 192, 349

Ebbesson, S. O. E. 52, 53, 54, 55, 101, 119, 121, 124, 139, 140, 141, 348, 349, 352, 357, 362, 376, 377, 380, 381
Ebner, F. F. 139, 141, 144, 145, 148, 246, 352, 358
Eckerman, D. A. 212, 213, 349
Edwards, C. A. 213, 385
Eikmanns, K. 125, 126, 352
Eimas, P. D. 96, 377
Elias, H. 247, 352
Elliott, R. C. 285, 352
Engelhardt, F. 71, 73, 74, 81, 95, 352
Epstein, R. 302, 353
Ermakova, T. V. 53, 380
Eskin, R. M. 82, 84, 85, 162, 353, 356
Estes, W. K. 342, 353
Etkin, M. W. 327, 353
Ettlinger, G. 251, 285, 310, 311, 347, 353, 361, 371
Evans, E. F. 185, 353
Everett, N. B. 36, 353
Ewert, J. P. 121, 122, 125, 126, 127, 353

Fallon D. 67, 68, 89, 90, 105, 349, 377
Fankhauser, G. 130, 133, 134, 353
Fantino, E. 69, 353
Farel, P. B. 132, 133, 353
Farris, H. E. 150, 153, 154, 353
Fearing, F. S. 191, 353
Ferguson, J. L. 177, 178, 347
Ferrier, R. J. 260, 346
Ferry, M. 89, 356
Fields, H. L. 181, 365
Finger, T. E. 55, 56, 353
Fitch, M. D. 216, 369
Flaherty, C. F. 93, 353
Fleshler, M. 196, 360
Fletcher, H. J. 274, 354
Flood, N. B. 103, 105, 106, 107, 108, 109, 110, 111, 354, 371
Foard, C. F. 294, 350
Fobes, J. L. 160, 363
Foree, D. D. 9, 354
Foskett, M. D. 85, 346
Foster, R. E. 147, 148, 354
Fowlkes, D. 284, 379
Fox, R. 182, 354
Francisco, E. W. 208, 374
Frank, A. G. 103, 106, 107, 111, 354
Frank, W. H. 130, 133, 134, 353
Franklin, S. R. 192, 359
Franzisket, L. 126, 354
Frey, K. 164, 380
Friedman, H. 271, 354
Frumin, N. 144, 358
Fuentes, I. 59, 356
Fuenzalida, C. E. 150, 151, 354
Fujita, O. 106, 354

Gaertner, I. 302, 303, 346
Gaffan, D. 325, 354
Galaburda, A. M. 317, 354
Galand, G. 122, 123, 365
Galkin, T. W. 267, 355
Gallagher, D. P. 162, 383
Gallon, R. L. 66, 354

Gallup, G. G. 191, 227, 233, 354, 367, 369
Gamble, H. J. 139, 143, 149, 165, 354, 355
Gamzu, E. 163, 354
Garbanati, J. A. 316, 377
Garcia, J. 305, 354
Gardner, B. T. 286, 294, 295, 296, 297, 298, 299, 301, 354, 355, 384
Gardner, R. A. 286, 294, 295, 296, 297, 298, 299, 301, 354, 355, 384
Gay, R. A. 91, 365
Geller, I. 60, 355
Geschwind, N. 285, 314, 317, 354, 355
Gibson, E. J. 146, 381
Girgus, J. S. 286, 358
Glanzer, M. 218, 355
Glanzman, D. L. 125, 126, 355
Glick, S. D. 315, 355
Goddard, G. V. 143, 355
Goff, D. M. 233, 349
Gojković, M. 54, 372
Goldberg, M. E. 181, 355
Goldby, F. 139, 143, 149, 165, 355
Goldman, P. S. 267, 355
Goldstein, A. C. 127, 128, 355
Gonzalez, R. C. 61, 72, 73, 76, 77, 82, 83, 84, 85, 86, 87, 89, 90, 91, 161, 162, 163, 197, 198, 203, 223, 259, 260, 355, 356, 372
Goodman, D. A. 125, 126, 356
Goodman, D. C 144, 348
Goodman, I. J. 233, 234, 356, 382
Goodrich, R. B. 157, 380
Gordon, D. 104, 356
Gordon, J. A. 282, 359
Gormezano, I. 59, 61, 128, 356
Gossette, M. F. 197, 223, 224, 272, 357
Gossette, R. L. 155, 159, 197, 200, 222, 223, 224, 270, 271, 272, 356, 357
Gould, S. J. 29, 30, 240, 357, 372
Graeber, R. C. 54, 56, 357
Graf, C. L. 162, 357
Graf, V. 164, 195, 200, 201, 357
Granda, A. M. 156, 157, 357
Grant, D. S. 219, 220, 375
Gray, J. A. 34, 108, 357
Grecian, E. A. 271, 348
Green, H. L. 189, 385
Greenwood, D. D. 277, 364
Gregory, R. L. 32, 357
Gross, C. G. 262, 263, 350, 357
Groves, P. M. 127, 379
Gruber, H. E. 286, 358
Gruber, S. H. 59, 358
Gruberg, E. R. 122, 358
Gruender, A. 61, 370
Grüsser, O.-J. 120, 358
Grüsser-Cornehls, U. 120, 358
Gubernick, D. J. 151, 152, 358
Guillery, R. W. 150, 157, 346
Gusel'nikov, V. I. 146, 358
Gutowski, K. A. 192, 194, 382
Guttman, N. 271, 352
Gygax, P. A. 41, 369

Hainsworth, F. R. 104, 105, 358
Hale, E. B. 102, 103, 358
Hall, G. 327, 351
Hall, W. C. 141, 144, 145, 147, 148, 180, 354, 358
Halpern, M. 120, 124, 143, 144, 358, 376
Hamilton, C. R. 315, 317, 358
Hankins, W. G. 305, 354
Hansen, G. 200, 368
Harlow, H. F. 59, 207, 209, 210, 213, 277, 282, 351, 358, 364, 368
Harman, P. J. 247, 248, 358
Harrison, S. N. 216, 364
Hayes, C. 293, 358
Hayes, K. J. 215, 283, 284, 358
Hayes, W. N. 146, 147, 150, 151, 152, 358, 359, 361
Hearst, E. 192, 193, 213, 359, 371, 385
Heath, J. E. 155, 370
Heckel, R. V. 130, 374
Hedges, M. 282, 283, 379
Heffner, R. 339, 359
Heier, P. 48, 49, 53, 359
Heimer, L. 54, 352
Henderson, T. B. 90, 348
Herman, L. M. 278, 280, 282, 324, 325, 359, 379
Herrick, C. J. 53, 57, 118, 359
Herrnstein, R. J. 216, 359
Hershkowitz, M. 129, 359
Hertzler, D. R. 146, 147, 151, 152, 358, 359
Hess, E. H. 152, 153, 348
Hicks, L. H. 283, 359
Hicks, R. E. 314, 359
Hinde, R. A. 57, 58, 191, 349, 359
Hineline, P. N. 9, 196, 359, 373
Hines, D. 220, 359
Hodos, W. 171, 173, 174, 178, 179, 180, 182, 183, 188, 231, 359, 360, 362, 371
Hoffman, H. S. 5, 196, 216, 360
Hoffmann, A. 124, 360
Hogan, D. E. 212, 213, 385
Hogan, J. A. 68, 344
Hogberg, D. K. 151, 152, 358
Holden, A. L. 181, 363
Holgate, V. 202, 366
Hollard, V. 234, 360, 385
Holman, J. G. 216, 369
Holmes, C. E. 156, 350
Holmes, N. K. 82, 83, 84, 356
Holmes, P. A. 159, 360

Hombach, A. 155, 159, 357
Honig, W. K. 204, 366
Honigmann, H. 214, 215, 360
Hood, P. 197, 200, 223, 224, 357
Hoogland, P. V. 141, 142, 360
Horel, J. A. 142, 319, 320, 360, 362
Horn, A. L. D. 132, 360
Horn, G. 132, 183, 232, 345, 347, 360, 365
Horridge, G. A. 133, 360
Hostetter, G. 32, 379
Hubel, D. H. 181, 360
Huber, G. C. 138, 142, 172, 246, 253, 362
Hull, C. L. 12, 42, 43, 306, 307, 360
Hume, D. 1
Humphrey, G. 150, 152, 360
Hunter, M. W. 209, 225, 226, 360, 361
Hunter, N. B. 192, 194, 382
Hunter, W. S. 274, 360
Hurst, J. W. 89, 383
Hutchison, L. V. 178, 365

Ichikawa, B. T. 150, 151, 354
Ince, L. P. 133, 360
Ingle, D. 121, 122, 125, 126, 127, 353, 360, 361
Inhelder, B. 286, 372
Inman, N. 223, 357
Ireland, L. C. 151, 152, 359, 361
Ito, H. 55, 361
Iwai, E. 56, 361

Jacobs, D. W. 66, 361
Jagoda, H. 96, 350
Jane, J. A. 54, 357
Jarvik, M. E. 327, 376
Jarvis, C. D. 182, 361
Jarvis, M. J. 285, 361
Jeeves, M. A. 8, 224, 271, 272, 341, 373
Jenkins, H. M. 60, 192, 193, 204, 347, 361, 368
Jennings, J. W. 281, 361
Jensen, A. R. 342, 361
Jerison, H. J. 15, 16, 25, 26, 27, 28, 29, 30, 31, 101, 102, 119, 165, 222, 241, 242, 243, 249, 250, 278, 335, 337, 338, 361
Johanson, D. C. 240, 361
Johnson, D. F. 212, 345
Johnson, J. I. 277, 278, 361
Jones, A. W. 174, 175, 361
Jones, C. N. 277, 278, 279, 352
Jones, T. B. 77, 210, 361
Julian, S. M. 157, 374
Juorio, A. V. 142, 361

Kaas, J. H. 339, 361
Kalat, J. W. 154, 375
Källén, B. 118, 141, 143, 174, 175, 177, 361
Kamil, A. C. 77, 209, 210, 211, 224, 225, 226, 360, 361, 362
Kang, D. S. 144, 363
Kaplan, H. 104, 105, 106, 108, 344, 362
Kapp, B. 327, 376
Kappers, C. U. A. 138, 142, 172, 246, 253, 362
Karamian, A. I. 122, 362
Karten, H. J. 55, 144, 171, 173, 174, 175, 176, 177, 178, 180, 182, 183, 184, 185, 187, 188, 189, 227, 231, 339, 349, 353, 359, 360, 362, 369, 374, 385
Katz, H. M. 281, 378
Kaufman, E. L. 214, 362
Kay, H. 281, 362, 384
Keating, E. G. 142, 362
Keefer, L. H. 281, 361
Keller, F. R. 195, 378
Kelley, D. B. 184, 362
Kellogg, L. A. 293, 362
Kellogg, W. N. 293, 294, 362
Kemp, F. D. 156, 362
Kemper, T. L. 317, 354
Kenigfest, N. B. 53, 380
Kennel, K. 287, 384
Kicliter, E. 120, 121, 122, 123, 124, 362
Killackey, H. 263, 363
Kimberly, R. P. 181, 363
Kimble, D. P. 125, 126, 363
Kimble, G. A. 271, 352
Kinsbourne, M. 314, 359
Kirk, K. L. 160, 161, 363
Kirkish, P. M. 160, 363
Kitt, C. A. 142, 177, 178, 347
Klara, J. W. 227, 233, 367
Kleinginna, P. R. 155, 363
Klempay, S. 192, 365
Kling, A. 143, 351
Klinman, C. S. 61, 363
Knapp, H. 120, 144, 363, 376
Knigge, K. M. 106, 351
Knight, L. 150, 151, 349
Knudsen, E. I. 55, 56, 185, 363
Koehler, O. 214, 215, 363
Köhler, W. 307, 363
Kokoros, J. J. 124, 363
Kolb, B. 266, 267, 363
Konishi, M. 181, 185, 363, 372, 385
Konjević, D. 54, 372
Konorski, J. 266, 267, 364
Kormann, L. A. 142, 362
Koronakos, C. 281, 363
Kosareva, A. A. 53, 146, 148, 149, 345, 380
Kovacević, N. 54, 372
Kral, P. A. 192, 383
Krasnegor, N. A. 233
Kraus, G. 270, 271, 357
Krayniak, P. F. 176, 177, 178, 363

Krekorian, C. O. 156, 363
Kremer, E. F. 327, 363
Krompecher, S. 30, 363
Kruger, L. 139, 148, 244, 363, 373
Krushinskaya, N. L. 233
Kuczka, H. 125, 363
Kuhl, P. K. 294, 364

Lacey, D. J. 140, 364
Laddin, L. H. 151, 361
Lancy, D. 69, 353
Landreth, G. E. 144, 370
Lanza, R. P. 302, 353
Lashley, K. S. 25, 42, 134, 255, 256, 257, 258, 259, 261, 364
Lawicka, W. 266, 267, 364
Le Gros Clark, W. E. 246, 247, 364
Lea, S. E. G. 216, 364, 369
Leakey, R. E. F. 240, 364, 381
Leary, R. W. 277, 283, 351, 364
Lehmkuhle, S. W. 182, 354
LeMay, M. 317, 354
Lemire, M. 55, 374
Lende, R. A. 245, 254, 255, 364, 382
Lenneberg, E. H. 294, 310, 311, 313, 347, 364
Leonard, C. M. 189, 253, 265, 267, 364, 370
Leppelsack, H.-J. 185, 364
Levi-Montalcini, R. 174, 175, 361
Levine, B. A. 224, 364
Levine, M. 208, 209, 225, 364
Levy, E. I. 220, 373
Lico, M. C. 124, 360
Lieberman, P. 293, 364, 365
Liege, B. 122, 123, 365
Likely, D. 93, 365
Limpo, A. J. 230, 231, 374
Linnaeus (Carl von Linné) 19
Lipak, J. A. 30, 363
Little, L. 78, 81, 93, 160, 198, 199, 201, 202, 203, 204, 205, 365, 366
Livesey, P. J. 271, 276, 277, 365
Lloyd Morgan, *see* Morgan, C. L.
Lloyd, T. 66, 382
Lockard, R. B. 16, 365
Lockhart, J. M. 210, 365
Lockhart, R. S. 322, 323, 350
Loeb, J. 2
Lohman, A. H. M. 139, 140, 141, 145, 149, 365
LoLordo, V. M. 9, 354
Longo, N. 61, 85, 192, 356, 365
Loop, M. S. 153, 154, 365
Lord, J. 81, 93, 160, 201, 202, 366
Lord, M. W. 214, 362
Losey, G. S. 65, 365
Lougee, M. 209, 210, 211, 361
Loveland, D. H. 216, 359
Lowes, G. 89, 365
Lubar, J. F. 266, 268, 370
Lubow, R. E. 205, 216, 365
Ludvigson, H. W. 91, 365
Lyons, J. 329, 365

Macadar, A. W. 178, 365
McCabe, B. J. 183, 232, 345, 360, 365
McCulloch, T. L. 213, 370
McDaniel, W. F. 263, 365
MacDonald, R. L. 191, 205, 349
McGill, T. E. 130, 365
McGonigle, B. O. 307, 308, 365
McHugh, T. G. 195, 367
McIlwain, J. T. 181, 365
Mackintosh, J. 95, 96, 366
Mackintosh, N. J. 11, 12, 13, 17, 61, 70, 71, 72, 73, 74, 75, 76, 78, 80, 81, 82, 84, 87, 88, 89, 90, 91, 92, 93, 94, 95, 96, 97, 99, 158, 160, 162, 196, 197, 198, 199, 201, 202, 203, 204, 205, 206, 270, 285, 327, 333, 351, 365, 366, 371, 379
McLaughlin, B. 313, 366
McNemar, Q. 341, 349
Macphail, E. M. 9, 164, 195, 218, 219, 230, 231, 232, 233, 234, 308, 366, 367, 377
McQuiston, M. D. 275, 368
Mahut, H. 321, 367
Maier, N. R. F. 3, 42, 274, 306, 341, 367
Mandriota, F. J. 67, 68, 367
Manley, J. A. 147, 367
Manning, F. J. 263, 266, 357, 368
Manocha, S. N. 277, 279, 367
Mantor, H. 156, 373
Margules, L. 65, 365
Markowitsch, H. J. 265, 268, 269, 367
Marshall, D. A. 271, 354
Marx, M. H. 84, 367
Maser, J. D. 227, 233, 367
Masterton, B. 269, 339, 359, 367
Matsumiya, Y. 156, 157, 357
Matthews, T. J. 195, 367
Matyniak, K. A. 206, 367
Mauldin, J. E. 77, 210, 361, 362
Mayer, J. 65, 375
Mayr, E. 19, 367
Mead, W. R. 230, 233, 373
Medin, D. L. 326, 367
Meibach, R. C. 315, 355
Meier, R. E. 179, 367
Melton, A. W. 219, 367
Meltzoff, A. N. 285, 367
Melzack, R. 57, 58, 191, 368, 375
Meyer, D. R. 61, 209, 214, 256, 280, 347, 368, 370
Meyer, P. M. 214, 256, 280, 347, 368
Meyers, W. J. 275, 368
Michels, K. M. 278, 361

Mihailović, J. 179, 367
Miles, C. G. 204, 368
Miles, R. C. 209, 274, 275, 368
Miller, J. D. 294, 364
Miller, J. T. 200, 368
Miller, R. R. 130, 131, 327, 368
Milner, B. 33, 267, 268, 319, 323, 368
Milstein, S. 84, 356
Mishkin, M. 244, 266, 267, 282, 319, 320, 364, 368, 380
Moffitt, M. 218, 378
Moise, S. L. 327, 368
Molina, E. J. 205, 382
Moody, D. B. 294, 316, 317, 372, 378
Moon, L. E. 213, 282, 368
Moore, A. R. 126, 129, 368
Moore, B. R. 192, 193, 194, 196, 361, 368
Moore, G. P. 148, 149, 150, 368
Moore, J. W. 323, 368
Moreines, J. 322, 349
Morenkov, E. D. 146, 358
Morgan, C. L. 1, 2, 303, 368
Morgan, M. J. 216, 369
Moriarty, K. M. 286, 384
Morita, Y. 55, 361
Morlock, H. C. 165, 369
Morruzzi, G. 188, 347
Morse, P. A. 294, 369
Mrosovsky, N. 157, 369
Mudry, K. M. 123, 369
Munn, N. L. 130, 271, 369
Muntz, W. R. A. 120, 369
Murillo, N. R. 162, 369
Myers, R. E. 143, 351

Nadel, L. 321, 323, 370
Narkiewicz, O. 265, 369
Nash, R. F. 191, 369
Nauta, W. J. H. 41, 174, 180, 188, 231, 265, 339, 362, 369
Nebes, R. D. 315, 369
Nelson, J. S. 44, 369
Nevin, J. A. 281, 369
Newcombe, F. 318, 369
Newman, J. D. 185, 384
Nieuwenhuys, R. 48, 49, 51, 52, 53, 56, 116, 117, 118, 119, 139, 369
Nigrosh, B. J. 281, 369
Nissen, C. H. 215, 283, 284, 358
Nissen, H. W. 213, 282, 369, 370
Noble, M. 61, 370
Nomokonova, L. M. 123, 380
Nonneman, A. J. 266, 267, 315, 363, 382
Norgren, R. 176, 370
North, A. J. 75, 159, 271, 370
Northcutt, R. G. 49, 50, 53, 101, 118, 119, 120, 124, 140, 144, 145, 146, 149, 155, 230, 233, 345, 348, 352, 363, 370, 373, 375
Nott, K. H. 233, 370
Nottebohm, F. 184, 189, 362, 370
Numan, R. 266, 268, 370

Oakley, D. A. 261, 370
Oeckinghaus, H. 184, 185, 351
Oi, S. 106, 354
O'Keefe, J. 321, 323, 370
Oldfield-Box, H. 281, 362
Olivier, A. 142, 177, 371
Olton, D. A. 324, 325, 370
Orbach, J. 257, 371
Ordy, J. M. 286, 380
Orrego, F. 146, 148, 371
Oscar, M. 80, 383
Otis, L. S. 60, 66, 371
Over, R. 285, 371
Overmier, J. B. 60, 102, 103, 104, 105, 106, 107, 108, 109, 110, 111, 354, 358, 371, 380

Pappas, B. A. 261, 351
Parent, A. 141, 142, 143, 175, 177, 371
Parker, D. M. 232, 376
Parks, T. E. 210, 365
Parsons, T. S. 22, 44, 45, 47, 114, 115, 137, 138, 139, 169, 170, 238, 375
Pasnak, R. 286, 371
Passingham, R. E. 247, 248, 251, 253, 254, 266, 267, 371
Pasternak, T. 183, 371
Pastore, N. 213, 371
Patterson, F. G. 294, 296, 297, 298, 299, 371
Pavlov, I. P. 10
Pearson, R. 171, 175, 221, 371
Peden, B. F. 193, 371
Peeke, H. V. S. 57, 108, 191, 371, 372
Peeke, S. C. 57, 108, 371, 372
Pepper, R. L. 280, 359
Pert, A. 161, 162, 163, 372
Peters, R. I. 133, 372
Petersen, M. R. 316, 317, 372
Petitto, L. A. 294, 296, 297, 298, 299, 379
Petrinovich, L. 191, 372
Petrović-Minić, B. 265, 367
Pettigrew, J. D. 181, 345, 372
Petty, F. 66, 381
Peyrichoux, J. 120, 372
Phillips, R. E. 175, 188, 189, 227, 233, 234, 372
Phillips, W. A. 220, 372
Piaget, J. 285, 286, 287, 308, 372
Pickett, J. M. 257, 372
Pietrewicz, A. 77, 210, 361
Pilbeam, D. 30, 240, 372

Pinckney, G. A. 68, 372
Pitcoff, K. 84, 89, 356
Pivovarov, A. S. 146, 358
Platel, R. 164, 165, 372
Platt, C. J. 54, 372
Plotnik, R. J. 224, 225, 226, 372
Poitras, D. 141, 371
Popper, A. N. 66, 361
Porter, R. H. 154, 350
Portmann, A. 221, 222, 373
Potter, M. C. 220, 373
Potts, A. 84, 89, 356
Powell, R. W. 156, 373
Powell, T. P. S. 187, 373
Powers, A. S. 89, 90, 91, 146, 147, 165, 217, 345, 356, 374
Premack, D. 287, 295, 299, 300, 303, 304, 305, 309, 310, 311, 373, 384
Prestrude, A. M. 293, 373
Pribram, K. H. 244, 373
Pridmore, P. A. 271, 376
Pritz, M. B. 145, 146, 147, 148, 149, 230, 233, 345, 373
Pritzel, M. 265, 268, 269, 367
Prockop, E. S. 282, 368
Przybylski, R. J. 53, 370
Pubols, B. H. 75, 159, 373
Putkonen, P. T. S. 189, 373

Quiring, D. P. 26, 27, 350

Rabinovitch, M. S. 103, 373
Rachlin, H. 9, 196, 359, 373
Rajalakshmi, R. 8, 224, 271, 272, 341, 373
Rakover, S. S. 66, 373
Ratcliff, G. 318, 369
Ratner, A. M. 5, 360
Rausch, L. J. 178, 365
Ray, R. S. 125, 126, 363
Raymond, B. 89, 373, 383
Razran, G. 59, 64, 373
Reboreda, J. C. 129, 376
Reese, H. W. 207, 373
Reese, T. W. 214, 362
Reid, R. L. 197, 373
Reiner, A. J. 144, 146, 147, 165, 374
Rensch, B. 29, 374
Repérant, J. 53, 55, 120, 144, 179, 372, 374, 380
Rescorla, R. A. 59, 64, 91, 92, 99, 110, 153, 154, 193, 204, 304, 374
Restle, F. 207, 208, 210, 211, 374
Revzin, A. M. 174, 180, 181, 182, 187, 188, 231, 362, 374
Reynolds, G. S. 230, 231, 374
Rhodes, J. 271, 348
Richardson, A. M. 153, 156, 157, 160, 345, 350, 363, 374, 380
Riddell, W. 197, 223, 357
Riege, W. H. 68, 374
Rieke, G. K. 177, 178, 188, 374
Riley, E. P. 93, 353
Rilling, M. 205, 380
Rio, J.-P. 144, 374
Riopelle, A. J. 208, 374
Riss, W. 120, 143, 144, 376
Robbins, D. O. 145, 374
Roberts, D. 130, 374
Roberts, W. A. 76, 203, 219, 220, 259, 260, 325, 356, 375
Robinson, D. L. 181, 355
Robinson, R. G. 316, 375
Roccaforte, P. 61, 98, 99, 383
Rodgers, W. L. 57, 58, 375
Roemer, R. A. 127, 379
Rollin, A. R. 281, 346
Romanes, G. J. 1, 2, 375
Romer, A. S. 21, 22, 44, 45, 47, 114, 115, 137, 138, 139, 169, 170, 238, 239, 375
Rose, M. 254
Rosen, G. D. 316, 377
Rosenfeld, S. A. 284, 375
Rosenkilde, C. E. 267, 375
Rosner, B. S. 294, 350
Rosvold, H. E. 103, 267, 282, 355, 368, 373
Rouse, J. E. 191, 375
Rovainen, C. M. 57, 59, 346, 375
Royce, G. J. 118, 119, 120, 370, 375
Rozin, P. 5, 65, 154, 375
Rudolph, R. L. 195, 375
Rudy, J. W. 326, 381
Rumbaugh, D. M. 295, 300, 301, 302, 375, 376
Rundus, D. 219, 326, 375
Runge, T. E. 184, 185, 351
Rusiniak, K. W. 305, 354
Russell, I. S. 243, 260, 261, 370, 375

Sacher, G. A. 251, 376
Safriel-Jorne, O. 95, 96, 366
Saiff, E. I. 150, 151, 359
Saito, S. 56, 361
Sakamoto, N. 55, 361
Salzen, E. A. 232, 234, 348, 376
Samuel, D. 129, 359
Samuels, I. 323, 352
Sanders, R. J. 294, 296, 297, 298, 299, 379
Sands, S. F. 325, 376
Sarich, V. M. 241, 385
Saunders, J. C. 271, 376
Savage, G. E. 102, 103, 104, 105, 106, 107, 108, 109, 110, 354, 371, 376
Savage-Rumbaugh, E. S. 301, 302, 376
Sawyer, J. R. 329, 384
Scalia, F. 52, 55, 120, 143, 144, 376
Scarborough, B. B. 60, 376

Schade, A. F. 198, 205, 376
Schapiro, H. 149, 376
Schein, M. W. 224, 344
Schiemann, K. 215
Schiller, P. H. 307, 376
Schmajuk, N. A. 129, 376
Schmidt, E. C. 125, 126, 355
Schneider, A. M. 327, 376
Schneider, G. E. 146, 182, 183, 377
Schneiderman, N. 59, 356, 358
Schneirla, T. C. 3, 42, 274, 367
Schoel, W. M. 74, 384
Schrier, A. M. 96, 210, 271, 273, 280, 377
Schroeder, D. M. 54, 377
Schuckman, H. 105, 108, 377
Schultz, W. J. 229, 230, 378
Schusterman, R. J. 77, 210, 272, 273, 377
Schutz, S. L. 85, 377
Schwartz, B. 163, 194, 354, 377
Schwartz, D. 247, 352
Schwartzkopff, J. 147, 382
Scobie, S. R. 60, 67, 68, 89, 90, 105, 346, 349, 377
Scoville, W. B. 33
Scull, J. W. 164, 377
Segaar, J. 56, 106, 377
Segal, J. R. 57, 58, 375
Segura, E. T. 129, 376
Seidenberg, M. S. 300, 379
Seifert, A. R. 268, 370
Seller, T. J. 190, 377
Sepinwall, J. 127, 128, 355
Sergeyev, B. F. 64
Settlage, P. H. 277, 364
Shaffer, W. O. 220, 377
Shapiro, S. 105, 108, 377
Sharma, S. C. 55, 377
Sheffield, V. F. 85, 377
Shepp, B. E. 96, 377
Sherman, G. F. 316, 377
Sherman, T. 219, 380
Shiffrin, R. M. 220, 377
Shimp, C. P. 201, 218, 377, 378
Shulman, R. I. 209, 210, 211, 361
Siegel, A. 176, 177, 178, 363
Sime, M. E. 281, 384
Simons, E. L. 240, 378
Simonsen, L. 150, 151, 349
Singer, W. 181, 378
Singh, R. K. 266, 267, 363
Sinnott, J. M. 294, 378
Skeen, L. C. 269, 367
Skinner, B. F. 42, 43, 302, 313, 353, 378
Slack, W. V. 130, 133, 134, 353
Sligar, C. M. 55, 141, 142, 381
Slotnick, B. M. 281, 369, 378
Smeets, W. J. A. J. 54, 55, 378
Smith, M. 150, 378
Smith, R. F. 195, 378
Smotherman, W. 286, 380
Snell, O. 29
Snowdon, C. T. 104, 105, 294, 358, 369
Snyder, M. 263, 363
Solhkhan, N. 61, 98, 99, 383
Solomon, R. L. 110, 374
Solomon, R. P. 323, 378
Spear, N. E. 93, 353
Spearman, C. E. 340, 341, 378
Spears, R. H. 263, 365
Speiss, J. 270, 357
Spencer, W. A. 58, 126, 270, 379
Spies, G. 127, 128, 355
Springer, A. D. 327, 368
Squier, L. H. 60, 378
Squire, L. R. 327, 378
Stahl, B. J. 44, 114, 238, 378
Stalling, R. B. 280, 359
Stanning, C. J. 220, 221, 382
Starkman, N. 60, 104, 105, 108, 110, 371
Stearns, E. M. 76, 378
Stebbins, W. C. 294, 316, 317, 372, 378
Stephan, H. 244, 247, 249, 250, 378
Stettner, L. J. 206, 229, 230, 231, 233, 367, 378
Stevens, D. A. 283, 351
Stewart, C. N. 277, 378
Stirling, C. E. 156, 157, 357
Stokes, T. M. 189, 370
Stollnitz, F. 210, 345
Stone, C. P. 341, 349
Storer, R. W. 169, 171, 378
Straub, R. O. 300, 379
Strong, P. N. 282, 283, 379
Sturm, T. 60, 379
Sugerman, R. A. 149, 379
Sussman, D. 105, 108, 377
Sutherland, N. S. 17, 72, 94, 95, 96, 205, 206, 366, 379
Swingland, I. R. 102, 107, 376

Tallarico, R. B. 224, 225, 226, 372
Tarr, R. S. 142, 143, 379
Tauxe, L. 240, 379
Taylor, S. 150, 151, 349
Ten Donkelaar, H. J. 149, 379
Tennant, W. A. 96, 97, 98, 99, 203, 204, 205, 350, 379
Terrace, H. S. 294, 296, 297, 298, 299, 300, 379
Testa, T. J. 305, 379
Teuber, H. L. 258, 319, 368, 379, 382
Teyler, T. J. 127, 379
Theios, J. 75, 159, 379
Thomas, D. R. 200, 368
Thomas, G. J. 32, 60, 371, 379
Thomas, R. K. 282, 284, 379

Thompson, C. R. 300, 379
Thompson, R. 283, 351
Thompson, R. F. 58, 126, 127, 270, 379
Thompson, R. K. R. 324, 325, 379
Thompson, R. L. 67, 68, 367
Thompson, R. W. 127, 128, 131, 384
Thompson, T. 60, 379
Thorndike, E. L. 2, 3, 12, 86, 379, 380
Thorpe, W. H. 131, 153, 193, 214, 215, 380
Tighe, L. S. 164, 380
Tighe, T. J. 164, 357, 380
Tinklepaugh, O. L. 165, 380
Trabasso, T. 217, 307, 308, 347
Tranberg, D. K. 205, 380
Trapold, M. A. 110, 380
Treichler, F. R. 214, 280, 368
Tsang, Y. C. 257, 380
Tschirgi, R. D. 148, 149, 150, 368
Tsukahara, S. 56, 361
Tucker, A. M. 105, 108, 377
Tucker, D. 117, 380
Turner, B. H. 244, 380
Tversky, B. 219, 380

Ueda, S. 55, 361
Ulinski, P. S. 149, 380
Ulrich, G. 150, 151, 354

Van Bergeijk, W. A. 129, 380
Van Hoesen, G. W. 284, 375
Van Houten, R. 195, 375
Van Sommers, P. 65, 155, 380
Van Woerden-Verkley, I. 139, 140, 141, 145, 149, 365
Vance, V. J. 156, 157, 363, 380
Vanegas, H. 55, 352, 380
Vaughter, R. M. 286, 380
Vernon, J. A. 130, 133, 134, 353
Vesselkin, N. P. 53, 122, 123, 362, 380
Vest, B. 267, 355
Vickery, J. D. 284, 379
Vinogradova, O. S. 233, 380
Vogt, M. 185, 364
Volkmann, J. 214, 362
Von Bonin, G. 241, 242, 247, 381
Voneida, T. J. 55, 139, 140, 141, 142, 352, 381

Wagner, A. R. 84, 85, 91, 92, 99, 204, 326, 374, 381
Walk, R. D. 146, 381
Walker, A. 240, 364, 381
Wallace, J. 60, 346
Ward, J. P. 263, 285, 344, 381
Warren, J. L. 66, 381
Warren, J. M. 70, 95, 96, 103, 111, 205, 210, 224, 268, 271, 272, 273, 274, 275, 276, 277, 278, 279, 280, 281, 282, 283, 315, 317, 344, 347, 352, 378, 381, 382
Warrington, E. K. 320, 322, 344, 382
Washburn, S. L. 240, 241, 382
Wasserman, E. A. 192, 194, 205, 382
Watson, J. B. 2
Weaver, G. E. 220, 221, 382
Weaver, L. M. 315, 355
Webster, D. B. 147, 351
Webster, D. M. 234, 385
Webster, W. G. 315, 317, 382
Weidner, C. 120, 372
Weigele, S. 69, 353
Weinberger, N. M. 125, 126, 356
Weinstein, S. 258, 379, 382
Weisbach, W. 147, 382
Weiskrantz, L. 264, 320, 321, 324, 325, 354, 382
Weitzman, R. A. 80, 382
Welch, J. C. 126, 129, 368
Welker, J. 57, 382
Welker, W. I. 57, 245, 246, 382
Wenzel, B. M. 177, 178, 365, 374
Werboff, J. 66, 382
Werner, T. J. 212, 349
Werz, M. A. 325, 370
Wesley, F. 214, 215, 283, 382
Wesp, R. 233, 382
Wessels, M. G. 193, 382
Westcott, P. W. 131, 347
Whitaker, H. A. 313, 351
White, T. D. 240, 361, 383
Whitfield, I. C. 185, 353
Whitlow, J. W. 326, 381
Wickelgren, W. O. 57, 59, 383
Wickens, D. D. 322, 383
Wiesel, T. N. 181, 360
Wiggins, S. L. 130, 374
Wike, E. L. 93, 348
Wilcoxon, H. C. 154, 192, 350, 383
Wild, J. M. 234, 385
Wildman, L. D. 263, 365
Williams, B. A. 200, 383
Williams, D. I. 204, 205, 383
Williams, D. R. 193, 194, 377, 383, 384
Williams, H. 193, 383
Williams, J. T. 156, 159, 160, 383
Williamson, A. J. 232, 376
Williston, J. S. 108, 372
Wilson, B. J. 225, 226, 383
Wilson, J. J. 84, 383
Wilson, M. 181, 262, 383
Wilson, W. A. 80, 383
Wirth, M. C. 133, 372
Wise, K. L. 286, 383
Wise, L. A. 286, 383
Wise, L. M. 162, 383

Wodinsky, J. 66, 70, 79, 84, 282, 283, 346, 383
Wolach, A. H. 61, 84, 89, 90, 98, 99, 347, 373, 383, 385
Wolff, J. L. 96, 384
Wollberg, Z. 185, 384
Wood, F. G. 302, 384
Wood, S. 286, 384
Woodard, W. T. 60, 62, 63, 67, 71, 73, 74, 80, 81, 90, 95, 105, 161, 348, 352, 384
Woodruff, G. 194, 287, 303, 384
Wright, A. A. 325, 376
Wright, J. W. 151, 152, 358
Wright, P. L. 281, 384

Yaremko, R. M. 127, 128, 131, 384
Yehle, A. L. 285, 381
Yeni-Komshian, G. H. 317, 384
Yerkes, R. M. 130, 165, 384
Yori, J. G. 156, 157, 384
Young, J. Z. 42, 46, 346, 384
Youngren, O. M. 189, 372
Yunis, J. J. 329, 384
Yutzey, D. A. 316, 377

Zagorulko, T. M. 122, 362
Zangwill, O. L. 255, 384
Zaretsky, M. D. 185, 385
Zecha, A. 188, 385
Zeier, H. 175, 176, 188, 227, 228, 229, 233, 385
Zeigler, H. P. 187, 188, 189, 225, 230, 233, 234, 385
Zeiler, M. D. 63, 385
Zentall, T. R. 212, 213, 385
Zihlman, A. L. 241, 385
Zimmerman, R. R. 286, 383
Zola, S. M. 321, 367
Zoloth, S. R. 316, 317, 372
Zych, K. A. 84, 385

Subject Index

The letters f, n, and t denote figure, footnote, and table respectively.

accessory olfactory bulb
absent in birds 171
in amphibians 116, 118
in reptiles 137, 139f
projections of, *see* olfactory system
acoustic confusion errors, in immediate recall 318
Actinopterygii 111
evagination not found in 48
pallial divisions in 53
phylogeny of 46–7
active avoidance learning
in amphibians 130
in birds 9, 195–6
following archistriatal lesions 227–8
in fish 66–8
following forebrain ablation 104–5, 107–9
in reptiles 156–7
Sidman
in birds 195
in fish 68
after-effects
and resistance to extinction 83, 85–7, 88, 92–4, 112, 162, 166–7
and single alternation 87–8, 92–3
and the successive negative contrast effect 90, 93–4, 112, 162, 166–7
and the use of strategies 78–9, 112
as memories 88, 217
sensory 85
see also sensory contrast
Age of Mammals 22t
Age of Reptiles 22t
aggression 6
in birds
following lobus parolfactorius lesions 233
following septal lesions 233
habituation of 191
in fish
following forebrain lesions 106
in mammals
cerebral lateralization of 316
following amygdalar lesions 143, 166
in reptiles
role of amygdala in 142–3, 166
Agnatha, phylogeny of 44–6; *see also* lampreys
albal stalk, and memory 319–20
Alligator spp
brain- and body-weight in 28f
discrimination learning in 156
forebrain of 139f
instrumental conditioning in 155
retinal projections in 144
serial reversal learning in 159, 167
somatosensory system in 148
see also Crocodilia
allocortex, in mammals 244, 320
allometry
defined 29
see also simple allometry
American chameleons (*Anolis carolinensis*)
habituation in 151
sensitization in 152
American Sign Language (Ameslan, or ASL)
acquisition by non-humans 294–9, 301
ammocaetes
telencephalon of 53
amnesia, *see* hippocampus, in mammals; memory
amnion 136
amniotes, definition of 136
amphibians
brain- and body-weight in 119, 335
classical conditioning in 127–9
classification of 114–15, 134
forebrain lesions and learning in 132–4
habituation in 125–7
instrumental conditioning in 129–32
sensory and motor organization of forebrain in 119–25, 339
structure of forebrain in 116–19
sun-compass orientation in 4–5
Amphisbaenia, phylogeny of 136–7
amygdala 40, 141
in amphibians 118–20, 124
in fish 49
in mammals 55, 120, 143, 166, 175–7, 245, 319–20
in reptiles 141–4, 166

see also archistriatum
analogy
contrasted with homology 39
analysers, *see* selective attention
Anapsida, phylogeny of 137
animal hypnosis, *see* tonic immobility
Annulata, phylogeny of 136-7
anoles, *see* American chameleons
ansa lenticularis, in birds 177
anterior commissure, in mammals 246
anteromedial cortex
in rats 267
anthracosaurs
as ancestors of reptiles 136, 114, 115f
anthropoids
brain size of 243, 249
classification of 238
neocortical volume and brain size of 248f, 249
progression indices of 249, 250f, 251
ratio of eulaminate to total neocortex in 253-4
Anura
auditory system of 122-3
habituation in 125-7
maze learning in 130
olfactory system of 119-20
phylogeny of 114-15
visual system of 120-2
see also frogs; toads
apes, *see* pongids
aphasia
and performance in intelligence tests 258
following cerebral lesions 314, 317
see also dysphasia
apodans, *see* Gymnophiona
Archaeopteryx 168-9
archicortex, *see* hippocampus
archistriatum 40
in birds 171, 173f, 175-6, 185, 186f, 187-9, 227-9, 233, 236
in reptiles 141
see also amygdala
archosaurs, phylogeny of 137, 138f, 168
area superficialis basalis, *see* olfactory tubercle
arguments 311
artificial intelligence, *see* intelligence
artificial languages
acquisition by chimpanzees 295, 299-301
use in communication by chimpanzees 301-2
and inductive reasoning in chimpanzees 305-6
and deductive reasoning in chimpanzees 309-11
artiodactyls, *see* ungulates
association areas 41, 252
and intelligence 254-5, 261-9
in an echidna 254, 255f
relative size of 253-5, 288
visual, *see* extrastriate visual cortex
association-formation
and communication between non-human animals 302-3
and deductive reasoning 306-9
and human language-acquisition 312-13
and inductive reasoning 304-6
and intentional deception by chimpanzees 304
and language analogues in non-humans 295-6, 300-1
and learning-set formation 207-8
as basis of all intellectual activity 3
distinct from reasoning 3
not required for habituation 10
see also classical conditioning; instrumental conditioning
associationists, *see* instrumental conditioning
attention, *see* selective attention
auditory neocortex
in mammals 148, 252, 255f
auditory nerve, projections of, *see* auditory system
auditory perception
of chimpanzees 293
see also auditory system
auditory system
in anurans 122-3, 134, 184
in birds 174, 184-5, 190, 233, 235
in fish 54, 56
in mammals 123, 148, 184-5
in reptiles 147-8, 166, 184
auditory tubercle, in reptiles 147
Australopithecus spp 240
brain size in 28f, 240
language in 329
autoshaping
and behavioural contrast 163-4
and communication between non-human animals 303
and interstimulus interval and stimulus duration 63
in birds 63, 205, 193-4, 303
in fish 60, 62-3, 65
in reptiles 153-4, 163-4
avoidance learning, *see* active avoidance learning; passive avoidance learning

babbling, in chimpanzees and children 298
bandicoots, brown short-nosed (*Isoodon obesulus*), one-trial reversals in 271
basal ganglia, in mammals, *see* striatum, in mammals
basal insectivores, *see* insectivores, basal

bats
  Microchiroptera, frontal cortex of 253
  vampire (Desmodontidae), brain- and body-weight of 26f
Beau gregory (*Pomacentrus leucostictus*), avoidance learning in 66
behavioural contrast 163–4
between-brain, *see* diencephalon
bichir (*Polypterus*)
  absence of pallial divisions in 53
  phylogeny of 46–7
binocular disparity units
  in birds and mammals 181
binomial system of nomenclature 19
bird song, physiology of 169, 184–5, 189–90, 235
birds
  behavioural contrast in 163
  biological intelligence in 15–16
  brain mechanisms and learning in 226–34, 236
  brain size in 26–9, 101, 221–2, 335–6
  classical conditioning in 192–3, 235
  classification of 168–9, 170f, 171, 234–5
  effects of reward shifts in 202–3
  habituation in 190–2, 235
  instrumental conditioning in 193–6, 235
  interspecies comparisons in 221–6, 236
  learning-set formation in 207–11
  mechanisms of attention in 74, 203–6
  memory in 216–21, 235, 324, 327–8, 335
  number concepts in 214–15
  phylogeny of 137, 138f, 168–9
  probability learning in 80, 95, 200–2
  sameness-difference concept in 211–14
  sensory and motor organization of forebrain in 178–90, 235
  serial reversal learning in 95, 197–200
  structure of forebrain in 171, 172f, 173f, 174–8, 235, 339
  visual concepts in 215–16
birds of paradise (Paradisaeidae), rank within passerines of 169
blackbirds, New World (Icteridae), rank within passerines of 171
blindsight, following striate cortical lesions 264
blocking
  in birds 204
  in fish 98–9, 112
  in mammals 98–9
bluegill, *see* sunfish
bony fish (Osteichthyes), phylogeny of 44–7
bowerbirds (Ptilonorhynchidae), rank within passerines of 169
Bowman's glands 116–17
brain lesions, technique of 32–4, 41
brain organization
  and intelligence 31–5, 338–40
  in vertebrates 35–42, 338–40
  see also forebrain, sensory and motor organization of; forebrain, structure of
brain size
  and body weight 26–31, 335–7
  and cortical folding 247
  and cortical volume 247, 248f, 249–51, 288
  and hippocampal volume 250–1
  and intelligence 16, 25–31, 221, 241–3, 258–9, 288, 335–8
  and neuron activity, connectivity and density, 25–6
  and sizes of brain subdivisions 251, 288
  evolution of 337–8
  in amphibians 119
  in birds 26–8, 221–2, 236
  in fish 26–8, 101–2, 113
  in reptiles 26–8, 164–6
  in mammals 26–8, 241–3, 247, 288
brain stimulation 34, 41
  and reinforcement
    and cerebral lateralization 315–16
    in birds 234
  see also forebrain, sensory and motor organization of
brain weight, *see* brain size
brainstem rest, in birds 221–2
Broca's area, in humans 314
Brontosaurus, *see* dinosaurs
bush-babies (*Galago senegalensis*)
  absence of sensory association cortex in 252
  cross-modal perception in 285
  extrastriate visual cortex of 263–4
  frontal cortex and delayed alternation in 269

cacomistles (*Bassariscus astutis*), serial reversal learning in 271
Caiman spp
  function of amydala in 142
  lateral olfactory tract projections in 144
  maze learning in 155–6
  probability learning in 160
  serial reversal learning in 159
  somatosensory system in 149
  see also Crocodilia
Cainozoic era 22t
Cambrian period 23t
canary (*Serinus canarius*)
  nervous control of singing in 184–5, 189
  sameness–difference concept in 213
Carboniferous period 22t
carnivores

brain size in 243
double alternation in 275
frontal cortex of 267
learning-set formation in 278–9
neocortical volume and brain size in 247, 248f, 249f
phylogeny of 239f
carp (*Cyprinus carpio*)
intra- versus extradimensional shifts in 96, 99
overshadowing in 98
catarrhines, classification of 238; *see also* monkeys
catecholaminergic terminals 141
categorical perception
of speech sounds 293–4
categorization, *see* encoding; categorical perception
catfish, bullhead (*Ictalurus nebulosus*), lateral line system of 55–6
cats (*Felis catus*)
absence of sensory association cortex in 252
brain size in 242, 243t, 277
delayed response in 274–5
double alternation in 276–7
frontal cortex of 265–9
lack of transfer from serial reversal learning to learning-set formation in 210, 272–3, 282
lateral geniculate nucleus cells in 181
learning-set formation in 277–9, 282
object permanence in 286–7
phylogenetic affinity to man 241
Reversal Index of 224, 272
sameness–difference concept in 282–3
split-brain, cerebral lateralization and paw preference in 315, 317, 328
striate cortex cells in 181
causality, *see* reasoning, inductive
cedar birds (*Nucifraga caryocataces*), effects of hippocampal lesions in 233
cephalization, *see* index of cephalization
cerebellum 36
in mammals, progression indices of 250f
cerebral dominance, *see* cerebral lateralization
cerebral hemispheres
in amphibians 116
in birds 172f, 188–9, 221–2, 233
in fish 48–50
in mammals 317
in reptiles 137, 139f, 149
cerebral lateralization
in humans 189–90, 258, 313, 315
and handedness 314
and intelligence 318
and language 318
and the critical age for language-acquisition 313
in non-human mammals 314–18, 328, 340
in song birds 189–90, 235
cetaceans
index of folding of 247
phylogeny of 239f
chaffinches (*Fringilla coelebs*)
habituation in 191
sensitization in 191
spontaneous recovery in 191
chameleons (Chamaeleonidae)
minute olfactory bulbs in 137
see also American chameleons
Chelonia
brain size in 164
olfactory and accessory olfactory bulbs in 137
phylogenetic relationships of 136–7
visual system in 144–7
chickens (*Gallus domesticus*)
archistriatal lesions and fear in 227
avoidance learning in 195
brain of 172f
forebrain lesions and punishment in 231
learning-set formation in 224–6
omission training in 194
overtraining and reversal learning in 206
probability learning in 201–2
relative brain size in 222
Reversal Index of 224
role of telencephalic structures in imprinting in 232–3, 183
self-stimulation of the brain in 234
serial reversal learning in 223–4
tecto-hyperstriatal projection in 183
children
acquisition of quantifiers by 310
babbling in 298
cross-modal perception in 285
deaf, intelligence of 291–2
dysphasic, intelligence of 291–2
feral, intelligence of 291
grammatical categories in 296
intra- versus extradimensional shifts in 96
language acquisition in 312–13
mean length of utterance in 298–9
number concepts in 284
reasoning and memory in 217
reversal versus extradimensional shifts in 164
speech comprehension without speech production in 294
stages of cognitive development in 285–7
transitive inference in 307–8
word order in 297–8
see also humans

chimpanzees (*Pan troglodytes*)
acquisition of quantifiers by 310–11
auditory perception of 293
brain- and body-weight in 28f, 243t
cerebral asymmetry in 317–18
chromosomes of, compared with human 329
communication between 301–2
complex visual concepts in 284
comprehension of logical operators by 309–10
concept of causality in 305–6
conservation in 287
cross-modal perception in 285
extent of frontal cortex in 253–4
insight in 307
intentional deception by 303–4
manual babbling in 298
mean length of utterance in 298–9
number concepts in 215, 283–4
object permanence in 286
phylogeny of 238, 241
production and comprehension of sentences in 296–301
production and comprehension of words in 293–6
ratio of eulaminate to total neocortex in 253
sameness–difference concept in 213, 282
transfer from reversal learning to learning-set formation in 210, 272–3
vocal capacities of 293
Chinchilla laniger, categorical perception in 294
cholinesterase 143, 175, 177
Chondrostei, phylogeny of 46–7
chordates
central nervous system of 20
percentage of known animal species 20
chromosomes, of man and chimpanzee 329
cingulate cortex 244
cingulate gyrus, in mammals, compared to ventromedial avian Wulst 174–5
Cladoselachii, ancestors of modern sharks 46
class, defined 19
classical conditioning 9–10
and interstimulus interval 61, 127
and reasoning 306
as simpler than instrumental conditioning 13, 194
distinguished from instrumental conditioning 10–12, 154, 193, 196
in amphibians 127–9, 134
in birds 192–3, 196, 235
in fish 58–64, 92
following forebrain lesions 102, 110, 113
in mammals 270
following neocortical lesions 260–1
in reptiles 153–5, 161, 164, 166
stimulus-substitution theory of 193
see also autoshaping; expectancies; omission training; poison-aversion learning; sensory preconditioning
classification
of animals 17–20
'natural' as opposed to 'artificial' methods of 18–19
of amphibians 114–15
of birds 168–171, 234–5
of fish 44–7, 331
of mammals 237–41, 331–2
of reptiles 136–7, 166
see also learning tasks
coelacanth (*Latimeria*)
brain and body weight in 28f
brain structure in 119
phylogenetic relationships of 46–7
Columbiformes, rank within birds of 170f, 171, 197
communication, in non-human animals 301–4, 328
comparative psychology, *see* intelligence, comparative approach to
complex learning
and classification of learning tasks 9
in birds 197–221
in fish 69–100
after forebrain ablation 103, 105, 111, 113
in mammals 270–87
in reptiles 158–65
computers
and the notions of mass action and equipotentiality 258
reason and consciousness in 343
simulation of chimpanzee language-production by 300–1
concept formation 4; *see also* number concepts; sameness–difference concept; stages of cognitive development; visual concepts
concrete operational stage, *see* stages of cognitive development
concrete operations period, *see* stages of cognitive development
conditional discriminations 200, 207, 212–13
and sentence-production by chimpanzees 300–1
delayed, and cerebral dominance in monkeys 316
conditioned response (CR) 10
conditioned stimulus (CS) 10
conditioned suppression
in birds, following archistriatal lesions

228–9
in fish 60
consciousness
and language 343
in a computer 343
in animals 1–2
conservation, *see* stages of cognitive development
consolidation
impaired in amnesic humans 321
following physiological treatments 327
see also rehearsal
context-specific capacities, *see* specific capacities
contextual variables
and avoidance learning 130, 156–7
and contrast effects 93
and delayed response 274
and discrimination learning 155
and instrumental conditioning 261
and learning-set formation 211, 226–7, 280–1
and maze learning 156
and measures of general intelligence 342
and probability learning 81, 95, 161, 202, 235
and resistance to extinction 93
and the sameness-difference concept 213–14, 236, 282–3
and serial reversal learning 71, 75–7, 95, 197–8, 200, 223–4, 235, 271
and single alternation 93
and the overtraining reversal effect 206
and transfer from serial reversal to learning-set formation 272–3
as influencing rate of learning 7–8, 14–15, 330–4, 337, 342
contrast effects, *see* behavioural contrast; simultaneous negative contrast effect; successive negative contrast effect
contrast mechanism
in birds 203
in mammals but not in fish 86
responsible for reward shift effects 85–7
responsible for the successive negative contrast effect 90
see also expectancies
convergence, concept of 19
core area, of cortical sensory field 253, 262
corpus callosum, in mammals 246, 288
corpus striatum, *see* striatum
corticoid layer, in birds 173f, 175, 177
corticospinal tract, *see* pyramidal tract
cortex 40
in birds 175
in reptiles 138–9, 161, 166
see also hippocampus; neocortex; pallium; piriform cortex
Corvidae
brain size in 222
high intelligence of 169, 171
rank within passerines of 169
cotylosaurs, phylogeny of 136–7, 138f
counting, *see* number concepts
cranial nerves 36
Crespi effect, *see* successive negative contrast effect
Cretaceous period 22t
critical age, for language-acquisition 313
crocodiles (*Crocodilus* spp)
instrumental conditioning in 155
serial reversal learning in 159
Crocodilia
auditory system in 148
brain size in 164
phylogenetic relationships of 137, 138f, 168
cross-modal perception, in mammals 285, 332
Crossopterygii, phylogeny of 46, 114, 115f, 134
crows (*Corvus* spp)
brain size in 28f
learning-set formation in 225–6
Reversal Index in 224
see also jackdaws; ravens; rooks
cueing technique, for amnesic humans 320–1
cyclostomes
brain of 47–9, 111
description 44
less intelligent than teleosts? 102

deafness, and intelligence 291–2
deception, intentional, in chimpanzees, 303–4
degeneration, *see* neuroanatomical techniques
Deinonychus 168
delayed alternation
in mammals 325–6
role of frontal cortex in 266, 268–9
delayed-matching-to-sample
in birds 219
in mammals 326–7
following frontal cortex lesions 266
delayed response, in mammals 273–5
following lesions of frontal cortex 266–9
delayed reward
influence on reward shift effects 93
similarity of after-effects to those of non-reward 94
depression effect, *see* successive negative contrast effect
developmental dysphasia, and intelligence 291–2
Devonian period 23t

diencephalon 35–6
in mammals, progression indices of 250f
digit span, *see* memory span
dinosaurs
as ancestors of birds 168
classification of 137, 168
evolution and extinction of 136
dishabituation
in amphibians 126
in birds 191–2
in fish 58
in reptiles 152
dogfish, brain of 53
dogs (*Canis familiaris*)
anecdotes about 2
brain size in 242, 243t
delayed response in 274
frontal cortex of 265–7
phylogenetic affinity to man 241
dolphins
brain size in 242, 243t
Delphinus bairdii
brain size in 247
index of folding in 247
Tursiops spp)
communication between 302–3
learning-set formation in 278, 280
sameness-difference concept in 282
short-term memory in 324–5
dominance, *see* cerebral lateralization
dopamine 177
dorsal cortex, in reptiles 138, 140–1, 145–8, 165; *see also* pallium, in reptiles
dorsal noradrenergic bundle, in mammals 141
dorsal ventricular ridge
in reptiles 140f, 141–3, 145, 147–9, 166, 246, 339
homologues of 118, 174–5, 178
double alternation, in mammals 275–7
doves 170f
Streptopelia risoria, serial reversal learning in 197–9
drugs 35, 327
Dryopithecus 240
ducks
brain of 172f
mallard (*Anas platyrhynchos*)
archistriatum and feeding in 189
archistriatum and fear in 227
habituation in 191
movement of decerebrate 188

ear advantage, *see* cerebral lateralization
easy-to-hard effect, in birds 204–5
echidnas
absence of corpus callosum in 246
Tachyglossus 237
brain size in 242
one-trial reversals in 271
telencephalon of 245
Zaglossus 237
echinoderms, as ancestors of chordates 21
ectostriatum, in birds 171, 173f, 174, 180–1, 182
ectothermic vertebrates 168
eels, moray (*Gymnothorax funebris*), olfactory system of 55
eighth nerve, *see* auditory system
Elasmobranchiomorphi, phylogeny of 44–6
elasmobranchs
brain weight in 101–2, 113
forebrain of 111
phylogeny of 45–6
electric organ, discharge rate as avoidance response 67–8
electroconvulsive shock, and memory, 327
electroencephalographic (EEG) technique 34–5
electrosensory system, in fish 54–5
elephants (Elephantidae)
brain- and body-weight in 28f, 30, 243t
thickness of cortex in 247
emu (*Dromiceius*) 169
encephalization 37–8
and the visual system 338
in amphibians 120–1, 134
in reptiles 166
in mammals 264
encephalization quotient
in mammals 242–3, 277–8
encoding, and memory 213–4
endbrain 35; *see also* forebrain
endocasts, *see* endocranial casts
endocranial casts, and brain size in extinct species 26, 337
endothermy 168
engrams, brightness 256
entorhinal cortex 244
enzyme concentrations 140, 174
Eocene epoch 22t
epithalamus 36
epochs, geologic 22t
equipotentiality
in a computer 258
within the neocortex 25, 256, 261
eras, geologic 22–3t
escape learning, *see* active avoidance learning
ethologists, interest in birds of 190
Eucinostomus gula (mojarras), habituation in 57
eulaminate cortex
defined 252
ratio to total neocortex 253–4
Euparkeria, as possible ancestors of birds 168
Eutheria, *see* placentals
evagination, hemispheric

illustrated 52f
in fish 48–9
eversion, hemispheric
illustrated 52f
in fish 51–2
evolution
and the geologic time scale 23
of language 329, 343
of learning capacity 1, 16, 342–3
of the brain 15
traditional view of 37–8
theory of 1
and the classification of animals 18
expectancies
absent in fish 13, 86, 92, 112
absent in reptiles 161, 164, 167
and behavioural contrast 164
and formation of classical associations 86, 92, 110
and resistance to extinction 86, 92, 112
and the generation of inhibition 91–2
and the simultaneous negative contrast effect 91–2, 112, 163
and the successive negative contrast effect 89–90, 92, 112
in birds 235
see also contrast mechanism
experimental material, *see* neuroanatomical techniques
extinction
in birds
and reward size 203
following hyperstriatal lesions 232
following partial reinforcement 203
in fish 59, 82
and amount of training 82–4
and reward size 84, 86–8, 93–4
following forebrain lesions 106–7
following partial reinforcement 84–8
in mammals 59, 82
and amount of training 82–4
and reward size 84, 86–8, 93–4
following frontal cortex lesions 266
following habituation to the unconditioned stimulus 153–4
following hippocampal lesions 319
following partial reinforcement 84–8
in reptiles
and reward size 161–2, 166
following habituation to the unconditioned stimulus 153–4
following partial reinforcement 162
extradimensional shifts (EDS), *see* intradimensional shifts (IDS) versus extradimensional shifts (EDS); reversal learning
extrastriate visual cortex, in mammals 55, 145, 181, 252, 262–4, 315, 339
face recognition
in rhesus monkeys 284
in human amnesics 323
factor analysis 341
family 19
fear
and frustration 34
in birds
and avoidance learning 196
following archistriatal lesions 175, 227–8, 236
following lobus parolfactorius lesions 233
following septal lesions 233
in fish
following forebrain lesions 108–11, 113
in mammals
cerebral lateralization of 316
following amygdalar lesions 143, 166
in reptiles
following amygdalar lesions 142–3, 166
see also habituation, in birds; habituation, in reptiles
feeding in birds, nervous control of 187–9, 234–5
ferrets (*Mustela furo*)
learning-set formation in 278–9, 281
serial reversal learning in 271
Field L, of avian neostriatum 184–5, 186f, 189, 233
finches, zebra (*Peophila guttata*)
auditorily-responsive cells in hyperstriatum of 185
see also chaffinches
fish
biological intelligence of 15
brain- and body-weight in 26–9, 101, 335–6
classical conditioning in 58–64
classification and phylogeny of 44–53, 111, 331
forebrain lesions and learning in 102–11, 113
habituation in 57–8
instrumental conditioning in 64–9
inter-species differences in learning in 100–2
mechanisms of attention in 94–9, 112–13
probability learning in 79–82, 111–12, 201
reward shift effects in 82–94, 112–13
sensory and motor organization of forebrain in 37, 53–6, 111, 117, 339
serial reversal learning in 70–9, 111–12, 332
structure of forebrain in 47–53, 111, 118–19

folding
of neocortex 246–7, 288
forebrain 35, 37–8
role in learning and memory of
in amphibians 127, 132–5
in birds 226–34
in fish 102–11
in reptiles 165
sensory and motor organization of 338–9
in amphibians 119–29, 134, 339
in birds 178–85, 186f, 187–90, 235
in fish 53–6, 339
in mammals 244–5
in reptiles 143–50
structure of 38–40
in amphibians 116–19, 134
in birds 171, 173f, 174–8, 235
in fish 47–53
in mammals 244–6
in reptiles 137–43
see also under forebrain subdivisions, and especially neocortex
forgetting, *see* proactive interference
formal operations stage, *see* stages of cognitive development
fornix, postcommissural
absent in birds 176
in mammals 320
free operant avoidance, *see* active avoidance learning, Sidman
frogs (*Rana* spp)
auditory system of 123
avoidance learning in 130
brain- and body-weight in 119
classical conditioning in 127–9
decerebrate, learning in 132
habituation in 125–6
internal clock in 4, 129
phylogeny of 114–5
sensitization in 127–8
spinal, learning in 132–3
strio-tegmental pathway in 124
sun-compass orientation in 4–5, 293
visual system in 120–2
frontal cortex, in mammals 34, 252–4
and perseveration 264–9
and spatial memory 264–9, 315
fronto-archistriatal tract
in birds 234
frustration 86
and behavioural contrast 164
and fear 34

Galliformes 170f, 197
brainstem rest weight in 221–2
ganglion cells, in amphibians 120
geckos
banded (*Coleonyx variegatus*), serial reversal learning in 160
Gekko gecko, telencephalon of 140f
general capacities, *see* intelligence
general cortex, *see* dorsal cortex
general intelligence, *see* intelligence
genus 19
geologic time scale 22–3t
gerbils, Mongolian *(Meriones unguiculates*), learning-set formation in 281
gibbons (*Hylobates*) 238
globus pallidus 40
homologue
in amphibians 118
in birds 177–8
in mammals 188, 245
Gnathonemus (an elephant fish), avoidance learning in 67
goldfish (*Carassius auratus*)
active avoidance learning in 66–8
amount of training and extinction in 82–4
autoshaping in 60, 62–3
blocking in 98–9
brain- and body-weight in 28f
dishabituation in 58
failure to obtain sensory preconditioning in 63–4
failure to obtain the overtraining reversal effect in 95–6
forebrain stimulation and arousal in 106
forebrainless
activity levels in 106
avoidance learning in 104–5, 107–9
classical conditioning in 102
extinction in 106–7
habituation in 108
instrumental conditioning in 102–3, 110
poison-aversion learning in 104
reversal learning in 103, 106
trace conditioning in 107
habituation in 57
intra- versus extradimensional shifts in 96–8
omission training in 62–3, 65
one-trial learning in 68
partial reinforcement and extinction in 84–5, 87, 93–4
probability learning in 80–2, 161
punishment in 68
reward size and extinction in 84, 93–4
second-order conditioning in 64
serial reversal learning in 70–9
simultaneous negative contrast effect in 90
single alternation in 87–8
successive negative contrast effect in 89–90, 93–4

gorillas (*Gorilla gorilla*)
brain- and body-weight in 28f
classification and phylogeny of 238, 241
production and comprehension of sentences by 296-9
production and comprehension of words by 294-6
grammar
generative 312
in apes 295-301, 328-30
universal 312, 329, 343
granule cell layer, in sensory cortex 252
great apes, *see* pongids
Great chain of being 17-18
guidance trials 80, 201
guppy (*Lebistes reticulatus*)
avoidance learning in 66
habituation in 57
gustatory system
in mammals 176
see also vomeronasal epithelium
Gymnophiona
cannot detect airborne sounds 123
phylogeny of 114-15
gyrencephalic brains 247
gyrus proreus, in carnivores 267

habituation 10, 58
in amphibians 125-7, 134
of visually-responsive cells 121-2
in birds 190-2
in fish 57-8
after forebrain lesions 108
in mammals 270
to the unconditioned stimulus prior to extinction 153-4
in reptiles 150-3, 166
to the unconditioned stimulus prior to extinction 153-4
hagfishes (Myxinidae) 44
hair cells, in the auditory and lateral line systems 56
hamsters (*Mesocricetus auratus*) effects of collicular lesions in 146, 182
handedness
in humans 315
and cerebral dominance 189, 314
in non-humans 315, 317
handling, and cerebral lateralization 316
hedgehogs (Erinaceidae)
as basal insectivores 249
Hemiechinus auritus, frontal cortex and delayed alternation in 269
Paraechinus spp
frontal cortex and delayed alternation in 269
interhemispheric connections of 246
visual system of 145
hemispheres, *see* cerebral hemispheres
hindbrain 35-6
in birds, projections to 177
hippocampal fissure 244
hippocampus 40
in amphibians 117f, 118, 122-4
in birds 173f, 174-7, 229, 231
and spatial information processing 233
in fish 48-9
in mammals 144, 176, 244-6
and cognitive mapping 323
and language 313-14
and memory 33, 318-24, 328, 340
and perseveration 319
and spatial information processing 319
progression indices of 250f, 251
volume related to brain volume 250-1
in reptiles 138, 140-1, 144, 148
histochemical analysis
of amphibian forebrain 118
of avian forebrain 174-5, 177
of mammalian forebrain 177
of reptilian forebrain 140-3
holocephalians, phylogeny of 45
Holostei, phylogeny of 46-7
homing, in birds 192
and the paleostriatum primitivum 178
Hominidae
brain size in 240
classification of 19, 238
hominoids, classification of 238
homology
contrasted with analogy 38-9
techniques for inferring 39-40
horseradish peroxidase 41, 183
horses (*Equus caballus*), brain size in 243
humans (*Homo* spp)
amnesic syndrome in 33, 318-24
brain size in 27f, 28f, 30, 240, 242-3, 247, 335
cerebral asymmetry in 317
cerebral dominance in 189-90, 258, 313-15, 317-18
chromosomes of, compared with chimpanzees' 329
classification and phylogeny of 19, 238, 240-1, 332
cortical lesions and intelligence in 258
delayed response in 274, 275f
effects of frontal cortex damage in 267-8
effects of hippocampal damage in 33, 318-24
evolution of language in 329, 343
extent of frontal cortex in 253-4
eyelid conditioning in 128

humans (*Homo* spp) (*cont.*)
general factor of intelligence in 340–2
handedness in 315
hippocampal volume and brain size in 251
index of cortical folding in 247
intelligence, contrasted with non-human 290–1, 328
intra- versus extradimensional shifts in 96
Korsakoff's psychosis in 319–20
language-acquisition device in 312–13
proactive interference and forgetting in 321
ratio of eulaminate to total neocortex in 253–4
short-term memory in 324–7
speech perception in 293–4
spinal, learning in 133
thickness of neocortex in 247
see also children
hylobatids 238
hyperstriatum, in birds 171, 173f, 174–5, 187, 229–34, 236, 246, 339
accessorium 173f, 174, 175, 232
see also nucleus, intercalated, of the hyperstriatum accessorium
dorsale 173f, 174, 180
homologue, in reptiles 174–5
ventrale 173f, 174, 178, 183–5, 189, 232
see also Wulst
hypnosis, *see* tonic immobility
hypopallium, *see* dorsal ventricular ridge
hypothalamus 36
in amphibians 120
in birds 176–7
in fish 55
in mammals 176, 319
in reptiles 142, 144
hypotheses, *see* strategies

ichthyostegids, phylogeny of 114, 115f
iguanas, desert (*Dipsosaurus dorsalis*)
avoidance learning in 157
escape learning in 156
extinction following partial reinforcement in 162
maze learning in 156
iguanids
auditory system of 148
size of olfactory bulbs in 137
imperative inference, as an account of instrumental learning 13
imprinting 192, 232
a comparable phenomenon in reptiles 152–3
role of telencephalic structures in, in birds 183, 232–3
index of cephalization 30, 222, 242, 258, 335–6
index of folding, of neocortex 247
index of somatic elaboration 336
induction, *see* reasoning, inductive
inference, *see* imperative inference; reasoning, deductive
inferior colliculus
homologue, in birds 184
in mammals 123
inferotemporal cortex, and discrimination learning 262–3, 315
inhibition
and expectancies 91–2
and the simultaneous negative contrast effect 91
and serial reversal learning 231
in birds, following hyperstriatal lesions 230–2
see also perseveration
innate, *see* specific capacities
insectivores
basal, as reference group for progression indices 249, 250f, 251
brain size in 243
frontal cortex of 253
phylogeny of 238, 239f, 241
insight, in chimpanzees 307
instinct, Darwin and the definition of 3
instrumental conditioning 9–10
as basic or early form of learning 13, 194
alternative interpretations of association formed 12–14
distinguished from classical conditioning 10–12
in amphibians 129–132, 134
in birds 193–6, 235
in fish 64–9
following forebrain lesions 102–3, 110
in mammals 270
following neocortical lesions 260–1
in reptiles 155–8, 166
insular cortex 244
intelligence
and body weight 28–9
and brain organization 31, 226, 246
and brain size 25–31, 221, 241–3, 259, 288, 335–8
and cerebral lateralization 318
and communication 301
and complex behaviour 4
and cortical association areas 254–5, 261–9
and general adaptiveness 332–5
and language 290–3, 314, 328–9
and laws of association-formation 307
and learning-set formation 279–81
and maze learning 257
and phylogeny, 331–2
in birds 169

in mammals 241, 288
and short-term memory, 217
and simple learning tasks 4, 342
and the amnesic syndrome 323
and the neocortex 244, 246, 255–61, 288
assessment of, in animals 3–9, 341–2
biological 15–16, 338
comparative approach to 3, 15–17, 100, 334
early history of 1–3
contrasted with artificial intelligence 4–5
contrasted with sensorimotor capacities 193
contrasted with specialized capacities 5–6, 192, 340–2
definition of 3–5
evolution of 342–3
general factor of 340–2
nature of 3, 16
physiological analysis of 25–35, 40–3, 226
intelligence tests 258, 340–2
non-verbal 292
see also learning tasks
intentions, in non-human animals 303–4, 328
internal clock, in amphibians 4, 129
intradimensional shifts (IDS) versus extra-dimensional shifts (EDS)
in birds 203–5
in fish 96–9, 112
in mammals 96
in reptiles 164
intraotic ganglion
in reptiles 147
introspectionism 1–2
intrusion errors, by amnesic patients 321
invertebrates
percentage of known animal species 20
relationship to vertebrates 21
isocortex, *see* neocortex
isthmus 35

jackdaws (*Corvus monedula*), number concepts in 215
Jacobson's organ, *see* vomeronasal epithelium
jays, blue (*Cyanocitta cristata*)
learning-set formation in 209–11, 224–6, 235
strategies formed in serial reversals in 77–8, 199–200, 235
Jurassic period 22t
juxtallocortex 244

kainic acid 188
kangaroos (*Macropus* spp)
classification of 237
corpus callosum in 246
poor learning in 271
kestrels, American (*Falco sparverius*), stereopsis in 182
kiwi (*Apteryx*), size of olfactory bulbs in 171
knowledge, in non-human animals 303
koala bear (*Phascolarctos*), classification of 237
Korsakoff's psychosis, and the amnesic syndrome 319–20, 322–3

labyrinthodonts, phylogeny of 114, 115f
laminae, in birds 174
lampreys (Petromyzonidae)
brain of 47–9
brain weight in 101, 113
phenomena resembling habituation in 57
sensitization, not classical conditioning in 59
language
acquisition
in humans 312–13
in non-humans 293–301, 331, 343
and cerebral lateralization 313, 318
and cross-modal perception 285
and depth of encoding 323
and hippocampal function 313
and intelligence 290–3, 328–9, 330, 342–3
and the concept of causality 306, 311
and the universal grammar 312–13
and transitive inference 308–9, 311
evolution of 329, 343
langurs, *see* monkeys, langurs
larynx, of chimpanzees 293
latent inhibition, in birds and mammals 205
lateral cortex, *see* piriform cortex
lateral forebrain bundle, in reptiles 147
lateral hypothalamus 34
in birds 177
in mammals 315–16
lateral line system 55–6
lateral pallium, *see* piriform cortex
learning-set formation
and short-term memory 208, 210, 221
in birds 207–11, 224–6, 235–6
in mammals 209–10, 277–82
learning tasks 7, 9–15
lemurs
classification of 238
Malagasy (*Lemur macaco*), learning-set formation in 278
lepidosaurs, phylogenetic relationships of 136–7
limbic archistriatum, in birds 176, 227–9
limbic cortex 174–5

limbic system 177
connections with frontal cortex 268-9
effects of damage to 33-4
lissencephalic brains 246-7
lizards (Lacertilia; Sauria)
auditory system of 148
Bengal monitor (*Varanus bengalensis*), autoshaping in 153-4
brain lesions and learning in 165
brain size in 164-5
collared (*Crotaphytus collaris*)
avoidance learning in 157-8
classical conditioning in 153
fence (*Sceloporus* spp)
effects of amygdalar lesions in 142-3
subdivisions of hippocampus in 140
instrumental conditioning in 155
Lacerta, forebrain of 139f
phylogeny of 136-7, 138f
retinal projections in 144
size of olfactory bulbs in 137
'taming' of 150
tegu (*Tupinambis* spp)
avoidance learning in 157
cortical layers in 139
thalamic projections in 141, 145
see also geckos; iguanas; slow worms
lobus parolfactorius,
in birds 173f, 176-8, 233
locus coeruleus, in mammals 141
logical operators, comprehension by a chimpanzee 309-10
long-term memory
and learning-set formation 209-10
and the hippocampus 33, 318-24
capacity of
and brain size 337
in fish 72
distinguished from short-term memory 217-18, 220-1, 235, 314, 322, 328
in birds 216-17
see also after-effects; proactive interference
loris family, classification of 238
lovebirds, peach-faced (*Agapornis roseicolis*), emotional responding following archistriatal lesions in 227
lungfishes (Dipnoi), phylogeny of 46-7

Macropodidae, classification of 237
magpies
Pica pica, brain size in 222
red-billed blue (*Urocissa occipitalis*), serial reversal learning in 223, 197
mallards, *see* ducks, mallard
mammals
as reference species in comparative studies 24
behavioural contrast in 163
biological intelligence in 15
brain- and body-weight in 26-9, 101, 241-3, 288
cerebral lateralization in 314-18, 328
classification and phylogeny of 19, 137, 138f, 237-41, 288, 331-2
communication in 301-4, 328
cross-modal perception in 285, 332
forebrain organization in 244-69, 288, 339
importance of instrumental conditioning in 194
interspecies differences in learning in 270-82, 288
language in 293-301, 311-13, 328-9
number concepts in 215, 283-4
organization of memory in 324-8
poison aversion learning in 154
probability learning in 80-1, 160-1
Reversal Index in 224, 271-2
role of hippocampus in 318-24, 328
reasoning in 304-11, 328
sameness-difference concept in 282-3, 288
stages of cognitive development in 285-7
visual concepts in 284-5
mammillary bodies, and the amnesic syndrome 319-20
maps, and hippocampal function 323
marmosets (*Callithrix*)
brain size of 243t
delayed response in 274-5
learning-set formation in 209, 277-8
ratio of eulaminate to total neocortex in 253, 254f
marsupials (Metatheria)
brain size in 242
classification and phylogeny of 237-8, 241, 288
corpus callosum of 246
cortical folding in 247
efficiency of learning in 271
one-trial reversals in 271
mass action 256
in a computer 258
in amphibians 134
of mammalian neocortex 25, 261
matching, *see* probability learning; sameness-difference concept
maximizing, *see* probability learning
maze learning (multi-unit)
in fish, following forebrain lesions 103
in reptiles 165

in birds 193
in mammals
and general intelligence 341–2
and memory span 325
following hippocampal lesions 319
following neocortical lesions 256–8
mean length of utterance
in chimpanzees and children 298–9
medial cortex, *see* hippocampus
medulla 36–7
in birds 184–5, 187
in fish 54
in reptiles 147
memory, *see* long-term memory; short-term memory
memory span
in humans 217
failure to find avian analogue of 218, 324, 327
failure to find non-human mammalian analogue of 324–5, 327
mesencephalic flexure 35–6
mesencephalon, *see* midbrain
Mesozoic era 22t
metencephalon 36
midbrain 35–6
in amphibians 124
in birds 177, 187
in mammals 245f
progression indices of 250f
mind–body problem, *see* consciousness
mink (*Mustela vison*)
learning-set formation in 278–9, 281
serial reversal learning in 271
minnows, golden shiner (*Notemigonus crysolecus*), classical conditioning in 60
Miocene epoch 22t
mobbing, in birds
habituation of 191
following archistriatal lesions 227
moles (Talpidae)
brain- and body-weight in 28f
classification of 238
mollies, green sailfin (*Mollienisia latipinna*), optimal inter-stimulus interval for conditioning in 61
monkeys
brain lesions and amnesia in 319–20
brain size in 335
Cebidae 270
Cebus spp
cross-modal perception in 285
learning-set formation in 280
Reversal Index in 224
sameness–difference concept in 282
serial reversal in 270–1
Cercopithecus spp
cerebral lateralization in 316–17
ratio of eulaminate to total neocortex in 253, 254f
double alternation in 275
extrastriate visual cortex in 262–4
frontal cortex in 264–9
hand preference in 315
intra- versus extradimensional shifts in 96
langurs (*Presbytis entellus*)
brain size of 278
learning-set formation in 277–9
learning-set formation in 207
Macaca spp
acquisition of quantifiers by 310–11
brain lesions and amnesia in 319–20
brain size in 243t, 277–8
categorical perception in 294
cerebral lateralization in 315–17
cerebral symmetry in 317–18
conservation in 287
cortical lesions and maze learning in 257
cross-modal perception in 285
delayed response in 274–5
double alternation in 276–7
face recognition in 284–5
learning-set formation in 210
number concepts in 283–4
object permanence in 286
one-trial reversals in 271
sameness-difference concept in 282
short-term memory in 326
transfer from serial reversals to learning-set formation in 273
squirrel (*Saimiri sciureus*)
brain size in 243t, 277
learning-set formation in 209, 277–8
number concepts in 284
object permanence in 286
Reversal Index of 224
sameness-difference concept in 282
serial reversal learning in 270–1
transitive inference in 307–9
unilateral cerebral lesions and handedness in 315
monotremes
as primitive mammals 237–8, 241
brain size in 242
one-trial reversals in 271
phylogeny of 237, 288
motor control system
in amphibians 124
in birds 174, 176, 178, 187–90, 234–5
in fish 54, 56, 111
in mammals 174, 188, 339
in reptiles 149, 166
motor cortex 174, 176, 252, 255f, 339
mouse (*Mus musculus*)
encephalization quotient of 243t

mouse (*Mus musculus*) (*cont.*)
modification of paw preference in 315
neuron density, dendritic tree length and brain weight in 26f
thickness of cortex in 247
mouthbreeders (*Tilapia* spp)
amount of training and extinction in 82
autoshaping in 60
failure to obtain serial reversal improvement in 70
forebrainless
avoidance learning in 104
escape behaviour in 105–7
matching in probability learning in 79
partial reinforcement and extinction in 84–5
preference dependent upon choice in 80
tectal brain grafts and serial reversal learning in 259
mudpuppy, *see* salamanders, mudpuppy
mullet (*Mugil cephalus*), autoshaping in 60
myelencephalon 36
mynah birds, greater hill (*Gracula religiosa*)
learning-set formation in 226
Reversal Index of 224
serial reversal learning in 197, 223

natural selection, *see* evolution
navigation, *see* specific capacities; homing
neanderthal man, brain size of 240
see also humans
negative contrast effect, *see* simultaneous negative contrast effect; successive negative contrast effect
neocortex 40, 244–69
and instrumental conditioning 260–1
and intelligence 255–61, 288
and reasoning 3
folding of 246–7, 288
homologues of 339
in amphibians 124, 134
in birds 174–6, 339
in fish 49, 111
in reptiles 140–1, 339
intra-mammalian variations in organization of 246–7, 249–55, 288
progression indices of 249, 250f
surface of, and brain size 247
thickness of 139, 246–7
volume of, and brain size 247, 248f, 249–51, 288
neostriatum 40, 174
in birds 171, 173f, 174, 177, 184–5, 186f, 187, 228, 233–4, 246, 339
in reptiles, *see* dorsal ventricular ridge
neural connectivity, and brain size 25–6
neural tube 20, 35, 40
neuroanatomical techniques 41–2; *see also* forebrain, sensory and motor organization of
neuron activity, independent of brain size 26
neuron density, and brain size 25–6
neuropil
in amphibians 119
in reptiles 144
neurotransmitters 35
New World monkeys 238; *see also* monkeys
newts (*Triturus* spp)
instrumental discrimination learning in 129
phylogeny of 114–15
triploid 133–4
Y-maze learning in 133–4
nictitating membrane 59, 125–9, 192, 196
non-reversal shifts
in birds, following hyperstriatal lesions 231–2
in reptiles 164
noradrenalin, and cerebral lateralization 316
normal material, *see* neuroanatomical techniques
nosebrain 53
nucleus (neclei)
accessory optic
in amphibians 120
in birds 179
in fish 55
in reptiles 144
accumbens
in amphibians 117f, 118
in birds 177
in reptiles 140f, 141
angularis
in birds 184, 186f
in reptiles 147
anterior olfactory 40
in birds 171
in fish 49
in mammals 244
in reptiles 141, 144
basalis, in birds 173f, 176, 185, 187–9, 234
caudate 40
in amphibians 118
in birds 177
in mammals 188, 245
see also striatum
central, of telencephalon, in sharks 49–50, 54
central, of torus semicircularis, *see* torus semicircularis, in reptiles
central pallial, in teleosts 51, 52f, 56
cochlear
in amphibians 123
in birds 184
cuneate, in birds 185

dorsal medullary, in anurans 123
dorsolateralis posterior, in birds 187
dorsolateralis tegmenti, in reptiles 141
dorsomedial, in mammals 252-3, 264-5
ectomammillaris, *see* nuclei, accessory optic, in birds
gracile, in birds 185
habenular, in birds 186f
lateral 177
intercalated, of the hyperstriatum accessorium, in birds 173f, 174, 179, 183, 231
intrapeduncularis, in birds 177
laminaris, in birds 184, 186f
lateral geniculate
in mammals 145, 180-1, 262
in reptiles 144-6
lateral posterior 55, 145, 180; *see also* pulvinar
lentiformis mesencephali, in birds 179
magnocellularis
in birds 185, 186f
in reptiles 147
medial geniculate, in mammals 148
medialis, in reptiles 147, 149
mesencephalicus lateralis dorsalis
in birds 184-5, 186f
of the basal optic root, *see* nuclei, accessory optic
of the diagonal band of Broca
in amphibians 118
in birds 176-7
in mammals 176
in reptiles 141
of the lateral lemniscus
in birds 184, 186f
in reptiles 147
of the oculomotor nerve 37
in birds 186f
of the trapezoid body, in reptiles 147
of the trochlear nerve 37
opticus tegmenti, *see* nuclei, accessory optic
ovoidalis, in birds 184, 186f, 189
posterocentral, in amphibians 123
posterolateral, in amphibians 122
preoptic, *see* preoptic area
prethalamicus, in fish 55
principal optic, of thalamus, in birds 179, 181, 183
principal sensory, of trigeminal ganglion
in birds 176, 185, 187-8
reuniens, in reptiles 147
rostral posteroventral, in amphibians 122
robustus, in birds 185, 189
rotundus
in birds 174, 180-3, 186f
in fish 55
in reptiles 145-9
septal, *see* septal area
sphaericus, in reptiles 141, 144
superficialis parvocellularis, in birds 187
superior olivary
in amphibians 123
in birds 184, 186f
in reptiles 147
suprachiasmaticus hypothalami, in birds 179
number concepts
in birds 214-15, 235, 284
in mammals 215, 283-4
numbers (of orders, genera, or species)
of amphibians 114
of birds 169
of fish 46
of mammals 238
of reptiles 136-7
nystagmus, in birds, habituation of 191

object permanence, *see* stages of cognitive development
obligate ranks, defined 19
occipitomesencephalic tract, in birds 187
oddity, *see* sameness-difference concept
Old World monkeys, classification of 238; *see also* monkeys
olfactory bulbs
in amphibians 116
in birds 171, 172f, 175
in fish 48-50
in mammals
progression indices of 250f
size not closely related to brain size 251
in reptiles 137, 139f
projections of, *see* olfactory system
olfactory nerve 36
and learning, in fish 107
in amphibians 116f
see also olfactory system
olfactory peduncle
in sharks 51
in reptiles 137, 141
olfactory system
in amphibians 119-20, 134
in birds 177-8, 190
in fish 53-6, 111, 117
in reptiles 142-4, 178
in mammals 55, 120, 142, 144, 244
olfactory tubercle 40
in amphibians 117f, 118
in birds 177
in fish 49-51, 119
in mammals 55, 244
in reptiles 141, 144
Oligocene epoch 22t

omission training 11–12, 193, 261
in birds 193–4, 196
compared with extinction 63
in fish 62–3, 65, 112
compared with extinction 63
see also active avoidance learning
one-trial learning
in amphibians 131
in birds 192, 195, 231
of reversals 197–8, 200, 223, 235
in fish 68
in mammals
following learning-set formation 207, 281–2
of reversals 75, 77, 159, 197, 271
in reptiles 154–5, 158
of reversals 159–60, 167
opossum, Virginia (*Didelphis virginiana*)
absence of corpus callosum in 246
as a primitive mammal 237, 241
brain size in 28f, 242
and neocortical volume 247, 249f
efficient learning in 271
optic nerve 36; *see also* visual system
optic tectum
in amphibians 120–2
in birds 172f, 179, 183, 186f
in fish 50, 54–6, 111, 259
in reptiles 144–7
see also superior colliculus
optokinetic response, in reptiles 151–2
optomotor response, in reptiles 152
orangutan (*Pongo pygmaeus*) classification and phylogeny of 238
order, defined 19
Ordovician period 23t
ornithischians, *see* dinosaurs
Oscar (*Astronotus ocellatus*), autoshaping in 60
Osteichthyes, phylogeny of 44–7
ostracoderms 44–6
ostrich (*Struthio*) 169
brain- and body-weight in 28f
overshadowing
in birds 204
in fish 98–9, 112
in mammals 98–9
overtraining reversal effect
failure to obtain in fish 95–6, 113
in birds 205–6
in rats 95–6, 205–6
owls
barn (*Tyto alba*), visual system of 181
burrowing (*Speotyto cunicularia*), projections from anterior Wulst in 174
large nucleus laminaris in 184

pain, *see* consciousness
Palaeozoic era 22–3t
Paleocene epoch 22t
paleocortex 40
in mammals, progression indices of 150f
see also piriform cortex
paleostriatum 40
in birds 171, 173f, 174–5, 177–8, 188, 233–4
in reptiles 141
pallial thickening, in reptiles 139–40
pallium
in amphibians 117f, 118, 120–1, 131
in birds 174–7
in fish 48–51, 119
in mammals 143, 175, 244–5
in reptiles 138–9, 149, 166, 175, 245
see also cortex
paradise fish (*Macropodus opercularis*)
failure to obtain serial reversal improvement in 70
failure to obtain the overtraining reversal effect in 95–6
forebrainless
maze learning in 103
reversal learning in 103
habituation in 57
parahippocampal area, in birds 173f, 176–7
parrots (Psittaciformes), 170f
Amazona ochrocephala, serial reversal learning in 223
brain size in 222
number concepts in 214
partial reinforcement extinction effect (PREE), *see* extinction
passerines 197
brain size in 222
evolution and classification of 169, 170f
nine-primaried, rank within passerines of 171
physiology of bird song in 184–5, 189–90
passive avoidance learning
in amphibians 131
without brain participation 132–3
in birds 195
following telencephalic lesions 231–2
in fish 68–9
following forebrain lesions 107
in mammals, following hippocampal lesions 319
in reptiles 157–8
patterning, *see* single alternation
pelycosaurs, phylogeny of 137, 237
penguins, *Spheniscus demersus*, 170f
brain size in 222
perception 4; *see also* visual system
periallocortex 244
periamygdalar region 244
periarchicortex 244

periectostriatal belt, in birds 180
periods, geologic 22–3t
peripaleocortex 244
perissodactyls, *see* ungulates
peristriate cortex 145, 181; *see also* extra-striate visual cortex
Permian period 22t
perseveration, in mammals
following frontal cortex lesions 266–9
following hippocampal lesions 319, 321
pharmacological techniques 35, 327
phylogeny, and intelligence 331–2; *see also* classification
phylum, defined 19
physiological psychology, *see* intelligence, physiological analysis of
pigeons (*Columba livia*)
autoshaping in 192–3
and inter-stimulus interval and stimulus duration 63
archistriatal lesions
and fear in 227–8
and feeding in 189
auditory system of 186f
avoidance learning in 195–6
blocking in 204
brain of 172f
brain size in 222
classification of 171
communication between 302–3
forebrain lesions and schedule performance in 228–9
habituation in 191
hippocampal lesions and reversal learning in 233
hyperstriatal lesions
and reversal learning in 230–2
and successive discrimination in 233
intra- versus extradimensional shift learning in 203–5
key-pecking in darkness in 195
latent inhibition in 205
learning of stimulus sequences in 300
learning-set formation in 225–6, 236
long-term memory in 216–17
number concepts in 215
omission training in 193
overshadowing in 204
partial reinforcement and extinction in 203
reward size and extinction in 203
Reversal Index in 224
sameness-difference concept in 212–13, 282
serial reversal learning in 70–5, 197–200, 223, 235
short-term memory in 218–20, 324–5
simultaneous negative contrast effect in 203
single alternation in 203
somatosensory system of 187
telencephalon of 173f
transfer along a continuum in 204–5
visual concepts in 215–16
visual system of 181–2
piriform cortex 40
in amphibians 117f, 118
in birds 173f, 175
in fish 48–9
in mammals 120, 244–6
in reptiles 138–9, 140f, 141, 144
placentals (Eutheria)
brain size in 242–3
classification and phylogeny of 237–8, 239f, 241, 288
corpus callosum in 246
placoderms, phylogeny of 44–5
planum temporale, in humans, asymmetry of 317
platypus (*Ornithorhynchus*)
brain size of 242
classification and phylogeny of 237
platyrrhines 238; *see also* monkeys
pleasure
not associated with twitters in chicks 234
see also consciousness
Pleistocene epoch 22t
Pliocene epoch 22t
poison-aversion learning 5
in birds 192
in fish
following forebrain lesions 104
in mammals 154, 305
in reptiles 154–5
pongids
classification and phylogeny of 238, 240–1, 288, 332
language-acquisition in 328–30
pons 36, 131, 176
porpoises (Delphinidae)
brain- and body-weight in 28f
index of cephalization in 30
positive behavioural contrast, *see* behavioural contrast
possums, brush-tailed (*Trichosurus vulpecula*)
brain size in 242
post-trial events, and rehearsal 326–7
preoperational period, *see* stages of cognitive development
preoptic area
in birds 177
in fish 48–9
prepiriform cortex
in birds 171, 178

prepiriform cortex (*cont.*)
in mammals 55, 244
presubicular cortex 244
pretectum
in amphibians 120
in birds 179
in fish 55
in reptiles 144
primacy effect (in list learning)
failure to obtain
in birds 219–20, 325
in humans 220
in non-human mammals 325
in humans 218
in non-human mammals 325
Primates
as most intelligent mammals 15–16
brain- and body-weight in 27f, 28f, 242–3, 288, 335
classification and phylogeny of 19, 238, 239f, 240–1, 288
cross-modal perception in 285
delayed response in 274
double alternation in 275–7
extent of frontal cortex in 253–4
generalized strategies in 272–3
hippocampal volume and brain size in 250–1
learning-set formation in 207–11
neocortical volume and brain size in 247, 248f, 249, 288
ratio of eulaminate to total neocortex in 253–4
sameness-difference concept in 282–3
see also individual subgroups, especially chimpanzees; monkeys
proactive inhibition, release from
in normals and amnesics 322
proactive interference
in birds 72–4, 197, 200, 219–20, 230–1
in fish 72–5, 79, 87–8
in human amnesics 320–1
in non-human mammals 72–4, 263
probability learning 69, 95
in birds 81–2, 200–2
in fish 79–82, 111–12
in mammals 79–81, 111–12
following cortical lesions 259
in reptiles 160–1, 166
procedural variables, *see* contextual variables
processing
depth of, and memory 321–4
levels of, and the amnesic syndrome 324
progression index, for mammalian brain structures 249, 250f, 251
proisocortex 244
propositional association, *see* imperative inference
prosencephalon 35
prosimians
brain size of 243
classification of 238
neocortical volume and brain size of 248f, 249
progression indices of 249, 250f, 251
Proterozoic era 23
Prototheria, phylogeny of 237
psycholinguistics 313, 323
pulvinar 55, 145, 252, 262, 264
punishment, *see* passive avoidance learning
putamen, in mammals 40, 188, 245
homologue
in amphibians 118
in birds 177
see also striatum
pyramidal cell layer, in motor cortex 252
pyramidal tract, in mammals 149, 174, 339

quail, bob white (*Colinus virginianus*)
hyperstriatal lesions and reversal learning in 229–31
overtraining and reversal learning in 206
Reversal Index of 224
serial reversal learning in 223, 197
quantifiers, acquired by non-humans 310–11
Quaternary period 22t
quinto-frontal structures, in birds 188
quokkas (*Setonix brachyurus*), report of poor learning in 271

rabbits (*Oryctolagus cuniculus*)
conditioning of nictitating membrane response in 59, 61, 128, 196
cross-modal perception in 285
double alternation in 276
encephalization quotient of 243t
neuron density, dendritic tree length and brain-weight in 26f
short-term memory in 326
raccoons (*Procyon* spp)
delayed response in 274
double alternation in 277
learning-set formation in 278
sameness–difference concept in 282
serial reversal learning in 270–1
Ramapithecus 240
random control procedure
in birds 192–3
in fish 59, 61
random matching, *see* probability learning
ranks, *see* obligate ranks
ratites 169
rats (*Rattus norvegicus*)
amount of training and extinction in 82–4
amygdala of 175

asymmetry of cerebral function in 315–16, 318, 328
brain- and body-weight in 28f
cortical lesions in
and brightness discrimination 256
and maze learning 256–7
and probability learning 259
and serial reversal learning 259–60
cross-modal perception in 285
delayed response in 174
detection of causal relationships by 305
direct projection from pons to telencephalon in 176
double alternation in 275–6
extrastriate visual cortex in 263
frontal cortex in 253, 265–9
habituation to the unconditioned stimulus preceding extinction in 153–4
intra- versus extradimensional shifts in 96, 112
learning-set formation in 277, 278f, 281
maze learning and general intelligence in 341
neuron density, dendritic tree length and brain-weight in 26f
number concepts in 284
one-trial reversals in 75, 159, 271
overtraining reversal effect in 95–6
partial reinforcement and extinction in 84–8, 92
phylogenetic affinity to man of 241
probability learning in 79–81, 111–13, 201–2
reversal versus extradimensional shifts in 164
reward size and extinction in 83
sameness–difference concept in 282–3
serial reversal learning in 70–4, 92, 112–13, 197–200
short-term memory in 325–7
simultaneous negative contrast effect in 90–3
single alternation in 87, 93
successive negative contrast effect in 89–90, 93
three-table reasoning problem in 306, 341
use of strategies by 77
ravens (*Corvus corax*), number concepts in 214
reasoning
absent in non-human animals 1, 3, 328
and language 311, 343
and memory 217
and stage of cognitive development 287
deductive, in non-human animals 306–11
in a computer 343
inductive
as the basic scientific method 17
in non-human animals 304–6, 311
see also imperative inference
recency effect (in list learning)
in humans 218, 220
in non-human mammals 325
recent memory, *see* short-term memory
recognition memory
in birds 218–20
in humans 220
following frontal cortex damage 267-8
in monkeys 325
following telencephalic lesions 319–20
see also face recognition
recording, of brain activity 34–5, 41
see also forebrain, sensory and motor organization of
reflexes, as instances of primitive brain organization 37–8
rehearsal
following physiological treatments 327
lack of evidence for, in birds 219–21, 235, 324
in humans 217–21, 318–19
in non-human mammals 325–7
reinforcer, *see* unconditioned stimulus
reptiles
biological intelligence of 15
brain- and body-weight in 26–9, 164–6, 335
classical conditioning in 153–5, 166
classification of 136–7, 166
forebrain lesions and learning in 165
habituation in 150–3, 166
instrumental conditioning in 155–8, 166
mechanisms of attention in 164–5
probability learning in 160–1
reward shift effects in 161–4
sensory and motor organization of forebrain in 143–50, 166
contrasted with avian organization 190
serial reversal learning in 159–60
structure of forebrain in 137–43, 166, 339
response inhibition, *see* inhibition; perseveration
retention, *see* short-term memory; long-term memory
retinal ganglion cells, in amphibians 120
retinal projections, *see* visual system
retinofugal pathway, *see* visual system
retroactive interference, in concurrent discriminations 263
retrosplenial cortex 244
Reversal Index 8
in birds 224
in mammals 224, 271–2

reversal learning 8
in amphibians 129
in birds
following hippocampal lesions 233
following hyperstriatal lesions 229–32, 236
in fish
following forebrain lesions 103, 106
in mammals
compared to extradimensional shifts 164
following frontal cortex lesions 266, 268–9
following hippocampal lesions 319, 321
in reptiles, compared to extradimensional shifts 164
see also overtraining reversal effect; serial reversal learning; Reversal Index
reward following, *see* strategies
reward shifts
in birds 202–3
in fish and mammals 13, 82–94
in reptiles 161–4
rhachitomes, phylogeny of 114, 115f
Rhea 169
rhinal fissure 244
rhombencephalon, *see* hindbrain
Rhynchocephalia, *see* tuatara
rodents
brain size of 243, 335
frontal cortex of 253, 267
learning-set formation in 281
neocortical volume and brain size in 247, 249f
phylogeny of 239f
see also individual subgroups, especially rats
rooks (*Corvus frugilegus*), learning-set formation in 225
ruling reptiles, phylogeny of 137, 138f, 168

salamanders
Ambystoma spp
habituation in 126
instrumental conditioning in 129
telencephalon of 117f, 124
mudpuppy (*Necturus maculosus*)
brain- and body-weight in 119
forebrain of 116
habituation in 125–6
phylogeny of 114–15
sameness–difference concept
in birds 211–14, 235
in mammals 282–3
Sarcopterygii
pallial divisions in 53
phylogeny of 46–7
saurischians, *see* dinosaurs
Scala naturae 17–18
schedules of reinforcement
differential reinforcement for low response rate (DRL)
in birds, following telencephalic lesions 228–9
in mammals, following frontal cortex lesions 266, 268–9
fixed-interval
in birds, following telencephalic lesions 230, 233
fixed-ratio
in birds, following telencephalic lesions 230
variable-interval
in birds, following telencephalic lesions 228–9, 233
sciatic nerve
in amphibians 123, 133
in reptiles 148
second language acquisition 313
second-order conditioning, in fish 64
secondary reinforcement, in forebrainless fish 109–10
selective attention
in birds 197, 203–6, 235
compared with rats 74, 95, 201–6
following hyperstriatal lesions 231–2
in fish 94–9
compared with rats 72, 74–5, 94–5, 97, 113
in mammals
following inferotemporal cortex lesions 262–3
in reptiles 164
self-stimulation, *see* brain stimulation, and reinforcement
semantic confusion errors, in delayed recall 318
sensitization
in amphibians 126–8
in birds 191
in fish 58–9, 63
following forebrain lesions 106
in reptiles 152
sensorimotor stage, *see* stages in cognitive development
sensory carry-over 85
sensory contrast
and the simultaneous negative contrast effect 91
sensory preconditioning (SPC)
and reasoning in rats 306
failure to obtain in fish 63–4, 92, 112
sentences, in apes 296–301, 311
septal area 40
in amphibians 117f, 118, 120, 124

in birds 173f, 176–8, 233–4
in fish 49, 51, 119
in mammals 144, 176, 245
progression indices of 250f
in reptiles 140f, 141, 144
septomesencephalic tract, in birds 188
serial reversal learning 69–70, 221
in birds 70–3, 197–200, 222–4, 236
and transfer to learning-set formation 210, 235
following hyperstriatal lesions 229–32
in fish 70–9, 95, 111–13, 199, 332
following tectal brain grafts 259
in mammals 70–2, 75, 111–13, 197, 270–3
and transfer to learning-set formation 210, 272–3
following cortical lesions 259–60, 203
in reptiles 159–60
sharks
brain size and intelligence in 101–2
lemon (*Negaprion brevirostris*)
classical conditioning in 59
instrumental conditioning in 64
Notorhynchus maculatus (cow sharks), brain of 50f
nurse (*Ginglymostoma cirratum*), instrumental conditioning in 64
phylogeny of 45–6
Scyliorhinus caniculus (cat sharks), telencephalon of 51f
sensory and motor organization of forebrain in 54
Sphyrna tiburo (hammerhead sharks) brain of 50
structure of forebrain in 49–50
short-term memory
and intelligence 217, 221, 313–14, 318
and language 318, 334–5
and learning-set formation 210
and the hippocampus 33, 318–24
distinguished from long-term memory 217–18, 220–1, 235, 314, 322, 328
in birds 217–21, 235
in fish, following forebrain lesions 107–8
in non-human mammals 324–7
see also after-effects
shrews (Soricidae)
as basal insectivores 249
classification of 238
siamangs (*Symphalangus syndactylus*) 238
Siamese fighting fish (*Betta splendens*)
avoidance learning in 66
discrimination using a classical procedure in 60
forebrainless
gill-cover responding and spontaneous movement in 106
habituation in 108
habituation in 57
punishment in 68–9
sensitization in 58
spontaneous recovery in 57
Sidman avoidance, *see* active avoidance learning, Sidman
sign order, *see* grammar
Silurian period 23t
simple allometry, equation of 29–31, 222, 242, 249, 335–6
simultaneous negative contrast effect
in birds 203
in fish 90–2, 112
in mammals 90–4, 112
in reptiles 163
single alternation
in birds 203
in fish 87–8
in mammals 87–8, 93–4
skin senses, *see* somatosensory system
skunks, striped (*Mephitis mephitis*)
learning-set formation in 278, 279f, 281
serial reversal learning in 270–1
sliders, *see* turtles, *Pseudemys* spp
slow worms (*Anguis fragilis*), 'taming' of 150
snakes (Ophidia; Serpentes)
brain size in 164–5
Eastern hognose (*Heterodon platyrhinus*), habituation in 150
garter (*Thamnophis* sp)
habituation in 150
olfactory projections in 144
passive avoidance learning in 158
poison-aversion learning in 154–5
indigo (*Drymarchon corais*), instrumental conditioning in 155
phylogeny of 136–7, 138f
rat (*Elaphe obsoleta*), escape learning in 156
rattlesnakes (*Crotalus*), habituation in 150
'taming' of 150
visual system of 144–5
water, diamond-backed (*Natrix rhombifera*), habituation in 150–1
somatic factor
in brain size 29–31, 221–2, 335–6
somatomotor archistriatum, in birds 176, 189, 227–9
somatosensory system
in amphibians 123–4, 134
in birds 176, 185, 187–90, 235
in fish 54–6
in reptiles 148–9, 166

song-birds, *see* passerines
song production, *see* bird song
spacing, and memory
  in birds 219
  in humans 219, 326
  in non-human mammals 325–6
sparrows
  house (*Passer domesticus*) 151
  brain of 172f
  olfactory bulbs of 171
  New World (Emberizidae), rank within passerines of 171
spatial information processing
  and cerebral lateralization, in rats 316
  and the hippocampus
    in birds 233
    in mammals 319, 321, 323
spatial memory
  following frontal cortex lesions 266–9
  see also spatial information processing
species, defined 19
species-specific (device, mechanism, response, symbol), *see* specific capacities
specific capacities
  and avoidance learning
    in birds 9, 195–6
    in fish 66
  and communication 301
  and counting 214
  and intelligence 5–6, 9, 192, 340
  and language 293, 312–3, 328, 330, 332, 334–5, 340
speech
  and cerebral dominance 189
  perception
    and Wernicke's area 314
    in humans and apes 293–4
  production
    and Broca's area 314
    in apes 293
  see also language
spinal cord 35
  and primitive brain organization 37
  isolated, learning in
    in amphibians 132–3, 135
    in humans 133
  see also motor system; somatosensory system
spiny anteaters, *see* echidnas
split-brain mammals 315
spontaneous recovery
  in amphibians 126
  in birds 191
  in fish 57, 59
  in mammals 59
Squamata
  phylogeny of 136–7
  retinal projections in 144
  serial reversal learning in 160
  size of olfactory bulbs in 137
  see also lizards; snakes
squirrels (*Sciurus*)
  brain size of 243t, 277
  learning-set formation in 277–8
stages of cognitive development, in mammals 285–7, 308
starling family (Sturnidae) 223, 226
starlings (*Sturnus vulgaris*), auditorily-responsive cells in Field L of 185
Stegosaurus, *see* dinosaurs
stem reptiles, *see* cotylosaurs
stereoscopic vision, in birds 181–2
sticklebacks (*Gasterosteus aculeatus*)
  forebrain lesions and motivation in 106
  telencephalon of 52f
stimulation, *see* brain stimulation
stimulus-response (S-R) reinforcement principle 86, 89
strategies 77, 80, 207–9
  in birds 77–8, 199–200, 209–11, 226, 235
  in humans
    following frontal cortex damage 268
  in non-human mammals 77, 271–3, 281–2
  in fish 77–9, 112
  in reptiles 159, 167
stria terminalis, in reptiles and mammals 142
striate cortex, in mammals 145, 180–3, 255–6, 260, 262, 264
striatum 40
  external 339
  in amphibians 117f, 118, 120–4
  in birds 171, 177
  in fish 48–9, 51, 119
  in mammals 142, 148, 166, 175, 188, 245
    progression indices of 250f
  in reptiles 139, 140f, 141–2, 147–9, 166
strict reinforcement principle 86, 89
subicular cortex, in mammals 176, 320
  avian homologue of 176
subitizing 214
substantia innominata, in rats 176
successive negative contrast effect 85
  in fish and rats 89–90, 92–4
  not obtained in reptiles 162, 166
sunfish (*Lepomis* spp)
  brain stimulation and nestbuilding in 106
  forebrainless, instrumental conditioning in 102–3
superior colliculus
  in mammals 55, 145, 180–2, 246, 262, 264
  see also optic tectum

superior temporal gyrus 316
superstition 11
supralaryngeal tract, of chimpanzees 293
suprarhinal cortex, in rats 267
syllogism 310–11
Sylvian fissure, in primates, asymmetry of 317
symbolic matching 212
syntax, *see* grammar
syrinx 169, 189–90
systematic variation 7–8, 333–4; *see also* contextual variables

taming, *see* fear; habituation, in reptiles
tarsiers (*Tarsius*)
  classification of 238
  ratio of eulaminate to total neocortex in 253–4
Tasmanian wolf (*Thylacinus*) 237
taste, *see* gustatory system
tectofugal pathway 338
  in amphibians 121–2
  in birds 180–3, 231
  in fish 55
  in mammals 55, 122, 145, 179–80
  in reptiles 145, 147, 166
tectum 36; *see also* optic tectum; superior colliculus
tegmentum 36
  in reptiles 144
telencephalon 35–8
  direct non-thalamic projections to
    in birds 176, 184–5, 187, 190, 235
    in mammals 176, 184
  medium, in fish 48–51
  see also forebrain
teleosts
  brain size in 28, 113
  forebrainless, learning in, 102–11, 113
  phylogeny of 46, 47f, 111, 115
  sensory and motor organization of forebrain in 54–6
  structure of forebrain in 51–3, 111
  see also fish; *and individual groups, especially* goldfish
temporal neocortex 319
temporal-parietal-occipital region, in birds 177
temporal stem, and memory 319–20
tenrecs (Tenrecidae), as basal insectivores 249
terrapins 136n
Tertiary period 22t
thalamofugal pathway
  in amphibians 121–2
  in birds 179–83, 231
  in fish 53
  in mammals 145, 166, 179–80
  in reptiles 145, 147, 166
thalamus 36–7
  dorsomedial, and the amnesic syndrome 319–20
  projections (from and to)
    in mammals 145, 175, 252–3, 263–5
    for other vertebrate classes, see forebrain, sensory and motor organization of
thecodonts 168
therapsids 237
Theria 237
toads
  Bufo spp
    avoidance learning in 130
    brain stimulation in 124
    classical conditioning in 128–9
    habituation in 125–7
    instrumental conditioning in 129
    one-trial learning in 131
  clawed (*Xenopus* spp)
    avoidance learning in 130–1
      following spinal section 132
  phylogeny of 114–15
  spadefoot (*Scaphiopus hammondi*), failure to learn avoidance by 130
  visual system of 121
tonic immobility, in birds 191–2
  following septal lesions 233
tortoises, phylogeny of 136–7
torus semicircularis
  in anurans 123
  in reptiles 147
  in teleosts 56
trace classical conditioning, in forebrainless goldfish 107
trace independence theory, of avian short term memory, 219–20
transfer along a continuum, in birds 204–5
transitive inference, *see* reasoning, deductive
tree-shrews (Tupaiidae)
  classification of 238
    Tupaia spp
    absence of sensory association cortex in 252
    extrastriate visual cortex of 263–4
    frontal cortex lesions in 266, 269
Triassic period 22t
Triceratops, *see* dinosaurs
truth-table 309
tuatara (*Sphenodon punctatus*)
  maze learning in 155
  phylogeny of 136–7, 138f
  retinal projections in 144
  telencephalon of 142f
turtles
  avoidance learning in 157–8
  Clemmys spp maze learning in 165

turtles (*cont.*)
Emys, forebrain of 139f
habituation in 150–3
instrumental conditioning in 155
painted (*Chrysemys picta*)
absence of successive negative contrast effect in 162
amygdalar divisions in 143
behavioural contrast in 163
brain lesions and visual performance in 146–7
dorsal cortex and learning in 165
extinction following partial reinforcement in 162
probability learning in 160–1
reversal versus extradimensional shifts in 164
reward size and resistance to extinction in 161–2
serial reversal learning in 159
simultaneous negative contrast effect in 163
phylogeny of 136–7, 138f
Pseudemys spp
avoidance learning in 156–8
classical conditioning in 153
extinction following partial reinforcement in 162
habituation in 150–2
habituation to the unconditioned stimulus preceding extinction in 153–4
instrumental conditioning in 155
properties of tectal cells in 145–6
somatosensory system in 148
side-necked (*Podocnemis unifilis*)
brain lesions and vision in 146–7
snapping (*Chelydra serpentina*)
development of food preference in 152–3
somatosensory system of 148–9
thalamic projections of 141
visual system in 144–7
two-process theories of learning 108
types of learning 9

'umweg' problems, in forebrainless fish 103
unconditioned response (UCR) 10
unconditioned stimulus (UCS) 10
understanding, in non-human animals 303
ungulates
brain size in 243
and neocortical volume 247, 249f
phylogeny of 239f
unnatural tasks, to assess intelligence 7, 9, 16
Urodela
cannot detect airborne sounds 123
habituation in 125–6
maze learning in 130
olfactory system of 119–20
phylogeny of 114–15
see also newts; salamanders

vallecula, in birds 173f, 174
ventral cerebral flexure 35–6
vertebrates
brain- and body-weight in 26–31
brain organization in 35–42
classification and phylogeny of 19–21
vervets, *see* monkeys, *Cercopithecus* spp
visual association cortex, *see* extrastriate visual cortex
visual cliff, in reptiles, following brain lesions 146–7
visual concepts
in birds 215–16, 235
in mammals 284–5
visual cortex
in an echidna 255f
see also extrastriate visual cortex; striate cortex
visual perception 4; *see also* visual system
visual system 338
in amphibians 120–2, 134, 178
in birds 121–2, 178–84, 190, 235
in fish 53–5, 111, 121
in mammals 121–2, 145, 166, 180–1, 252–7, 260–4
in reptiles 121–2, 144–7, 166, 178
vocal cords, of chimpanzees 293
vocal tract, of chimpanzees 293
vomeronasal epithelium
in amphibians 116–18
in reptiles 137
vomeronasal nerve
in a frog 116f
see also olfactory system

wallabies
bridled nail-tailed (*Onychogalea frenata*) brain size in 242
classification of 237
warblers, New World (Parulidae)
rank within passerines of 171
weasel family (Mustelidae), *see* ferrets; mink; skunks
Wernicke's area, in humans 314, 317
homologue, in monkeys 316
'Wh' questions, put to apes and children 295–6
whales
Atlantic pilot (*Globicephala melaena*), index of folding in 247
blue (*Balaenoptera musculus*), brain size in 28f, 30
brain size in 243t

win–stay, lose–shift, *see* strategies
word-order, *see* grammar
words, in apes 293–6, 311
worm-lizards (Amphisbaenia), phylogeny of
  136–7
Wulst, in birds 174–5, 187–8
  visual 179–81, 183
    columnar arrangement of 181
  see also hyperstriatum, accessorium; hyperstriatum, dorsale

zona limitans
  in amphibians 117f, 118
  in fish 49, 51